Lecture Notes in Artificial Intelligence   11625

Subseries of Lecture Notes in Computer Science

Series Editors

Randy Goebel
    University of Alberta, Edmonton, Canada
Yuzuru Tanaka
    Hokkaido University, Sapporo, Japan
Wolfgang Wahlster
    DFKI and Saarland University, Saarbrücken, Germany

Founding Editor

Jörg Siekmann
    DFKI and Saarland University, Saarbrücken, Germany

More information about this series at http://www.springer.com/series/1244

Seiji Isotani · Eva Millán ·
Amy Ogan · Peter Hastings ·
Bruce McLaren · Rose Luckin (Eds.)

Artificial Intelligence in Education

20th International Conference, AIED 2019
Chicago, IL, USA, June 25–29, 2019
Proceedings, Part I

 Springer

Editors
Seiji Isotani 🆔
University of Sao Paulo
Sao Paulo, Brazil

Eva Millán 🆔
University of Malaga
Málaga, Spain

Amy Ogan
Carnegie Mellon University
Pittsburgh, PA, USA

Peter Hastings 🆔
DePaul University
Chicago, IL, USA

Bruce McLaren
Carnegie Mellon University
Pittsburgh, PA, USA

Rose Luckin
University College London
London, UK

ISSN 0302-9743   ISSN 1611-3349 (electronic)
Lecture Notes in Artificial Intelligence
ISBN 978-3-030-23203-0   ISBN 978-3-030-23204-7 (eBook)
https://doi.org/10.1007/978-3-030-23204-7

LNCS Sublibrary: SL7 – Artificial Intelligence

This Springer imprint is published by the registered company Springer Nature Switzerland AG
The registered company address is: Gewerbestrasse 11, 6330 Cham, Switzerland

Preface

The 20th International Conference on Artificial Intelligence in Education (AIED 2019) was held during June 25–29, 2019, in Chicago, USA. AIED 2019 was the latest in a longstanding series of now yearly international conferences for high-quality research in intelligent systems and cognitive science for educational applications.

The theme for the AIED 2019 conference was "Education for All in the XXI Century." Inequity within and between countries continues to grow in the industrial age. Education that enables new economic opportunities plays a central role in addressing this problem. Support by intelligent information technologies have been proposed as a key mechanism for improving learning processes and outcomes, but may instead increase the digital divide if applied without reflection. The collective intelligence of the AIED community was convened to discuss critical questions, such as what the main barriers are to providing educational opportunities to underserved teachers and learners, how AI and advanced technologies can help overcome these difficulties, and how this work can be done ethically.

As in several previous years, the AIED 2019 events were co-located with a related community, the Learning at Scale (L@S 2019) conference. Both conferences shared a reception and a plenary invited talk by Candace Thille (Stanford University, USA). Also, three distinguished speakers gave plenary invited talks illustrating prospective directions for the field with an emphasis on accessibility, equity, and personalization: Jutta Treviranus (Ontario College of Art and Design University, Canada); Nancy Law (University of Hong Kong, SAR China); and Luis von Ahn (Carnegie Mellon University, USA).

There were 177 submissions as full papers to AIED 2019, of which 45 were accepted as long papers (ten pages) with oral presentation at the conference (for an acceptance rate of 25%), and 43 were accepted as short papers (four pages) with poster presentation at the conference. Of the 41 papers directly submitted as short papers, 15 were accepted. Apart from a few exceptions, each submission was reviewed by three Program Committee (PC) members. In addition, submissions underwent a discussion period (led by a leading reviewer) to ensure that all reviewers' opinions would be considered and leveraged to generate a group recommendation to the program chairs. The program chairs checked the reviews and meta-reviews for quality and, where necessary, requested for reviewers to elaborate their review more constructively. Final decisions were made by carefully considering both meta-reviews (weighed more heavily) scores and the discussions. Our goal was to conduct a fair process and encourage substantive and constructive reviews without interfering with the reviewers' judgment. We also took the constraints of the program into account, seeking to keep the acceptance rate within the typical range for this conference.

Beyond paper presentations and keynotes, the conference also included:

– A Doctoral Consortium Track that provided doctoral students with the opportunity to present their emerging and ongoing doctoral research at the conference and receive invaluable feedback from the research community.
– An Interactive Events session during which AIED attendees could experience first-hand new and emerging intelligent learning environments via interactive demonstrations.
– An Industry and Innovation Track, intended to support connections between industry (both for-profit and non-profit) and the research community.

The AIED 2019 conference also hosted ten half-day workshops with topics across a broad spectrum of societal issues, such as: life-long learning; educational data mining; multi-modal multi-channel data for self-regulated learning; ethics; informal learning; human-centered AI products design; standardization opportunities; team tutoring; intelligent textbooks and using AI to teach AI in K12 settings.

We especially wish to acknowledge the great efforts by our colleagues at DePaul University for hosting this year's conference.

Special thanks goes to Springer for sponsoring the AIED 2019 Best Paper Award and the AIED 2019 Best Student Paper Award. We also want to acknowledge the amazing work of the AIED 2019 Organizing Committee, the PC members, and the reviewers (listed herein), who with their enthusiastic contributions gave us invaluable support in putting this conference together.

May 2019

Seiji Isotani
Eva Millán
Amy Ogan
Peter Hastings
Bruce McLaren
Rose Luckin

Organization

General Conference Chairs

Bruce M. McLaren   Carnegie Mellon University, USA
Rose Luckin   London Knowledge Lab, UK

Program Chairs

Amy Ogan   Carnegie Mellon University, USA
Eva Millán   Universidad de Málaga, Spain
Seiji Isotani   University of Sao Paulo, Brazil

Local Organization Chair

Peter Hastings   DePaul University, USA

Workshop and Tutorial Chairs

Mingyu Feng   WestEd, USA
Ma. Mercedes T. Rodrigo   Ateneo de Manila University, Philippines

Industry and Innovation Track Chairs

Elle Yuan Wang   ASU EdPlus, USA
Ig Ibert Bittencourt   Federal University of Alagoas, Brazil

Doctoral Consortium Chairs

Janice Gobert   Rutgers Graduate School of Education, USA
Mutlu Cukurova   University College London Knowledge Lab, UK

Poster Chairs

Alexandra Cristea   Durham University, UK
Natalia Stash   Eindhoven University, The Netherlands

Awards Chairs

Tanja Mitrovic   University of Canterbury, New Zealand
Julita Vassileva   University of Saskatchewan, Canada

International Artificial Intelligence in Education Society

Organization

President

Bruce M. McLaren   Carnegie Mellon University, USA

Secretary/Treasurer

Benedict du Boulay   University of Sussex, UK
(Emeritus)

Journal Editors

Vincent Aleven   Carnegie Mellon University, USA
Judy Kay   University of Sydney, Australia

Membership Chair

Benjamin D. Nye   University of Southern California, USA

Publicity Chair

Erin Walker   Arizona State University, USA

Finance Chair

Vania Dimitrova   University of Leeds, UK

Executive Committee

Ryan S. J. d. Baker   University of Pennsylvania, USA
Tiffany Barnes   North Carolina State University, USA
Min Chi   North Carolina State University, USA
Cristina Conati   University of British Columbia, Canada
Ricardo Conejo   Universidad de Málaga, Spain
Sidney D'Mello   University of Notre Dame, USA
Vania Dimitrova   University of Leeds, UK
Neil Heffernan   Worcester Polytechnic Institute, USA
Diane Litman   University of Pittsburgh, USA
Rose Luckin   University College London, UK
Noboru Matsuda   Texas A&M University, USA
Manolis Mavrikis   University College London Knowledge Lab, UK

Tanja Mitrovic	University of Canterbury, New Zealand
Amy Ogan	Carnegie Mellon University, USA
Zachary Pardos	University of California, Berkeley, USA
Kaska Porayska-Pomsta	University College London, UK
Ido Roll	University of British Columbia, Canada
Carolyn Penstein Rosé	Carnegie Mellon University, USA
Julita Vassileva	University of Saskatchewan, Canada
Erin Walker	Arizona State University, USA
Kalina Yacef	University of Sydney, Australia

Program Committee

Esma Aimeur	University of Montreal, Canada
Patricia Albacete	University of Pittsburgh, USA
Vincent Aleven	Human-Computer Interaction Institute, Carnegie Mellon University, USA
Ivon Arroyo	Worcester Polytechnic Institute, USA
Nilufar Baghaei	OPAIC, New Zealand
Ryan Baker	University of Pennsylvania, USA
Gautam Biswas	Vanderbilt University, USA
Ig Ibert Bittencourt	Federal University of Alagoas, Brazil
Emmanuel Blanchard	IDÛ Interactive Inc., Canada
Nigel Bosch	University of Illinois Urbana-Champaign, USA
Jesus G. Boticario	UNED, Spain
Kristy Elizabeth Boyer	University of Florida, USA
Bert Bredeweg	University of Amsterdam, Netherlands
Christopher Brooks	University of Michigan, USA
Geiser Chalco Challco	ICMC/USP, Brasil
Maiga Chang	Athabasca University, Canada
Mohamed Amine Chatti	University of Duisburg-Essen, Germany
Min Chi	NC State University, USA
Andrew Clayphan	The University of Sydney, Australia
Cristina Conati	The University of British Columbia, Canada
Mark G. Core	University of Southern California, USA
Scotty Craig	Arizona State University, Polytechnic, USA
Mutlu Cukurova	University College London, UK
Ben Daniel	University of Otago, New Zealand
Diego Dermeval	Federal University of Alagoas, Brazil
Tejas Dhamecha	IBM, India
Barbara Di Eugenio	University of Illinois at Chicago, USA
Daniele Di Mitri	Open Universiteit, Netherlands
Vania Dimitrova	University of Leeds, UK
Peter Dolog	Aalborg University, Denmark
Fabiano Dorça	Universidade Federal de Uberlandia, Brazil
Mingyu Feng	SRI International, USA
Rafael Ferreira	Federal Rural University of Pernambuco, Brazil

Carol Forsyth	Educational Testing Service, USA
Davide Fossati	Emory University, USA
Reva Freedman	Northern Illinois University, USA
Dragan Gasevic	Monash University, Australia
Isabela Gasparini	UDESC, Brazil
Elena Gaudioso	UNED, Spain
Janice Gobert	Rutgers University, USA
Ashok Goel	Georgia Institute of Technology, USA
Ilya Goldin	2U, Inc., USA
Alex Sandro Gomes	Universidade Federal de Pernambuco, Brazil
Art Graesser	University of Memphis, USA
Monique Grandbastien	LORIA, Université de Lorraine, France
Gahgene Gweon	Seoul National University, South Korea
Jason Harley	University of Alberta, Canada
Andreas Harrer	University of Applied Sciences and Arts Dortmund, Germany
Peter Hastings	DePaul University, USA
Yuki Hayashi	Osaka Prefecture University, Japan
Tobias Hecking	University of Duisburg-Essen, Germany
Neil Heffernan	Worcester Polytechnic Institute, USA
Tsukasa Hirashima	Hiroshima University, Japan
Ulrich Hoppe	University Duisburg-Essen, Germany
Sharon Hsiao	Arizona State University, USA
Paul Salvador Inventado	California State University Fullerton, USA
Seiji Isotani	University of São Paulo, Brazil
Sridhar Iyer	IIT Bombay, India
G. Tanner Jackson	Educational Testing Service, USA
Patricia Jaques	UNISINOS, Brazil
Srecko Joksimovic	Teaching Innovation Unit and School of Education, University of South Australia, Australia
Pamela Jordan	University of Pittsburgh, USA
Sandra Katz	University of Pittsburgh, USA
Judy Kay	The University of Sydney, Australia
Fazel Keshtkar	St. John's University, USA
Simon Knight	University of Technology Sydney, Australia
Tomoko Kojiri	Kansai University, Japan
Amruth Kumar	Ramapo College of New Jersey, USA
Rohit Kumar	Raytheon BBN Technologies, Cambridge, MA, USA
Jean-Marc Labat	Université Paris 6, France
Sébastien Lallé	The University of British Columbia, Canada
H. Chad Lane	University of Illinois at Urbana-Champaign, USA
Nguyen-Thinh Le	Humboldt Universität zu Berlin, Germany
Blair Lehman	Educational Testing Service, USA
James Lester	North Carolina State University, USA
Chee-Kit Looi	National Institute of Education, Singapore
Yu Lu	Beijing Normal University, China

Kazuhisa Seta	Osaka Prefecture University, Japan
Lei Shi	University of Liverpool, UK
Sergey Sosnovsky	Utrecht University, Netherlands
Pierre Tchounikine	University of Grenoble, France
Maomi Ueno	The University of Electro-Communications DFKI, Japan
Carsten Ullrich	GmbH, Germany
Kurt Vanlehn	Arizona State University, USA
Julita Vassileva	University of Saskatchewan, Canada
Felisa Verdejo	UNED, Spain
Rosa Vicari	Universidade Federal do Rio Grande do Sul, Brazil
Erin Walker	Arizona State University, USA
Elle Wang	Arizona State University, USA
John Whitmer	Blackboard, Inc., USA
Beverly Park Woolf	University of Massachusetts, USA
Marcelo Worsley	Northwestern University, USA
Kalina Yacef	The University of Sydney, Australia
Elle Yuan Wang	Arizona State University, USA
Diego Zapata-Rivera	Educational Testing Service, USA
Jingjing Zhang	Beijing Normal University, China
Gustavo Zurita	Universidad de Chile, Chile

Additional Reviewers

Afzal, Shazia
Anaya, Antonio R.
Andrews-Todd, Jessica
Arevalillo-Herráez, Miguel
Arroyo, Ivon
Botelho, Anthony F.
Chavan, Pankaj
Chen, Chen
Chen, Penghe
Choi, Heeryung
Cochran, Keith
D'Mello, Sidney
Deep, Anurag
Deitelhoff, Fabian
Doberstein, Dorian
du Boulay, Benedict
Erickson, John
Galafassi, Cristiano
Gaweda, Adam
Gerritsen, David

Gitinabard, Niki
Harrison, Avery
Hartmann, Christian
Hayashi, Yusuke
Herder, Tiffany
Horiguchi, Tomoya
Hulse, Taylyn
Hutchins, Nicole
Hutt, Stephen
Ishola, Oluwabukola
Ju, Song
Kay, Judy
Kent, Carmel
Kojima, Kazuaki
Landers, Richard
Lawson, Marylynne
Lelei, David Edgar
Lin, Tao Roa
Madaio, Michael
Maehigashi, Akihiro

Malkiewich, Laura
Mao, Ye
Matsumuro, Miki
Mavrikis, Manolis
Mcbroom, Jessica
McNamara, Danielle
Memon, Muhammad Qasim
Minn, Sein
Mishra, Shitanshu
Mittal, Anant
Molenaar, Inge
Morita, Junya
Munshi, Anabil
Negi, Shivsevak
Nikolayeva, Iryna
Oertel, Catharine
Okoilu, Ruth
Patikorn, Thanaporn
Praharaj, Sambit
Rajendran, Ramkumar
Rodriguez, Fernando
Saha, Swarnadeep
Shahriar, Tasmia
Shen, Shitian

Shimmei, Machi
Smith, Hannah
Smith, Karl
Snyder, Caitlin
Stewart, Angela
Strauss, Sebastian
Sánchez-Elvira Paniagua, Angeles
Tan, Hongye
Thompson, Craig
Toda, Armando
Tomoto, Takahito
Tsan, Jennifer
Vanlehn, Kurt
Wang, April
Wiggins, Joseph
Yamamoto, Sho
Yang, Xi
Yett, Bernard
Yi, Sherry
Ying, Kimberly
Yokoyama, Mai
Zhang, Ningyu
Zhou, Guojing

Abstracts of Keynotes

Learning to Learn Differently

Jutta Treviranus

OCAD University in Toronto, Canada
jtreviranus@ocadu.ca

Abstract. Our data-driven decision processes reduce diversity & complexity. All data is about the past. This leads to bias against outliers, small minorities, and novel changes. Most artificial intelligence amplifies and automates this pattern. This leads to disparity and blind spots in education and research. How can we design intelligence that recognizes, understands and works for diverse learners and educators?

Human Development and Augmented Intelligence

Nancy Law

The University of Hong-Kong, SAR China
nlaw@hku.hk

Abstract. Records of human civilizations date back to more than five millennia. The history of human civilization is deeply intertwined with its history of technological advancement. While humans are not alone in their ability to create tools for augmented performance, humans are the only species that create and use technology to connect minds over time and space. Hence human society has been able to advance not only through evolution, but more importantly, through learning. The twentieth century has brought a major technological breakthrough in creating machines that learn, machines that provide humans with augmented intelligence. Scientific investigations of human intelligence and human learning have inspired and benefitted from technological advances in artificial intelligence from the start of these efforts. Drawing on current studies on human development in the digital age, this talk explores how human development may be affected ontologically in the increasingly digitally connected and augmented world that we are in, and its social implications, particularly for education.

Duolingo: Free Language Education for the World

Luis von Ahn

Carnegie Mellon University, USA
biglou@cs.cmu.edu

Abstract. Duolingo is the free language education platform that has motivated over 300 million people worldwide to learn a language. The platform's digital-native experience, intuitive design and data-based approach to optimizing education has lead to its selection by Apple as iPhone App of the Year by Google, as "Best of the Best Android App" 2 years in a row. Luis will talk about the company's trajectory and mission, the future of education, and the role of computer science in optimizing the learning process in ways that were previously impossible.

Contents – Part I

Contents – Part II

Doctoral Consortium

Industry Papers

Workshop Papers

Towards the Identification
of Propaedeutic Relations in Textbooks

Giovanni Adorni, Chiara Alzetta, Frosina Koceva, Samuele Passalacqua$^{(\boxtimes)}$,
and Ilaria Torre

Department of Informatics, Bioengineering, Robotics and Systems Engineering,
University of Genoa, Genoa, Italy
{giovanni.adorni,ilaria.torre}@unige.it,
{chiara.alzetta,frosina.koceva}@edu.unige.it,
samuele.passalacqua@dibris.unige.it

Abstract. As well-known, structuring knowledge and digital content
has a tremendous potential to enhance meaningful learning. A straightforward approach is representing key concepts of the subject matter and
organizing them in a knowledge structure by means of semantic relations.
This results in hypergraphs with typed n-ary relationships, including the
so-called prerequisite or propaedeutic relations among concepts. While
extracting the whole concept graph from a textbook is our final goal,
the focus of this paper is the identification of the propaedeutic relations
among concepts. To this aim, we employ a method based on burst analysis and co-occurrence which recognizes, by means of temporal reasoning,
prerequisite relations among concepts that share intense periods in the
text. The experimental evaluation shows promising results for the extraction of propaedeutic relations without the support of external knowledge.

Keywords: Relation extraction · Knowledge structure ·
Temporal reasoning

1 Introduction

Concept Maps (CM) are a way for representing key concepts of the subject matter and organizing them in a knowledge structure by means of semantic relations.
They can be used to automatically generate lesson plans [2] and evaluation tests
[37]. While several methods exist (e.g. [10,34]) to face the issue of automatic
concept extraction, the identification of prerequisite relations among concepts is
still an open research problem, even though there is a long-standing interest on
learning hierarchies since 1971 Gagné's work [18].

In [1] we proposed an approach for prerequisite relation (**PR relation**, hence-
forth) extraction from an educational textbook, by combining two methods:
(1) using temporal ordering and co-occurrence of concepts, (2) using the structure of the textbook and the relevance of the terms. In this paper we propose

S. Isotani et al. (Eds.): AIED 2019, LNAI 11625, pp. 1–13, 2019.
https://doi.org/10.1007/978-3-030-23204-7_1

an alternative approach to method 1: instead of using only co-occurrence of concepts and temporal ordering, we propose the use of *burst analysis* [22] *based on co-occurrence and combined with temporal reasoning.* Burst analysis has already been used in text mining for summarization [31] and relation extraction [36]. It is based on the idea that terms in a text have bursting intervals, i.e. portions of text where they are particularly prominent. Relations between pairs of concepts are derived by observing how pairs of burst intervals that belong to different terms are positioned in the text flow.

Co-occurrence based methods are the core of many approaches for PR relation identification [20,24]. However, while co-occurrence is an intuitive condition for PR, high co-occurrence is not necessarily a measure of PR strength, since it could identify other types of relations, such as associations, taxonomic relations and co-requisites among others.

For the experimental evaluation we compared our method for PR relation extraction, based on burst analysis and co-occurrence, against a baseline method based exclusively on co-occurrence. Therefore we asked domain experts to manually annotate the relations automatically extracted by both methods and compared the results. The experimental evaluation provides promising results in terms of Accuracy of PR identification and Precision of top identified relations. In particular, preliminary results suggest the effectiveness of burst analysis to filter out relationships between concepts that co-occur frequently but are not relevant for the educational purpose in terms of PR relations.

The main contribution of this paper to the literature is the improvement of prerequisite extraction through an unsupervised and domain independent approach, that exploits only the unstructured content of a digital textbook, i.e. without using external resources such as Wikipedia links [32] or other knowledge bases [29]. Notice that the algorithms and the prerequisite datasets (both manually annotated and automatically generated) are available for use in future research (teldh.dibris.unige.it/projects/).

2   Background and Related Work

Relationship extraction is a well-known task of Information Extraction. Its main goal is to identify relations between entities in a document (see [12,30] for comprehensive surveys) in order to give a structured representation of the information in the text. Many relationship types can be identified, such as temporal [26] and lexical-semantic [9] only to name a few. We notice a growing interest towards the analysis of scientific and educational content [5,17,19]. Pedagogical relations are in fact of great interest in the AIED community for automatic construction of ontologies and CM [14].

In this line of research [38] and, more recently, [33] retrieves relations exploiting syntactic analysis of sentences in a text and use them to automatically build CM. More similarly to our approach, [23,36] use burst analysis to recognize relationships between concepts and draw them as links in a CM. Contrary to us, their methods extract all the possible relations between pairs of concepts, while our effort is to identify specifically PR relations.

The PR relation is a dependency relation defining precedence between two concepts t_u and t_v. From a cognitive perspective, it represents what a learner must know/study (concept t_u) before approaching concept t_v. In CM a concept is an atomic piece of knowledge of a domain. Contrary to [34], who used latent semantic analysis, we address concept extraction from educational materials as a task of terminology extraction (i.e. the most relevant terms of a document are the domain concepts). By definition, the main properties of a PR relation are the followings: (1) binary relation: it involves pairs of concepts; (2) anti-reflexive relation: concept t_u cannot be a prerequisite of itself; (3) transitive relation: if $t_u \prec t_v$ and $t_v \prec t_z$, than $t_u \prec t_z$ (see for instance the CM of Fig. 1: *browser* $\prec$ *HTTP* and *HTTP* $\prec$ *WWW*, hence *browser* $\prec$ *WWW*); (4) anti-symmetric relation: if concept $t_u \prec t_v$, than $t_v \prec t_u$ must not hold (in the map below, *network* $\prec$ *internet*, so *internet* cannot be prerequisite of *network*). These conditions imply that CM is acyclic.

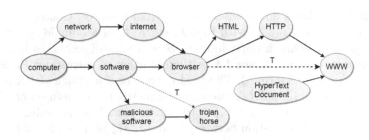

Fig. 1. Small Concept Map.

Extracting PR from educational materials is a relatively new field of research. Methods for PR extraction may rely on graph analysis [8,21], structured representations of knowledge, such as Wikipedia [32] or other knowledge bases [29] or, more similar to our approach, on linguistic information of textual documents [20], often enriched with external knowledge [25,35]. In [27] the authors propose two approaches based on feature extraction and machine learning to map courses from different universities onto a space of concepts. Likewise, in [29] the authors define various features and train a classifier that can identify PR relations from video transcripts. Both methods use semantics and context based features. [19] introduces a weak ontology driven approach: they extract lexical and semantic features and apply machine learning techniques for detecting a PR between learning objects. Contrary to them, we do not need an annotated dataset for feature extraction, which makes our approach not only faster but also more domain independent. Another popular approach is based on the RefD metric [24], which measures the prerequisite relation between two concepts by computing how differently they refer to each other. This co-reference might be intended as a co-occurence of terms in a window of textual context or as an explicit reference, like a citation. Contrary to RefD, that works well on medium-large sets of documents, our approach works also on small data, like a book chapter. We perform PR extraction starting from the educational material where concepts are described since a PR relation strictly depends on the writer's communicative intent and teaching style. As an example, consider top-down and bottom-up approaches, both used in textbooks: the former tends to present a topic starting from broad concepts and definitions, while the latter starts from examples. Choosing one approach over the other influences the direction of the PR relation in the text (i.e. from a general concept to a specific one, or vice versa). In other cases, the relationship identification is made easier by the hierarchical lexical relation between terms representing concepts: *network* and *Local Area Network* not only share a PR relation, but also the lexical head. In the same way, non-hierarchical lexical relations are useful to spot misleading prerequisite pairs, like co-requisites, which are usually presented together for providing complementary knowledge and have a non-hierarchical nature. To clarify, imagine a possible description of the HTTP protocol: the *client*, who sends a request to a *server*, is not a prerequisite of *server*, nor vice versa.

Notation. We define a document D as a textual resource. Concept extraction returns a terminology $T \in D$ with $t_u \in T$, where t_u is a domain-specific term. Following [36], we define a burst interval B as a slice of sentences in D where the occurrences of a term t_u are denser than in other segments and $B_{t_u}[i]$ is the *i-th* bursting period of term t_u. The final output of concepts and PR relations extraction is a concept graph G represented, similarly to [35], as a set of triples in the form $G = \{(t_u, t_v, p) | t_u, t_v \in T, p \geq 0\}$, where p is a positive value indicating the strength of the PR relation between t_u and t_v (t_u prerequisite of t_v).

3   Proposed Method

In this paper we propose a new approach for building the concept graph. As mentioned above, co-occurrence is not always a satisfying measure of PR relation, since it often overestimates PR relations, including other kinds of relations between terms that frequently co-occur. Moreover, deciding which concept plays the role of prerequisite in a pair, by considering only their temporal order of appearance in the text, may result in a PR relation with wrong direction, where the prerequisite has been extracted as consequent and vice versa. Actually, concepts in educational textbooks may appear with different scopes along the text flow: first they might be just mentioned or introduced, then used inside their definition and later recalled to explain some new information. Therefore, by viewing the textbook as a stream of sentences, one could analyze these changes and better understand how the relation between two concepts evolves in the document.

Kleinberg formally defines and models the periods of an event along a time series (e.g., a stream of documents such as e-mails or news articles) as a two state automaton in which the event is in the first state if it has a low occurrence, but then it moves to the second state if its occurrence rises above a certain threshold, and eventually it goes back to the first state if its occurrence goes below the threshold [22]. These transitions are repeated along the entire duration of a time series and the periods in which the event remains in the second state are called *burst intervals*. If applied to a single document rather than a set, Kleinberg's algorithm can be used to detect the bursting intervals of keywords [23,36]. Intuitively, a rising of *bursting activity* associated with a concept signals its appearance or re-appearance in the flow of the discourse, revealing that certain features, mainly the frequency of the concept in that interval, are sharply rising [22] and suggesting that the concept has become more prominent. With the burst detection we gain not only the intervals in which a concept t_u is "bursty" (i.e., $B_{t_u}[i]$), but also the hierarchical level of "burstiness" of these intervals. In fact, bursts associated with an event form a nested structure, with a long burst of low intensity potentially containing several shorter bursts of higher intensity inside it [22]. Moreover, the employ of burst analysis also allows us to analyze different types of temporal patterns established by two concepts while they are used in the text. Our interest in applying burst analysis largely arises from the temporal nature of the PR relation: instead of using co-occurrence as criterion for the extraction, we propose to extract burst intervals of the concepts and then apply spatial-temporal reasoning on the extracted patterns in order to identify PR relations. As a matter of fact, the comparison of temporal patterns allows a richer analysis: by analyzing the pairs of intervals between two different concepts t_u and t_v, we can exploit Allen's interval algebra [3] to capture and formalize their temporal relations. Among Allen's basic relations, we used only a subset of temporal patterns, for which we could recognize some meaningful interpretation with respect to the PR relation. Consistently with our main assumption of co-occurrence as a necessary (though not sufficient) condition for a PR relation, all adopted patterns imply a co-occurrence of two terms within a temporal window. Even the *before* relation, if detected by applying a maximum gap between two intervals, entails co-occurrence. Our selection is shown in Fig. 2. For simplicity, t_u and t_v are referred as X and Y, and $B_X[i]$ as $B_{X,i}$.

Allen's Relations. Contrary to [23,36], where combinations of burst intervals were used for identifying generic relationships between concepts, we seek to recognize the PR relation. To this aim, we make the following assumptions in order to give a prerequisite interpretation to Allen's relations.

$B_{X,i}$ rel $B_{Y,j}$	pattern	$B_{X,i}$ rel $B_{Y,j}$	pattern
equals	|-- $B_{X,i}$ --| |-- $B_{Y,j}$ --|	overlap	|--- $B_{X,i}$ ---| 　|---- $B_{Y,j}$ ----|
before	|-- $B_{X,i}$ --| 　　　|-- $B_{Y,j}$ --|	meets	|-- $B_{X,i}$ --| 　　|-- $B_{Y,j}$ --|
starts	|-- $B_{X,i}$ --| |----- $B_{Y,j}$ -----|	finishes	|-- $B_{X,i}$ --| |---- $B_{Y,j}$ -----|
includes	|----- $B_{X,i}$ -----| 　|-- $B_{Y,j}$ --|		

Fig. 2. Burst relations interpretation

- $B_{X,i}$ *equals* $B_{Y,j}$. This pattern emphasizes the relatedness of two concepts without necessarily implying the existence of a PR relation. In these cases, some kind of relation between X and Y is highly probable, but we cannot say whether X is a prerequisite of Y or vice versa. Moreover, this pattern may not reveal a PR relation at all, since *equal* is very common when two concepts are co-requisites. Consequently, we assume *equal* has a low potential to reveal a prerequisite.

- $B_{X,i}$ *before* $B_{Y,j}$. Since a prerequisite commonly precedes its subsidiary concept, in this pattern X could be probably a prerequisite of Y. We do not consider pairs of bursts with a *before* pattern when their gap exceeds a certain number of sentences, since in such cases the two concepts are almost certainly too far to establish a direct PR relation.

- $B_{X,i}$ *overlaps* $B_{Y,j}$. If concept X is prerequisite of Y, in the text we would expect at least some cases where X is first explained and shortly after Y is introduced, with a certain area of overlapping. Thus, this pattern is highly informative for the existence of a PR relation.

- $B_{X,i}$ *meets* $B_{Y,j}$. Here the bursting period of concept X stops exactly when concept Y begins to be more intense in the text. The two concepts are too near to completely disregard the possibility of a PR relation, and yet, as already mentioned, the proximity is not per se a sufficient condition for a PR relation. Hence, we assume this pattern has a moderate force to suggest a prerequisite.

- $B_{X,i}$ *starts* $B_{Y,j}$. The *starts* pattern can be representative of situations where two concepts emerge almost simultaneously (most likely because they are highly related), but then the author temporarily abandons one of the two concepts while he further develops the other. According to this observation, there is a moderate/high chance that X is prerequisite of Y.

- $B_{X,i}$ *includes* $B_{Y,j}$. This pattern shows a concept being discussed within the span of a more long-standing concept, with the longer one that totally encompasses the smaller one. The nested concept can be very likely a specification of the embedding concept at a more fine-grained level (and thus a PR relation can be appropriately traced), or sometimes it could represent a detour from the main line of discussion (and thus disclosing a learning content that is suggested for a deeper analysis). For these reasons, *includes* is highly informative.

- $B_{X,i}$ *finishes* $B_{Y,j}$. Compared to other patterns, here a PR between X and Y is harder to assume, since $B_{Y,j} \prec B_{X,i}$. Nevertheless, a low weight should be still considered to deal with cases of bottom-up explanations.

Algorithm Description. The algorithm is structured in three phases (the pseudocode is available at teldh.dibris.unige.it/projects/): a burst extraction phase (*ExtractBursts*), a temporal pattern detection phase (*DetectTemporalPatterns*) and a prerequisite extraction phase (*ExtractPrereqs*). The burst extraction phase takes as inputs a document D containing the full text to analyze, a terminology T consisting in a list of terms appearing in D and a set of parameters for constructing the Markov's chain according to Kleinberg's description (the base s of the exponential distribution used for modeling the event frequencies, the coefficient γ for the transition costs between states, and the desidered level l within the hierarchy of the extracted intervals). D is transformed into an ordered list of sentences by means of sentence splitting, and the result is $\mathcal{Q}_D = \{q_1, q_2, ..., q_i\}$, where q_i is the i-th sentence of D. A dictionary $\mathcal{O}$ is built for mapping each concept in T with the indexes of sentences where it occurs. Burst intervals are identified for every concept t given its list $\mathcal{O}_{t_u}$ of sentence indexes, e.g. the burst intervals of t_u are: $B_{t_u} = \{[b_{starts_1} - b_{ends_1}], [b_{starts_2} - b_{ends_2}], ..., [b_{starts_i} - b_{ends_i}]\}$. The function $kleinberg(\mathcal{O}[t], s, \gamma, l)$ involves the construction of an infinite hidden Markov model as described in [22]. For this particular procedure we relied on an implementation of Kleinberg's algorithm available for Python[1] that needs to be fed with $\mathcal{O}[t]$. In addition, two parameters, s and γ, need to be set in advance: the former controls the exponential distribution from which an event is assumed to be drawn (i.e., how frequent an event must be in order to trigger the detection of a burst); the latter modifies the transition cost to a higher state. Higher values of s increase the strictness of the algorithm's criterion for how dramatic an increase of activity has to be in order to be considered as a burst; higher values of γ mean that a burst must be sustained over longer periods of time in order to be recognized [6]. During the tuning of these parameters we opt for minimal permitted values ($s = 1.05$, $\gamma = 0.0001$) with the aim of maximizing the extraction of bursting intervals. In the phase $DetectTemporalPatterns$, every pair of bursts $B_{t_u}[i]$ and $B_{t_v}[j]$ (belonging to two distinct concepts t_u and t_v) are compared, and temporal relations are identified by performing pattern matching. A weight W_r is therefore assigned to the identified Allen's relation r, according to the considerations described in Section $Allen's$ $relations$. Similarly to [23], we also follow the idea that adding a tolerance gap is necessary in this stage. As a matter of fact, by considering only the exact starting/ending/meeting point of two bursts, we can hardly find a complete match, while by adding a tolerance gap the method becomes more permissive during the identification of temporal patterns. The result of the current phase is a square matrix $\mathcal{P}$ of size $|\mathcal{B}| \times |\mathcal{B}|$, where $|\mathcal{B}|$ is the total number of extracted bursts, reporting a weight for each pair of bursts as resulted from the pattern matching procedure (the weight is zero only for bursts pairs with a distant $before$ relation and for bursts pairs belonging to the same concept). In the PR extraction phase, the matrix obtained from the previous step is taken as a basis for constructing an undirected square matrix $\mathcal{M}$ of size $|T| \times |T|$: for each two distinct concepts t_u and t_v, all the weights associated to the burst pairs belonging to t_u and t_v are combined and normalized by means of the PR formula below, i.e. a modified version of the normalized relation weight (NRW) formula described in [23]. The resulting weight is stored both in $\mathcal{M}_{t_u,t_v}$ and $\mathcal{M}_{t_v,t_u}$. Given $X, Y \in T$ and $X \neq Y$, we compute $PR_{X,Y}$ as the sum of the relation weights W_r assigned to the recognized Allen's patterns, then we normalize this value by taking into account the frequency f of X and Y in their respective intervals $B_{X,i}$ and $B_{Y,j}$, the total length (measured in sentences) of

---

[1] Library $pybursts$, https://pypi.org/project/pybursts/0.1.1/.

all bursts of X and Y, and also the number of these bursts[2]. $\mathcal{M}$ is therefore converted into a direct matrix, and the direction is given by comparing the first bursts of the concepts in the pair. A directed graph $\mathcal{G}$, with concepts as nodes and PR relations as edges, can be finally built from $\mathcal{M}$.

$$PR_{X,Y} = \sum_i \left(W_r \; \frac{f(X, B_{X,i}) \times |\mathbf{B_X}|}{\sum_i |B_{X,i}|} \; \frac{\sum\limits_{j \in rel\,(B_{Y,j}, B_{X,i})} f(Y, B_{Y,j}) \times |\mathbf{B_Y}|}{\sum_j |B_{Y,j}|} \right)$$

4 Experimental Evaluation

Goal. The method we proposed for PR relation identification is based on the assumption that co-occurrence of concepts is likely a condition for the existence of PR relation between two concepts. In general, high co-occurence frequency is a good indicator of relations (as shown in previous works [11,24]), thus it can also underpin other kinds of relations besides PR. The goal of our evaluation is to investigate the following two hypotheses: (**HP1**) burst analysis, as in our proposed method, could perform better than methods based only on co-occurrence frequency, reducing false positive and false negative PR relations; (**HP2**) burst-based method for PR identification could reduce false positive PR relations when two high co-occurring concepts are related by a relation that is not PR.

Methodology. We tested our Burst-based method on a chapter of a computer science textbook, "*Computer Science: An Overview*" [7]. The output of the algorithm is compared against a method based on co-occurrence of terms in a window of context. Both methods are manually evaluated and compared by domain experts. In the following we call such methods respectively Burst-based method (**BM**) and Frequency-based method (**FM**). To test HP1, we computed the Accuracy [28] of BM and FM on a set of 150 randomly selected relations from the results of BM and FM. To test HP2, we computed the Precision [28] of the Top 150 PR relations returned by the algorithms and therefore analyzed the types of error. Details are in the following.

Corpus and Concept Extraction. For the evaluation we used Sect. 4 "Networking and the Internet" of the above mentioned textbook [7] (20,378 tokens, distributed over 751 sentences). Concept extraction is addressed by relying on Text-To-Knowledge platform [13]. The extracted terminology contained both single nominal structures (e.g. *computer*) and complex nominal structures with modifiers (e.g. *hypertext transfer protocol*). The set of extracted terms was manually revised by three experts and missing terms were added. The final terminology consists of 125 terms, for a total of 15,500 pairs of distinct terms (representing the candidate PRs), excluding symmetric pairs.

---

[2] Note that the current formula takes into account all the relations where an Allen's pattern is recognized, while we are working on an improved version that limits them to relations where the subsidiary concept exhibits high burstiness.

PR Relation Extraction. We ran the burst algorithm as described above, assigning W_r weights to burst relations (according to the assumptions discussed in Sect. 3) on a 10pt scale, and thus we obtained a direct matrix. On the other hand, the FM method computes how many times two terms of the terminology appear together in a three sentences span (i.e. the one where a term appears, the preceding and the following). The output of FM is a direct matrix as well: values represent co-occurrence frequencies of each term pairs and the direction is given by the order of first occurrence of the terms.

Experts' Annotation. In order to evaluate **HP1** and **HP2**, we asked domain experts to annotate the extracted pairs of concepts, in line with [15,20,24,25]. For the first hypothesis (**HP1**), we created a sample by randomly selecting, for each method, 150 pairs of concepts using the following criteria: (1) 50 pairs identified by the method as having PR relationship, (2) 50 pairs identified by the method as not having PR relation, (3) 50 pairs (among those not selected for the other partitions) regardless of whether they have been identified with a PR relation or not. The third set was done to make the sample more homogeneous with respect to the algorithms' outputs, which are significantly unbalanced (i.e. only 5.11% of the pairs obtained a PR label for the BM method and 4.07% for the FM method). To evaluate the second hypothesis (**HP2**), we selected the Top 150 relations returned by each method, ordered according to their weight.

In both cases (**HP1** and **HP2** evaluations), two domain experts were asked to annotate the pairs in the two samples, assigning what they believed to be the correct label for that pair. The guidelines for evaluation explicitly dictated to read the textbook and assign labels based on how concepts are addressed in the text. Moreover, a third expert was asked to analyze cases of disagreement between annotators in order to check if disagreement was due to annotators' subjectivity or to annotation errors (e.g., distraction, misinterpretation of guidelines or misinterpretation of the text). The risk of errors is well-known in the literature as well as the disagreement due to annotators' subjectivity [4,15,16].

4.1 Results and Discussion

In order to test **HP1** (i.e. if our BM method could produce less false positives and false negatives compared to co-occurrence-only based methods), we computed the Accuracy, as defined in [28], of BM and FM. To this aim, we compared the output of the algorithms for the 150 randomly selected pairs of concepts against each expert's annotation, and then we took their average score. Notice that in this evaluation our aim was to assess the correctness of the PR relation identification, not its strength. Results in Fig. 3 show that BM Accuracy is 0.84, slightly outperforming FM, whose Accuracy is 0.77.

Considering **HP2** (i.e. BM could reduce false positives in cases of frequently co-occurring concepts connected by a relation that is not a PR relation), we took into account the Top 150 relations returned by each method. Obviously, for the FM method such relations are those whose concept pairs have the highest co-occurrence frequency. First of all, we computed Precision (as defined in [28]) of both methods against the experts' annotations. As displayed in Fig. 3, Precision of BM outperforms FM with a decreasing trend, suggesting that BM method works better especially in cases of high co-occurrence frequency. To deepen this analysis, we also performed a qualitative analysis on the error types occurring in the top 85 relations (as shown in the chart, at point 85 we have the minimum distance between BM and FM Precision). The third expert was asked to classify the errors as: (a) very distant relation, (b) relation different from PR, (c) inverse relation, (d) no relation. As can be seen in Fig. 3, *relations different from PR* are 12% with BM method and 41% with FM method, confirming that BM method allows to reduce the identification of non-PR relations between high co-occurring concepts. These preliminary results seem promising if we also consider that, by tuning the weights of Burst relations, we could further improve the outcomes.

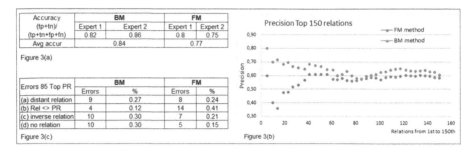

Fig. 3. Results: (a) Accuracy, (b) Precision top 150 PR, (c) Errors top 85 PR

5   Conclusion and Limits

In this paper we presented a new method based on Burst analysis, co-occurrence, and temporal reasoning for PR relation extraction from educational materials. Contrary to most approaches for prerequisite extraction, we developed an unsupervised method able to extract knowledge from unstructured text, without exploiting external structured knowledge and with light need of training. This represents our main contribution to the literature, with the goal to provide a method that is not heavily domain-dependent and not complex to be applied.

Current results in terms of Accuracy and Precision are encouraging, even though further evaluations are needed to draw conclusions. In this respect, our future work includes refining the PR formula, applying the method to different domains and types of resources, and evaluating it on larger corpora.

Moreover, we are confident that a major improvement of the BM method can be obtained by taking into account more complex patterns and not only making one-by-one comparisons of pairs of intervals. By taking benefit from this knowledge we could enhance the method with a much deeper analysis of how the relation between two concepts evolves across time.

Further research directions are analyzing and interpreting the annotators agreement and improving the burst-based algorithm with machine learning methods: starting from [4], we are collecting a gold dataset of educational resources annotated with PR relations that can be used to tune the weights of the temporal patterns.

References

1. Adorni, G., Dell'Orletta, F., Koceva, F., Torre, I., Venturi, G.: Extracting dependency relations from digital learning content. In: Serra, G., Tasso, C. (eds.) IRCDL 2018. CCIS, vol. 806, pp. 114–119. Springer, Cham (2018). https://doi.org/10.1007/978-3-319-73165-0_11
2. Adorni, G., Koceva, F.: Designing a knowledge representation tool for subject matter structuring. In: Croitoru, M., Marquis, P., Rudolph, S., Stapleton, G. (eds.) GKR 2015. LNCS (LNAI), vol. 9501, pp. 1–14. Springer, Cham (2015). https://doi.org/10.1007/978-3-319-28702-7_1
3. Allen, J.F.: Maintaining knowledge about temporal intervals. Commun. ACM **26**(11), 832–843 (1983)
4. Alzetta, C., Koceva, F., Passalacqua, S., Torre, I., Adorni, G.: PRET: prerequisite-enriched terminology. A case study on educational texts. In: Proceedings of the Fifth Italian Conference on Computational Linguistics (2018)
5. Augenstein, I., Das, M., Riedel, S., Vikraman, L., McCallum, A.: SemEval 2017 task 10: scienceIE-extracting keyphrases and relations from scientific publications. In: Proceedings of the 11th International Workshop on Semantic Evaluation (SemEval-2017), pp. 546–555 (2017)
6. Binder, J.: Package 'bursts': Markov model for bursty behavior in streams. R package version 1.0-1 (2014)
7. Brookshear, G., Brylow, D.: Networking and the Internet. In: Computer Science: An Overview, Global Edition. Pearson Education Limited (2015). Chapter 4
8. Brusilovsky, P., Vassileva, J.: Course sequencing techniques for large-scale web-based education. Int. J. Continuing Eng. Educ. Life long Learn. **13**(1/2), 75–94 (2003). https://www.inderscience.com/info/inarticletoc.php?jcode=ijceell&year=2003&vol=13&issue=1/2
9. Camacho-Collados, J., et al.: SemEval-2018 task 9: hypernym discovery. In: Proceedings of the 12th International Workshop on Semantic Evaluation, pp. 712–724 (2018)
10. Cimiano, P., Völker, J.: Text2Onto. In: Montoyo, A., Muñoz, R., Métais, E. (eds.) NLDB 2005. LNCS, vol. 3513, pp. 227–238. Springer, Heidelberg (2005). https://doi.org/10.1007/11428817_21
11. Clariana, R.B., Koul, R.: A computer-based approach for translating text into concept map-like representations. In: Proceedings of the First International Conference on Concept Mapping, pp. 14–17 (2004)
12. de Abreu, S.C., Bonamigo, T.L., Vieira, R.: A review on relation extraction with an eye on Portuguese. J. Braz. Comput. Soc. **19**(4), 553 (2013)
13. Dell'Orletta, F., Venturi, G., Cimino, A., Montemagni, S.: T2k^2: a system for automatically extracting and organizing knowledge from texts. In: Proceedings of the Ninth International Conference on Language Resources and Evaluation (LREC-2014) (2014)
14. Devedzic, V.: Education and the semantic web. Int. J. Artif. Intell. Educ. **14**(2), 165–191 (2004)

15. Fabbri, A.R., et al.: TutorialBank: a manually-collected corpus for prerequisite chains, survey extraction and resource recommendation. In: ACL (2018)
16. Fort, K., Nazarenko, A., Rosset, S.: Modeling the complexity of manual annotation tasks: a grid of analysis. In: International Conference on Computational Linguistics, pp. 895–910 (2012)
17. Gábor, K., Buscaldi, D., Schumann, A.-K., QasemiZadeh, B., Zargayouna, H., Charnois, T.: SemEval-2018 Task 7: semantic relation extraction and classification in scientific papers. In: Proceedings of International Workshop on Semantic Evaluation (SemEval-2018), New Orleans, LA, USA (2018)
18. Gagné, R.M.: Learning hierarchies. In: Merrill, M.D. (ed.) Instructional Design: Readings (Englewood Clis, NJ: Prentice-Hall, 1968), pp. 118–131 (1971)
19. Gasparetti, F., Medio, C.D., Limongelli, C., Sciarrone, F., Temperini, M.: Prerequisites between learning objects: automatic extraction based on a machine learning approach. Telematics Inform. 35(3), 595–610 (2018)
20. Gordon, J., Zhu, L., Galstyan, A., Natarajan, P., Burns, G.: Modeling concept dependencies in a scientific corpus. In: Proceedings of the 54th Annual Meeting of the Association for Computational Linguistics (Volume 1: Long Papers), vol. 1, pp. 866–875 (2016)
21. John, R.J.L., McTavish, T.S., Passonneau, R.J.: Semantic graphs for mathematics word problems based on mathematics terminology. In: EDM (Workshops) (2015)
22. Kleinberg, J.: Bursty and hierarchical structure in streams. Data Min. Knowl. Disc. 7(4), 373–397 (2003)
23. Lee, S., Park, Y., Yoon, W.C.: Burst analysis for automatic concept map creation with a single document. Expert Syst. Appl. 42(22), 8817–8829 (2015)
24. Liang, C., Wu, Z., Huang, W., Giles, C.L.: Measuring prerequisite relations among concepts. In: EMNLP pp. 1668–1674 (2015)
25. Liang, C., Ye, J., Wang, S., Pursel, B., Giles, C.L.: Investigating active learning for concept prerequisite learning. In: Proceedings of EAAI (2018)
26. Ling, X., Weld, D.S.: Temporal information extraction. In: AAAI, vol. 10, pp. 1385–1390 (2010)
27. Liu, H., Ma, W., Yang, Y., Carbonell, J.: Learning concept graphs from online educational data. J. Artif. Intell. Res. 55, 1059–1090 (2016)
28. Olson, D.L., Delen, D.: Advanced Data Mining Techniques. Springer, Heidelberg (2008). https://doi.org/10.1007/978-3-540-76917-0
29. Pan, L., Wang, X., Li, C., Li, J., Tang, J.: Course concept extraction in MOOCs via embedding-based graph propagation. In: Proceedings of the Eighth International Joint Conference on Natural Language Processing (Volume 1: Long Papers), vol. 1, pp. 875–884 (2017)
30. Sarawagi, S., et al.: Information extraction. Found. Trends Databases 1(3), 261–377 (2008)
31. Subasic, I., Berendt, B.: From bursty patterns to bursty facts: the effectiveness of temporal text mining for news. In: Proceedings of ECAI 2010: 19th European Conference on Artificial Intelligence, pp. 517–522 (2010)
32. Talukdar, P.P., Cohen, W.W.: Crowdsourced comprehension: predicting prerequisite structure in Wikipedia. In: Proceedings of the Seventh Workshop on Building Educational Applications Using NLP, pp. 307–315. Association for Computational Linguistics (2012)
33. Thenmozhi, D., Aravindan, C.: An automatic and clause-based approach to learn relations for ontologies. Comput. J. 59(6), 889–907 (2016)

34. Villalon, J., Calvo, R.A.: Concept extraction from student essays, towards concept map mining. In: Ninth IEEE International Conference on Advanced Learning Technologies, ICALT 2009, pp. 221–225. IEEE (2009)
35. Wang, S., et al.: Using prerequisites to extract concept maps from textbooks. In: Proceedings of the 25th ACM International on Conference on Information and Knowledge Management, pp. 317–326. ACM (2016)
36. Yoon, W.C., Lee, S., Lee, S.: Burst analysis of text document for automatic concept map creation. In: Ali, M., Pan, J.-S., Chen, S.-M., Horng, M.-F. (eds.) IEA/AIE 2014. LNCS (LNAI), vol. 8482, pp. 407–416. Springer, Cham (2014). https://doi.org/10.1007/978-3-319-07467-2_43
37. Žitko, B., Stankov, S., Rosić, M., Grubišić, A.: Dynamic test generation over ontology-based knowledge representation in authoring shell. Expert Syst. Appl. **36**(4), 8185–8196 (2009)
38. Zouaq, A., Nkambou, R.: Building domain ontologies from text for educational purposes. IEEE Trans. Learn. Technol. **1**(1), 49–62 (2008)

Investigating Help-Giving Behavior in a Cross-Platform Learning Environment

Ishrat Ahmed[1]($\boxtimes$), Areej Mawasi[2], Shang Wang[2], Ruth Wylie[2],
Yoav Bergner[3], Amanda Whitehurst[4], and Erin Walker[1]

[1] University of Pittsburgh, Pittsburgh, USA
{isa14,eawalker}@pitt.edu
[2] Arizona State University, Tempe, USA
{amwassi,Ruth.Wylie,swang158}@asu.edu
[3] New York University, New York, USA
ybb2@nyu.edu
[4] STEMteachersPHX, Gilbert, USA
amanda@stemteachersphx.org

Abstract. A key promise of adaptive collaborative learning support is the ability to improve learning outcomes by providing individual students with the help they need to collaborate more effectively. These systems have focused on a single platform. However, recent technology-supported collaborative learning platforms allow students to collaborate in different contexts: computer-supported classroom environments, network based online learning environments, or virtual learning environments with pedagogical agents. Our goal is to better understand how students participate in collaborative behaviors across platforms, focusing on a specific type of collaboration - help-giving. We conducted a classroom study (N = 20) to understand how students engage in help-giving across two platforms: an interactive digital learning environment and an online Q&A community. The results indicate that help-giving behavior across the two platforms is mostly influenced by the context rather than by individual differences. We discuss the implications of the results and suggest design recommendations for developing an adaptive collaborative learning support system that promotes learning and transfer.

Keywords: Adaptive collaborative learning support ·
Intelligent collaborative support · Help-giving-behavior ·
Motivation

1 Introduction

Adaptive collaborative learning support (ACLS) provides intelligent support to enhance collaborating students' learning outcomes [10]. There are many ACLS systems that have been applied in different contexts, from face-to-face classroom environments [1,8,19] to online learning [6]. For example, [6] designed adaptive support in the form of strategic prompts (e.g., request an explanation, offer

© Springer Nature Switzerland AG 2019
S. Isotani et al. (Eds.): AIED 2019, LNAI 11625, pp. 14–25, 2019.
https://doi.org/10.1007/978-3-030-23204-7_2

assistance, encourage collaboration) to provide a structured and extended discussion in an online collaborative learning environment. [19] developed a system to support help-giving in a classroom peer tutoring context by providing timely and appropriate help. While these technologies show promise, they focus on supporting students within a single activity in a given context, and do not take into account that students are often collaborating across multiple educational platforms. We build on this prior work to design a cross-platform ACLS to support student collaborative activity across multiple platforms and improve learning.

Designing ACLS for multiple platforms is important because of the need to understand how students' interactions and skills transfer across different contexts. First, students might behave differently on different platforms, and focusing interaction within a single context limits the potential effectiveness of the ACLS. Student behavior in a synchronous collaborative learning environment (e.g., text-based communication) might be different than in an asynchronous collaborative learning environment (e.g., online threaded discussion). For example, [4] assessed graduate student participation in a synchronous chat and asynchronous threaded discussion environment, and reported more responding and reacting statements in the synchronous environment compared to the asynchronous one. [12] explored the social and cognitive presence of graduate students in a synchronous and a asynchronous tool within the same online environment. In this work, we are interested specifically in the quality of collaborative support within a learning activity distributed across multiple platforms. Second, as students learn how to collaboratively construct knowledge on one platform, they will hopefully transfer these skills to a second platform. Their collaborative activity could be informed by one platform to the other. So, facilitating the transfer of skills using ACLS might ultimately enhance students' collaborative learning abilities beyond a single context. However, one of the challenges in supporting collaboration is modeling the collaborative behaviors of students [19]. For multiple platforms, our first research question is: **How do individual students' collaborative interactions vary across different learning platforms?**

In this paper, we examine student help-giving behavior in a mathematics classroom. Help-giving is defined as an activity where students interact with their peers, give explanations to one another, and provide feedback and examples [20]. While doing this, students clarify, elaborate, articulate their own understanding, justify their reasoning, and organize concepts to explain their idea [16,21]. These behaviors contribute to the co-construction of knowledge, which help students learn and transfer their knowledge in multiple contexts. To support student collaborations across multiple contexts, we need to understand their patterns of collaborative interactions. As student collaborates with each other, their motivation in help-giving can affect their participation during collaboration. Thus our second research question is: **How does student collaborative behavior across different platforms predict learning and motivation?** Here, we explore motivation in the context of expectancy-value theory [22]. Both expectancy (whether an individual expects to be able to perform a task) and value are essential motivational factors that might help us to understand student help-giving behaviors and design ACLS systems for multiple platforms.

In addition to the influence of different learning environments, students' collaborative help-giving behaviors can also be influenced by their motivations. Much work in CSCL has analyzed the influence of motivation on students' contributions to online discussions, learning activities, and knowledge acquisition during classroom collaboration [15,17,24]. We considered both student attitude towards mathematics (ATM) and self-efficacy (SE) as motivational factors in this paper. Self-efficacy, a concept developed by Bandura [2], refers to a students' beliefs about their capacity to accomplish certain tasks which affect human motivation, efforts, persistence, and achievement. Attitude refers to student beliefs whether a "task is important, enjoyable, or difficult" [9]. Both ATM and self-efficacy play an important role in how students learn mathematics. To support a student with a cross-platform ACLS, we need to explore how these individual motivational differences influence student interaction across different educational platforms. Our third research question is: **How do individual differences predict student interactions across multiple platforms?**

The purpose of this study is to develop an understanding of students' collaborative interactions in a middle school mathematics classroom where students have different collaborators and use different learning environments. In order to explore the concept of a cross-platform ACLS, we have chosen two platforms: (1) Modelbook, an interactive digital textbook, and (2) Khan Academy, an online question answering platform. Modelbook allows synchronous communication through different tools promoting collaboration mainly in the form of discussion and text-based chat. On the other hand, in Khan Academy students participate in a collaborative activity through answering questions. We investigate our research questions within the context of these two platforms.

2   System Description

Our learning environment leverages a 5-day curriculum which we co-designed with an expert consultant fluent in Modeling pedagogy, a type of instructional pedagogy where students collaborate in small groups to answer a problem [7]. The curriculum focused on a ratio and proportions unit in middle school mathematics, including the topics of proportional relationships, lines, and linear equations. On Day 1, students were asked to devise a method for mixing blue and red paint in the perfect ratio to create purple. On Day 2, students iterated on their models and discussed the definition of a proportion, and then on Day 3 students looked at other examples of proportions. In Days 4 and 5, students received hands on experience applying their understanding of ratios and proportions to modeling the speed of a moving car. Following modeling pedagogy, on each day students alternated between individual problem-solving, small-group activities, and whole class discussions, creating multiple opportunities for collaboration.

Modelbook incorporates several components to facilitate student interaction. For example, once students have completed their small-group activities, the teacher asked students to upload photos of their work to a gallery. The students discussed each image, providing feedback to others (see Fig. 1 left). Modelbook also has a chat feature where students can engage in general discussion. With the guidance of the teacher, students were encouraged to perform help-giving interactions in the gallery and general chat.

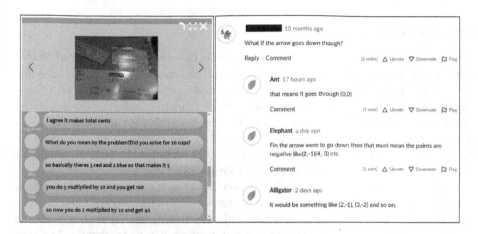

Fig. 1. Left: Gallery image thread with student discussion, Right: Khan Academy discussion thread

ModelBook was built using the Django web framework. The front end of the application is implemented using HTML, jQuery, and CSS. Templates within Django contain the static parts of the desired HTML output as well as provision for inserting dynamic content. For each tool in Modelbook, there are icons on the left hand of the application, which, when clicked, triggers a jQuery event to dynamically load related interface on the right-hand side of the application. All discussion threads were implemented using the "Pusher" service - a hosted API for quickly, easily and securely adding a real-time bi-directional connection. We have used the default SQLite database that accompanies the Django framework. All user activity (e.g., uploaded images, messages) is stored in the database.

The other collaborative platform used is Khan Academy. While most known for its instructional videos, Khan Academy allows asynchronous collaboration with geographically distributed learners in a question and answer environment under each video (www.khanacademy.org). For our curriculum, students posted responses to Khan Academy questions four times over the 5-day period. To facilitate this activity, students were given a homework sheet that included instructions to watch a related Khan Academy video, and were asked to look for two questions posted by other people and provide a response. Students' responses were then discussed in class the following day, and they were encouraged to post in class if they had not done their homework. An example of student participation in Khan Academy is shown in Fig. 1 (right).

3 Method

We conducted a five-day design study to explore how student interactions differed between ModelBook and Khan Academy, whether individual characteristics predicted how they interacted, and whether their interactions on the two platforms predicted their learning outcomes.

3.1 Participants

We conducted the study at a middle school in the southwestern United States with a minority enrollment of 47%. The study was conducted as part of regular classroom practice in an eighth-grade mathematics class with a class of 28 students. We received parental permission for 20 students and thus excluded the other 8 from the analysis. Student ages ranged from 12–14; there were 11 boys, 8 girls, and 1 student who selected "Other" on the question. Participant self-reported ethnicity was as follows: Hispanic (10), White (3), African American (2), Native American (1), Asian (1), and Other (3).

Of the 20 students who consented to participate in the study, only 16 students participated in all elements of the study: two students incorrectly filled out the motivation questionnaire (i.e., selected multiple items), one student was absent for the pre-survey, and one student did not post on Khan Academy. Thus, the final data analysis was done with 16 students.

3.2 Procedure

Domain pretests and a motivation survey were given to students on the Friday before the intervention week. Over the five days of the intervention, students followed the curriculum and engaged in multiple types of activities and interactions, such as: receiving direct instruction from the facilitator, working in small groups of two or three, participating in classroom discussions, completing Modelbook activities, and answering questions on Khan Academy (both in class and for homework). While the classroom teacher was present for each day of the study, the activities were facilitated by one of the authors who was a former teacher and was also our expert consultant in modeling curriculum. On the Wednesday following the five intervention days, students took a domain posttest and a motivation post-survey.

3.3 Measures

Domain Assessment. The pretest and posttest consisted of two isomorphic forms designed to assess students' ability to solve proportional and ratio problems and relevant proportional definitions. It was based on district benchmarks and co-designed with the expert consultant. Each test form included 12 items, with eleven items assessing students' mastery of the domain concepts, and one item asking students to provide an explanation. Forms were counterbalanced across participants (i.e., half the participants received form A for the pretest and B for the posttest, and half received form B for the pretest and A for the posttest). After giving the tests in the study, we noticed that a multi-part question (consisting of 3 items) on Form B was unclear to students and resulted in a disproportionate number of incorrect responses. We excluded that question from the test analysis, along with the corresponding question on Form A. Thus, a total of 8 items were summed to assess student domain learning, with 1 item used to assess student explanatory skill.

Motivation Pre-measure. We surveyed students about their attitudes towards math and mathematical self-efficacy. The instrument consisted of 22 five-level Likert-type items. Value and enjoyment of mathematics were assessed using a portion of the Attitudes Towards Math Scale [18], modified by reversing some items to balance positive and negative statements. To examine students' mathematics self-efficacy, we adapted items from the Motivated Strategies for Learning Questionnaire (MSLQ) [5,14]. The MSLQ scale is generic, so we modified the items to be specific to mathematics. An example item is, "I believe I will receive an excellent grade in math class."

Motivation Post-measure. The post-intervention motivation scale consisted of 15 questions based on Expectancy-Value Theory, with 5 equivalent questions for each platform (ModelBook, Khan Academy, and face-to-face interaction). We wanted to assess whether students perceptions of the tasks differed between platforms and varied based on their experiences during the intervention. The scale was modified from [3] to reflect students' motivation towards help-giving in math. Two example items are: "I'm certain I can make others understand the most difficult material presented in the question" (expectancy), and, "I enjoy helping others with their math questions" (value).

Coding of Interactions. We coded the digital interaction data using a coding scheme based on [21] with the following dimensions: (1) Level of Relevance to the content (LOR), (2) Level of Elaboration (LOE), and (3) Social factors (S). LOR was coded using three categories: *General* (information on the content but not enough to call it a explanation; e.g., "I agree because my board also was not an exact pattern."), *Specific* (information specific to the content; e.g., "I think the unit rate is not 2/3 but it is 2:3"), *Offtopic* (irrelevant to the domain content). LOE coded for on-topic (general & specific) utterances has two categories: *Non-Elaborated* (answer without example or explanation; e.g., "I agree our car also did not go in a straight line.") and *Elaborated* (answer with example, proper explanation with reasoning and justification; e.g., "if we have 2 cups+3 cups that would = five but we need 20 cups"). Finally, we classified an utterance as social if it had at least one of the following four factors: *praise* ("the graph is good"), *apologetic* ("No offense but this makes no sense to me, sorry."), *polite* ("Thank you"), and *encouragement* ("Just do your best"). A second rater independently coded 17% of the dialogues with LOE (kappa = .805), LOR (kappa = .954) and Social (kappa = 1.0). Disagreements were resolved through discussion.

4 Results

For the analysis, we computed both the total numbers of each code dimension as well as student-level percentages with respect to the total utterances for each dimension. Table 1 shows the means and standard deviations for N = 16 for Modelbook (MB) and Khan Academy (KA):

Table 1. M and SD for each coding category

	LOE				LOR					
	Non-Elaborated		Elaborated		General		Specific		Offtopic	
Platform	M	SD	M	SD	M	SD	M	SD	M	SD
MB	8.5	4.89	1.06	1.237	5.81	4.490	3.75	3.0	1.375	2.156
KA	2.438	1.364	0.81	1.223	0.25	0.683	3.0	1.155	0	0

1. How does student interaction differ between Modelbook and Khan Academy?

Table 2 shows mean percentages and standard deviations of categories elaborated, specific, and social utterances for both Modelbook and Khan Academy with respect to the total utterances for each dimension (i.e., LOE, LOR, and S).

Table 2. M and SD for distinct types of utterances

	Modelbook		Khan Academy	
	M	SD	M	SD
Elaborated	10.7	12.5	22.9	34.2
Specific	44.0	26.7	92.7	20.2
Social	21.9	18.2	0	0

To investigate differences in interaction between platforms, a repeated measures MANOVA was conducted with percent elaborated, percent specific, and percent social as dependent variables, and platform (Modelbook or Khan Academy) as an independent variable. The overall model was significant, $F(3, 13) = 32.136$, $p < .001$. Univariate tests revealed that while percent elaborated was not significantly different between conditions [$F(1, 15) = 2.480$, $p = .136$], percent specific was [$F(1, 15) = 45.226$, $p < .001$], as was percent social [$F(1, 15) = 23.122$, $p < .001$]. It should be noted that interaction on Khan Academy followed a fairly uniform pattern, with nearly all on-topic utterances being specific, and no utterances being social.

As students gave both elaborated help and specific help in Modelbook and Khan Academy, we computed correlations between elaborated help across both platforms and specific help across both platforms. Elaborated help in Modelbook was not significantly correlated with elaborated help in Khan Academy [$r(16) = 0.433, p = 0.094$]; and specific help in Modelbook was not significantly correlated with specific help in KA [$r(16) = 0.261, p = 0.328$]. Interestingly, specific help in Modelbook was correlated with elaborated help in Khan Academy [$r(16) = .746, p = .001$]. This analysis demonstrates that not only was interaction different in general across the different platforms, but for each individual student, interaction on one platform did not predict their interaction on another platform.

Table 3. M and SD for post-motivational measure on help-giving behavior

	Modelbook		Khan Academy	
	M	SD	M	SD
Self-efficacy	3.25	1.065	2.88	1.147
Importance	3.50	1.095	3.38	1.408
Interest	3.19	.911	3.13	1.147
Utility	3.62	.957	3.38	1.025
Cost	3.75	.931	3.44	1.031

While behaviors were different across the different platforms, perceptions of students' own interactions in the platforms were not. A repeated measures MANOVA was conducted with each of the motivational post-measures (self-efficacy, importance, interest, utility, and cost) as the dependent variables and platform (Modelbook or Khan Academy) as an independent variable. The overall model was not significant [$F(5, 11) = 1.082, p = .422$] and there were no significant univariate effects. Table 3 summarizes the result.

2. How does help-giving behavior predict learning and motivation?

Table 4 shows means and standard deviations of the pre-test and post-test scores. We conducted a repeated-measures ANOVA and found that learning was not significantly different from pretest to posttest. Despite the overall lack of learning gains, we still look at predictors that may contribute to learning for individuals.

Table 4. M and SD for domain assessment and pre-motivational measures

Measures	Pretest	Posttest	Self-efficacy	Value	Enjoyment
M	4.563	4.687	2.742	3.312	2.617
SD	1.4127	1.4477	.8773	.7288	.975

We did a stepwise multiple regression analysis with percent elaborated in both Modelbook and Khan Academy, percent specific in both Modelbook and

Khan Academy, percent social in Modelbook, pre self-efficacy, attitude towards math score, and pre-test score as predictor variables, with post-test score as the dependent variable. The model that emerged from the stepwise analysis contained only percent elaborated in Modelbook ($\beta = 0.584$; $p = 0.003$) and pretest score ($\beta = .488$; $p = 0.010$) as significant predictors, together explaining 67% of the total variance (Adjusted R-square $= 0.619$; $F(2, 13) = 13.181$; $p = 0.001$). Thus, the only behavioral variable that predicted posttest score was the level of elaborated help in Modelbook.

3. How does motivation and prior domain knowledge predict student help-giving behavior across the two platforms?

Table 4 shows the means and standard deviations of the pre-motivational measures: math self-efficacy, value, and enjoyment. To determine how motivation and prior knowledge predicts student help-giving behaviors, we conducted two multivariate regressions. The first analysis was done for Modelbook behaviors. We used percent elaborated, percent specific and social as dependent variables with pre-test score, average self-efficacy, and average attitude towards math score as predictors. No significant model emerged from it. Univariate tests also did not show any significant results; $F(3, 10) = .471$, $p = .709$, for pre-test score; $F(3, 10) = 1.046$, $p = .414$ for average pre self-efficacy, and $F(3, 10) = 1.007$, $p = .430$ for average attitude towards math score. Multivariate analysis done for Khan Academy behaviors with percent elaborated, percent specific as dependent variables with pre-test score, average self-efficacy, and average attitude towards math score as predictors also demonstrated similar result. Univariate tests didn't show any significant results, $F(2, 11) = .618$, $p - .557$, for pre-test score; $F(2, 11) = .596$, $p = .568$ for average pre self-efficacy, and $F(2, 11) = .286$, $p = .756$ for average attitude towards math score. Students' motivation prior to the intervention did not have an effect on their behaviors during the intervention.

5   Discussion and Conclusion

To design adaptive support for collaboration, student activity history in the collaboration contexts and current engagement in collaborative activities are essential [13]. In this paper, we examined whether student interactions differed across different technological platforms, how their interactions predicted learning and motivation, and how their interactions were informed by their individual characteristics. We found that students displayed better help-giving behavior in Khan Academy compared to Modelbook, but only help-giving behaviors in Modelbook predicted student learning. Individual characteristics like prior knowledge and math motivation did not predict how students gave help.

One interesting finding from this work was that while students gave more high-quality help in Khan Academy than in Modelbook, only the elaborated help in Modelbook was predictive of student posttest scores (controlling for pretest). The affordances of Khan Academy (asynchronous communication with an external community) may have led students to take more time to formulate their response [23], leading to more specific help and more elaborated help. In contrast,

Modelbook represented synchronous, informal communication with peers, leading to overall less high-quality help but more social behaviors (which have shown in other work to be beneficial for learning [11]). In Khan Academy, because of the increased pressure of asynchronous public posts, students may have engaged in knowledge-telling behaviors [16], where they gave help on concepts they had already mastered. This may have led to less learning than their more off-the-cuff interactions in Modelbook, which may have represented knowledge-building, where they construct their knowledge as they are constructing their explanations. One implication of this finding for the design of adaptive support is that to improve outcomes from help-giving, it may be sufficient to encourage more elaborated help in Modelbook. However, in Khan Academy, it may be necessary to directly scaffold students in constructing the elaborated help so that they engage in reflective knowledge-building behaviors.

Another critical element of our results is that while context dictated how students gave help, individual differences did not. Students' help-giving behavior was more elaborate and specific in Khan Academy compared to their behavior in Modelbook, and for individual students, these behaviors were not correlated with each other. This indicates that student behavior in one platform does not inform how they will behave in another platform; rather, the different platforms influenced how students will help each other. Additional support for this finding is provided by the fact that neither prior knowledge, math self-efficacy, nor attitude towards math predicted how students gave help in either platform. This finding implies that a model of student help-giving on one platform is unlikely to generalize to the same student's help-giving behaviors on a different platform, and context thus needs to be part of any knowledge-tracing model of help-giving.

This study has a number of limitations. First, the sample size was small. Second, the number of interactions was greater in Modelbook compared to Khan Academy due to the design of the curriculum. To adapt to this limitation, we used student-level percentages to compute the results rather than absolute counts of student interactions. Third, students did not learn as a whole from pretest to posttest, possibly because the intervention time was too short or our assessment wasn't sensitive enough to detect changes in student knowledge.

Nevertheless, the present research has important implications for computer-supported collaborative learning and ACLS. Students' interactions in different platforms can be used to design individualized support that facilitates productive communication across collaborative learning environments. This goal will require a cross-platform student interaction model along with a domain knowledge model and motivation model for each student. Investigation is required to understand how to make predictions about student behavior within a single platform using this cross-platform interaction model, whether and how to encourage students to participate in platforms they are less comfortable with, whether and how to encourage students to transfer their skills from one platform to a different platform, and whether and how the same student should be given different kinds of support on different platforms.

In this paper, we examined students' help-giving behavior across Modelbook and Khan Academy. This paper takes a step towards establishing the need for understanding cross-platform collaborative behavior, and based on our findings, we are currently building a cross-platform help-giving model. We believe this approach will ultimately enhance peer collaboration as students move between platforms of interaction.

Acknowledgements. This work is supported by the National Science Foundation under Grant No 1736103.

References

1. Baghaei, N., Mitrovic, A., Irwin, W.: Supporting collaborative learning and problem-solving in a constraint-based CSCL environment for UML class diagrams. Int. J. Comput. Support. Collaborative Learn. **2**(2–3), 159–190 (2007)
2. Bandura, A.: Self-efficacy: toward a unifying theory of behavioral change. Psychol. Rev. **84**(2), 191 (1977)
3. Berger, J.L., Karabenick, S.A.: Motivation and students' use of learning strategies: evidence of unidirectional effects in mathematics classrooms. Learn. Instr. **21**(3), 416–428 (2011)
4. Davidson-Shivers, G.V., Muilenburg, L.Y., Tanner, E.J.: How do students participate in synchronous and asynchronous online discussions? J. Educ. Comput. Res. **25**(4), 351–366 (2001)
5. Duncan, T.G., McKeachie, W.J.: The making of the motivated strategies for learning questionnaire. Educ. Psychol. **40**(2), 117–128 (2005)
6. Gweon, G., Rose, C., Carey, R., Zaiss, Z.: Providing support for adaptive scripting in an on-line collaborative learning environment. In: Proceedings of the SIGCHI Conference on Human Factors in Computing Systems, pp. 251–260. ACM, April 2006
7. Jackson, J., Dukerich, L., Hestenes, D.: Modeling instruction: an effective model for science education. Sci. Educ. **17**(1), 10–17 (2008)
8. Kumar, R., Rosé, C.P., Wang, Y.C., Joshi, M., Robinson, A.: Tutorial dialogue as adaptive collaborative learning support. Front. Artif. Intell. Appl. **158**, 383 (2007)
9. Liu, X., Koirala, H.: The effect of mathematics self-efficacy on mathematics achievement of high school students (2009)
10. Magnisalis, I., Demetriadis, S., Karakostas, A.: Adaptive and intelligent systems for collaborative learning support: a review of the field. IEEE Trans. Learn. Technol. **4**(1), 5–20 (2011)
11. Ogan, A., Finkelstein, S., Walker, E., Carlson, R., Cassell, J.: Rudeness and rapport: insults and learning gains in peer tutoring. In: Cerri, S.A., Clancey, W.J., Papadourakis, G., Panourgia, K. (eds.) ITS 2012. LNCS, vol. 7315, pp. 11–21. Springer, Heidelberg (2012). https://doi.org/10.1007/978-3-642-30950-2_2
12. Oztok, M., Zingaro, D., Brett, C., Hewitt, J.: Exploring asynchronous and synchronous tool use in online courses. Comput. Educ. **60**(1), 87–94 (2013)
13. Paramythis, A.: Adaptive support for collaborative learning with IMS learning design: are we there yet. In: Proceedings of the Workshop on Adaptive Collaboration Support, Held in Conjunction with the 5th International Conference on Adaptive Hypermedia and Adaptive Web-Based Systems, Hannover, Germany, pp. 17–29, July 2008

14. Pintrich, P.R., De Groot, E.V.: Motivational and self-regulated learning components of classroom academic performance. J. Educ. Psychol. **82**(1), 33 (1990)
15. Rienties, B., Tempelaar, D., Van den Bossche, P., Gijselaers, W., Segers, M.: The role of academic motivation in computer-supported collaborative learning. Comput. Hum. Behav. **25**(6), 1195–1206 (2009)
16. Roscoe, R.D., Chi, M.T.: Understanding tutor learning: knowledge-building and knowledge-telling in peer tutors' explanations and questions. Rev. Educ. Res. **77**(4), 534–574 (2007)
17. Schoor, C., Bannert, M.: Motivation in a computer-supported collaborative learning scenario and its impact on learning activities and knowledge acquisition. Learn. Instr. **21**(4), 560–573 (2011)
18. Tapia, M., Marsh, G.E.: An instrument to measure mathematics attitudes. Acad. Exch. Q. **8**(2), 16–22 (2004)
19. Walker, E., Rummel, N., Koedinger, K.R.: Adaptive intelligent support to improve peer tutoring in algebra. Int. J. Artif. Intell. Educ. **24**(1), 33–61 (2014)
20. Webb, N.M., Farivar, S.: Promoting helping behavior in cooperative small groups in middle school mathematics. Am. Educ. Res. J. **31**(2), 369–395 (1994)
21. Webb, N.M., Mastergeorge, A.: Promoting effective helping behavior in peer-directed groups. Int. J. Educ. Res. **39**(1–2), 73–97 (2003)
22. Wigfield, A., Eccles, J.S.: The development of achievement task values: a theoretical analysis. Develop. Rev. **12**(3), 265–310 (1992)
23. Wu, D., Hiltz, S.R.: Predicting learning from asynchronous online discussions. J. Asynchronous Learn. Netw. **8**(2), 139–152 (2004)
24. Xie, K., Ke, F.: The role of students' motivation in peer-moderated asynchronous online discussions. Br. J. Educ. Technol. **42**(6), 916–930 (2011)

Predicting Academic Performance: A Bootstrapping Approach for Learning Dynamic Bayesian Networks

Mashael Al-Luhaybi$^{(\boxtimes)}$, Leila Yousefi, Stephen Swift, Steve Counsell, and Allan Tucker

Intelligent Data Analysis Laboratory, Brunel University London, London, UK
{Mashael.Al-luhaybi,Leila.yousefi,stephen.swift,steve.counsell,
Allan.Tucker}@brunel.ac.uk

Abstract. Predicting academic performance requires utilization of student related data and the accurate identification of the key issues regarding such data can enhance the prediction process. In this paper, we proposed a bootstrapped resampling approach for predicting the academic performance of university students using probabilistic modeling taking into consideration the bias issue of educational datasets. We include in this investigation students' data at admission level, Year 1 and Year 2, respectively. For the purpose of modeling academic performance, we first address the imbalanced time series of educational datasets with a resampling method using bootstrap aggregating (bagging). We then ascertain the Bayesian network structure from the resampled dataset to compare the efficiency of our proposed approach with the original data approach. Hence, one interesting outcome was that of learning and testing the Bayesian model from the bootstrapped time series data. The prediction results were improved dramatically, especially for the minority class, which was for identifying the high risk of failing students.

Keywords: Performance prediction · Bayesian networks ·
Resampling · Bootstrapping · EDM

1 Introduction

According to the Higher Education Statistics Agency (HESA) in the UK [1], the drop-out rate among undergraduate students has increased in the last three years. The statistics published by the HESA reveal that a total of 26,000 students in England in 2015 dropped out from their enrolled academic programmes after their first year. Also, the statistics show that the higher education (HE) qualifications obtained by students for all levels, including undergraduate and postgraduate levels, decreased from 788,355 in 2012/13 to 757,300 in 2016/17.

Supported by the Intelligent Data Analysis Research Laboratory, Brunel University London, United Kingdom.

S. Isotani et al. (Eds.): AIED 2019, LNAI 11625, pp. 26–36, 2019.
https://doi.org/10.1007/978-3-030-23204-7_3

The growing availability of such reports provides new opportunities to educational researchers. A large body of research has investigated the issues associated with students' learning and academic performance. For instance, educational data mining (EDM) researchers have attempted to analyse and evaluate student data to enhance their education and provide solutions to failure issues using the state-of-the-art Artificial Intelligence (AI) methods.

Predicting student performance is a major area of interest within the field of EDM in terms of ascertaining accurately what, as yet, unknown knowledge regarding this performance, such as final grades [2], will transpire to be. However, it is a very difficult task as it is influenced by social, environmental and behavioral factors [3,4]. Thus, machine learning algorithms are increasingly being used to discover the relationships between these factors and the academic performance of students. There are different educational predictive models for student assistance aimed at helping them to achieve an improvement in their studies. What is interesting about these models is that they model students' achievement by using Bayesian networks (BNs) to handle uncertainty as well as representing the student's knowledge. For instance, [5] used Dynamic Bayesian networks (DBNs) to analyze students' cognitive structure over time. BNs [6] involve a classification approach based on probability theory [7] and are considered the best predictors. Such probability predicts the membership of all student-related factors and the class factor by assuming that the independency of the latter is based on the associated values with the other attributes in the prediction model [8].

However, from a practical point of view, a common issue with classifying students is that the educational datasets usually contain imbalanced data, especially for high risk or failed students compared to the excellent or medium performance ones. Because of this, we exploited a resampling method on students' obtained grades and other students related attributes, namely bootstrapping, to ensure that more states of student overall performance are obtained than using the original time series datasets.

This paper provides, first, a novel DBN approach for predicting university students' academic performance from time series educational data using a probabilistic modeling approach. Secondly, it explores the use of a bootstrapping method to resample the educational datasets in order to improve the learning of the BN structure, whilst also enhancing the detection of the students of the minority class, who are those at high risk, as early as possible.

2   Related Work

A considerable number of studies have been conducted to predict the performance of the students based on the Bayesian method. For instance, [8] compared the Bayesian approach with other classification approaches to identify useful patterns that could be extracted from students' personal and pre-university data in order to predict students' performance at the university. Similarly, authors of [9] used the same approach for a comparative study of classification algorithms but their aim was to classify the students and identify the most influencing attributes on students' failure. Though some researchers were attempting to use the BNs for characterizing performance and exploring the correlation between the performance and the attributes, others were interested in detecting and modeling students' learning styles [10,11]. For example, [12] conducted a study to analyze demographic, social and assessment data to predict the slow learning students in order to improve their performance and reduce failure rate prior to the exam.

However, there is not much-related work in the educational system that handles the issues with the imbalanced educational data through exploiting the bootstrap approach [13,14]. Feng and co-authors [15] utilized and validated their statistical results by using bootstrapping with logistic regression to evaluate students' learning based on different educational interventions. Similarly, a study has been conducted by [16] to evaluate students' understanding of statistical inference with a bootstrapping approach while did not consider time. To the best of our knowledge, there is no previous work in this field that applied DBNs on the bootstrapped "time series" dataset of students progression data. Hence, this work is a first attempt to use them, with the aim of achieving an improvement in student performance overall.

3   Method

As with many prediction issues, educational datasets usually include imbalanced data, because the number of students in Low, Medium and High risk classes is not equally balanced (see Fig. 5(A)). We focus here on first predicting students performance at university based on the original educational time series records. Then, we investigate our resampling approach in order to have some insights into the current problems with the imbalanced educational time series records. Hence, the issue of predicting students' performance based on imbalanced data can be determined using our resampling approach. For this purpose, a DBN model was learned from student's temporal data, taking into consideration the imbalance issue of the predictive classes (see Fig. 1).

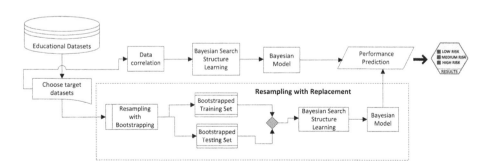

Fig. 1. This diagram presents Bayesian structure learning and the resampling strategy used for learning the DBN model.

3.1   Datasets

The datasets used in this paper were collected from Brunel University admissions and computer science databases. This consists of 377 records for students' progression and other student-related data in 2014, 2015 and 2016, respectively. The datasets contained the following data categories:

- **Admissions dataset:** includes students' application data when entering the university. such as: nationality, ethnicity, country of birth, disability, been in care, socioeconomic class ...etc.;
- **Progression dataset:** this includes student final grades for all Year 1 and Year 2 modules at the university for measuring students' overall academic performance;
- **Engagement attitude:** this includes students' attitude towards turning up to classes and labs in Year 1 and Year 2 at the university;
- **Online temporal assessment profiles:** this includes a student online assessment profile based on their online time-series assessment trajectories. These profiles were obtained using dynamic time warping (DTW) and hierarchical clustering algorithms.

3.2   Resampling with Bootstrapping

Resampling strategies are fundamental approaches in the pre-processing phase, which are used to change the distribution of data in a dataset [17]. After we had discretised the students' overall grade bands from (A, B, C, D, E and F) to qualitative states of low, medium and high risk students, we still encountered an imbalance issue especially for the high risk students (see the confusion matrix in Fig. 5(A)). As aforementioned, imbalance data is a very common issue in educational datasets, which affects learning the predictive models as well as making difficulties in identifying the cases of the minority classes. The minority class in this work is the high risk class, which is the class assigned for those students who obtained low grades (D, E and F) in most of the modules. We exploited here a resampling approach on the datasets to obtain reliable accuracy of the prediction results using bootstrapping.

We exploited the bootstrapping approach using bootstrap aggregation (bagging) [18] to sample the original data with replacement. This approach was also applied to estimate the accuracy of the BN in predicting more student records of all classes and to avoid overfitting. To implement the bootstrap approach, we used the REPTree algorithm with the classification and regression tree algorithm (CART) in the WEKA [19] mining tool. We decided to resample the data using the decision tree algorithm as it is widely used for the low bias and high variance models. To this end, we spilt the dataset into 60% training set and 40% testing set. We tested the model on the training data through 10 iterations for bagging using the same size as the original dataset. It is important to mention that, we could have resampled any size from the dataset, but we decided to obtain the same number of students' records as in the original dataset for better comparisons to the imbalanced data and decision making using our proposed approach.

To validate the bootstrapped data, we computed the mean μ of the distribution to give an 95% bootstrap confidence interval. The mean was $x = 0.67$ for students overall performance, which we used as an estimated value of the mean for the underlying distribution. To calculate the confidence interval we needed to measure the difference between the distribution of x around the mean μ, as follows:

$$\delta = \overline{x} - \mu \tag{1}$$

To get this distribution, we found the standard deviation for the entire student records $\delta.1$ and $\delta.9$, the 0.1 and 0.9, which are critical values of δ to achieve a 95% confidence interval of $[\overline{x} - \delta.1, \overline{x} - \delta.9]$. The StdDev for the full data was obtained from the following equation:

$$P(\delta.9 \leq \overline{x} - \mu \leq \delta.1|\mu) = 0.95 \quad \Longleftrightarrow \quad P(\overline{x} - \delta.9 \geq \mu \geq \overline{x} - \delta.1|\mu) = 0.95 \tag{2}$$

However, the bootstrap offers a direct approach to obtain the distribution of δ, which can be measured by the distribution of:

$$\delta^* = \overline{x}^* - \overline{x} \tag{3}$$

where, $\overline{x}^*$ indicates the mean of the bootstrap data. We generated one bootstrapped sample of size of 377, which was the size of the original data.

3.3   Bayesian Structure Learning

The learning of the BN structure was performed using GeNIe [20], software implemented for learning and modelling Bayesian Networks (BNs) and Dynamic Bayesian networks (DBNs). For learning the BN structure, we used a Bayesian search on two datasets: the original and bootstrapped datasets. We trained the BNs on these two datasets as we wanted to obtain a very accurate and reliable predictive model. The BNs with temporal links inferred from the admissions and students historical grades, are represented in three time slots (t, t+1 and t+2) (see Fig. 2). In our discrete time BNs, three-time slots are observed to identify the correlation between students' overall performance and other related attributes, such as module grades, online temporal assessment profiles and students' engagement attitude. For example, Fig. 3 shows that the disability attribute at time (t) affects the states of some grades at Year 1 and Year 2.

3.4   Bayesian Parameter Learning

The principle goal of learning a DBN is to find the posterior distribution that is adapted to students' progression data, which allows for identifying the states of all students' attributes as well as overall performance. The parameters of the Dynamic Bayesian model were learned using the expectation maximization (EM) algorithm [21] with the bootstrapped data. We implemented this algorithm to estimate the posterior distribution of students' attributes in time slots t, t+1 and t+2. We used the EM algorithm as it performs the maximum likelihood for temporal data, which supports learning from time series data.

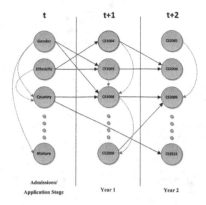

Fig. 2. Dynamic Bayesian network proposed approach for three time slots (t, t+1 and t+2).

4   Experimental Results

The results provide an evaluation of our proposed DBN approach in predicting third year university students' performance. Two key experiments were undertaken: learning from the original data and learning from the bootstrapped data. We set up these two experiments as we wanted to show improvement in predicting the performance, especially for high risk students, who belonged to the minority class in these experiments. In the learning process, we learned the BN structure from students' data in three different time slots. These were students' admissions attributes (time slot t), their obtained grades at Year 1 (time slot t+1) and Year 2 (time slot t+2). In Fig. 3, we present the discovered correlations between students' admission and progression data (grades). It is interesting to note that some of the admission nodes influence students' achievement in some of the Year 1 and Year 2 modules. In addition, students' performance in Year 2 was mainly influenced by their grades in CS1811 (Fundamental Programming Assessment), CS2003 (Usability Engineering) and CS2005 (Networks and Operating Systems), which are compulsory modules for computer science students.

To examine the bootstrapping on improving prediction of the high risk students and the other classes, we provide a comparison between the two approaches used, as shown in Fig. 4. The accuracy results were obtained using the 10 fold cross validation for predicting the academic performance in time slot t+2. Figure 4 shows a significant improvement in identifying the low, medium and high risk students for the bootstrapped data. For example, the accuracy obtained for the high risk class using the bootstrapped data was 0.94, whilst when using the original data it was only 0.63.

The confusion matrices in Fig. 5 indicate the predicted low, medium and high risk students using the original and the proposed bootstrapped data approach. It also reveals the percentage of classification accuracy for each predicted class using the original dataset (A) and bootstrapped dataset (B).

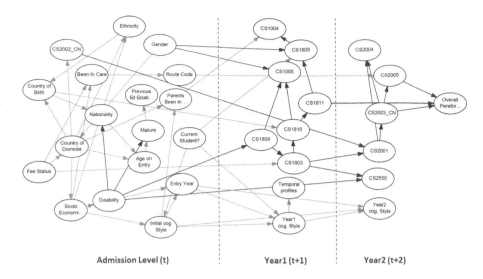

Fig. 3. Dynamic Bayesian structure learned from the bootstrapped temporal educational data. The strong relationships between students' attributes were coloured in blue. (Color figure online)

For evaluating the performance of the BN model, we performed sensitivity and specificity analysis on the cohort of students who were predicted to be at low, medium and high risk. To this end, we visualized the Receiver Operating Characteristics curve (ROC) and the Area under the Curve (AUS), as shown in Fig. 6. We used these two performance measurements as we had a multi class predictive model. It can be seen from the ROC curves in Fig. 6 that for the low risk prediction (A) and high risk prediction (C) are very close to 100% sensitivity and 100% specificity, which means a perfect discrimination of the overall prediction accuracy based on the bootstrapped data.

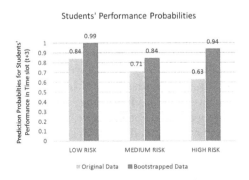

Fig. 4. Prediction probabilities for students' overall performance using two approaches (original and bootstrapped data). It represents the accuracy results for the three classes.

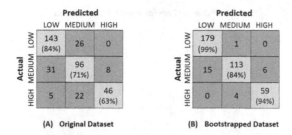

Fig. 5. Academic performance confusion matrices comparing prediction results for the class attribute using the original dataset (A) and bootstrapped dataset (B).

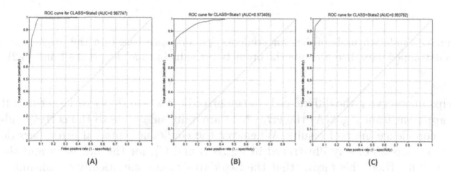

Fig. 6. ROC curves of students' overall performance for the three states, these being: (A) state 0 for the low risk, (B) state 1 for the medium risk and (C) state 2 for the high risk students.

We then examined the validation of our DBN approach in predicting the performance in Year 3 using supplied test sets, as shown in Fig. 7. Firstly, we predicated students' performance based on admissions data only at time slot t for the two datasets, which were the original and the bootstrapped data. Secondly, we added more data, which were students' progressions and final grades at Year 1, to see how better we can predict using the temporal approach. After that, we predicted performance using all students' attributes. It is apparent from Fig. 7 that, the prediction was improved in time slot t+1 when we added Year 1 grades. This improvement was due to the direct relation between students' achieved grades in Year 1 with their overall performance.

5   Confidence Interval Results

This section presents the influence of using bootstrapping to improve learning a BN model. The plotted chart in Fig. 8(A–C) compares the accuracy, the precision and the sensitivity results for predicting the performance of students, among 377 students' time series records for two different datasets (original, bootstrapped). We also show the error bars with a 95% confidence interval, which

Students' Overall Pereformance Probabailties

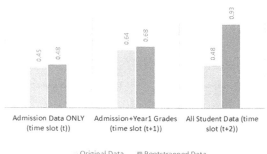

Fig. 7. Validation probabilities for students' performance based on the original and bootstrapped data. It represents the prediction results in time slots (t, t+1 and t+2).

helps in observing the difference between error bars where they overlap or not. It is apparent from Fig. 8(A) the error bars are quite small due to the corresponding confidence interval results. Whenever the confidence interval error bars do not overlap, as clearly illustrated in Fig. 8(B and C), for the precision and the sensitivity, then, this means that the two datasets are statistically significant.

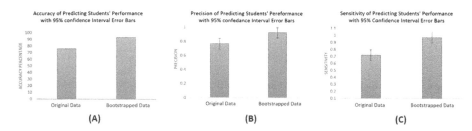

Fig. 8. Confidence interval (CI) error bars charts for the accuracy (A), the precision (B) and the sensitivity (C) for predicting the performance of the students based on the original and the bootstrapped data approaches.

6   Conclusion and Future Work

The prediction of the performance of students has been increasingly emerging in the educational field as it is now possible to transform huge amounts of data into useful knowledge. However, this is a very difficult task because of the issues associated with data. Usually, educational datasets have missing, inaccurate, imbalanced data and so forth, which are also very common issues in all the other research fields. Learning from imbalanced data requires approaches and techniques to transform such data into useful knowledge [22]. To this end, a resampling approach was explored in this paper for learning DBNs using bootstrap aggregating (bagging). This approach was adopted to tackle the imbalance issue with educational datasets.

The objective of this paper was to model a DBN for predicting the performance of students and the early detection of students at risk of failing or dropping a module based on their time series data. For this purpose, we used students' admission, Year 1 and Year 2 grades in conjunction with other attributes to predict the performance at Year 3, taking into consideration the imbalanced issue of the educational data. A set of two BNs were learned from the educational time series data. The first was learned from the original data, whereas the second model was learned from the resampled data (via bootstrapping). We evaluated the obtained BN models in terms of predicting more states of students' overall performance from temporal educational data using the two different approaches.

Important analytically relevant findings were found when comparing the two approaches used for learning the DBNs. The results show that more states of student's overall performance were achieved when learning from the bootstrapped data, especially for the minority class which was for detecting the high-risk students. We have also demonstrated how the bootstrapped resampling approach enhances the overall prediction of student performance using DBNs. These findings have significant implications for developing education and enhancing students' learning using artificial intelligence. We intend to use these findings to differentiate between the different cohorts of students who perform with similar dynamics and therefore, simplify them to obtain a better understanding of students' performance.

Further experimental works are needed to explore the extension of these Bayesian models with the investigation of latent attributes, with the aim of capture hidden factors that may influence the dynamics of students' academic performance. In our future works, we attempt to compare the proposed methodology especially DBNs with other classification approaches. Moreover, we try to compare other balancing methods with the Bootstrap approach and being more precise in bootstrapping time series grades.

Acknowledgment. This work was partially funded through an internal Brunel Student Assessment and Retention grant (STARS Project).

References

1. The Higher Education Statistics Agency (HESA) Website, HE student enrolments by level of study. https://www.hesa.ac.uk/data-and-analysis/sfr247/figure-3. Accessed 30 Jan 2019
2. Romero, C., López, M.I., Luna, J.M., Ventura, S.: Predicting students final performance from participation in on-line discussion forums. Comput. Educ. **68**, 458–472 (2013)
3. Bhardwaj, B.K. and Pal, S.: Data mining: a prediction for performance improvement using classification. arXiv preprint arXiv:1201.3418 (2012)
4. Araque, F., Roldán, C., Salguero, A.: Factors influencing university drop out rates. Comput. Educ. **53**(3), 563–574 (2009)

5. Seffrin, H.M., Rubi, G.L. Jaques, P.A.: A dynamic bayesian network for inference of learners' algebraic knowledge. In: Proceedings of the 29th Annual ACM Symposium on Applied Computing, pp. 235–240. ACM (2014)
6. Pearl, J.: Probabilistic Reasoning in Intelligent Systems: Networks of Plausible Inference. Elsevier (2014)
7. Witten, I.H., Frank, E., Hall, M.A., Pal, C.J.: Data Mining: Practical Machine Learning Tools and Techniques. Morgan Kaufmann, Burlington (2016)
8. Kabakchieva, D.: Predicting student performance by using data mining methods for classification. Cybern. Inf. Technol. **13**(1), 61–72 (2013)
9. Kaur, G., Singh, W.: Prediction of student performance using weka tool. Int. J. Eng. Sci. **17**, 8–16 (2016)
10. García, P., Amandi, A., Schiaffino, S., Campo, M.: Evaluating Bayesian networks' precision for detecting students' learning styles. Comput, Educ. **49**(3), 794–808 (2007)
11. Carmona, C., Castillo, G., Millán, E.: Designing a dynamic bayesian network for modeling students' learning styles. In: 2008 Eighth IEEE International Conference on Advanced Learning Technologies, pp. 346–350 (2008)
12. Kaur, P., Singh, M., Josan, G.S.: Classification and prediction based data mining algorithms to predict slow learners in education sector. Procedia Comput. Sci. **57**, 500–508 (2015)
13. Beal, C., Cohen, P.: Comparing apples and oranges: computational methods for evaluating student and group learning histories in intelligent tutoring systems. In: Proceedings of the 12th International Conference on Artificial Intelligence in Education, pp. 555–562 (2005)
14. McLaren, B.M., Koedinger, K.R., Schneider, M., Harrer, A., Bollen, L.: Bootstrapping novice data: semi-automated tutor authoring using student log files. In: Proceedings of Workshop on Analyzing Student-Tutor Interaction Logs to Improve Educational Outcomes, Proceedings of the 7th International Conference on ITS-2004: Intelligent Tutoring Systems (2004)
15. Feng, M., Beck, J.E., Heffernan, N.T.: Using Learning Decomposition and Bootstrapping with Randomization to Compare the Impact of Different Educational Interventions on Learning. International Working Group on Educational Data Mining (2009)
16. Pfannkuch, M., Forbes, S., Harraway, J., Budgett, S., Wild, C.: Bootstrapping students' understanding of statistical inference. Summary research report for the Teaching and Learning Research Initiative (2013)
17. Chawla, N.V., Bowyer, K.W., Hall, L.O., Kegelmeyer, W.P.: SMOTE: synthetic minority over-sampling technique. J. Artif. Intell. Res. **16**, 321–357 (2002)
18. Moniz, N., Branco, P. and Torgo, L.: Resampling strategies for imbalanced time series. In: 2016 IEEE International Conference Data Science and Advanced Analytics (DSAA), pp. 282–291. IEEE (2016)
19. Hall, M., Frank, E., Holmes, G., Pfahringer, B., Reutemann, P., Witten, I.H.: The WEKA data mining software: an update. ACM SIGKDD Explor. Newslett. **11**(1), 10–18 (2009)
20. Druzdzel, M.J.: GeNIe: a development environment for graphical decision-theoretic models (1999)
21. Moon, T.K.: The expectation-maximization algorithm. IEEE SIgnal Process. Mag. **13**(6), 47–60 (1996)
22. He, H., Garcia, E.A.: Learning from imbalanced data. IEEE Trans. Knowl. Data Eng. **9**, 1263–1284 (2008)

The Impact of Student Model
Updates on Contingent Scaffolding
in a Natural-Language Tutoring System

Patricia Albacete[1(✉)], Pamela Jordan[1], Sandra Katz[1],
Irene-Angelica Chounta[2], and Bruce M. McLaren[3]

[1] Learning Research and Development Center,
University of Pittsburgh, Pittsburgh, PA, USA
palbacet@pitt.edu
[2] Institute of Education, University of Tartu, Tartu, Estonia
[3] Human Computer Interaction Institute, Carnegie Mellon University,
Pittsburgh, PA, USA

Abstract. This paper describes an initial pilot study of Rimac, a natural-language tutoring system for physics. Rimac uses a student model to guide decisions about *what content to discuss next* during reflective dialogues that are initiated after students solve quantitative physics problems, and *how much support to provide* during these discussions—that is, domain contingent scaffolding and instructional contingent scaffolding, respectively. The pilot study compared an experimental and control version of Rimac. The experimental version uses students' responses to pretest items to initialize the student model and dynamically updates the model based on students' responses to tutor questions during reflective dialogues. It then decides what and how to discuss the next question based on the model predictions. The control version initializes its student model based on students' pretest performance but does not update the model further and assigns students to a fixed line of reasoning level based on the student model predictions. We hypothesized that students who used the experimental version of Rimac would achieve higher learning gains than students who used the control version. Although we did not find a significant difference in learning between conditions, the experimental group took significantly less time to complete the pilot study dialogues than did the control group. That is, the experimental condition led to more efficient learning, for both low and high prior knowledge level learners. We discuss this finding and describe future work to improve the tutor's potential to support student learning.

Keywords: Dialogue-based tutoring systems · Student modeling · Contingent scaffolding

1 Introduction

The key features of instructional scaffolding, as described by [12], include *contingency*, *fading* and, correspondingly, the gradual *transfer of responsibility* for learning and successful performance to the learner. "Contingency" refers to the adaptive nature of

© Springer Nature Switzerland AG 2019
S. Isotani et al. (Eds.): AIED 2019, LNAI 11625, pp. 37–47, 2019.
https://doi.org/10.1007/978-3-030-23204-7_4

scaffolding and is believed to be its core feature, from which the other two features stem. Instructors dynamically adjust their degree of control over the learning task according to their diagnosis of the student's current level of understanding or performance [14]. "Fading" refers to the gradual release of this support so that scaffolding can achieve its ultimate aim: to shift responsibility for successful performance to the student.

Wood and Wood [14] distinguished between three types of contingency during human tutoring sessions: *temporal*, *domain*, and *instructional contingency* (see also [13]). Temporal contingency is concerned with deciding *when* to intervene versus letting the learner struggle for a while or request help. Domain contingency is concerned with choosing appropriate content to address during an intervention, while instructional contingency is concerned with deciding how to address focal content—for example, in how much detail and through which pedagogical strategies (e.g., modeling, hinting, explaining, question asking)?

For the Rimac natural-language tutor [1, 2, 5, 9], we developed an Instructional Factors student model [4] that dynamically updates throughout the tutorial dialogue in order to represent the student's current level of understanding. The student model is used during decision-making about domain and instructional contingency. We compared this version of Rimac to a version that uses a static representation of the student's understanding based solely on the student's pretest performance, i.e., to a version that uses an array of knowledge components initialized with pretest scores as a student model, to make decisions about domain and instructional contingency. We predicted that classroom students who interacted with the version of Rimac that incorporates the adaptive student model would show greater learning gains than those who interacted with a version of Rimac that incorporates a simple static representation of a student's level of understanding. A student model that reflects students' progress should lead to more appropriate decisions regarding domain and instructional contingency. To our knowledge, this is the first real-time test of an Instructional Factors Model (IFM) being used by an ITS to tutor students in the classroom.

2   Rimac: An Adaptive Natural-Language Tutoring System

Rimac is a dialogue-based tutoring system that engages high school students in conceptual discussions after they solve quantitative physics problems (e.g., [1, 2, 10]). These dialogues are developed using an authoring framework called *Knowledge Construction Dialogues* (KCDs) (e.g., [6, 7, 11]). KCDs present a series of carefully ordered questions known as a *Directed Line of Reasoning* (DLR) [6], which guide students in responding to complex conceptual questions (reflection questions, or RQs). When the student makes an error at a particular step in the DLR, the tutor initiates a remedial sub-dialogue to address that error. Figure 1 shows the system's interface which presents, in the left pane, the problem statement along with a sample solution to a quantitative problem that students watch as a video and, in the right pane, an excerpt of a reflective dialogue between the system and the student which addresses conceptual knowledge associated with the quantitative problem.

Rimac adapts its instruction to students' ever evolving knowledge by incorporating a student model that is updated as the student engages in the dialogues and by implementing policies that, with the help of the student model predictions, allow it to choose the next question to ask at the appropriate level of granularity and with adequate support. The granularity level refers to domain contingency—that is, how much content is explicitly discussed with the student (e.g. discuss all the steps in the reasoning vs skip over some steps that the student can likely infer on her own). Adequate support refers to instructional contingency—that is, how much detail should be provided in questions and hints about the selected content.

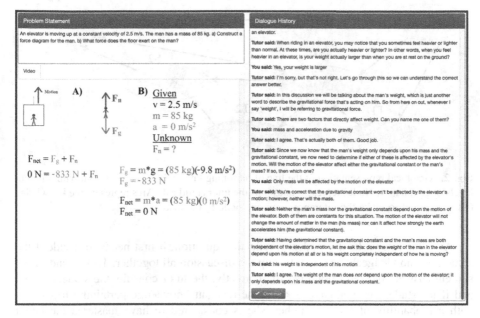

Fig. 1. Rimac interface. Problem statement shown in upper left pane, worked example video in lower left pane, and dialogue excerpt in right pane.

An individual learner's student model is built in two steps: first, using the results of the student's pretest, a clustering algorithm classifies the student as low, medium, or high. The purpose of this initial clustering is to increase the accuracy of the student model's predictions. Second, the student is assigned a cluster-specific regression equation that is then personalized with the results of the student's pretest. The regression equation assigned to the student represents an implementation of an Instructional Factor Analysis Model (IFM), as proposed by [4]. This student model uses logistic regression to predict the probability of a student answering a question correctly as a linear function of the student's proficiency in the relevant knowledge components (KCs). Additionally, as the student progresses through the dialogues, her student model is dynamically updated according to the correctness of her responses to the tutor's questions [5].

To be able to vary the level at which the tutorial discussions are conducted, for each reflection question (RQ), we developed dialogues at three different levels of granularity: an expert level (P—primary) which only includes the essential steps of the reasoning, a medium level (S—secondary), and a novice level (T—tertiary) which includes more basic knowledge such as definitions of concepts and laws. Figure 2 shows a graphic representation of an excerpt of a line of reasoning (if the net force on an object is zero then the object's velocity is constant) at three different levels of granularity.

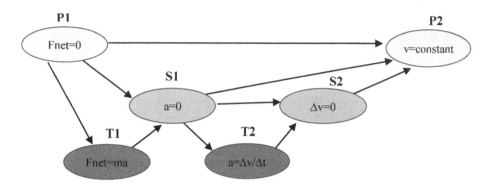

Fig. 2. Graphical representation of the line of reasoning Fnet = 0 → v = constant with different levels of granularity. Nodes represent questions the tutor could ask. Arcs represent the knowledge (KCs) required to make the inference from one node to the next.

After the tutor asks the student a reflection question, it first needs to decide if the student is knowledgeable enough to skip the discussion all together. To this end, if the student answers the reflection question correctly, the tutor consults the student model and if the student is predicted to know the relevant knowledge pertaining to the RQ with a probability of 80% or higher, she is considered to have mastered the target knowledge and is allowed to skip the RQ. On the other hand, if the student either does not answer the RQ correctly or has not mastered its relevant knowledge, the tutor engages in a reflective dialogue with the learner. At each step of this discussion, the tutor needs to decide at what level of granularity it will ask the next question in the line of reasoning (LOR) (or in a remedial sub-dialogue if the previous question was answered incorrectly) in order to proactively adapt to the student's changing knowledge level. It performs this adaptation by following policies aimed at driving the student to reason in an expert-like manner while providing adequate scaffolding. Hence, the tutor will choose a question in the highest possible granularity level that it deems the student will respond to correctly or that it perceives will be in the student's zone of proximal development (ZPD)—"a zone within which a child can accomplish with help what he can later accomplish alone" [3].

To make this choice, Rimac consults the student model, which predicts the likelihood that the student will answer a question correctly. The tutor interprets this probability in the following way: if the probability of the student responding correctly is higher than 60% then the student is likely to be able to respond correctly, and if it is lower than 40% the student is likely to respond incorrectly. However, as the prediction gets closer to 50%, there is greater uncertainty since there is a 50% chance that she will be able to answer correctly and a 50% chance that she will answer incorrectly. This uncertainty on the part of the tutor about the student's ability could indicate that the student is in her ZPD with regards to the relevant knowledge. Hence the tutor perceives the range of probabilities between 40% and 60% as a model of the student's ZPD [5]. Thus, the tutor will choose to ask the question in the highest possible level of the LOR that has a predicted probability of at least 40% of being answered correctly [2]. The exception to this policy is for questions belonging to the expert level LOR. For those questions, the tutor takes a more cautious approach and only asks them if it is quite certain that the student will answer them correctly, i.e., if the predicted probability of the student answering the expert level question is equal to or greater than 60%.

The expression of each question within the LOR is adapted to provide increased support as the certainty of a correct answer decreases [9]. For example, the tutor can ask a question directly with little support such as, "What is the value of the net force?" or with more support by expressing it as "Given that the man's acceleration is zero what is the value of the net force applied on the man?" In the latter case, the object is named concretely and a relevant hint ("Given that the man's acceleration is zero") is included, making this second version of the question less cognitively demanding.

3   Testing the System

3.1   Conditions

Two versions of the system were developed to use as control and experimental conditions. The control version used a "poor man's" student model that consisted of an array of KCs initialized with a score based on the student's pretest performance and that score did not vary throughout the study. Additionally, when students started a reflection question, they were assigned to a fixed LOR level (expert, medium, or novice) based on the correctness of their response to the RQ and on their KC scores according to the algorithm shown in Fig. 3.

The experimental condition used the adaptive version of the system described in previous sections, which embeds a student model that updates its estimates as the dialogue progresses and implements domain and instructional contingent scaffolding.

3.2   Participants

Students from a high school in Pittsburgh, Pennsylvania, in the U.S. were recruited to participate in the study. They were taking a college preparatory class (though not honors or Advanced Placement) that covered the topics discussed in the system. Students were randomly assigned to the control and experimental conditions and used the system as an in-class homework helper, hence the system was used after the material had been covered in class. A total of 73 students participated in the study; N = 42 were in the control condition and N = 31 in the experimental condition. The imbalance in the number of participants was due to students missing school and hence not completing the study (a t-test revealed no pretest difference between students who completed the study and those that did not, $p = .471$).

3.3   Materials

Using the experimental and control versions of the system, students solved 5 problems with 3–5 reflection questions per problem on the topic of dynamics. A pretest and isomorphic posttest (i.e., the pretest and corresponding posttest items only differed in their cover stories) were developed. The tests consisted of 35 multiple-choice test items that were presented online and automatically graded, though students did not receive feedback on the correctness of their answers. The test items were conceptual questions that tested the KCs associated with tutor's reflection questions but were not similar to the homework problems which required quantitative solutions as seen in the sample problem solution in Fig. 1. Students were given 30 min to complete the tests.

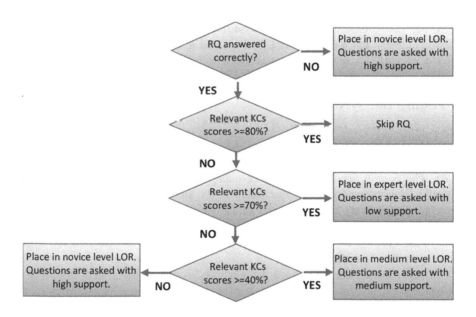

Fig. 3. Flow chart showing behavior of control condition

3.4   Protocol

Students started by taking the online pretest. After the pretest, they interleaved solving homework problems on paper with using the system in the following way: First, students solved on paper the quantitative homework problem presented by the system; second, they viewed a video of a sample solution to that problem on the system as feedback (the video contained no discussion of conceptual material); third, students engaged in conceptual dialogues with the tutorial system which addressed the conceptual aspects of the quantitative problem they had just attempted to solve. After all problems were completed, students took the online posttest and a short satisfaction survey. The entirety of the study was performed in class over the course of 4 days. All students took the pretest on Day 1 and the posttest on Day 4 and worked on the homework problems at their own pace on Days 1–3.

3.5 Results

Our main hypothesis is that students in the experimental condition would learn more than those in the control condition due to the system's proactive adaptation of scaffolding to students' evolving needs. To test this hypothesis, we started by evaluating whether students in each condition learned from interacting with the system. Then we compared the mean learning gains between conditions and checked for an aptitude treatment interaction. Finally, we compared time on task between conditions.

Did students in each condition learn from interacting with the system? To answer this question a paired-samples t-test was performed comparing the mean scores of the pretest to those of the posttest in each condition. The tests revealed a statistically significant difference between mean pretest scores and mean posttest scores for students in both conditions suggesting that students learned from interacting with the system. Table 1 shows the results.

Table 1. Pretest vs. Posttest scores

Condition	Pretest mean SD	Posttest mean SD	$t(n)$	p	Cohen's d
Experimental	M = .505 SD = .093	M = .592 SD = .091	$t(30) = 6.540$	<.001	1.2
Control	M = .503 SD = .091	M = .615 SD = .089	$t(41) = 7.565$	<.001	1.2

Did students in one condition learn more than in the other? To investigate whether one version of the system fostered more learning than the other, we first performed an ANCOVA with Condition as fixed factor, prior knowledge (as measured by pretest) as covariate, and Posttest as the dependent variable. The results of this test suggest that condition had no statistically significant effect on posttest when controlling for the effects of prior knowledge, $F(1,70) = 1.770$, $p = .19$ Additionally, we performed an independent samples t-test comparing the mean gain from pretest to posttest between conditions. No statistically significant difference was found between the mean gain of the experimental condition (M = .087, SD = .074) and the mean gain of the control condition (M = .112, SD = .096), t(71) = 1.226, $p = .22$. The results of the t-test and ANCOVA suggest that students in both conditions learned equally. We also evaluated whether the incoming knowledge—as measured by pretest score—of students in each condition was comparable. An independents sample t-test revealed no statistically significant difference in students' prior knowledge between conditions t(71) = .127, $p = .90$.

Did the effectiveness of the treatment vary depending on students' prior knowledge? In other words, was there an aptitude-treatment interaction? To study this issue, we performed a regression analysis using Condition, Pretest, and Condition*Pretest (interaction term) as independent variables and gain as the dependent variable. The regression coefficient of the interaction term was not significant suggesting no aptitude-treatment interaction $F(1,69) = 1.456$, $p = .23$.

Was one version of the system more efficient than the other? To investigate this possibility, we compared the mean time that students spent working on the system[1] between conditions by performing an independent samples t-test. The test revealed that the mean time on task of the experimental condition (M = 51.26 min, SD = 12.44 min) was significantly shorter than the mean time on task of the control condition (M = 71.52 min, SD = 16.42 min), $t(71) = 5.754$, $p < .001$, Cohen's $d = 1.4$.

A closer look at time on task: Was the experimental system more efficient than the control system for students of *all* incoming knowledge levels? In a prior study where we compared a version of Rimac that used a "poor man's" student model (similar to the control condition of this study) to a version of Rimac that did not have a student model and had all students go through the novice LOR, we found the system with the student model was significantly more efficient than the system without the student model, but *only* for high prior knowledge students [8]. Hence, we decided to investigate if in the current study the experimental version was more efficient than the control for students of *all* levels of incoming knowledge. To this end, we partitioned the students in each condition into those with high incoming knowledge and those with low incoming knowledge using a median split. We then compared the time on task of high prior knowledge students in the control and experimental groups. To that end we performed an ANOVA which revealed that the mean time of task of high pretesters in the experimental group was 31% (20.8 min) shorter than in the control group, a statistically significant difference. Similarly, when comparing time on task for low prior knowledge students between conditions, an ANOVA revealed a 27% time on task difference in favor of the experimental condition which was statistically significant. See results in Table 2 and Fig. 4.

Table 2. Comparison of time on task (TOT) between conditions for high and low incoming knowledge students

Student prior kw	Condition	N	Mean TOT (min)	SD TOT (min)	F	p	Cohen's d
Low	Control	21	74.72	14.82	F(1,35) = 18.29	<.001	1.4
	Experimental	16	54.78	12.95			
High	Control	21	68.33	17.66	F(1,34) = 16.201	<.001	1.4
	Experimental	15	47.51	11.09			

---

[1] Time on task did not include the time students spent solving the problems on paper. Additionally, any inactivity longer than three minutes while a student worked on the system was not counted towards the time on task estimate since it could be indicative that the student had taken a break from the learning activity.

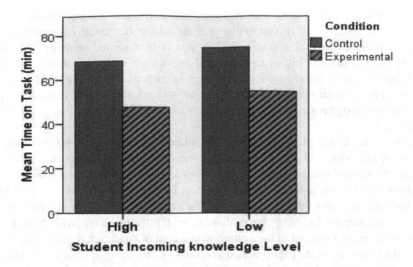

Fig. 4. Comparison of time on task between conditions for High and Low prior knowledge students.

4   Discussion and Future Work

In this paper we report on the comparison of two versions of Rimac to explore the effectiveness of incorporating a student model that is dynamically updated throughout the interaction to enable domain and instructional contingency during tutorial dialogues. One version of Rimac (experimental version) proactively adapts the content it discusses as well as the amount of support it provides during its interaction with the student by using the predictions of a student model that dynamically updates its assessment of students' understanding of particular KCs as the student progresses through the dialogues. The second version of Rimac (control version) sets the student on a fixed line of reasoning, rather than adapting to the students' evolving knowledge needs, based on the student's initial response to the reflection question under consideration and on the predictions of a static student model that only considers the student's pretest performance. We found that students in both conditions learned equally well. One possible reason this may have occurred is that regardless of the level of line of reasoning at which students are placed in the control system, if they lack the necessary knowledge to answer a question correctly, they are presented with a remedial sub dialogue that covers the knowledge subsumed in the lower level LORs. Hence, it is possible that the fixed LOR with its remediations were enough for students to have comparable knowledge gains as in the more adaptive, experimental condition.

The key finding of this work is that students who used the system with the dynamic student model (i.e., the experimental system) learned more efficiently, that is, in less time, than those who used the system with the static student model (i.e., control version). Of particular interest is the discovery that students with low incoming knowledge in the experimental condition were able to go through all the dialogues 27% faster (on average, experimental condition: 55 min, control condition: 75 min) than those in the control condition. This suggests that a dynamic student model is more effective than a static one in supporting domain and instructional contingency. The dynamic student model is able to effectively adjust to the students' evolving knowledge allowing them to traverse higher level lines of reasoning—which are shorter—as their knowledge improves, thereby saving them time. In contrast, a static student model will keep the granularity of the discussions with the students at the level defined by their incoming knowledge regardless of improvements in their knowledge that occur during the dialogues.

In future work, we plan to compare the adaptive system with two less adaptive versions of the system to try to separate on the one hand, the effect on learning of updating the student model during the dialogues and, on the other hand, the effects of providing domain and instructional contingency. In the first study, we will perform a more in-depth analysis of the impact that the student model's dynamic updates have on students' learning by isolating the evaluation of this feature. We will compare the current experimental version of the system with a control condition that would perform exactly the same way as the experimental version—i.e., deciding at what level to ask the next question and with how much support to express it rather than placing students in a fixed LOR—except that it would choose the next question based on the predictions of the static KC scores derived from the pretest rather than on the dynamically updated model. In the second study, we will evaluate more precisely the value of performing domain and instructional contingency (i.e., deciding what to ask and how to ask it on each step of the dialogue) by comparing the current version of the experimental condition with a control condition that improves on the flexibility of the one presented in this paper by placing students in fixed low, medium or high levels of lines of reasoning not just when the student answers the reflection question correctly (as in the current study) but also when the student answers it incorrectly. This may allow Rimac to place a student who may have slipped when answering the RQ in a more appropriate LOR level. The comparison of these versions of Rimac might provide additional evidence of the value of implementing scaffolding that contains domain and instructional contingency.

Acknowledgments. We thank Sarah Birmingham, Dennis Lusetich and Scott Silliman. This research was supported by the Institute of Education Sciences, U.S. Department of Education, through Grant R305A150155 to the University of Pittsburgh. The opinions are those of the authors and do not represent views of the Institute or the U.S. Department of Education.

References

1. Albacete, P., Jordan, P., Katz, S.: Is a dialogue-based tutoring system that emulates helpful co-constructed relations during human tutoring effective? In: Conati, C., Heffernan, N., Mitrovic, A., Verdejo, M.F. (eds.) AIED 2015. LNCS (LNAI), vol. 9112, pp. 3–12. Springer, Cham (2015). https://doi.org/10.1007/978-3-319-19773-9_1
2. Albacete, P., Jordan, P., Lusetich, D., Chounta, I.A., Katz, S., McLaren, B.M.: Providing proactive scaffolding during tutorial dialogue using guidance from student model predictions. In: Penstein Rosé, C., et al. (eds.) AIED 2018. LNCS (LNAI), vol. 10948, pp. 20–25. Springer, Cham (2018). https://doi.org/10.1007/978-3-319-93846-2_4

3. Cazden, C.: Peekaboo as an Instructional Model: Discourse Development at Home and at School. Stanford University Department of Linguistics, Palo Alto (1979)
4. Chi, M., Koedinger, K.R., Gordon, G.J., Jordon, P., VanLehn, K.: Instructional factors analysis: a cognitive model for multiple instructional interventions. In: Pechenizkiy, M., Calders, T., Conati, C., Ventura, S., Romero, C., Stamper, J. (eds.) EDM 2011, pp. 61–70 (2011)
5. Chounta, I.-A., Albacete, P., Jordan, P., Katz, S., McLaren, B.M.: The "Grey Area": a computational approach to model the zone of proximal development. In: Lavoué, É., Drachsler, H., Verbert, K., Broisin, J., Pérez-Sanagustín, M. (eds.) EC-TEL 2017. LNCS, vol. 10474, pp. 3–16. Springer, Cham (2017). https://doi.org/10.1007/978-3-319-66610-5_1
6. Evens, M., Joel, M.: One-on-One Tutoring by Humans and Computers. Psychology Press, New York (2006)
7. Graesser, A.C., Lu, S., Jackson, G.T., et al.: AutoTutor: a tutor with dialogue in natural language. Behav. Res. Methods 36, 180–192 (2004)
8. Jordan, P., Albacete, P., Katz, S.: Adapting step granularity in tutorial dialogue based on pretest scores. In: André, E., Baker, R., Hu, X., Rodrigo, M.M.T., du Boulay, B. (eds.) AIED 2017. LNCS (LNAI), vol. 10331, pp. 137–148. Springer, Cham (2017). https://doi.org/10.1007/978-3-319-61425-0_12
9. Katz, S., Albacete, P., Jordan, P., Lusetich, D., Chounta, I.A., McLaren, B.M.: Operationalizing Contingent Tutoring in a Natural-Language Dialogue System. Nova Science Publishers, New York (2018)
10. Katz, S., Albacete, P.: A tutoring system that simulates the highly interactive nature of human tutoring. J. Educ. Psychol. 105(4), 1126–1141 (2013)
11. Rosé, C., Jordan P., Ringenberg, M., Siler, D., VanLehn, K., Weinstein, A.: Interactive conceptual tutoring in Atlas-Andes. In: AIED 2001, pp. 151–153(2001)
12. van de Pol, J., Volman, M., Beishuizen, J.: Scaffolding in teacher–student interaction: a decade of research. Educ. Psychol. Rev. 22, 271–296 (2010)
13. Wood, D.: The why? what? when? and how? of tutoring: the development of helping and tutoring skills in children. Literacy Teach. Learn. 7(1/2), 1–30 (2003)
14. Wood, D., Wood, H.: Vygotsky, tutoring and learning. Oxford Rev. Educ. 22(1), 5–16 (1996)

Item Ordering Biases in Educational Data

Jaroslav Čechák[✉] and Radek Pelánek

Masaryk University, Brno, Czech Republic
{xcechak1,pelanek}@fi.muni.cz

Abstract. Data collected in a learning system are biased by order in which students solve items. This bias makes data analysis difficult and when not properly addressed, it may lead to misleading conclusions. We provide clear illustrations of the problem using simulated data and discuss methods for analyzing the scope of the problem in real data from a learning system. We present the data collection problem as a variant of the explore-exploit tradeoff and analyze several algorithms for addressing this tradeoff.

1 Introduction

As students solve items in a learning system, the system can collect data on their performance. This data can be used to reason about student knowledge and learning and properties of items. Based on the results of data analysis we can take actions to improve student learning; the action can be either automatic (algorithmic) or manual (based on insight obtained from analysis) [1].

However, data from learning systems are biased by data collection, and these biases can lead to misleading conclusions [15]. A specific data collection issue is item ordering. Students often solve items in a similar order, specifically "from easier to more difficult". This ordering bias makes the analysis of data difficult as it confounds increase in item difficulty and student learning [4,13,15]. Moreover, data are often influenced by attrition bias—students leave a system in non-random order, e.g., due to mastery learning or self-selection bias. Samples of population solving different items are thus not representative. This issue has been studied previously in learning curves research [7,10,11].

For data analysis purposes it is beneficial to have some degree of randomization in the data collection process [12,15]. In this way, we may overcome some of the biases. However, randomization usually goes against pedagogical principles, e.g., the progression "from easier to more difficult" is usually pedagogically desirable. We thus have a specific case of the explore-exploit tradeoff [2]: we want to present items to students in such a way that the collected data enable us to understand student learning and properties of items (*exploration*) and at the same time we want students to learn efficiently in the system (*exploitation*). The explore-exploit tradeoff in learning system has been in different variants studied in several recent works [3,8,9].

In this work, we focus specifically on biases related to item ordering. Although item ordering biases have been noted before, it was only as a sidenote of another

© Springer Nature Switzerland AG 2019
S. Isotani et al. (Eds.): AIED 2019, LNAI 11625, pp. 48–58, 2019.
https://doi.org/10.1007/978-3-030-23204-7_5

analysis [4] or as one example of a wide discussion of biases [15]. We provide a deep analysis of the issue, which is relevant to many learning systems.

We use simulated data to provide a clear illustration of the item ordering bias and to discuss under what circumstances the problem is important and why. We also discuss methods for understanding data collected from real students (with respect to item ordering) and illustrate the application of these methods on real data from learning systems. Finally, we describe and compare several algorithms for addressing the explore-exploit tradeoff with respect to item ordering in learning systems.

2   Background

The basic setting for our analyses and discussions are problem solving exercises. These exercises are typically presented in an interactive form such as a web page or a mobile app. Students are presented with problem solving exercises with varying difficulty. The students may be working on the problem either in their free time or as a part of the class assignment. Examples of these problems include but are not limited to simple programming exercises, math problems, and logic puzzles.

The student performance is measured as the time required to solve the exercise. To be precise, we use a logarithm of time since problem solving times in learning systems are typically log-normally distributed [6]. While there are other possible performance measures, solving time is universally applicable to various exercise types.

2.1   Used Data

The analyses are performed on data collected from systems Umíme[1] and RoboMission[2]. There are four datasets in total, three from the former and one from the latter. These datasets cover three problem solving exercises: a logic puzzle with marble in a maze (*Marble* and *Marble2*), Sokoban logic puzzle (*Sokoban*) and block-based programming tasks (*RoboMission*).

These datasets contain student interaction logs where each entry represents an attempt of a student at solving an item (a single problem solving exercise). Each entry also holds information whether the student's attempt was successful or not. Overall statistics can be seen in Table 1. The distribution of the number of items seen by a given student and the number of students that have seen a given item is roughly geometric.

2.2   Simulations

To better understand the behavior of the system and possible biases we used simulations. Simulations give us the possibility to control every aspect of the

---

[1] https://www.umimeto.org/, available only in Czech.

[2] https://en.robomise.cz/.

Table 1. Basic statistics of the used datasets.

Dataset	Students	Items	Interactions
Marble	17868	110	258848
Marble2	2364	98	28810
Sokoban	12384	109	182979
RoboMission	4174	85	62534

Table 2. An overview of the simulation scenarios used in this paper.

Name	Skill distribution	Learning	Skill Δ	Attrition
Incremental	$\mathcal{N}(-3, 0.1)$	Incremental	6	No
Incr. small	$\mathcal{N}(-0.5, 0.1)$	Incremental	1	No
Step	$\mathcal{N}(-3, 0.1)$	Step	6	No
Steep	$\mathcal{N}(-3, 0.1)$	Steep	6	No
Diverse	$\mathcal{N}(-3, 3)$	Incremental	6	No
diverse + att.	$\mathcal{N}(-3, 3)$	Incremental	6	Yes
const. + att.	$\mathcal{N}(0, 3)$	No	0	Yes

system and also provide us ground truth for observation comparison. The main function of our simulation framework can be simplified as take a student, while there is an available item, pick an item and generate solving time using a model. The source code of the framework is publicly available[3].

The student solving time model is defined as follows: $t = b + a\theta + \epsilon$. In this equation, t is a logarithm of solving time, b is an item difficulty, a is an item discrimination factor, θ is a student skill, and $\epsilon = \mathcal{N}(0, c^2)$ is Gaussian noise. In-depth description of the model is in [16]. To keep the simulation simple parameters a and c were fixed to values -1 and 1 respectively.

The framework is modular to allow modeling of different scenarios. They do not reflect reality but deliberately exaggerate some aspect to illustrate a specific bias. We choose to keep item difficulties fixed for better comparison. An overview of the scenarios used later in this paper can be found in Table 2.

The column *name* refers to a label we will use for the given scenario later in the paper. In the description of simulation results, we will also add a suffix indicating which item selection strategy was used.

Skill distribution describes how student skills were generated. We sampled the skill values from a normal distribution with parameters given in the table.

The column *learning* contains information about the type of student learning. Incremental means that a student is learning in small steps with each solved item. After each item, the student skill is increased by $\frac{skill\ \Delta}{\#items}$. Step learning models a situation where students do not understand the concept, but they can

---

[3] https://github.com/adaptive-learning/simulations-aied2019.

"learn" the concept with some probability after solving an item. The learning happens by increasing the student skill by skill Δ with 5% probability after each solved item. The skill of a given student is increased at most once. We do not model knowledge deterioration, e.g., forgetting. Once the concept is mastered, the student keeps the gained skill. Steep learning is a situation where a student is learning rapidly and only a few items are enough to master the concept. Real life example would be learning of the user interface. In our simulation, it takes 10 items for a student to completely learn the concept.

The *skill* Δ states the maximal amount of skill the student can gain after solving every item. For example, if the student has a skill of -3.2 at the beginning and the skill Δ is 6. Then after solving all items, the student's skill will be 2.8.

The last column refers to whether the attrition is modeled or not. When there is no attrition, all students solve every item. When the attrition is modeled, all students have a fixed time budget they can spend solving items. For better illustration, it is set so that only a few students reach the last item. In our simulations, this happens to be at 0.6 times a sum of all items difficulties.

3   Item Ordering Bias

The data collected from student interactions in a learning system that uses a fixed item order may be biased. These biases manifest themselves as dispro-portions between the estimated item difficulty and the actual item difficulty. We use simulations to explore and clearly illustrate estimates in scenarios from Table 2. We also compare situations when students solve items from easier to more difficult (*ordered*) and in a random order (*random*).

Figure 1 shows plots of different scenarios and their effect on estimated item difficulties. In this context, true item difficulty is equal to solving time of an average skilled student, i.e., the one with skill value of 0. The y-axis shows item difficulty. The x-axis corresponds to item order when sorted by their true diffi-culty. Curves for different scenarios show estimates of item difficulties computed as mean solving times. The translucent bands around them signify 95% confi-dence interval (computed by bootstrapping) for the mean.

Figure 1(a) illustrates the importance of the relationship between student learning and item difficulty increase. We consider *incremental* and *incr. small.* scenarios in the *ordered* variant. In the *incremental* scenario, student learning dominates item difficulty increases, and the estimates of item difficulty are almost reversed, i.e., easier items are estimated as harder than truly hard items. In the *incr. small* scenario the item difficulty increases dominate. While easier items are still being overestimated and harder item underestimated, the relative ordering is reasonably well estimated.

Figure 1(c) illustrates the results of the analysis under different assumptions about student learning. We keep the *incremental* scenario and compare it with *step* and *steep*. Students again solve items in the fixed order. The curve shapes differ substantially though in all three the difficulties of the first few items are highly overestimated. Real life example of this behavior could be introductory items teaching students how to use and work with the system.

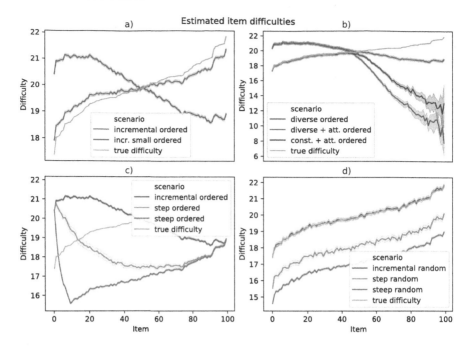

Fig. 1. Illustration of problem ordering bias. Note that in (b) and (d) the *true difficulty* (partially) overlaps with other curves.

Figure 1(d) illustrates that the fixed item order causes problems with estimates. If we let students solve items in a random order, estimates become accurate in the sense of relative ordering. The absolute values of estimates and true difficulties differ as the expected value of student skill for a random student differs between the scenarios.

Figure 1(b) illustrates the effect of attrition. In the *diverse* and *diverse + att.* scenarios, the student population is changing as students are learning in the process of solving items. In addition, the attrition leads to a change in a population sample for different items. Only the students with high skill are able to solve many items in a fixed amount of time. Consequently, later items are solved only by outstanding students. The similar result is achieved even in *const. + att.* scenario. In *diverse + att.* the learning amplifies this effect.

4   Understanding Data

Experiments with simulated data show that the fixed item order can have a significant impact on the analysis of data. To asses how relevant the problem is in a particular system, we need to understand the data. In what order do students solve items within a system? In some systems, the order is completely fixed. In such cases the potential bias caused by the ordering is high, but at least it is quite clear what is happening. In many realistic systems, however, the ordering of items

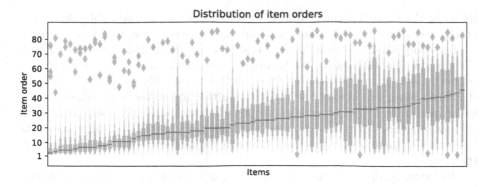

Fig. 2. Analysis of item ordering in RoboMission dataset.

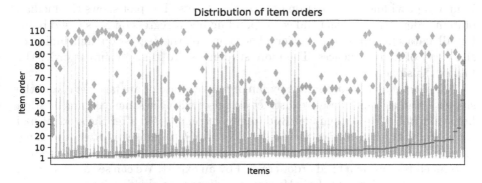

Fig. 3. Analysis of item ordering in Marble dataset.

is neither completely fixed nor random. One common setting is the following: items are presented to students in some specific default order, but students do not necessarily have to follow this order exactly. The provided ordering can be chosen manually or algorithmically; often it is explicitly or implicitly from easier items to more difficult.

To understand the data, we propose to use the following type of graph. For each student, we construct a student's item solving sequence by ordering the items the student has seen. The items are sorted from the first seen to the last seen. Aggregating over all students gives us a set of observed orders for any given item. We then visualize the distribution of these orders, sorting items by their median order.

An example of this visualization for the RoboMission dataset is in Fig. 2. The distributions for individual items are visualized using the "letter value" plot [5] (an extension of boxplot). Items with tall boxes usually have fewer observations (solutions). This visualization provides an overview of student behavior in the system. In the specific case illustrated in Fig. 2, it is clear that although students do not proceed through the system in a completely identical order, there are strong regularities, particularly with respect to items solved at the beginning.

Items in the RoboMission system are divided into levels and sublevels; in the graph, we can see "steps", where flat part are the items from the same sublevel.

For comparison, we include the same analysis for Marble dataset in Fig. 3. We can see that the median orders are slightly lower and they do not increase between items as much resulting in a "flatter" look. The former is caused by students solving on average fewer items resulting in shorter item sequences and lower item orders. The latter is caused by the randomness of the student item solving sequences. If orders of items are identically distributed on some finite interval, i.e., there is no precedence between items, the medians of all items are the same.

To better grasp the item ordering effect, it is useful to provide comparative analysis across different datasets and to quantify the effect. Figure 4 presents analysis analogical to Figs. 2 and 3, but showing only the median order for each item and providing a comparison of several datasets. The plot shows the median of item orders (translucent) and corresponding linear regression lines (solid bold) for all datasets introduced earlier. The slope of the regression line indicates similarity of item sequences. The more students solve items in similar order the higher the slope.

For *Marble* and *Sokoban*, the median orders of all items are fairly low, and the slope is small as well. These are distinct exercises with the same recommendation behavior. In both a random unsolved item is recommended after the student had successfully finished solving the item. The *Marble2* dataset is from the same exercise as *Marble* with a change in the recommender behavior. The system recommends items in a fixed order created by an expert. We can see a substantial increase in the slope. In *RoboMission* the items are divided into nine levels and each level is further divided into three sublevels. The recommended item is always taken randomly from the easiest not yet mastered sublevel. Hence the recommendations are ordered in terms of sublevels while random within the sublevel. The slope is between a fixed and random ordering as one would expect from the semi-ordered items. Note that similar type of analysis (using slightly different visualization) was used by [4].

In all four cases, the students may ignore the recommendation and choose any item from an item overview. Moreover, students typically solve only a few items. Consequently, most median item orders are low. The items in the overview are presented in some kind of lattice resulting in an implicit ordering that some students may choose to follow. Due to these factors, the extreme values of the slope are unlikely in real life scenarios.

Slope lines may not intersect the origin as evident from the Fig. 4. The slope intersecting x-axis to the right of the origin indicates that students choose items randomly in the beginning and later begin to follow a common trend. The slope intersecting on the left-hand side of the origin indicates that students solve more randomly chosen items later as they progress through the system.

One may choose to interpolate the item medians with polynomials or different kinds of curves. This is a viable option, and by definition, the fit will be tighter. Linear regression has the benefit of being non-parametric. Figure 4 illustrates

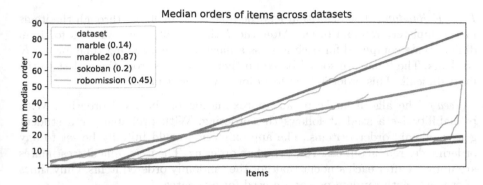

Fig. 4. Analysis of ordering from real systems. The regression line slopes for each dataset are located in parentheses in the plot legend.

that even the simple slope analysis can differentiate between various system behaviors. It could, in turn, be used to asses the relevance of the item ordering related biases in a system.

5 Algorithms for Dynamic Item Ordering

In learning systems, we would like the items to be ordered well for efficient student learning, i.e., from easier to more difficult. We call it the ideal ordering. At the same time though, we would like "good" orderings for the analysis. As evident from the earlier simulations that would be a randomized ordering. We can perceive the situation as a variation on exploit-explore tradeoff. The problem lies in balancing exploitation (students are learning efficiently) and exploration (obtaining unbiased data about items).

It would be possible to avoid the use of randomization and to fully focus on exploitation if we take student learning into account in the analysis. The shown bias arises from student learning, and the modeling of student learning removes the bias. However, this approach is valid only when the used model of student learning fits very well the actual progress of student learning. There are many different student modeling approaches [14], and it is nontrivial to choose a suitable model for a particular situation. Moreover, all models are simplifications, and even the use of an appropriate type of model can cause distortions in model analysis. It thus makes sense to explore "model-free" methods for the analysis.

5.1 Algorithms

We have devised three algorithms that combine exploration with exploitation in various ways. The algorithms are quite simple and definitely could be improved upon. Our main aim is to illustrate the methodology that could be used to compare such algorithms.

First K Random. The algorithm lets the first K students go through the items in a completely random order. After the Kth student, the estimates for item difficulty are computed for each item as a mean problem solving time for the K students. The items are sorted based on their estimates from the easiest to the most difficult. This order is then fixed for any later student.

ε-greedy. The algorithm keeps an approximation of the ideal ordering. With probability 1-ε a student follows this ordering. With probability ε a student gets randomly ordered items. The approximation could initially be set to any ordering desired, in our case some random ordering. The item difficulty estimates are updated after each student that got the randomly ordered items. Only times of students with random passes are used for estimates.

Adaptive Periodic K. The algorithm keeps an ordering based on estimates of item difficulties. The initial ordering is random, and the estimates are recomputed after K students have gone through the system since the last update. Times of all the previous students are used in the computation of estimates.

5.2   Experiments

We once again used the simulations to get insight into the properties of these algorithms in different scenarios. We are mainly interested in convergence properties, i.e., can the algorithm find the ideal ordering. The metric we choose is the Spearman's correlation coefficient with ideal ordering. The Fig. 5 shows the progression of the correlation as new students are coming to the system. Each data point is a mean of correlations for ten consecutive students. This step is done to smooth out the curves yet still retain some local variations.

In the case Fig. 5(a) we can see that all three algorithms are performing well. This is the scenario where problem solving time estimates follow true item difficulties. Figure 5(b) shows results for a scenario where the ordering bias is highly present. We can see that *Adaptive periodic 20* algorithm is struggling and oscillating around a non-ideal ordering. Even in this scenario, 20 random student passes are enough to obtain a reasonable approximation of the ideal ordering as illustrated by *First 20 random* algorithm. The scenario (c) adds attrition on top. We can see that the performance of all algorithms is impacted. The *0.05-greedy* algorithm is still able to slowly approach the ideal ordering. We can compare the scenario (d) to others to gain some insight into the interplay of attrition and learning. Compared to (c) we can see that all algorithms are performing better. *Adaptive periodic 20* is doing even better than in scenario (b) with learning and no attrition. Other algorithms are affected by limited information gained from student solving only a subset of items.

It is worth noting that the robustness of *ε-greedy* algorithm is achieved at the expense of some students having randomly ordered items. That is also the reason for the ruggedness of its curve in the plots.

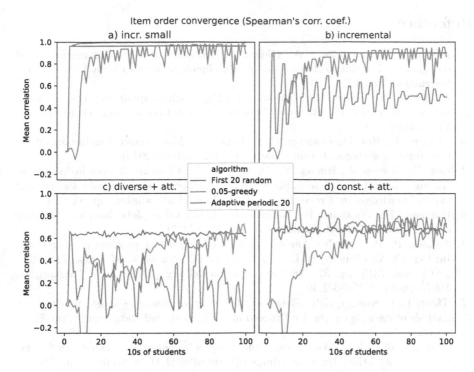

Fig. 5. Convergence of ordering computed as the Spearman's correlation coefficient.

6   Discussion

Students often solve items in a specific order, particularly in a progression from easier to more difficult. Such regularities in student behavior lead to biases in data collected from learning systems. When not taken into account in the analysis, these biases may lead to misleading conclusions. We have illustrated the potential biases using simulated data, showing how the importance of biases depends on the speed and nature of student learning (relative to the increases in item difficulty). We have also proposed a technique for analyzing the degree of ordering bias in data from real systems.

To overcome biases caused by fixed ordering, we can utilize dynamic item orderings. An interesting research question is how to realize such a dynamic ordering. In this work, we analyzed three simple algorithms for a dynamic ordering of items. Using simulated data we showed that their performance depends on the specific application scenario (what is the speed and nature of student learning, what is the distribution of item difficulties). The results show that under some settings, naive use of automatic adaptation can lead to worse results than solutions based on simple randomization. An interesting direction for future work is to consider more complex combinations of adaptation and randomization.

References

1. Aleven, V., McLaughlin, E.A., Glenn, R.A., Koedinger, K.R.: Instruction based on adaptive learning technologies. In: Handbook of Research on Learning and Instruction, Routledge (2016)
2. Audibert, J.Y., Munos, R., Szepesvári, C.: Exploration-exploitation tradeoff using variance estimates in multi-armed bandits. Theoret. Comput. Sci. **410**(19), 1876–1902 (2009)
3. Clement, B., Roy, D., Oudeyer, P.-Y., Lopes, M.: Multi-armed bandits for intelligent tutoring systems. J. Educ. Data Min. **7**(2), 20–48 (2015)
4. González-Brenes, J., Huang, Y., Brusilovsky, P.: General features in knowledge tracing: applications to multiple subskills, temporal item response theory, and expert knowledge. In: Proceedings of Educational Data Mining, pp. 84–91 (2014)
5. Hofmann, H., Wickham, H., Kafadar, K.: Letter-value plots: boxplots for large data. J. Comput. Graph. Stat. **26**(3), 469–477 (2017)
6. Jarušek, P., Pelánek, R.: Analysis of a simple model of problem solving times. In: Cerri, S.A., Clancey, W.J., Papadourakis, G., Panourgia, K. (eds.) ITS 2012. LNCS, vol. 7315, pp. 379–388. Springer, Heidelberg (2012). https://doi.org/10.1007/978-3-642-30950-2_49
7. Käser, T., Koedinger, K.R., Gross, M.: Different parameters - same prediction: an analysis of learning curves. In: Proceedings of Educational Data Mining, pp. 52–59 (2014)
8. Lan, A.S., Baraniuk, R.G.: A contextual bandits framework for personalized learning action selection. In: Proceedings of Educational Data Mining, pp. 424–429 (2016)
9. Liu, Y.E., Mandel, T., Brunskill, E., Popovic, Z.: Trading off scientific knowledge and user learning with multi-armed bandits. In: Proceedings of Educational Data Mining, pp. 161–168 (2014)
10. Murray, R.C., et al.: Revealing the learning in learning curves. In: Lane, H.C., Yacef, K., Mostow, J., Pavlik, P. (eds.) AIED 2013. LNCS (LNAI), vol. 7926, pp. 473–482. Springer, Heidelberg (2013). https://doi.org/10.1007/978-3-642-39112-5_48
11. Nixon, T., Fancsali, S., Ritter, S.: The complex dynamics of aggregate learning curves. In: Proceedings of Educational Data Mining (2013)
12. Papoušek, J., Stanislav, V., Pelánek, R.: Evaluation of an adaptive practice system for learning geography facts. In: Gasevic, D., Lynch, G., Dawson, S., Drachsler, H., Rosé, C.P. (eds.) Proceedings of Learning Analytics & Knowledge, pp. 40–47. ACM (2016)
13. Pavlik, P.I., Yudelson, M., Koedinger, K.R.: A measurement model of microgenetic transfer for improving instructional outcomes. Int. J. Artif. Intell. Educ. **25**(3), 346–379 (2015)
14. Pelánek, R.: Bayesian knowledge tracing, logistic models, and beyond: an overview of learner modeling techniques. User Model. User-Adap. Inter. **27**(3), 313–350 (2017)
15. Pelánek, R.: The details matter: methodological nuances in the evaluationof student models. User Model. User-Adap. Inter. (2018)
16. Pelánek, R., Jarušek, P.: Student modeling based on problem solving times. Int. J. Artif. Intell. Educ. **25**(4), 493–519 (2015)

A Comparative Study
on Question-Worthy Sentence Selection
Strategies for Educational Question
Generation

Guanliang Chen[1]([⊠]), Jie Yang[2], and Dragan Gasevic[1]

[1] Monash University, Melbourne, Australia
{guanliang.chen,dragan.gasevic}@monash.edu
[2] University of Fribourg, Fribourg, Switzerland
jie@exascale.info

Abstract. Automatic question generation, which aims at converting sentences in an article to high-quality questions, is an important task for educational practices. Recent work mainly focuses on designing effective generation architectures based on deep neural networks. However, the first and possibly the foremost step of automatic question generation has largely been ignored, i.e., identifying sentences carrying important information or knowledge that is worth asking questions about. In this work, we (i) propose a total of 9 strategies, which are grounded on heuristic question-asking assumptions, to determine sentences that are question-worthy, and (ii) compare their performance on 4 datasets by using the identified sentences as input for a well-trained question generator. Through extensive experiments, we show that (i) *LexRank*, a stochastic graph-based method for selecting important sentences from articles, gives robust performance across all datasets, (ii) questions collected in educational settings feature a more diverse set of source sentences than those obtained in non-educational settings, and (iii) more research efforts are needed to further improve the design of educational question generation architectures.

Keywords: Educational question generation · Sentence selection · Deep neural network

1 Introduction

In education, automatically generating high-quality questions for learning practices and assessment has long been desired. Previous studies have indicated that reading is one of the most frequent strategies adopted by students to learn [30]. To assess students' understanding, instructors and teachers need to design corresponding assessment questions about the reading material [3,28]. However,

© Springer Nature Switzerland AG 2019
S. Isotani et al. (Eds.): AIED 2019, LNAI 11625, pp. 59–70, 2019.
https://doi.org/10.1007/978-3-030-23204-7_6

such a question creation process is usually time-consuming and cognitively-demanding. Therefore, *automatic question generation*, which aims at automating the creation process through computational techniques, has attracted much research attention [5, 8, 10, 15].

One common strand of work in automatic question generation is generating questions in a rule-based manner, in which educational experts are recruited to carefully define a set of syntactic rules to turn declarative sentences into interrogative questions [1, 15, 23]. In recent years, deep neural networks have emerged as a more promising approach to question generation [8–11, 33, 35]. In contrast to rule-based methods, neural question generation methods can capture complex question generation patterns from data without handcrafted rules, thus being much more effective and scalable. Typically, neural question generation methods tackle the generation process as a sequence-to-sequence learning problem, which directly maps a piece of text (usually a sentence) to a question [10].

Table 1. Question-worthy sentence in a paragraph.

Have you ever dropped your swimming goggles in the deepest part of the pool and tried to swim down to get them? It can be frustrating because the water tries to push you back up to the surface as you're swimming downward. **This upward force exerted on objects submerged in fluids is called the buoyant force.**

Given a paragraph or an article, often there are only a limited number of sentences that are worth asking questions about, i.e., those carrying important concepts. An example is shown in Table 1, where the last sentence defines the most important concept "buoyant force". We, therefore, argue that selecting question-worthy sentences is of critical importance to the generation of high-quality educational questions.

Existing studies, however, pay little attention to this step: they either assume that the question-worthy sentences have been identified already [10] or simply take every sentence in an article as input for the question generator. For instance, [15] assumes that all sentences in an article are question-worthy and thus generate one question for each sentence and select high-quality ones based on their linguistic features. To our knowledge, [8] is the only study that explicitly tackles the question-worthy sentence selection problem. It uses a bidirectional LSTM network [19] to simultaneously encode a paragraph and calculate the question-worthiness of a sentence in the paragraph. However, training such a network relies on a large amount of ground-truth labels of question-worthy sentences (e.g., tens of thousands). Obtaining these labels is a long, laborious, and usually costly process. Furthermore, the proposed deep neural network was only validated in short paragraphs instead of the whole article. Considering that reading materials can be much longer and deep neural networks can fail at processing long sequence data due to the vanishing gradient problem [18], it remains an open question whether the proposed method can handle long articles.

Instead of developing a novel neural network architecture that simultaneously does sentence selection and question generation (like [8] does), this work takes one step back and focuses extensively on question-worthy sentence selection. We aim at achieving a better understanding of the effectiveness of different textual features in identifying question-worthy sentences from an article, so as to clarify the main criteria in selecting question-worthy sentences, and to adequately inform question generator design.

To this end, we first propose a total of 9 strategies for question-worthy sentence selection, which cover a wide range of possible question-asking patterns inspired by both low-level and high-level textural features. For instance, we represent our assumption that *informative* sentences are more likely to be asked about by leveraging low-level features such as sentence length and the number of concepts as informativeness metrics; our assumption that *important* sentences are more worth asking about is represented by leveraging semantic relevance between sentences, which can be measured by using summative sentence identification techniques [4, 13]. To evaluate the effectiveness of the proposed strategies, we apply them to identify question-worthy sentences on 4 question generation datasets, i.e., *TriviaQA* [20], *MCTest* [31], *RACE* [21] and *LearningQ* [5]. Among the four considered datasets, only RACE and LearningQ are collected in educational settings and they consist of questions covering various cognitive levels. In contrast, TriviaQA and MCTest are collected to advance the development of machine reading comprehension and they mainly contain questions seeking for factual details. By including all of these datasets, we expect to identify the specific characteristics of question-worthy sentences for the task of educational question generation. We use the sentences identified by the proposed strategies as input for a well-trained question generator and evaluate the effectiveness of sentence selection strategies by comparing the quality of the generated questions.

To the best of our knowledge, our work is the first one that systematically studies question-worthy sentence selection strategies across multiple datasets. Through extensive experiments, we find that *LexRank*, which identifies important sentences by calculating their eigenvector centrality in the graph representation of sentence similarities, gives the most robust performance across different datasets among the nine selection strategies. Furthermore, we demonstrate that questions collected for human learning purposes usually feature a more diverse set of sentences, including those that are most informative, important, or contain the largest amount of novel information, while non-learning questions (e.g., those seeking for factual details from Wikipedia articles in TriviaQA) are often positioned at the start of sentences. Lastly, we show that there is a large improvement space for existing educational question generation architectures.

2 Methodology

Our research methodology is depicted in Fig. 1. In the following, we first describe the nine strategies we developed for question-worthy sentence selection and then introduce our method for evaluating these strategies, including the question generator that takes the selected sentences as input for question generation, the

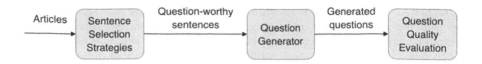

Fig. 1. Research methodology.

experimental datasets, the automatic evaluation metrics, and the human study. We evaluate the effectiveness of the proposed sentence selection strategies by comparing the quality of the questions generated by applying those sentence selection strategies.

2.1   Sentence Selection Strategies

In the following, we describe in detail our proposed sentence selection strategies based on different question-asking assumptions and sentence properties measured by different textual features.

Random Sentence (Random). As the baseline, we randomly select a sentence and use it as input for the question generator.

Longest Sentence (Longest). This strategy selects the longest sentence in an article. The assumption is that people tend to ask questions about sentences containing a large amount of information, which, intuitively, can be measured by their lengths.

Concept-Rich Sentence (Concept). In contrast to *Longest*, this strategy assumes that the amount of information can be better measured by the total number of entities in a sentence. The more entities a sentence contains, the richer the information it has.

Concept-Type-Rich Sentence (ConceptType). This strategy is a variant of *Concept*. It calculates the total number of entity types in a sentence to measure the informativeness of a sentence.

The above three strategies approximate question-worthiness of a sentence by *informativeness*, which is further measured by different textual features. In contrast, the following two strategies approximate question-worthiness of a sentence by *difficulty* and *novelty*, respectively.

Most Difficult Sentence (Hardest). This strategy is built on the assumption that difficult sentences can sometimes bring the most important messages that should be questioned and assessed. Therefore, it chooses the most difficult sentence in an article as the question-worthy sentence. We calculate the Flesch Reading Ease Score [6] of sentences as their difficulty indicators.

Novel Sentence (Novel). Unlike *Hardest*, this strategy assumes that sentences with novel information that people do not know before are more question-worthy.

We calculate the number of words that never appear in previous sentences as a sentence's novelty score [34] and select the most novel one.

Finally, we introduce three strategies that approximate question-worthiness of a sentence by the relative *importance* of the sentence with respect to the remaining ones in an article. The importance is either measured by the relative position of a sentence or its centrality represented by semantic relevance with other sentences.

Beginning Sentence (Beginning). In the research of text summarization, a common hypothesis about sentence positions is the importance of a sentence decreases with its distance from the beginning of the article [26], and therefore less question-worthy. This strategy selects the first sentence in an article as the most question-worthy sentence.

Centroid Based Important Sentence (LexRank). In line with *Beginning*, this strategy also assumes that question-worthy sentences should be selected from those of greater importance. The difference here is that the sentence importance is measured by the centroid-based method, *LexRank* [13], which calculates sentence importance based on eigenvector centrality in a graph of sentence similarities.

Maximum Marginal Relevance Based Important Sentence (MMR). In contrast to *LexRank*, this strategy computes sentence importance by considering a linear trade-off between relevance and redundancy [4]. That is, the strategy selects the sentence that is most relevant but shares least similarity with the other sentences as the most important sentence.

2.2   Evaluation Method

To evaluate the proposed strategies, we feed the selected sentences to a popular question generator and evaluate the effectiveness on four benchmarking datasets through both automatic evaluation and human evaluation.

Question Generator. There have been several studies working on constructing effective question generators [10,11,33,35], most of which assume that the answer to a question (usually a span of text in the input sentence) is determined prior to the generation of the question and use both the sentence and the answer as input for the question generator. Our focus in this work is to develop effective sentence selection strategies for educational question generation without observing the answers to questions beforehand. We, therefore, adopt the question generator proposed in [10], which only takes a sentence as input to generate a question, as our testbed to evaluate the effectiveness of the proposed sentence selection strategies. The question generator is based on an attention-based sequence-to-sequence learning framework [2], which maps a sentence from an article to a question by using an LSTM encoder and decoder [19]. Particularly, the decoder incorporates the attention mechanism over the encoder hidden states, which enables the question generator to focus on important concepts in the input sentence during question generation.

Table 2. Statistics of the 4 experimental datasets.

Data source	TriviaQA	MCTest	RACE	LearningQ
	Wikipedia articles	Human-generated stories	Language learning articles	Lecture transcripts
# Articles	138,537	660	23,530	1,102
# Questions	138,537	2,640	64,120	7,621
# Avg. sentence/article	202.39 ± 173.12	18.93 ± 7.25	19.05 ± 7.05	42.89 ± 25.57
# Avg. words/sentence	27.19 ± 24.92	12.84 ± 6.47	17.36 ± 10.12	19.76 ± 12.48
# Avg. questions/article	1.00 ± 0.00	4.00 ± 0.00	2.73 ± 1.11	6.92 ± 1.80

Datasets. Generally, all datasets with questions and the corresponding articles, which the questions are about, can be used to evaluate the selection strategies. Our work aims at identifying the unique characteristics of sentences that can be used to generate high-quality questions for educational practices, i.e., questions that are natural and readable to people and contain pedagogical value. We, therefore, select experimental datasets that contain natural questions designed by humans instead of search queries [12,24] or cloze-style questions [16,17,25]. We also include datasets that are collected for not only educational purposes but also non-educational purposes, e.g., those designed for the advancement of machine reading comprehension, so as to underline the difference between selecting sentences for the generation of learning questions and non-learning questions. To summarize, we include the following datasets for experiments.

- **TriviaQA** [20] contains questions from trivia and quiz-league websites and evidence articles gathered from web search and Wikipedia. Here we only consider questions with evidence articles collected from Wikipedia, which results in 138K question-article pairs.
- **MCTest** [31] consists of 660 stories written by crowd-workers and 2,000 associated questions about the stories.
- **RACE** [21] collects 23,000 reading comprehension articles for English learning and 64,000 assessment questions.
- **LearningQ** [5] contains both instructor-designed questions gathered from TED-Ed and learner-generated questions gathered from Khan Academy. As the learner-generated questions can be redundant about the same knowledge concepts (i.e., same sentences), to avoid concept bias, we only include the 7,000 instructor-designed questions for experiments.

TriviaQA and MCTest are collected in non-educational settings: TriviaQA questions mainly seek for factual details and the answers can be found as a piece of text in the source article from Wikipedia, and MCTest questions are designed for young children. RACE and LearningQ are collected in educational settings: RACE questions are mainly used to assess students' knowledge level of English, whereas LearningQ covers a diverse set of educational topics, more complex articles, and the questions require higher-order cognitive skills to solve. Descriptive statistics of these datasets are given in Table 2.

Automatic Evaluation Metrics. We adopt *Bleu1*, *Bleu2*, *Bleu3*, *Bleu4*, *Meteor* and *Rouge$_L$* for evaluation (following [10]). Bleu-n scores rely on the maximum n-grams for counting the co-occurrences between a generated question and a set of reference questions; the average of Bleu is employed as the final score [27]. Meteor computes the similarity between the generated question and the reference questions by taking synonyms, stemming and paraphrases into account [7]. Rouge$_L$ reports the recall rate of the generated question with respect to the reference questions based on the longest common sub-sequence [22].

Table 3. Examples of generated Questions and the corresponding source Sentences.

	Grammaticality	Clarity	Usefulness
S1: As other sources of natural gas decrease, the costs of non-renewable energies rise, and cutting-edge technologies make renewable energies so accessible **Q1:** *What is the source of natural gas decrease?*	2	1	0
S2: The brain can solve complicated problems, and grasp concepts such as infinity or unicorns. **Q2:** *What problems can the brain solve?*	3	3	0
S3: When testing a new headache medication, a large pool of people with headaches would be randomly divided into two groups, one receiving the medication and another receiving a placebo **Q3:** *How is a new headache medication tested?*	3	3	1

Human Study. To gain better insights about the quality of the generated questions, we recruited three native speakers of English to rate the quality of 200 randomly-selected questions (along with the corresponding source sentence and article) generated from RACE and LearningQ, respectively. Specifically, we considered three metrics here: *Grammaticality*, *Clarity* and *Usefulness*. Firstly, we only presented a question to the evaluators and asked them to rate the grammatical correctness of the question on a scale of $[1, 3]$, with 3 being exactly correct. Then, the corresponding sentence from which the question was generated was presented to the evaluators and the evaluators were asked to specify how clear the question was and to what extent the question was making sentence given the input sentence on a scale of $[1, 3]$, with 3 being very clear. Lastly, the evaluators were presented with the source article (i.e., the article from which the source sentence was selected) and rate the usefulness of the question for learning (e.g., enabling a better understanding of the article, assessing students' knowledge) on a scale of $[0, 1]$ with 1 being useful and 0 being not useful at all. The ratings given by the three evaluators for each metric were averaged as the final rating for a question. Three examples of the generated questions along with the source sentences and the corresponding human evaluation ratings are given in Table 3. Due to the limited space, we omit the source articles in Table 3.

3   Experiments

This section presents and compares the performance of our proposed strategies for question-worthy sentence selection as evaluated across the considered datasets.

3.1   Experimental Setup

To our knowledge, *SQuAD* [29] is the only dataset which contains ground-truth labels for over 97,000 sentence-question pairs. In line with [10], we also use the labeled input sentences and the corresponding questions in SQuAD for training the question generator. We set the hyper-parameters as suggested in [10] and use beam search ($N = 3$) to generate a question.

Articles can be of different lengths and thus possibly contain different numbers of question-worthy sentences (shown in Table 2). During experiments, the number of selected sentences should be dependent on the number of ground-truth questions gathered about an article: different ground-truth questions are seeking for different details about the article, i.e., based on different question-worthy sentences. We therefore evaluate each of the questions generated by different selected sentences against all the ground-truth questions of a article and consider the result with the best performance as an indication of the selected sentence matched with the ground-truth question.

3.2   Results and Analysis

Table 4 reports the results of the proposed sentence selection strategies on four datasets. We highlight the top-3 strategies for each dataset. Based on these results, several interesting findings are observed as follows.

For TriviaQA, *Beginning* achieves the best performance, indicating that most questions in TriviaQA are about the first sentence in the source article. Considering that the articles of TriviaQA are collected from Wikipedia, such a result can be interpreted by the fact that the first sentences of Wikipedia paragraphs/articles often contain the most important information worth asking about [36]. This observation can be further verified by the well-performing results given by *LexRank* and *MRR* – ranking at the 2nd and 3rd position, respectively – which also identifies important sentences but uses a different method. Overall, these results show that importance-based strategies are more effective than informativeness-based (e.g., *Longest, Concept*), difficulty-based (i.e., *Hardest*), or novelty-based ones (i.e., *Novel*).

For the two datasets collected in educational contexts, namely RACE and LearningQ, *Longest, LexRank*, and *Novel* generally show better performance than the other strategies. Such a result suggests that questions in learning related datasets are relevant to a more diverse set of sentences, i.e., those informative, important, or contain novel information, a result is likely due to the diverse learning goals related to the questions. We further observe big gaps between these three strategies and the remaining ones. For example, *Longest, LexRank*, and

Table 4. Experimental results of automatic evaluation on TriviaQA, MCTest, RACE and LearningQ. The top three results in each metric are in bold.

Datasets	Strategies	Bleu1	Bleu2	Bleu3	Bleu4	Meteor	Rouge$_L$
TriviaQA	Random	6.69	2.07	0.70	0.31	**6.21**	8.29
	Beginning	**9.42**	**3.67**	**1.51**	**0.74**	**6.66**	**10.70**
	Longest	3.37	1.21	0.45	0.21	3.57	8.67
	Hardest	1.99	0.66	0.23	0.11	2.37	6.84
	ConceptMax	0.73	0.19	0.06	0.02	1.63	4.04
	ConceptTypeMax	1.94	0.57	0.19	0.08	2.94	6.02
	LexRank	**8.79**	**3.11**	**1.14**	**0.52**	**5.54**	**9.81**
	MMR	**7.13**	**2.44**	**0.92**	**0.42**	5.06	**8.93**
	Novel	3.25	1.12	0.42	0.20	3.47	8.28
MCTest	Random	4.18	1.41	0.55	0.21	7.04	16.46
	Beginning	4.69	1.56	**0.63**	**0.27**	7.93	**17.36**
	Longest	**5.75**	**1.99**	**0.79**	**0.29**	**9.95**	**17.96**
	Hardest	4.41	1.42	0.51	0.18	7.53	16.73
	ConceptMax	3.92	1.48	0.60	**0.27**	5.72	16.71
	ConceptTypeMax	4.01	1.49	0.60	**0.28**	5.89	16.66
	LexRank	**5.24**	**1.85**	**0.70**	0.22	**8.55**	**18.13**
	MMR	4.53	1.53	0.57	0.22	7.52	17.20
	Novel	**4.92**	**1.58**	0.57	0.20	**9.15**	17.14
RACE	Random	4.24	1.28	0.45	0.20	5.74	11.47
	Beginning	4.50	1.33	0.43	0.18	5.95	11.82
	Longest	**6.48**	**2.08**	**0.74**	**0.33**	**7.83**	**12.84**
	Hardest	4.36	1.35	0.47	0.21	5.51	11.60
	ConceptMax	2.74	0.86	0.34	0.16	3.41	10.69
	ConceptTypeMax	2.78	0.88	0.35	0.17	3.45	10.71
	LexRank	**5.47**	**1.73**	**0.63**	**0.29**	6.79	**12.59**
	MMR	4.45	1.39	0.51	0.24	5.78	11.75
	Novel	**5.89**	**1.80**	**0.61**	**0.26**	**7.59**	**12.32**
LearningQ	Random	5.66	1.48	0.43	0.14	5.55	14.83
	Beginning	5.02	1.29	0.37	0.13	5.13	14.53
	Longest	**6.34**	**1.81**	**0.57**	**0.22**	**9.10**	**16.86**
	Hardest	5.92	1.60	**0.52**	**0.21**	5.77	15.48
	ConceptMax	4.57	1.25	0.40	0.16	4.77	14.20
	ConceptTypeMax	4.75	1.29	0.41	0.16	4.91	14.24
	LexRank	**6.74**	**1.91**	**0.62**	**0.26**	7.44	**16.40**
	MMR	5.86	1.53	0.47	0.17	5.72	15.12
	Novel	**6.00**	**1.64**	0.50	0.17	**8.93**	**16.28**

Novel are the only strategies achieving *Bleu1* scores greater than 5 and *Meteor$_L$* scores greater than 6 on RACE. This observation reveals that sentence selection strategies based on similar sentence properties however measured through different textual features (e.g., *Longest* vs. *Concept* and *LexRank* vs. *Beginning*) can have big variance in terms of performance. This highlights the importance of selecting appropriate textual features in question-worthy sentence selection.

Similar results also hold on the MCTest dataset: *Longest*, *LexRank*, and *Novel* generally achieve good performance, which suggests that questions in MCTest are also relevant to a diverse set of sentences. On the other hand, strategies such as *Beginning* and *ConceptMax* also perform well on several metrics, signifying that different measures of sentence properties (e.g., informativeness using *Longest* and *ConceptMax*) do not necessarily lead to highly different sentence selection results on MCTest. Despite this, we can observe that *LexRank* is the only sentence selection strategy consistently ranking in top-3 across all the 4 datasets, demonstrating its robustness against all the other compared strategies.

Table 5. Experimental results of human evaluation on RACE and LearningQ.

	Grammaticality	Clarity	Usefulness
RACE	2.12	1.92	0.53
LearningQ	1.87	1.34	0.22

Table 5 reports the human evaluation results on 200 questions randomly selected from RACE and LearningQ, whose source sentences were selected by applying the best-performing strategies, i.e., *Longest* and *LexRank*, respectively. Compared to LearningQ questions, RACE questions have higher ratings across all metrics, which indicate that RACE questions are more readable and making better sense to people. This can be explained by the fact that RACE only consists of articles and questions used for English learning, while LearningQ covers a wide range of subjects ranging from arts and humanities to science and technology; as well LearningQ articles are much longer and contain relatively more diverse sentence and question patterns. Correspondingly, this poses more challenges to the question generator to deliver high-quality questions. Noticeably, in terms of *Usefulness*, the ratings are 0.53 and 0.22 on the 0–1 scale for RACE and LearningQ, respectively. This indicates that only about 50% and 20% of the generated RACE and LearningQ questions respectively can be used for educational practices. This is in line with previous findings from [5] and demands further investigations to improve the existing architectures for educational question generation.

4   Conclusion and Future Work

Question-worthy sentence selection is an important however largely ignored topic in automatic generation of educational questions. This paper presented a systematic study on nine sentence selection strategies inspired by different question-asking heuristics. Extensive experiments showed that *LexRank*, which selects important sentences from articles by calculating eigenvector centrality in a graph of sentence similarities, gave robust performance across multiple datasets. Our experimental results also revealed that the beginning sentence in an article is often worth questioning about in non-educational settings, while questions in educational contexts feature a more diverse set of source sentences that are informative, important, or contain novel information. Also, we demonstrated that there is quite some improvement space for developing effective educational question generation architectures. These findings inspire our future research to combine multiple strategies for selecting question-worthy sentences in learning contexts and improve the design of existing educational question generation architectures by applying techniques such as reinforcement learning [32] and generative adversarial networks [14].

References

1. Adamson, D., Bhartiya, D., Gujral, B., Kedia, R., Singh, A., Rosé, C.P.: Automatically generating discussion questions. In: Lane, H.C., Yacef, K., Mostow, J., Pavlik, P. (eds.) AIED 2013. LNCS (LNAI), vol. 7926, pp. 81–90. Springer, Heidelberg (2013). https://doi.org/10.1007/978-3-642-39112-5_9
2. Bahdanau, D., Cho, K., Bengio, Y.: Neural machine translation by jointly learning to align and translate. arXiv preprint arXiv:1409.0473 (2014)
3. Bahrick, H.P., Bahrick, L.E., Bahrick, A.S., Bahrick, P.E.: Maintenance of foreign language vocabulary and the spacing effect. Psychol. Sci. **4**(5), 316–321 (1993). https://doi.org/10.1111/j.1467-9280.1993.tb00571.x
4. Carbonell, J., Goldstein, J.: The use of MMR, diversity-based reranking for reordering documents and producing summaries. In: SIGIR, pp. 335–336 (1998)
5. Chen, G., Yang, J., Hauff, C., Houben, G.J.: LearningQ: a large-scale dataset for educational question generation. In: ICWSM (2018)
6. Collins-Thompson, K.: Computational assessment of text readability: a survey of current and future research. ITL-Int. J. Appl. Linguist. **165**(2), 97–135 (2014)
7. Denkowski, M., Lavie, A.: Meteor universal: language specific translation evaluation for any target language. In: SMT (2014)
8. Du, X., Cardie, C.: Identifying where to focus in reading comprehension for neural question generation. In: Proceedings of the 2017 Conference on Empirical Methods in Natural Language Processing, pp. 2067–2073 (2017)
9. Du, X., Cardie, C.: Harvesting paragraph-level question-answer pairs from Wikipedia. In: Association for Computational Linguistics (ACL) (2018)
10. Du, X., Shao, J., Cardie, C.: Learning to ask: neural question generation for reading comprehension. In: ACL (2017)
11. Duan, N., Tang, D., Chen, P., Zhou, M.: Question generation for question answering. In: Proceedings of the 2017 Conference on Empirical Methods in Natural Language Processing, pp. 866–874 (2017)
12. Dunn, M., Sagun, L., Higgins, M., Guney, V.U., Cirik, V., Cho, K.: SearchQA: a new Q&A dataset augmented with context from a search engine. arXiv preprint arXiv:1704.05179 (2017)

13. Erkan, G., Radev, D.R.: LexRank: graph-based lexical centrality as salience in text summarization. J. Artif. Intell. Res. **22**, 457–479 (2004)
14. Goodfellow, I., et al.: Generative adversarial nets. In: Advances in Neural Information Processing Systems, pp. 2672–2680 (2014)
15. Heilman, M., Smith, N.A.: Good question! statistical ranking for question generation. In: HLT-NAACL (2010)
16. Hermann, K.M., et al.: Teaching machines to read and comprehend. In: NIPS, pp. 1693–1701 (2015)
17. Hill, F., Bordes, A., Chopra, S., Weston, J.: The goldilocks principle: reading children's books with explicit memory representations. CoRR abs/1511.02301 (2015)
18. Hochreiter, S., Bengio, Y., Frasconi, P., Schmidhuber, J., et al.: Gradient flow in recurrent nets: the difficulty of learning long-term dependencies (2001)
19. Hochreiter, S., Schmidhuber, J.: Long short-term memory. Neural Comput. **9**(8), 1735–1780 (1997)
20. Joshi, M., Choi, E., Weld, D.S., Zettlemoyer, L.: TriviaQA: a large scale distantly supervised challenge dataset for reading comprehension. In: ACL, July 2017
21. Lai, G., Xie, Q., Liu, H., Yang, Y., Hovy, E.: RACE: large-scale reading comprehension dataset from examinations. In: EMNLP (2017)
22. Lin, C.Y.: Rouge: a package for automatic evaluation of summaries. In: ACL (2004)
23. Mitkov, R., Ha, L.A.: Computer-aided generation of multiple-choice tests. In: HLT-NAACL (2003)
24. Nguyen, T., et al.: MS MARCO: a human generated machine reading comprehension dataset. arXiv preprint arXiv:1611.09268 (2016)
25. Onishi, T., Wang, H., Bansal, M., Gimpel, K., McAllester, D.A.: Who did what: a large-scale person-centered cloze dataset. In: EMNLP (2016)
26. Ouyang, Y., Li, W., Lu, Q., Zhang, R.: A study on position information in document summarization. In: COLING. pp. 919–927 (2010)
27. Papineni, K., Roukos, S., Ward, T., Zhu, W.J.: BLEU: a method for automatic evaluation of machine translation. In: ACL (2002)
28. Prince, M.: Does active learning work? a review of the research. J. Eng. Educ. **93**(3), 223–231 (2004)
29. Rajpurkar, P., Zhang, J., Lopyrev, K., Liang, P.: Squad: 100, 000+ questions for machine comprehension of text. In: EMNLP (2016)
30. Rayner, K., Foorman, B.R., Perfetti, C.A., Pesetsky, D., Seidenberg, M.S.: How psychological science informs the teaching of reading. Psychol. Sci. Public Interes. **2**(2), 31–74 (2001)
31. Richardson, M., Burges, C.J., Renshaw, E.: MCTest: a challenge dataset for the open-domain machine comprehension of text. In: EMNLP, pp. 193–203 (2013)
32. Sutton, R.S., Barto, A.G.: Introduction to Reinforcement Learning, vol. 135. MIT Press, Cambridge (1998)
33. Tang, D., Duan, N., Qin, T., Yan, Z., Zhou, M.: Question answering and question generation as dual tasks. arXiv preprint arXiv:1706.02027 (2017)
34. Tsai, F.S., Tang, W., Chan, K.L.: Evaluation of novelty metrics for sentence-level novelty mining. Inf. Sci. **180**(12), 2359–2374 (2010)
35. Wang, T., Yuan, X., Trischler, A.: A joint model for question answering and question generation. arXiv preprint arXiv:1706.01450 (2017)
36. Yang, Y., Yih, W.t., Meek, C.: WikiQA: a challenge dataset for open-domain question answering. In: EMNLP, pp. 2013–2018 (2015)

Effect of Discrete and Continuous Parameter Variation on Difficulty in Automatic Item Generation

Binglin Chen[1]([✉]) [iD], Craig Zilles[1] [iD], Matthew West[2] [iD], and Timothy Bretl[3] [iD]

[1] Department of Computer Science,
University of Illinois at Urbana-Champaign, Urbana, USA
{chen386,zilles}@illinois.edu
[2] Department of Mechanical Engineering,
University of Illinois at Urbana-Champaign, Urbana, USA
mwest@illinois.edu
[3] Department of Aerospace Engineering,
University of Illinois at Urbana-Champaign, Urbana, USA
tbretl@illinois.edu

Abstract. Automatic item generation enables a diverse array of questions to be generated through the use of question templates and randomly-selected parameters. Such automatic item generators are most useful if the generated item instances are of either equivalent or predictable difficulty. In this study, we analyzed student performance on over 300 item generators from four university-level STEM classes collected over a period of two years. In most cases, we find that the choice of parameters fails to significantly affect the problem difficulty.

In our analysis, we found it useful to distinguish parameters that were drawn from a small number (<10) of values from those that are drawn from a large—often continuous—range of values. We observed that values from smaller ranges were more likely to significantly impact difficulty, as sometimes they represented different configurations of the problem (e.g., upward force vs. downward force). Through manual review of the problems with significant difficulty variance, we found it was, in general, easy to understand the source of the variance once the data were presented. These results suggest that the use of automatic item generation by college faculty is warranted, because most problems don't exhibit significant difficulty variation, and the few that do can be detected through automatic means and addressed by the faculty member.

Keywords: Automatic item generation · Item models

1 Introduction

In classroom settings, computerized assessment offers many advantages compared to paper-based assessment, in both formative and summative assessment. In both contexts, it enables automatically grading a wide range of problem types [26], which reduces grading workloads, and provides students with

© Springer Nature Switzerland AG 2019
S. Isotani et al. (Eds.): AIED 2019, LNAI 11625, pp. 71–83, 2019.
https://doi.org/10.1007/978-3-030-23204-7_7

immediate feedback [3], which has been shown to improve learning [22]. In formative assessment, computerized assessment enables mastery-oriented pedagogies [6,18] where students can repeat problem types until mastery is demonstrated, and it permits assigning different problems to each student to discourage plagiarism [23]. In summative assessment, computer-based testing enables more authentic item types (e.g., programming exams on computers where compilers and debuggers are available), reduces the overhead of exam administration, especially for large classes [27] and permits the use of adaptive testing [19].

Many of these applications of computer-based assessment benefit from large pools of problems. One proposed method of generating a large collection of problems is automatic item generation (AIG) [13,17], where item instances (items) are generated by instantiating parameterizable problem templates with specific values (see Sect. 2.1). AIG has been used across a broad range of disciplines [15] and has proven useful for generating large pools of items.

AIG is most useful, however, if the item instances produced by an item generator are of similar difficulty, or at minimum of a predictable difficulty. Previous research has shown that generally there is variation in psychometric properties of the item instances (items) produced by an item generator (item model), but that variation tends to be smaller than the variation between generators [1,2,7,10–12,16,20,21,24]. There is a growing consensus that calibration should be done at the level of the item generator, using a multi-level strategy where item instances are nested within item generator [7,9,11,14,25].

This previous work, however, has largely focused on questions designed by psychometricians for standardized testing. For wide spread adoption of AIG across educational contexts, it is important to understand the extent to which disciplinary experts that are not experts in test construction can construct AIGs.

In this paper, we study AIG item difficulty variance in an ecologically-valid higher-education setting where university faculty members have written AIGs for use in both computerized homework and exams in large-enrollment STEM courses. We analyze student exam performance across a two-year time period on a collection of over 300 AIGs. We believe this also represents the largest reported study of AIGs to date. We describe our experimental data in detail in Sect. 2.

Our analysis focuses on how the choice of parameters for a given problem affects its difficulty. Methodologically, we found it necessary to break our analysis of parameters into two separate groups. In the first group, the number of unique values of the parameter was small (<10), which meant that we had sufficiently many samples for each parameter to use the hybrid Fisher's exact test to check for significant variation in difficulty between the parameters. The analysis of these *discrete* parameters is presented in Sect. 3. In the other group, the number of unique parameters was large enough that only a few samples were available for many of the parameter values, and for some parameters every student had a unique value. For this *continuous* parameter group, we used the Kolmogorov-Smirnov test, as described in Sect. 4.

In this analysis, we find that the vast majority of the AIGs studied can be characterized as *uniform generators*, in that their generated instances are of equal difficulty (or at least any differences in difficulty are too small to be detected). In our university-level STEM context, the few *non-uniform generators*—where some instances are significantly harder than others—that we found, were more frequently due to discrete parameters that represented different "configurations" of the problem. University STEM students may be largely insensitive to the specific numbers in problems because they are allowed to use calculators. Our paper makes the following contributions:

1. We demonstrate the utility of using the Kolmogorov-Smirnov test to analyze the difficulty variance in AIG with many unique parameter values,
2. We provide evidence that faculty that are disciplinary content experts, but not experts in test construction or psychometrics, are largely capable of writing AIGs without significant difficulty variation, and
3. We find that once an AIG has been identified as having significant difficulty variance, it is generally easy to understand the source of the variance. This suggests that an automated analysis tool could be used by disciplinary faculty to correct item generators or divide an item generator into a collection of item generators that each have a stable difficulty.

2   Experimental Data

The data was collected at a large R1 university in the US from Fall 2016 to Summer 2018. The four courses studied are drawn from introductory sequences in Electrical & Computer Engineering (electronics) and Mechanical Engineering (statics, dynamics, and solid mechanics). These courses administered a significant part of their summative assessments in a Computer-Based Testing Facility (CBTF) via the PrairieLearn system. With IRB approval we obtained all the PrairieLearn data collected in the CBTF.

2.1   PrairieLearn and the Computer-Based Testing Facility

In their simplest conception, two components comprise AIGs: a *template* (or *model*) providing the structure of the question and a computation that randomly parameterizes this template (see Fig. 1) [4]. AIGs, however, can be quite sophisticated, if enabled by the authoring tool. The courses studied in this paper use the PrairieLearn LMS [26], which permits authors to write an arbitrary piece of server-side code and use the full power of HTML5/JavaScript to render a question. This enables questions to not just populate text templates, but also programmatically generate images (e.g., pictures of beams with forces in different places) and provide client-side interaction, e.g., providing in-browser CAD tools for students to design finite-state machines.

PrairieLearn can be used for both online homework and for exams. The data for this paper was drawn from exams, which we believe—due to their higher stakes and better security—to be more reliable. The exams in question were taken in a Computer-Based Testing Facility (CBTF) [27], which is a proctored

```
a = random.randint(5, 10)        // random parameter generation
b = random.randint(5, 10)
solution = a + b                 // compute solution from params
```
```
Compute the sum:  {{a}} + {{b}}                  // template
```
```
Compute the sum:  9 + 7                          // problem instance
```

Fig. 1. A simple automatic item generator involving a computation that generates a pair of random numbers which are used in a template to generate an item instance.

physical computer lab where the tests are available. To support classes with hundreds of students with an 85-seat computer lab, CBTF exams are asynchronous; students can choose when to take their CBTF exam during a 3-day period. As such, each student's exam is uniquely generated when they take the exam; PrairieLearn supports both AIGs for random problem parameterization and randomly selecting from a pool of AIGs for a given slot on the exam. PrairieLearn exams are not adaptive, so the sub-population receiving any given problem is an unbiased random sample of the class's population.

2.2 Detailed Data Specification

For each class in each semester, we obtained the information of all generated items in the form (**class ID, semester ID, exam ID, generator ID, problem parameters, student ID, score**). The class, semester, exam, generator, and student IDs are unique identifiers that differentiate between classes, semesters, exams within a semester (retained across semesters), generators, and students. The problem parameters are a list of parameter-value pairs used by the item generator to generate an item. The value of each parameter can be an integer, a real number, a string, or a container (dictionary or array) containing further parameters. The score is an integer that is 1 if the student's submitted answer was correct and 0 if the submitted answer was incorrect. For questions in PrairieLearn that allowed multiple attempts, we only extracted the first scored student submission. Course details can be found in Table 1.

Table 1. The four courses studied have, in aggregate, 378 unique AIGs.

Course	Number of item generators	Number of discrete parameters	Number of continuous parameters	Number of exams	Number of semesters	Number of students	Number of item instances	Number of unique item instances
Class A	92	213	69	3	4	1,670	40,496	12,063
Class B	116	243	269	7	4	1,532	31,538	23,353
Class C	93	398	234	6	7	1,473	58,157	46,799
Class D	77	153	269	7	4	940	21,430	18,792

To facilitate the analysis, we performed the two preprocessing steps. First, we dropped the semester ID to merge data across semesters to obtain the best statistical power possible. We did not, however, drop exam ID (giving us 432 unique combinations of class ID, exam ID and generator ID) as some AIGs were used in multiple exams within a semester, and students' performance on an AIG might vary across the different exams. Second, we flattened containers (dictionaries or arrays) with less than 10 elements to facilitate studying the parameters in the containers independently. The 10-element cutoff prevents flattening arrays that contain hundreds of real values. After this flattening, the AIGs had a total of 1,848 parameters (4.88 parameters/AIG).

Some of these parameters had a small number of discrete values, while other were drawn from large, often continuous, ranges (e.g., floating-point numbers). These two classes of parameters required different statistical approaches and so were analyzed independently. As our AIGs, in general, had both discrete and continuous parameters, we have chosen to formulate the analysis in terms of parameters to identify the parameters that have a significant impact on difficulty.

3 Discrete Parameters

3.1 Analysis Method for Discrete Parameters

For each unique combination of (class ID, exam ID, generator ID), we first considered all of the discrete parameters of the item generators. For each discrete parameter, we computed a $2 \times n$ contingency table between the score and the parameter values, where n is the number of unique values of the parameter. Table 2 shows an example contingency table where $n = 2$. With the contingency table, we applied the hybrid Fisher's exact test to obtain a p-value describing whether the score is dependent on the parameter values. The hybrid Fisher's exact test is a combination of Fisher's exact test and the chi-squared test. Specifically, a chi-squared test is performed when Cochran's rule [8] (no cell has expected counts less than 1 and more than 80% of the cells have expected counts at least 5) is satisfied, otherwise Fisher's exact test is performed. The null hypothesis of the test states that the score is independent of the parameter values. A small p-value would indicate that the null hypothesis is not likely true and thus we may want to reject it.

Table 2. Example contingency table.

Score	Parameter value		Total
	Left	Right	
0	399	275	674
1	96	250	346
Total	495	525	1,020

With the above method, the data resulted in 1,223 contingency tables. Such a large number of hypothesis tests introduces the multiple testing problem, which states that running large number of hypothesis tests without correction will unavoidably end up with rejected null hypotheses (discoveries) that shouldn't have been rejected. For example, if we run 1,000 hypothesis tests with $\alpha =$ 0.05, even if all null hypotheses are true, we still expect about 50 ($= 1,000 \times$ 0.05) of them to be rejected. To address this issue, we employed the Benjamini-Yekutieli procedure [5] which provides a bound on the percentage of rejected null hypotheses that shouldn't have been rejected, regardless of the dependency structure of these tests.

3.2   Results for Discrete Parameters

We conducted the hybrid Fisher's test on the 1,223 contingency tables and obtained the same number of p-values. We sorted the p-values from the smallest to the largest and plotted them in this sorted order in Fig. 2. The way the Benjamini-Yekutieli procedure works is that it draws a line passing through the origin with slope $\alpha/(NH_N)$ where N is the number of hypothesis tests to be performed and H_N is the partial sum of the first N terms in the harmonic series. The Benjamini-Yekutieli procedure states that the points below this straight line correspond to significant results where the points above correspond to insignificant results. Figure 2 shows there are 17 points below the line, which suggests that there are 17 significant cases (that is, 17 discrete parameters causing non-uniform generators).

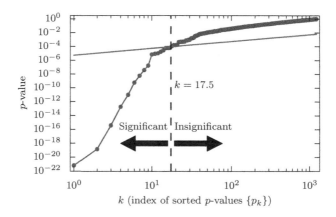

Fig. 2. Distribution of p-values for discrete parameters, log log version.

With the significant cases identified, we computed the non-uniform parameter fraction for discrete parameters to be 1.39% (95% CI [0.73, 2.05]), and plotted it as the first bar of Fig. 3a. We also computed the average maximum difference in correct rate for discrete parameters, which is calculated by iterating over each non-uniform parameter discovered by discrete parameter analysis, finding the pair of parameter values that gives the maximum difference in correct rate, and then averaging this over all non-uniform parameters. The result is 25.97% (95% CI [21.14, 30.80]), and we plotted it as the first bar of Fig. 3b.

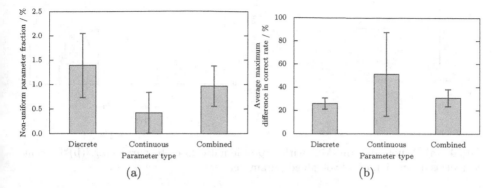

Fig. 3. (a) The non-uniform parameter fraction for each parameter type. (b) The average maximum difference in correct rate for parameters that we detected as non-uniform.

3.3   Manual Analysis of Outliers for Discrete Parameters

Figure 4a shows one of the seventeen non-uniform AIGs with a problematic discrete parameter. This AIG asked students to find either the x or y component of the velocity of a particle, given its speed, a description of its path, and its direction—either "left" or "right"—along this path. Figure 4b shows that students are significantly more likely to answer the item generator incorrectly when the direction of motion is "left." This result is easily explained by noting that a common mistake is neglecting to flip the sign of the velocity—from positive to negative—when the particle is moving to the left. Post-hoc analysis verified this hypothesis. As shown in Fig. 4b, percent incorrect was originally 80.6% for left and 52.4% for right. If we treat all cases in which students made a sign error as correct, then these results become 53.0% for left and 50.3% for right—in other words, the significant variation disappears. Difficulty variation in the sixteen other non-uniform generators could be similarly explained.

4   Continuous Parameters

4.1   Analysis Method for Continuous Parameters

Similar to the discrete parameter case, we considered the continuous parameters for each unique combination of (class ID, exam ID, problem ID). We excluded any string– or container–valued parameters from the analysis. For each parameter, we grouped the parameter values based on the corresponding score. This divides parameter values into two groups (those answered correctly and those answered incorrectly). We then applied the Kolmogorov-Smirnov test to the two groups

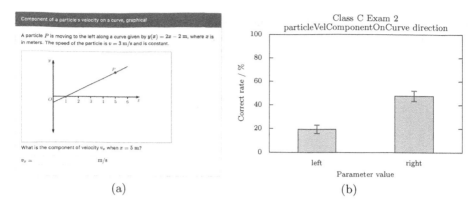

(a)   (b)

Fig. 4. (a) A non-uniform AIG with a problematic discrete parameter. (b) Percent correct as a function of a two-valued parameter.

of parameter values to examine if the two groups of parameter values are likely to be drawn from the same distribution. If the p-value from the Kolmogorov-Smirnov test is small, then it is likely that the two groups of parameter values are from different distributions, which indicates that this particular parameter can alter the difficulty of the item generated.

With the above method, the data resulted in 947 hypotheses tests to be performed. We again applied the Benjamini-Yekutieli procedure to resolve the multiple testing problem as in the discrete parameter case.

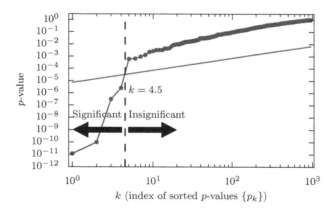

Fig. 5. Distribution of p-values for continuous parameters, log log version.

4.2   Results for Continuous Parameters

We again plotted the sorted p-values against the index and the straight line used by the Benjamini-Yekutieli procedure in Fig. 5. As the figure shows, there are 4 points below the line, which indicates that there are 4 significant cases (that is, 4 continuous parameters producing non-uniform generators).

Similarly to the discrete parameter case, we computed the non-uniform parameter fraction for continuous parameters to be 0.42% (95% CI [0.01, 0.84]), and plotted it as the second bar of Fig. 3a. We also computed the average maximum difference in correct rate by treating the continuous parameters as discrete and using the procedure in Sect. 3.2, to obtain 51.37% (95% CI [15.21, 87.53]), and we plotted this as the second bar of Fig. 3b.

With the number of significant cases computed for both discrete and continuous parameters, we performed a t-test to examine if there is a difference in non-uniform parameter fraction for these two types of parameters. The p-value for this test is 0.0146, which indicates that there is a statistically significant difference in the probability of a discrete parameter producing a non-uniform generator versus the probability of a continuous parameter doing so. We also performed a t-test on the average maximum difference in correct rate between the two types of parameters, and the p-value is 0.1093, which suggests that there isn't enough evidence to conclude that one type of parameter is more damaging to fairness than the other.

To understand the overall behavior of parameters, we also computed the non-uniform parameter fraction for the two types of parameters combined, which is 0.97% (95% CI [0.56, 1.38]). We plotted it as the third bar of Fig. 3a. We also computed the average maximum difference in correct rate for both type of parameters combined, which is 30.81% (95% CI [23.58, 38.04]). We plotted it as the third bar of Fig. 3b. In total, there were 20 non-uniform generators (17 discrete and 4 continuous non-uniform parameters, with two of them appearing on a single generator).

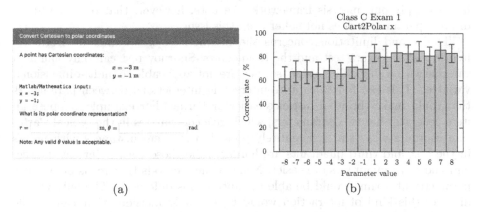

(a)   (b)

Fig. 6. (a) One of the non-uniform AIGs we found where a continuous parameter significantly affects difficulty. (b) Percent correct as it varies with this continuous parameter—note the change in difficulty between positive and negative values.

4.3   Manual Analysis of Outliers for Continuous Parameters

Figure 6a shows one of the four non-uniform AIGs with a problematic continuous parameter. This generator asked students to find the polar coordinates (r, θ) of a point, given its Cartesian coordinates (x, y). Figure 6b shows, in particular, that students are significantly more likely to answer this AIG correctly when $x > 0$. This result is easily explained by noting that θ must be computed with an arc-tangent and that a common mistake is not to use the four-quadrant arc-tangent function. The arc-tangent and the four-quadrant arc-tangent are the same for precisely those points for which $x > 0$. Post-hoc analysis verified this hypothesis. As shown in Fig. 6b, percent incorrect was originally 32.8% when $x < 0$ and 16.8% when $x > 0$. If we treat all cases in which students used the arc-tangent instead of the four-quadrant arc-tangent as correct, then these results become 20.8% when $x < 0$ and 16.8% when $x > 0$—in other words, the significant variation disappears. The continuous parameter variation in difficulty indicated by our analysis in the other three item generators can be similarly explained.

5   Limitations

There are three limitations in the current work. The first limitation is that the continuous parameter analysis only considered parameters with numerical values. This was done so that there was a natural ordering of parameter values, which is necessary for the Kolmogorov-Smirnov test to be applied (the K-S test needs to construct a cumulative density function). For parameters that have string-valued or complex-object-valued parameters, it is not immediately clear how to apply our analysis framework. We note, however, that our analysis of discrete parameters does not suffer from this issue.

The second limitation concerns the interactions of parameters. Both the hybrid Fisher's exact test and the Kolmogorov-Smirnov test can only handle a variable with a single dimension, so they are not applicable to multi-dimensional variables. This can fail to capture interesting interactions between parameters that could make the item instance easier or harder. For example, suppose an item generator has two parameters, which give coordinates in the typical Cartesian coordinate system. It is entirely possible that item instances with points in the first and third quadrants are harder while those with points in the second and forth quadrants are easier. None of the methods that focus on a single parameter at a time would be able to capture this difference. The only way to discover this kind of interaction would be to apply methods that can handle multi-dimensional data. Unfortunately, the increase of dimensions can result in the curse of dimensionality, since the increase of dimensionality means that more data points are necessary for any method to draw a solid conclusion. This would be an interesting direction for future research.

The third limitation relates to the study's generalizability. Since the data is collected in introductory engineering courses at an R1 institution in the US, it is unclear how well these results apply to K-12 education and other institution types in other countries. It is also unclear how applicable these results are to non-STEM disciplines, as the questions in the courses studied were highly numerical.

6 Conclusion

Of the 378 AIGs that we studied, we found only 20 of them (5.3%) to have parameters that led to statistically significant difficulty variation. From this study, we conclude that there is reason to be cautiously optimistic about the potential for university STEM faculty to author uniform AIGs. We believe that AIGs have the potential to improve the efficiency and effectiveness of many educational contexts, and this result suggests that instructors can be trusted to develop AIGs.

Furthermore, we found that the source of the variation of non-uniform AIGs was frequently very easy to comprehend when the parameter that led to the variance was identified. This suggests to us that an automated analysis run after every exam, much in the way that standard psychometrics tests are run, can be used to bring problematic generators to the faculty member's attention. All of the instances we observed could be fixed by one of three methods: (1) removing a particularly problematic parameter value (e.g., 0), (2) splitting the generator into multiple generators (e.g., one for quadrants 1 and 2 and the other for 3 and 4), or (3) shifting the range of a given parameter. We did have one generator that required a few hours to track down the source of the difficulty variance, and it came down to an incorrect problem solving strategy working for part of the parameter range and not for the rest; this could be addressed by excluding the problematic part of the range from the question.

Finally, we observed that discrete parameters were more likely to be the cause of non-uniform generators than continuous parameters. One hypothesis for this observation is that, in the engineering problems we studied, the continuous parameters often represented the real-numbered values of forces or voltages or distances. Our students generally use calculators when working with these numbers and the equations necessary to solve the problems, making the precise numerical values less important in determining question difficulty.

References

1. Arendasy, M.E., Sommer, M.: Using automatic item generation to meet the increasing item demands of high-stakes educational and occupational assessment. Learn. Individ. Differ. **22**(1), 112–117 (2012)
2. Attali, Y.: Automatic item generation unleashed: an evaluation of a large-scale deployment of item models. In: Penstein Rosé, C., et al. (eds.) AIED 2018. LNCS (LNAI), vol. 10947, pp. 17–29. Springer, Cham (2018). https://doi.org/10.1007/978-3-319-93843-1_2
3. Attali, Y., Powers, D.: Immediate feedback and opportunity to revise answers to open-ended questions. Educ. Psychol. Measure. **70**(1), 22–35 (2010)

4. Bejar, I.I.: Generative testing: from conception to implementation. In: Irvine, S., Kyllonen, P. (eds.) Item Generation for Test Development. Lawrence Erlbaum Associates (2002)
5. Benjamini, Y., Yekutieli, D.: The control of the false discovery rate in multiple testing under dependency. Ann. Stat. **29**, 1165–1188 (2001)
6. Bloom, B.S.: Learning for mastery. Evaluation Comment **1** (1968)
7. Cho, S.J., De Boeck, P., Embretson, S., Rabe-Hesketh, S.: Additive multilevel item structure models with random residuals: item modeling for explanation and item generation. Psychometrika **79**(1), 84–104 (2014)
8. Cochran, W.G.: Some methods for strengthening the common χ^2 tests. Biometrics **10**(4), 417–451 (1954)
9. Embretson, S.: Generating items during testing: psychometric issues and models. Psychometrika **64**(4), 407–433 (1999). https://doi.org/10.1007/BF02294564
10. Enright, M.K., Morley, M., Sheehan, K.M.: Items by design: the impact of systematic feature variation on item statistical characteristics. App. Measure. Educ. **15**(1), 49–74 (2002)
11. Geerlings, H., Glas, C.A., van der Linden, W.J.: Modeling rule-based item generation. Psychometrika **76**(2), 337–359 (2011)
12. Geerlings, H., van der Linden, W.J., Glas, C.A.: Optimal test design with rule-based item generation. Appl. Psychol. Measure. **37**(2), 140–161 (2013)
13. Gierl, M., Haladyna, T.: Automatic Item Generation: Theory and practice. Routledge, Abingdon (2013)
14. Glas, C.A., van der Linden, W.J.: Computerized adaptive testing with item cloning. Appl. Psychol. Measure. **27**(4), 247–261 (2003)
15. Haladyna, T.M.: Automatic item generation: a historical perspective. In: Gierl, M., Haladyna, T. (eds.) Automatic Item Generation· Theory and Practice, pp. 23–35. Routledge, Abingdon (2013)
16. Hively II, W., Patterson, H.L., Page, S.H.: A "universe-defined" system of arithmetic achievement tests. J. Educ. Measure. **5**(4), 275–290 (1968)
17. Irvine, S., Kyllonen, P.: Item Generation for Test Development. LawrenceErlbaum Associates, Mahwah (2002)
18. Kulik, C.L.C., Kulik, J.A., Bangert-Drowns, R.L.: Effectiveness of mastery learning programs: a meta-analysis. Rev. Educ. Res. **60**, 265 (1990)
19. van der Linden, W.J., Glas, C.A.: Elements of Adaptive Testing. Springer, Heidelberg (2010). https://doi.org/10.1007/978-0-387-85461-8
20. Macready, G.B.: The use of generalizability theory for assessing relations among items within domains in diagnostic testing. Appl. Psychol. Measure. **7**(2), 149–157 (1983)
21. Meisner, R., Luecht, R., Reckase, M.: The comparability of the statistical characteristics of test items generated by computer algorithms. Technical report ACT Research Report Series Nop. 93–9, ACT, Inc. (1993)
22. Opitz, B., Ferdinand, N.K., Mecklinger, A.: Timing matters: the impact of immediate and delayed feedback on artificial language learning. Front. Hum. Neurosci. **5**, 8 (2011)
23. Rasila, A., Havola, L., Majander, H., Malinen, J.: Automatic assessment in engineering mathematics: evaluation of the impact. In: ReflekTori 2010 Symposium of Engineering Education, pp. 37–45 (2010)
24. Sinharay, S., Johnson, M.S.: Use of item models in a large-scale admissions test: a case study. Int. J. Test. **8**(3), 209–236 (2008)

25. Sinharay, S., Johnson, M.S.: Statistical modeling of automatically generated items. In: Gierl, M., Haladyna, T. (eds.) Automatic Item Generation: Theory and Practice, pp. 183–195. Routledge, Abingdon (2013)
26. West, M., Herman, G.L., Zilles, C.: PrairieLearn: mastery-based online problem solving with adaptive scoring and recommendations driven by machine learning. In: 2015 ASEE Annual Conference and Exposition. ASEE Conferences, Seattle, Washington, June 2015
27. Zilles, C., West, M., Mussulman, D., Bretl, T.: Making testing less trying: lessons learned from operating a computer-based testing facility. In: 2018 IEEE Frontiers in Education (FIE) Conference, San Jose, California (2018)

Automated Summarization Evaluation (ASE) Using Natural Language Processing Tools

Scott A. Crossley[1](✉), Minkyung Kim[1], Laura Allen[3],
and Danielle McNamara[2]

[1] Department of Applied Linguistics/ESL,
Georgia State University, Atlanta, GA 30303, USA
{scrossley,mkim89}@gsu.edu
[2] Department of Psychology, Arizona State University, Tempe, AZ, USA
dmcnamara@asu.edu
[3] Department of Psychology, Mississippi State University,
Starkville, MS 39762, USA
lka22@msstate.edu

Abstract. Summarization is an effective strategy to promote and enhance learning and deep comprehension of texts. However, summarization is seldom implemented by teachers in classrooms because the manual evaluation of students' summaries requires time and effort. This problem has led to the development of automated models of summarization quality. However, these models often rely on features derived from expert ratings of student summarizations of specific source texts and are therefore not generalizable to summarizations of new texts. Further, many of the models rely of proprietary tools that are not freely or publicly available, rendering replications difficult. In this study, we introduce an automated summarization evaluation (ASE) model that depends strictly on features of the source text or the summary, allowing for a purely text-based model of quality. This model effectively classifies summaries as either low or high quality with an accuracy above 80%. Importantly, the model was developed on a large number of source texts allowing for generalizability across texts. Further, the features used in this study are freely and publicly available affording replication.

Keywords: Natural language processing · Summarization · Discourse · Writing · Machine learning · Summary scoring

1 Introduction

There are a number of different strategies to teach students to comprehend and produce text, including text summarization [1]. Indeed, a recent meta-analysis indicated that summarization techniques improve comprehension in over 90% of studies [2]. Text summarization is unique in that it allows students to read and comprehend short texts and then demonstrate their understanding of those texts in writing. Thus, summarization taps into both reading and writing skills, essentially *writing to learn*. Research has

© Springer Nature Switzerland AG 2019
S. Isotani et al. (Eds.): AIED 2019, LNAI 11625, pp. 84–95, 2019.
https://doi.org/10.1007/978-3-030-23204-7_8

demonstrated that summarization effectively promotes learning, enhances deeper understanding of domain topics [3–5] and helps students practice critical academic and life skills including distinguishing relevant from irrelevant material and integrating new information with prior knowledge [6]. Written summaries also provide students with the opportunity to practice writing skills including writing objectively, communicating main ideas, paraphrasing, and developing cohesive structures [7–10].

While multiple studies have shown that summarization techniques benefit students, learning to write summaries requires practice. Students often write too much, write too little, copy verbatim, or fail to appropriately synthesize information [11]. Fortunately, summarization strategies can be taught and are effective for a wide range of students, including less skilled readers [12, 13], language learners [14], and students with learning disabilities [15]. Indeed, a meta-analysis reported that the average weighted effect size for summarization instruction for adolescent learners (Grades 4–12) was quite large [16].

Instruction and practice enhance summarization skills; however, it is challenging for teachers to implement summarization tasks because evaluating summaries requires effort and time [17, 18]. In response, a number of methods have been developed for automated summarization evaluation (ASE) in order assess textual elements related to summaries including integrated content, accuracy of content, language use, and text coherence [17, 19–24]. However, many of these approaches rely on specific information outside of the text to predict summarization quality (e.g., expert summarizations of the source text) or depend on text features that are not publicly available. Such limitations make these approaches less generalizable, difficult to replicate, and problematic to implement in a dynamic learning environment. Thus, the purpose of this study is to develop an automated linguistic model of text summarization quality that is founded on natural language processing (NLP) features that are publicly available and do not require topic specific data. We do so by analyzing over 1,000 summaries collected through crowd-sourced and traditional techniques. The summaries were written on over 30 prompts providing some assurance that an algorithm gleaned from them will be generalizable. In addition, the bases for our algorithm are linguistic features that are freely and publicly available to all researchers to facilitate replication of our results. Our goal is to develop an ASE model that can be used to inform on-line tutoring and feedback systems and in turn provide strategy instruction to students as well as formative and summative feedback to writers regarding the quality of their summaries.

1.1 Summarization

Good summaries provide a concise overview of the most important content in a given passage. To do so, individuals need to construct a coherent, accurate mental model of the passage and paraphrase it using concise statements [25, 26]. Thus, successful summarization depends on (at least) two processes: comprehending source material and reproducing the key elements of that material. Generally speaking, this is achieved through the processes of reading and writing. Successful summarization depends on identifying main ideas and their supporting ideas as well as detecting the rhetorical and organizational structures of the source text [3, 27]. Once these textual elements have been identified, the content of the source text needs to be reproduced by communicating the main points of the text, generally in writing, by using inferences or generalizations, omitting unimportant details, rephrasing portions of the text, using objective language, and developing cohesive structures [8, 9, 11, 27].

Summary writing enhances reading comprehension [3, 4] by encoding and strengthening the retention of information [28–30]. Summarizing texts may also improve the retention of information to a greater degree than other educational tasks including argumentative writing [12], short-answer questions [31], and both fill-in-the-blank and multiple-choice questions [32]. Summary writing also helps develop writing skills, requiring students to convey information in a succinct manner in one's own words and build coherence through well connected sentences [25, 33].

1.2   Automated Summarization Evaluation (ASE)

Research has demonstrated the benefits of using summarizations in education settings, but the adoption of summarization tasks is not widespread. Manually scoring summaries is time-consuming and potentially subjective [17, 18]. Automatically assessing summary quality can help alleviate time constraints and, as a result, has become an important component of many educational technologies. Specifically, researchers are interested in how summarizations can be scored automatically using linguistic and semantic features of the summary and the source text and how these same features can be used to provide feedback to learners in online educational systems [17].

In assessing summary quality, the most important criterion is the inclusion of complete and accurate content that is consistent with the source text [17, 22–24]. There are a number of different approaches that have been used to automatically assess summary quality including overlap of key n-grams between source and summaries, examining lexical and semantic overlap between source texts and their summaries (or expert summaries), and the use of rhetorical devices in summaries including connectives. Perhaps the most common approach is the use of Latent Semantic Analysis (LSA) to assess overlap between summaries and their sources. LSA is a mathematical method to represent the meaning of words and text segments based on large text corpora, and in turn, the extent to which documents are semantically related to one another [20, 21]. Early work by Landauer et al. [34] focused on assessing students' reading skills by their ability to summarize short articles or excerpts from leveled readers within a system called Summary Street. Studies on Summary Street indicated that students receiving feedback from the system performed better than those in control groups [5, 35]. Other research has examined the potential for LSA to assess summarization quality using semantic overlap between student summaries and the source text as well as between student summaries and high-quality summaries written by peers and experts. These studies have shown that that semantic similarity with the expert rater summary [17, 19] and highly rated peer summaries [17] explain significant amounts of variance in human judgments of summarization quality.

Other researchers have relied on n-gram (i.e., multi-word phrases) overlap between summaries and source texts and linguistic features within summaries to predict summarization quality. For instance, Madnani et al. [22] successfully predicted summary scores using n-gram overlap between the summary and the source text, lexical and phrasal overlap between the student summary against a set of model (or reference) summaries that received the highest score, the ratio of n-grams copied from the source text, number of sentences in the summary, and incidence of discourse connector terms in the summary. In a later study, Sladoljev-agejev et al. [24] examined automated scoring of summaries for two source texts in the assessment of college-level writing in English as a second language using six analytic rubrics related to accuracy, completeness, relevance, coherence, cohesion and text organization. They used linguistic features related to n-gram overlap between summaries and source texts and between summaries and reference summaries along with linguistic indices computed by the NLP tool Coh-Metrix [36, 37]. They also found that accuracy was related to n-gram overlap and that scores for the other analytic features were only predicted by indices related to connectives and referential cohesion.

1.3 Current Study

In the current study, our objective is to improve upon previous studies by developing a an AES model that depends solely on textual features within the source text or in the summary – in essence, any text any time. In practice, this means that none of the features used to predict summarization quality are based on sources outside of the text allowing us to develop a model that can be extended to new source texts. To make potential extensions more reliable, we develop our model using a wide variety of source texts ($n = 30$). Further, the text features we use to develop our model are freely and publicly available ensuring that replications are possible.

2 Method

2.1 Data

A total of 1,023 summaries were collected from adult participants in the United States. Among them, 792 summaries of 30 different source texts were collected using the Amazon Mechanical Turk (MTurk) online research service. The MTurk workers in this study were asked to each write one summary on three different topics. The remaining 231 summaries were produced by adults who read at less than ninth grade level (i.e., adult literacy population). These participants were asked to each write a summary on two different source texts. Source texts were on unrelated topics ranging from child safety to internet shopping. The source texts were given by the California Distance Learning Project (CDLP), with permission from the Sacramento County Office of Education. The CDLP texts are simplified news stories that are developed to be read by low-literate adults to improve their comprehension skills [18]. Each of the CDLP texts was between four and eight paragraphs and ranged from 128 to 452 words ($SD = 73.900$ words). Flesch-Kincaid grade level was between 4th and 8th grade ($SD = 1.100$) for all texts. On average, each source text was summarized by 31 participants ($SD = 27.644$). The minimum number of participant summaries per text was 20 while the maximum was 129. Participants were given instruction on how to write a summary, shown a text, and then asked to summarize the text. No demographic or individual difference information was collected for the MTurk workers, while this information was collected for the adult literacy population.

2.2 Summary Rating

The summaries were scored by two expert raters using an analytic scoring rubric that focused on inclusion of main ideas, accuracy of main ideas, and appropriate length. All analytic features were on a 0–3 scale. Before rating the summarization, the raters examined the original texts independently and identified main ideas. Through adjudication, the raters created a finalized list of main ideas for each source text. During scoring, the raters referenced this list of main ideas. For the current analysis, we only focus on the inclusion of main ideas in the summary. Raters gave the summary a score of 3 if the summary included all main ideas, a score of 2 if the summary included most of the main ideas, a score of 1 if the summary included some main ideas, and a score of 0 if the summary included no main ideas. Expert raters were first trained with an independent set of summaries that were not part of the final corpus of summary. Once the raters were normed, they then scored the entire corpus of summaries independently.

2.3 Linguistic Features

Four NLP tools were used to collect information about the lexical sophistication, syntactic complexity, lexical diversity, and cohesion of the summaries. These tools were the Tool for the Automatic Analysis of Lexical Sophistication (TAALES) [38], and the Tool for the Automatic Analysis of Syntactic Sophistication and Complexity (TAASSC) [39], the Tool for the Automatic Analysis of Lexical Diversity (TAALED, a beta version), and the Tool for the Automatic Analysis of Cohesion (TAACO) [40]. These tools were selected because they are freely available, open-source, and well documented: all features that allow for replicability. Each is briefly discussed below.

TAALES. TAALES calculates lexical and phrasal features, such as lexical frequency (i.e., how often a word occurs in a reference corpus), psycholinguistic word information (e.g., human ratings of familiarity, imageability, and concreteness), and *n*-gram (i.e., sequences of contiguous words) frequency (i.e., how often an *n*-gram occurs in a reference corpus). In calculating indices related to frequency, various reference corpora are used, such as the SUBTLEXus corpus of subtitles [41] and the Corpus of Contemporary American English (COCA) [42]. Indices are calculated using all words, content words, or function words.

TAACO. TAACO computes indices related to textual features and text cohesion, such as type-token ratios (TTRs; the number of types [unique words] divided by the number of tokens [total running lemmas]), sentence overlap, paragraph overlap, and connectives (e.g., *moreover* and *nevertheless*). Overlap indices are calculated based on lemma overlap and semantic overlap (e.g., using LSA and vector-representation-of-words [word2vec]). TACCO also provides source-text similarity indices using LSA, latent dirichlet allocation (LDA), and word2vec.

TAALED. TAALED calculates lexical diversity indices, such as the measure of textual lexical diversity (MTLD; the mean length of sequential word strings in a text that maintain a given TTR value) [43] and Moving-average TTR (MATTR). Indices are calculated using all lemmas, content lemmas, function lemmas, or bi-grams.

TAASSC. TAASSC computes indices related to clausal and phrasal indices of syntactic complexity and indices related to complexity of verb-argument constructions (defined as a main verb plus all of its direct dependents). These features include clausal complexity (e.g., average numbers of particular structures per clause and dependents per clause) and noun-phase complexity (e.g., standard deviations of dependents per each noun phrase).

2.4 Statistical Analysis

Based on human ratings, the summary texts were grouped into two categories: a low-quality group in which summary texts were scored 0 or 1 ($n = 432$) and a high-quality group in which summary texts were scored 2 or 3 ($n = 591$). To par down the number of linguistic features and control for statistical assumptions, a series of pre-analytic pruning steps were undertaken. First, linguistic features for which correlations with summary scores were lower than |.20| were removed. Our threshold of .20 ensured that only variables with meaningful relations were included in the analysis. These linguistic indices were then controlled for multicollinearity (defined as $r > .700$).

To predict summary scores, a generalized linear mixed model (GLMM) was used. GLMMs combine linear mixed models (which handle both fixed and random effects) and generalized linear models (which address non-normal data, such as binomial distributions) to develop predictive model. In our GLMM model, the response/dependent variable was a binomial response defined as either high-quality or low-quality summarizations. The fixed effects in the analysis were the linguistic features calculated in each summary text. The random effects in the analysis quantified variation across source texts and participants. Thus, GLMMs can measure the effects of the linguistic features on the response variable (i.e., high-quality or low-quality of summary texts) while accounting for prompt effects and the repeated testing of the same participants. The GLMM model developed for this study using backward selection of the fixed effects, such that only significant fixed effects ($t > 1.96$ at a .05 significance level) were retained. We also tested interaction effects among the significant fixed effects. We then included random slope adjustments of source texts for each significant fixed effect because the effects of linguistic measures on summary scores are likely to differ depending on the source text.

The data were randomly divided into a training set and a test set using a 67/33 split [44]. The GLMM was created using the training set ($n = 685$), and then applied into the test set ($n = 338$) to evaluate how well the model classified an independent set of summaries. The test set contained approximately 33% of the summaries from each of the participant groups - 264 summaries from the MTurk participants (164 high quality; 100 low quality) and 74 summaries from the adult literacy participants (17 high quality; 57 low quality). Descriptive statistics indicated that the adult literacy participants produced more low quality summaries than the MTurk participants.

For data analysis, we used *R* (R Core Team, 2016) and the *lme4* package [45] to construct a GLMM model. We used the *LMERConvenienceFunctions* package [46] to perform backward selection of fixed effects and the *MuMIn* package [47] to calculate a marginal *r*-squared (i.e., variance explained by fixed effects only) and a conditional *r*-squared (i.e., variance explained by both fixed and random effects).

3 Results

After pruning the data using test for multicollinearity and effect size, 21 linguistic indices were retained and used to develop the GLMM model. Details on these features are reported in Table 1.

Table 1. Correlations between summary scores and computational indices

Index	Feature	Tool	r
Type-token ratio (AW)	Cohesion	TACCO	.469
Frequency (COCA spoken, AW)	Lexical sophistication	TAALES	−.435
Source similarity (word2vec)	Cohesion	TAACO	.406
Adjacent sentence similarity (word2vec)	Cohesion	TAACO	.362
MATTR (FW)	Lexical diversity	TAALED	−.324
Number of CW tokens	Lexical diversity	TAALED	.318
MTLD (FW)	Lexical diversity	TAALED	.309
MTLD bi-grams (FW)	Lexical diversity	TAALED	.303
MTLD (AW)	Lexical diversity	TAALED	.292
MTLD (CW)	Lexical diversity	TAALED	.292
Word frequency (COCA spoken, AW)	Lexical sophistication	TAALES	−.282
Repeated content lemmas and pronouns	Cohesion	TAACO	.282
Lexical density (Percentage of CWs)	Cohesion	TAACO	.254
Type-token ratio (CW)	Cohesion	TAACO	−.247
Frequency (COCA spoken, FW)	Lexical sophistication	TAALES	−.246
SD of dependents per nominal subject	Syntactic complexity	TAASSC	.232
Binary adjacent sentence overlap (FW)	Cohesion	TAACO	.229
Word frequency (SUBTLEXus, CW)	Lexical sophistication	TAALES	−.225
Word frequency (SUBTLEXus, AW)	Lexical sophistication	TAALES	−.216
SD of dependents per clause	Syntactic complexity	TAASSC	.215
SD of dependents per object of the preposition	Syntactic complexity	TAASSC	.204

AW = All words, CW = Content words, FW = Function words, SD = Standard Deviation

Using the training set ($n = 685$) to develop a baseline GLMM, a random intercept model was created including the prompts and the participants as random intercepts. This model explained 45.198% of the variance in the human ratings of summarization. Performing backward selection of fixed effects, the GLMM included four significant linguistic features (see Table 2). No significant interaction effects were revealed. The results indicated that high-quality summary texts tended to have higher type-token ratios of all words (i.e., less repetition of words/more unique words; $t = 7.424$, $p < .001$), greater similarity with source texts as measured by word2vec ($t = 3.388$, $p < .001$), words that occur less frequently in the COCA spoken corpus ($t = -3.376$, $p < .001$), and lower type-token ratios for content words (i.e., greater repetition of content words; $t = -2.592$, $p < .01$). This model explained 53.412% of the variance using the fixed factors (i.e., the linguistic features) and 80.756% of the variance using both fixed and random factors.

Table 2. Results of the generalized linear mixed model (GLMM)

Fixed effect	Feature	Estimate	Standard error	t	p
(Intercept)		−6.975	2.605	−2.678	<.010
Type-token ratio for all words	Cohesive	1.596	.246	6.493	<.001
Source similarity (word2vec)	Cohesive	7.424	2.191	3.388	<.001
Frequency for all words (COCA spoken)	Lexical	−.001	.001	−3.376	<.001
Type-token ratio for content words	Cohesive	−4.433	1.710	−2.592	<.010

The final GLMM was used to examine how accurately it classified high- and low-quality summaries in the training set ($n = 685$; see Table 3). The GLMM correctly allocated 597 of the 685 summaries in the training set for an accuracy of 87.153%. The precision scores (i.e., the ratio of correctly classified hits into the high-quality group to all hits classified into the high-quality group; 374/426) and recall scores (i.e., the ratio of correctly classified hits into the high-quality group to hits incorrectly classified into the low-quality group plus hits correctly classified into the high-quality group; 374/410) were .878 and .912, respectively. The combined accuracy of the model (F1) was .895.

Table 3. Confusion matrix for classifying high- and low-quality summaries in the training set

Actual group	Predicted group	
	Low-quality	High-quality
Low-quality	223	52
High-quality	36	374

The GLMM classification model was next extended to the test set ($n = 338$) to assess classification accuracy (see Table 4). The GLMM correctly allocated 276 of the 338 summaries in the test set for an accuracy of 81.657%. The precision scores (161/203) and recall scores (161/181) were .793 and .890, respectively. The F1 was .839. These results provide strong evidence that linguistic features as found in summaries can be used to classify summaries in terms of the inclusion of main ideas.

Table 4. Confusion matrix for classification of high- and low-quality summaries in the test set

Actual group	Predicted group	
	Low-quality	High-quality
Low-quality	115	42
High-quality	20	161

To assess generalization of the model for the two groups, we separately examined the accuracy of the model on the MTurk and Adult Literacy participant populations. The GLMM correctly allocated 216 of the 264 summaries in the MTurk summaries in the test set for an accuracy of 81.818%. The precision and recall scores were .826 and .896, respectively. The F1 was .860. Similarly, the GLMM correctly allocated 60 of the 74 summaries in the adult literacy populations. This yielded an accuracy of 81.081% with precision and recall scores of .560 and .824, respectively. The F1 was .667. Overall, these results are similar to those of the combined test set, suggesting that the model was generalizable to the two populations.

4 Discussion

Summarization is considered an effective strategy to promote and enhance learning and deep comprehension of texts. However, this strategy is seldom implemented by teachers in classrooms because the manual evaluation of students' summaries requires time and effort. This problem has led to the development of automated models of summarization quality. However, these models often rely on features outside of the student or source texts and are therefore not generalizable to summarizations of new texts. Furthermore, many of the models rely of proprietary tools that are not freely or publicly available, rendering replications difficult.

In this study, we developed and presented an ASE model that depends strictly on features of the source text or the summary, allowing for a purely text-based model of quality. Summaries ($n = 1,023$) were collected from adult participants in the United States. Among them, 792 summaries of 30 different source texts were collected using the Amazon Mechanical Turk online research service while other summaries were collected from adult literacy participants with low reading skills. The summaries were scored by two expert raters using an analytic scoring rubric that focused on inclusion of main ideas, accuracy of main ideas, and appropriate length Additionally, four NLP tools were used to collect information about the lexical sophistication, syntactic complexity, lexical diversity, and cohesion of the summaries.

Our derived model was able to classify summaries as either low or high quality with an accuracy above 80%. Importantly, the model was developed on a large number of source texts allowing for generalizability across texts. Moreover, the model generalized well to summaries produced by two different populations (i.e., MTurk and Adult Literacy populations) for different text genres (i.e., science texts and adult literacy texts). However, results were stronger for the MTurk population indicating that the model is better suited for summaries written by writers that are not low literacy. Further, the features used in this study are freely and publicly available affording replication. These results provide strong evidence that linguistic features as found in summaries can be used to classify summaries in terms of the inclusion of main ideas. Specifically, the model reported that higher rated summaries had greater lexical repetition of all word types (i.e., both function and content words), contained more infrequent words, had greater semantic overlap with the source text, and contained fewer repetitions of content words. These features support the notion that more lexically sophisticated summaries that have greater source overlap and more repetition of words in general (although fewer repetitions of content words) are scored higher.

5 Conclusion

The current study provides a strong foundation for future research on the automated scoring of student summaries and writing more broadly. Specifically, in terms of summarization scoring, we are in the process of collecting additional summaries from different prompts to extend the number and types of prompts in our database. We also plan to examine expert ratings that go beyond source integration and include text cohesion, the use of objective language, paraphrasing, and language sophistication. We are also in the process of refining newer NLP tools that may provide insight into how textual features are predictive of summarization quality.

The overarching objective of this work is to develop models of writing assessment that are open and free for educators, students, and researchers. These models will be integrated within the Writing Assessment Tool (WAT) which is currently under development. WAT will provide automated writing evaluation (AWE) for persuasive essays, source-based essays, and ASE for summaries. WAT will also provide access to validated linguistic and semantic features that characterize writing quality to researchers so that features such as those identified in this study are readily available to researchers. In doing so, we hope to increase access to tools that can improve writing research and writing education.

Acknowledgments. This research was supported in part by the Institute for Education Sciences (IES R305A180261). Ideas expressed in this material are those of the authors and do not necessarily reflect the views of the IES. We would also like to express thanks to Amy Johnson, Kristopher Kopp, and Cecile Perret for their help in collecting the data.

References

1. Marzano, R.J., Pickering, D.J., Pollock, J.E.: Classroom Instruction That Works: Research-Based Strategies for Increasing Student Achievement. Association for Supervision and Curriculum Development, Alexandria (2008)
2. Graham, S., Herbert, M.A.: Writing to Read: Evidence for How Writing Can Improve Reading: A Carnegie Corporation Time to Act Report. Alliance for Excellent Education, Washington (2010)
3. Spirgel, A.S., Delaney, P.F.: Does writing summaries improve memory for text? Educ. Psychol. Rev. **28**, 171–196 (2016)
4. van Dijk, T.A., Kintsch, W.: Strategies of Discourse Comprehension. Academic Press, New York (1983)
5. Wade-Stein, D., Kintsch, E.: Summary street: Interactive computer support for writing (2004). http://www.tandfonline.com/doi/abs/10.1207/s1532690xci2203_3
6. Rinehart, S.D., Stahl, S.A., Erickson, L.G.: Some effects of summarization training on reading and studying. Read. Res. Q. **21**, 422–438 (1986)

7. Brown, A.L., Campione, J.C., Day, J.D.: Learning to learn: on training students to learn from texts. Educ. Res. **10**, 14–21 (1981)
8. Brown, A.L., Day, J.D.: Macrorules for summarizing texts: the development of expertise. J. Verbal Learn. Verbal Behav. **22**, 1–14 (1983)
9. van Dijk, T.A., Kintsch, W.: Strategies of Discourse Comprehension. Academic, New York (1977)
10. Westby, C., Culatta, B., Lawrence, B., Hall-Kenyon, K.: Summarizing expository texts. Top. Lang. Disord. **30**(4), 275–287 (2010)
11. Jones, R.: Strategies for reading comprehension: Summarizing
12. Gil, L., Bråten, I., Vidal-Abarca, E., Strømsø, H.I.: Summary versus argument tasks when working with multiple documents: which is better for whom? Contemp. Educ. Psychol. **35**, 157–173 (2010)
13. Perin, D., Lauterbach, M., Raufman, J., Kalamkarian, H.S.: Text-based writing of low-skilled postsecondary students: relation to comprehension, self-efficacy and teacher judgments. Read. Writ. **30**, 887–915 (2017)
14. Chiu, C.-H.: Enhancing reading comprehension and summarization abilities of EFL learners through online summarization practice. J. Lang. Teach. Learn. **5**(1), 79–95 (2015)
15. Rogevich, M.E., Perin, D.: Effects on science summarization of a reading comprehension intervention for adolescents with behavior and attention disorders. Except. Child. **74**, 135–154 (2008)
16. Graham, S., Perin, D.: A meta-analysis of writing instruction for adolescent students. J. Educ. Psychol. **99**(3), 445–476 (2007)
17. Li, H., Cai, Z., Graesser, A.C.: Computerized summary scoring: crowdsourcing-based latent semantic analysis. Behav. Res. Methods **50**(5), 2144–2161 (2018)
18. Ruseti, S., et al.: Scoring summaries using recurrent neural networks. In: Nkambou, R., Azevedo, R., Vassileva, J. (eds.) ITS 2018. LNCS, vol. 10858, pp. 191–201. Springer, Cham (2018). https://doi.org/10.1007/978-3-319-91464-0_19
19. Jorge-Botana, G., Luzón, J.M., Gómez-Veiga, I., Martín-Cordero, J.I.: Automated LSA assessment of summaries in distance education: some variables to be considered. J. Educ. Comp. Res. **52**(3), 341–364 (2015)
20. Landauer, T.K., Dumais, S.T.: A solution to Plato's problem: the latent semantic analysis theory of acquisition, induction and representation of knowledge. Psychol. Rev. **104**, 211–240 (1997)
21. Landauer, T.K., McNamara, D.S., Dennis, S., Kintsch, W.: Handbook of Latent Semantic Analysis. Lawrence Erlbaum, Mahwah (2007)
22. Madnani, N., Burstein, J., Sabatini, J., O'reilly, T.: Automated scoring of a summary writing task designed to measure reading comprehension. In: Proceedings of the Eighth Workshop on Innovative Use of NLP for Building Educational Applications, pp. 163–168 (2013)
23. Mani, I.: Automatic Summarization. John Benjamins Publishing, Amsterdam (2001)
24. Sladoljev-agejev, T., Snajder, J., Analysis, T.: Using analytic scoring rubrics in the automatic assessment of college-level summary writing tasks in L2. In: Proceedings of the 8th International Joint Conference on Natural Language Processing, pp. 181–186 (2017)
25. Dole, J.A., Duffy, G.G., Roehler, L.R., Pearson, P.D.: Moving from the old to the new: research on reading comprehension instruction. Rev. Educ. Res. **61**(2), 239–264 (1991)
26. Kintsch, W., Van Dijk, T.A.: Toward a model of text comprehension and production. Psychol. Rev. **85**, 363–394 (1978)
27. Kintsch, W., Welsch, D., Schmalhofer, F., Zimny, S.: Sentence memory: a theoretical analysis. J. Mem. Lang. **29**, 133–159 (1990)
28. Hinze, S.R., Rapp, D.N.: Retrieval (sometimes) enhances learning: performance pressure reduces the benefits of retrieval practice. Appl. Cogn. Psychol. **28**(4), 597–606 (2014)

29. Butler, A.C., Karpicke, J.D., Roediger III, H.L.: The effect of type and timing of feedback on learning from multiple-choice tests. J. Exp. Psychol. Appl. **13**(4), 273–281 (2007)

30. Stewart, T.L., Myers, A.C., Culley, M.R.: Enhanced learning and retention through "writing to learn" in the psychology classroom. Teach. Psychol. **37**(1), 46–49 (2009)

31. Shokrpour, N., Fotovatian, S.: Effects of consciousness raising of metacognitive strategies on EFL students' reading comprehension. ITL – Int. J. Appl. Linguist. **157**, 75–92 (2009)

32. Mok, W.S.Y., Chan, W.W.L.: How do tests and summary writing tasks enhance long-term retention of students with different levels of test anxiety? Instruct. Sci. **44**(6), 567–581 (2016)

33. Delaney, Y.A.: Investigating the reading-to-write construct. J. Engl. Acad. Purp. **7**, 140–150 (2008)

34. Landauer, T.K., Lochbaum, K.E., Dooley, S.: A new formative assessment technology for reading and writing. Theor. Pract. **48**(1), 44–52 (2009)

35. Franzke, M., Kintsch, E., Caccamise, D., Johnson, N., Dooley, S.: Summary street: computer support for comprehension and writing. J. Educ. Comput. Res. **33**, 53–80 (2005)

36. Graesser, A.C., McNamara, D.S., Louwerse, M.M., Cai, Z.: Coh-Metrix: analysis of text on cohesion and language. Behav. Res. Meth. Ins. C. **36**, 193–202 (2004)

37. McNamara, D.S., Graesser, A.C., McCarthy, P.M., Cai, Z.: Automated Evaluation of Text and Discourse with Coh-Metrix. Cambridge University Press, Cambridge (2014)

38. Kyle, K., Crossley, S., Berger, C.: The tool for the automatic analysis of lexical sophistication (TAALES) version 2.0. Behav. Res. Methods **50**(3), 1030–1046 (2018)

39. Kyle, K.: Measuring syntactic development in L2 writing: fine grained indices of syntactic complexity and usage-based indices of syntactic sophistication. Doctoral Dissertation (2016). http://scholarworks.gsu.edu/alesl_diss/35

40. Crossley, S.A., Kyle, K., McNamara, D.S.: The tool for the automatic analysis of text cohesion (TAACO): automatic assessment of local, global, and text cohesion. Behav. Res. Methods **48**(4), 1227–1237 (2016)

41. Brysbaert, M., New, B.: Moving beyond Kucera and Francis: a critical evaluation of current word frequency norms and the introduction of a new and improved word frequency measure for American English. Behav. Res. Methods **40**(4), 977–990 (2009)

42. Davies, M.: The 385+ million word Corpus of Contemporary American English (1990–2008+): design, architecture, and linguistic insights. Int. J. Corpus Linguist. **14**, 159–190 (2009)

43. McCarthy, P.M., Jarvis, S.: MTLD, Vocd-D, and HD-D: a validation study of sophisticated approaches to lexical diversity assessment. Behav. Res. Methods **42**(2), 381–392 (2010)

44. Witten, I.A., Frank, E., Hall, M.A.: Data mining: Practical Machine Learning and Techniques. Elsevier, San Francisco, CA (2011)

45. Bates, D., Maechler, M., Bolker, B., Walker, S.: lme4: linear mixed-effects models using Eigen and S4. R Packag. Version **1**(7), 1–23 (2014)

46. Tremblay, A., Ransijn, J.: LMERConvenienceFunctions: a suite of functions to back-fit fixed effects and forward-fit random effects, as well as other miscellaneous functions. R Packag. Version **2**, 919–931 (2013)

47. Barton, K., Barton, M.K.: Package MuMIn. Model selection and model averaging based on information criteria (2018)

The Importance of Automated Real-Time Performance Feedback in Virtual Reality Temporal Bone Surgery Training

Myles Davaris[1], Sudanthi Wijewickrema[1(✉)], Yun Zhou[1], Patorn Piromchai[1,2], James Bailey[1], Gregor Kennedy[1], and Stephen O'Leary[1]

[1] The University of Melbourne, Melbourne, Australia
swijewickrem@unimelb.edu.au
[2] Khon Kaen University, Khon Kaen, Thailand

Abstract. Virtual reality (VR) is increasingly being used as a training platform in many fields including surgery. However, practice on VR simulators alone is not sufficient to impart skills. Provision of performance feedback is essential to enable skill acquisition by ensuring that mistakes are identified and corrected, strengths are reinforced, and insights into consequences of actions are provided. As such, for a simulation system to be an effective training platform and to enable self-directed learning, it is imperative that automated performance feedback is provided by the system. Although there has been increased interest in the development of feedback methodologies in VR-based surgical training in recent years, their effectiveness in practice has rarely been investigated. In this paper, we investigate the impact of performance feedback in a VR-based surgical training platform with respect to skill acquisition and retention through a randomized controlled trial. We show that feedback during training is essential for both acquisition and retention of surgical skills.

Keywords: Virtual reality surgical training ·
Automated performance feedback · Temporal bone surgery

1 Introduction

Virtual reality (VR) has become the go-to technology when developing education systems in recent years. VR simulators are accepted as ideal for this task, as they offer risk-free, interactive, immersive, repeatable, and easily accessible platforms, using which standardized training programs can be developed. The effectiveness of VR-based systems in teaching skills and knowledge has been tested in different application domains, but the results have been mixed.

For example, Gamito et al. [13] showed that a VR-based serious games application can be used to significantly improve attention and memory functions in patients in cognitive rehabilitation. Mao et al. [27] discussed how VR can be used with a robot to improve the gait of subacute stroke patients. A case study on the

© Springer Nature Switzerland AG 2019
S. Isotani et al. (Eds.): AIED 2019, LNAI 11625, pp. 96–109, 2019.
https://doi.org/10.1007/978-3-030-23204-7_9

use of VR in American football training [21] showed a 30% average improvement in scores after VR training. Sacks et al. [35] found that, in training construction safety, VR is better for stone cladding work and for cast-in-situ concrete work, but not for general site safety. Wijewickrema et al. [49] showed that there was no significant difference in knowledge after training on a VR ear anatomy simulator, when compared to training on the same content in the form of a presentation. In a comparison between virtual training and physical training for teaching a bimanual assembly task, Murcia-Lopez and Steed [30] found that there was no significant difference between groups.

This trend continues in the field of surgery as well, where some studies showed that VR-based training was better than traditional training, while others found no significant differences. In Hamilton et al. [16], Seymour et al. [40], and Grantcharov et al. [14] it was found that the skill level of surgical residents performing laparoscopic cholecystectomy improved with VR training. Ost et al. [32], Rowe et al. [34], and Blum et al. [5] showed that fellows and residents performed a bronchoscopy task faster and more skilfully after VR training than untrained controls. Sedlack et al. [37,38] observed that residents and fellows performed better in colonoscopy after VR training. Ahlberg et al. [2] observed that medical student performance with VR training for laparoscopic appendectomy was no better than that of non-trained controls. In Hogle et al. [20] it was seen that there was no significant difference in performance in surgical residents when performing cholecystectomy between intervention and control groups.

These mixed results show that it is impractical to form sweeping conclusions as to the effectiveness of VR in surgical education. As such, we also need to consider other factors such as the task being trained, the skill level of the student, the level of instruction, and even the design of the evaluation study that may affect evaluation results. In this paper, we explore one of these factors, namely, the effect of performance guidance/feedback in VR-based surgical simulation.

It has become evident that the sole availability of a surgical simulator is not sufficient for a meaningful educational experience and an appropriate curriculum should be available to utilize its full potential [12,44]. One important aspect when designing an effective surgical curriculum is the provision of performance feedback [44,45]. Feedback is essential for effective skill acquisition, and must be both timely and contextually relevant [10,28]. Its purpose is to reinforce strengths, address weaknesses, and foster improvements in the learner by providing insights into the consequences of their actions and by highlighting the differences between intended and actual results [44]. It was shown in Hattie & Timperley [19] that the most effective forms of feedback provide cues or reinforcement to learners; are in the form of video-, audio-, or computer-assisted instructional feedback; and/or relate to goals.

In recent years, research in developing automated feedback systems for VR simulation has grown. In temporal bone surgery, most simulators today provide some form of guidance/feedback to students during and/or after training. Bhutta [4] in a recent review of simulation platforms in temporal bone surgery identified several VR simulators that have in-built guidance systems. For example, the

VR temporal bone surgery simulator developed by Stanford University [29,41] provides interactive feedback on maintaining proper technique in the form of coloured dots [39]. The VOXEL-MAN TempoSurg simulator [24,33] provides step-by-step procedural guidance for performing an operation by showing the desired end product of each step and a textual explanation in a separate panel.

The virtual temporal bone simulator developed by Ohio State University [47, 48] has an integrated intelligent tutor which provides functions such as structure identification and an expert demo mode (replaying of a pre-recorded expert procedure with customizable viewing parameters). The Visible Ear simulator [42, 43] supports an integrated tutor function which provides step-by-step procedural guidance through the green-lighting of steps on the temporal bone and a separate panel with information about the current step and the end-product view.

The University of Melbourne VR Temporal Bone Surgery Simulator [31] provides both technical and procedural guidance during training. For example, step-by-step guidance on how to perform a cortical mastoidectomy is presented as highlighted areas on a temporal bone [54]. Copson et al. [7] discuss a similar implementation of procedural guidance, based on the same simulator, where visual cues are presented one step at a time along with verbal explanations of each step. Methods of providing (verbal, auditory) feedback on surgical technique for this platform are discussed in Zhou et al. [57,58] and Ma et al. [25,26].

In most validation studies, the simulation system as a whole, inclusive of the automated guidance system, has been evaluated, usually with respect to a control group [47,56] or in a pre-post comparison [7,11]. A few studies exist that test the effect of feedback. For example, Wijewickrema et al. [50,53] evaluated how feedback on technique and procedure respectively affected the performance of medical students performing a cortical mastoidectomy. Here, the control groups received no feedback while the intervention groups did. No post-tests were conducted and performance was measured during training. As such, whether surgical skills were properly acquired and retained were not tested. In another study, Wijewickrema et al. [52] compared the effect of two different feedback generation methods. However, it has not been clearly shown what improvements can be expected with respect to surgical skill acquisition and retention, by providing automated guidance/feedback.

Here, we aim to bridge this gap through a user study comparing the performance of students trained on a VR simulator with and without automated guidance/feedback. To this end, we use the previously developed guidance/feedback system of the University of Melbourne temporal bone surgery simulator [52], which provides guidance/feedback on different forms of surgical skill.

2   Background

The VR platform used in this research is the University of Melbourne temporal bone surgery simulator (see Fig. 1). The virtual operating space consists of a model of a temporal bone and a surgical drill. The virtual model is generated using a segmented micro-CT scan of a human cadaveric temporal bone.

The virtual drill reflects the movements of a haptic device which also provides tactile feedback to the user. The impression of depth is achieved through NVIDIA 3D vision technology. A MIDI controller is used as a convenient input device to change environment variables such as magnification level and burr size. Using the VR simulator, surgeons can perform ear operations to remove disease and improve hearing. This often involves removing parts of the temporal bone or operating on the middle or inner ears and requires safe navigation around anatomical structures such as the facial nerve, sigmoid sinus, and dura.

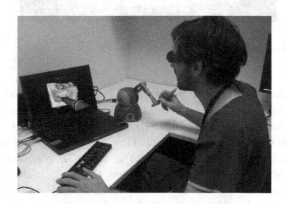

Fig. 1. The University of Melbourne VR temporal bone surgery simulator.

As surgical skills are multi-faceted, when expert surgeons teach trainees, they provide guidance/feedback on different aspects of the same. To emulate this, the simulation system considers four main aspects of skill that need to be acquired: procedural knowledge, knowledge of landmarks/surgical limits, manipulation of environmental variables, and drill handling/technical skills. The following sections discuss how guidance/feedback is provided in order to teach these skills.

Procedural Guidance: Procedural guidance is provided using the step-by-step guidance method of Wijewickrema et al. [54]. Each step of the surgery is highlighted on the temporal bone and the next step is only provided once the current step is completed. Figure 2 illustrates how the second step of a cortical mastoidectomy is highlighted (in green) after the first step has been drilled.

Advice on Landmarks/Surgical Limits: There are inherent cues (for example, changes in colour and smoothness of the bone) that inform surgeons when they are nearing an anatomical structure. However, these cues may be too subtle for a novice to detect. Therefore, it is necessary to provide more obvious warnings to this effect. Following the work of Wijewickrema et al. [51], the system currently provides verbal warnings when a trainee is drilling within a specified distance of a structure. Further, to enable learning of the anatomical structures, functionality to make the temporal bone transparent, so that the underlying structures can be viewed, is also available.

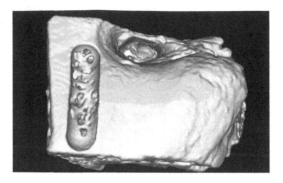

Fig. 2. Presentation of step-by-step procedural guidance. (Color figure online)

Feedback on Environmental Settings: The ideal values for environmental settings such as magnification level and burr size differ according to which area of the temporal bone is currently being drilled. For example, at the start of the procedure, in the central area, an overall view of the surgical space is required, and therefore, a lower magnification level is used. In contrast, when drilling deeper insider the mastoid, for example, near the facial nerve, more magnification will be required to get a better view. We use the method of generating regions in a temporal bone using morphological operations such as dilation and erosion as discussed in Wijewickrema et al. [51]. Once the regions are defined, to identify the valid ranges for each region, pre-collected expert data is used. Then, in real-time, if an environment setting is outside the pre-defined range of the relevant region, verbal auditory feedback is provided to inform the student of this. Figure 3 illustrates how these regions are defined for the purpose of providing feedback on environmental settings.

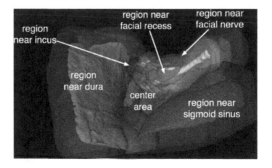

Fig. 3. Definition of regions where surgical technique is considered to be uniform.

Feedback on Drill Handling/Technical/Motor Skills: Typically, surgical technique adopted at one stage of a surgery is different to that of another.

For example, long strokes with a higher force can be used when drilling in an open area. However, more caution is warranted when drilling near a critical anatomical structure such as the facial nerve, and to avoid damage, less force should be used. Wijewickrema et al. [51] identified that the stages of a procedure is related to the regions being drilled. As such, in our system, different behaviour models are developed to identify poor technical skills per region in the following process. The same regions that were defined to provide feedback on environmental settings (as shown in Fig. 3) are used for this purpose.

The method based on random forests (RFs) discussed in Zhou et al. [57] is used here to generate feedback on drill handling. First, strokes are identified from the surgical trajectory using the method introduced in Hall et al. [15]. This enables the extraction of meaningful segments of the trajectory that can be used to define the quality of drill handling.

Once strokes are extracted, the values of metrics that define the quality of a stroke (such as speed and force) can be calculated for each stroke. We used such stroke metrics calculated for pre-collected expert and trainee data as features to train behaviour models that identified expert and trainee skill. Note that we made some modifications to the methods discussed in the original paper [57], when integrating them into our simulator. First, only the motion-based metrics (stroke length, duration, speed, acceleration, straightness, and force) were used as features when training the RF classifier. In contrast, in Zhou et al.'s work, they used motion-based metrics as well as environmental settings (simulator parameters) and proximity metrics (distance to structures). We separate the environmental settings from this and use a simple rule-based method of providing feedback on these, as discussed above. The distance measures are not required, once the regions around anatomical structures are defined. Second, in the original paper, the procedure was divided into stages, which were predicted using a pre-trained classifier, and behaviour models were trained for each stage. The prediction errors introduced by the stage detection was avoided here by using pre-defined regions instead, as discussed above.

If a stroke is classified as a trainee stroke, advice has to be provided on how to improve it so that expert-level behaviour can be learned. For this, we used the voting-based scheme used in Zhou et al. [57]. First, the expert stroke closest to the current trainee stroke was selected (from the pre-collected expert data) using a nearest neighbour strategy. In order to choose the specific feedback feature, the current trainee stroke and the closest expert stroke were classified by each tree in the RF. In a given tree, provided both strokes have been classified correctly, we computed the first feature (and direction: increase or decrease) on which the strokes were split into different branches and this feature received one vote. The feedback was then considered to be the feature and direction that received the most votes (for example, increase stroke length, decrease force etc.). This feedback is presented by the system to the user as verbal auditory instructions.

3   Methodology

We conducted a randomized controlled trial of 40 medical students, with no prior surgical experience, to evaluate the importance of providing automated guidance/feedback in VR-based temporal bone surgery. This study was approved by the University of Melbourne Human Ethics Committee (#1135497.3). On the first day, they were first shown a video tutorial on how to perform a simple temporal bone surgery (cortical mastoidectomy) on our VR simulator. Then, they performed this surgery on the VR simulator with no automated guidance (pre-test). The pre-test was performed in order to gauge their initial skill level, to account for individual variations in skill acquisition and retention. Next, they were randomly allocated to one of two groups: control or feedback. Then, they underwent a training session, performing the same procedure on the simulator. The feedback group received automated guidance/feedback during this procedure, while the control group did not. On the second day, they underwent another training session similar to the previous day. After this, on the same day, the participants performed a post-test: the same procedure without guidance/feedback. They came back a week later (on the ninth day), and performed a cortical mastoidectomy without guidance/feedback as a retention test. We recorded all the procedures conducted by participants using screen capture software. The study design in illustrated in Fig. 4.

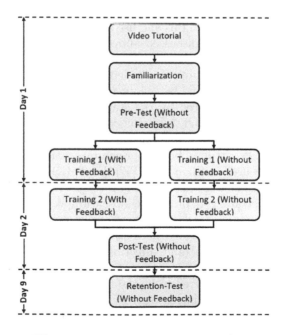

Fig. 4. Design of the evaluation study.

A blinded expert surgeon evaluated videos of the pre-, post- and retention tests based on a validated assessment scale [23]. This assessment scale was designed specifically for cortical mastoidectomy. It has been validated to be a feasible tool with high inter-rater agreement. This scale comprises two parts: checklist and global instruments, and assesses competency of the surgeon in performing the surgery as a whole. This takes into consideration all aspects of surgical skill, for example, knowledge of landmarks and procedure as well as technical skills.

The checklist instrument consists of 22 items (in 7 categories: initial bone cuts, defining anatomic limits, open antrum, digastric dissection, thin posterior EAC cortex (translucent), open facial recess, and posterior atticotomy). Each item is scored on a Likert scale ranging from 1 (unable to perform), through 3 (performs with minimal prompting), to 5 (performs easily with good flow). As such the minimum and maximum scores for this instrument are 22 and 110 respectively. However, as our study participants were medical students with no prior surgical experience, we did not teach them how to perform the 2 latter parts of the surgery (open facial recess and posterior atticotomy). As such only 13 items were relevant for this study, with minimum and maximum scores of 13 and 65 respectively. The global instrument comprises 10 items: understanding of objectives of surgery, interpretation of preoperative tests, use of otologic drill, knowledge of instruments, use of microscope, respect for surgical limits, time and motion, knowledge of specific procedure, Flow of operation, and overall surgical performance. The scoring is based on a 5-point Likert scale similar to that of the checklist instrument. As such, the minimum score is 10 and the maximum possible score is 50.

4 Results

First, we tested if the initial skill levels of the two groups were significantly different, using analysis of variance (ANOVA). We found that although the initial skill level of the feedback group was higher than that of the control group, these differences were not significant for either scores: checklist or global. Therefore, we can infer that the randomization procedure was successful.

To compare the level of surgical skill acquired after training by the two groups (post-test scores), taking into consideration the initial skill level of participants (pre-test scores) to account for individual aptitude, we performed an analysis of covariance (ANCOVA). A similar analysis was conducted to test for skill retention. Significant differences were observed between groups for both skill acquisition and retention, with the feedback group showing larger improvements in performance. The comparison results are shown in Table 1. In these ANCOVA analyses it was also tested what the effect of initial skill level was on skill acquisition and retention. It was seen that initial skill level was a significant factor in skill acquisition ($p = 0.026$ and $p = 0.007$ for checklist and global scores respectively), but not in skill retention.

To test if there were skill improvements within groups before and after training, we used paired t-tests to compare the post- and retention test scores with

Table 1. Between-group analysis of skill acquisition and retention, taking initial skill level as a covariate. Statistically significant results are shown in bold.

Score	Group	Adjusted mean	F(1,37)	p > F
Skill acquisition				
Checklist	Feedback	37.08	27.73	**<0.001**
	Control	20.37		
Global	Feedback	33.13	24.84	**<0.001**
	Control	16.57		
Skill retention				
Checklist	Feedback	28.46	13.73	**<0.001**
	Control	18.84		
Global	Feedback	25.17	13.24	**<0.001**
	Control	14.83		

pre-test scores. Cohen's d was used to calculate the effect sizes. Significant improvements with large effect sizes were observed in the feedback group with respect to both skill acquisition and retention, while the improvements in the control group were not significant with low effect sizes. Table 2 shows the results.

5   Discussion

From the results of the between group analyses in Table 1, we observe that the group that received automated feedback during training performed significantly

Table 2. Within-group analysis of performance with respect to skill acquisition and retention. Statistically significant results are shown in bold.

Group	Score	P-Value	Effect size (d)
Skill acquisition			
Feedback	Checklist	**<0.001**	1.74
	Global	**<0.001**	1.56
Control	Checklist	0.239	0.35
	Global	0.306	0.28
Skill retention			
Feedback	Checklist	**0.004**	0.90
	Global	**0.006**	0.85
Control	Checklist	0.386	0.27
	Global	0.552	0.19

better than the control group with respect to both skill acquisition and retention. The within group analysis (Table 2) shows that the learning and retention rates of the control group was not significant with low effect sizes. In contrast, the feedback group had significant levels of skill acquisition and retention with high effect sizes. These results imply that, at the level of the participants (complete novices with no prior experience in surgery), task demonstration is not sufficient to teach surgical skills, and performance feedback on different aspects of surgical skill is essential during training to ensure that skills are acquired and retained. These findings are in accordance with the principle of deliberate practice [10] which states that practice and relevant feedback are both important in gaining expertise. They also support other studies suggesting the benefit of real-time automated feedback in training [6, 8, 18, 55].

The significantly better performance after training with automated feedback further indicates that it was presented in such a way that it could be easily understood, thus retaining the cognitive load (burden placed on the human cognitive processing system [46]) at a manageable level to enable greater learning and performance [18]. That these results contradicts the guidance hypothesis which states that concurrent (real-time) feedback may lead to over-reliance and diminished performance [36], also indicates that the right amount of feedback was provided that supported learning, but discouraged over-reliance. However, as the students gain more experience in surgical procedures, it may be necessary to reduce the level of instruction, to avoid over-reliance [9, 22].

From the between group test results, we see that in addition to the presence or absence of feedback, the individual aptitude (as measured in the pre-test) also plays a significant role in the acquisition of surgical skills, but not in their retention. That individual differences affect complex skill acquisition seems obvious, but to what extent this is true is dependent on other factors as well. For example, Ackerman [1] discussed the ability-performance relationship as a function of task complexity, degree of task practice, and consistency of information-processing demands. This seems to indicate that with more practice, perhaps the significance of individual aptitude may lessen.

In this study, the training and evaluation of performance were both conducted on our VR simulation environment. As such, the results are not indicative of how participants would perform in a real-world scenario. However, there is evidence to suggests that skills learned in VR are transferable to real-world applications [3, 17]. This should be explored in future studies.

6 Conclusion

Here, we evaluated the effect of performance feedback on skill acquisition and retention in a VR-based surgical training platform. We observed that task demonstration and repeated practice to emulate the task on a VR simulator is not sufficient to acquire and retain surgical skills. Real-time feedback is not only helpful, but essential in the acquisition and retention of skills. Although in this paper, we established this for the test case of temporal bone surgery, the findings are in line with educational principles such as that of deliberate practice. This raises an important point which is often overlooked in VR education system design and development: the importance of appropriate feedback during training to ensure that the right skills are acquired and retained.

References

1. Ackerman, P.L.: Determinants of individual differences during skill acquisition: cognitive abilities and information processing. J. Exp. Psychol. Gen. **117**(3), 288 (1988)
2. Ahlberg, G., Heikkinen, T., Iselius, L., Leijonmarck, C.E., Rutqvist, J., Arvidsson, D.: Does training in a virtual reality simulator improve surgical performance? Surg. Endosc. Other Intervent. Techn. **16**(1), 126–129 (2002)
3. Beyer-Berjot, L., Aggarwal, R.: Toward technology-supported surgical training: the potential of virtual simulators in laparoscopic surgery. Scand. J. Surg. **102**(4), 221–226 (2013)
4. Bhutta, M.: A review of simulation platforms in surgery of the temporal bone. Clin. Otolaryngol. **41**(5), 539–545 (2016)
5. Blum, M.G., Powers, T.W., Sundaresan, S.: Bronchoscopy simulator effectively prepares junior residents to competently perform basic clinical bronchoscopy. Ann. Thorac. Surg. **78**(1), 287–291 (2004)
6. Chang, J.Y., Chang, G.L., Chien, C.J.C., Chung, K.C., Hsu, A.T.: Effectiveness of two forms of feedback on training of a joint mobilization skill by using a joint translation simulator. Phys. Ther. **87**(4), 418–430 (2007)
7. Copson, B., et al.: Supporting skill acquisition in cochlear implant surgery through virtual reality simulation. Cochlear Implants Int. **18**(2), 89–96 (2017)
8. Day, T., Iles, N., Griffiths, P.: Effect of performance feedback on tracheal suctioning knowledge and skills: randomized controlled trial. J. Adv. Nurs. **65**(7), 1423–1431 (2009)
9. Driscoll, M.P., Driscoll, M.P.: Psychology of learning for instruction (2005)
10. Ericsson, K.A.: Deliberate practice and the acquisition and maintenance of expert performance in medicine and related domains. Acad. Med. **79**(10), S70–S81 (2004)
11. Francis, H.W., et al.: Technical skills improve after practice on virtual-reality temporal bone simulator. Laryngoscope **122**(6), 1385–1391 (2012)
12. Fried, G.M.: Lessons from the surgical experience with simulators: incorporation into training and utilization in determining competency. Gastrointest. Endosc. Clin. **16**(3), 425–434 (2006)
13. Gamito, P., et al.: Cognitive training on stroke patients via virtual reality-based serious games. Disabil. Rehabil. **39**(4), 385–388 (2017)
14. Grantcharov, T.P., Kristiansen, V., Bendix, J., Bardram, L., Rosenberg, J., Funch-Jensen, P.: Randomized clinical trial of virtual reality simulation for laparoscopic skills training. Br. J. Surg. **91**(2), 146–150 (2004)
15. Hall, R., et al.: Towards haptic performance analysis using k-metrics. In: Pirhonen, A., Brewster, S. (eds.) HAID 2008. LNCS, vol. 5270, pp. 50–59. Springer, Heidelberg (2008). https://doi.org/10.1007/978-3-540-87883-4_6
16. Hamilton, E., et al.: Comparison of video trainer and virtual reality training systems on acquisition of laparoscopic skills. Surg. Endosc. Other Intervent. Tech. **16**(3), 406–411 (2002)

17. Hammoud, M.M., et al.: To the point: medical education review of the role of simulators in surgical training. Am. J. Obstet. Gynecol. **199**(4), 338–343 (2008)
18. Hatala, R., Cook, D.A., Zendejas, B., Hamstra, S.J., Brydges, R.: Feedback for simulation-based procedural skills training: a meta-analysis and critical narrative synthesis. Adv. Health Sci. Educ. **19**(2), 251–272 (2014)
19. Hattie, J., Timperley, H.: The power of feedback. Rev. Educ. Res. **77**(1), 81–112 (2007)
20. Hogle, N., et al.: Validation of laparoscopic surgical skills training outside the operating room: a long road. Surg. Endosc. **23**(7), 1476–1482 (2009)
21. Huang, Y., Churches, L., Reilly, B.: A case study on virtual reality American football training. In: Proceedings of the 2015 Virtual Reality International Conference, p. 6. ACM (2015)
22. Koedinger, K.R., Corbett, A.T., Perfetti, C.: The knowledge-learning-instruction framework: bridging the science-practice chasm to enhance robust student learning. Cogn. Sci. **36**(5), 757–798 (2012)
23. Laeeq, K., et al.: Pilot testing of an assessment tool for competency in mastoidectomy. Laryngoscope **119**(12), 2402–2410 (2009)
24. Leuwer, R., et al.: Voxel-man temposurg a virtual reality temporal bone surgery simulator. J. Jpn. Soc. Head Neck Surg. **17**(3), 203–207 (2008)
25. Ma, X., et al.: Adversarial generation of real-time feedback with neural networks for simulation-based training. arXiv preprint arXiv:1703.01460 (2017)
26. Ma, X., Wijewickrema, S., Zhou, Y., Zhou, S., O'Leary, S., Bailey, J.: Providing effective real-time feedback in simulation-based surgical training. In: Descoteaux, M., Maier-Hein, L., Franz, A., Jannin, P., Collins, D.L., Duchesne, S. (eds.) MICCAI 2017. LNCS, vol. 10434, pp. 566–574. Springer, Cham (2017). https://doi.org/10.1007/978-3-319-66185-8_64
27. Mao, Y., et al.: Effects of virtual reality with robot training on the gait of subacute stroke patients. Ann. Phys. Rehabil. Med. **61**, e180 (2018)
28. McGaghie, W.C., Issenberg, S.B., Petrusa, E.R., Scalese, R.J.: A critical review of simulation-based medical education research: 2003–2009. Med. Educ. **44**(1), 50–63 (2010)
29. Morris, D., Sewell, C., Blevins, N., Barbagli, F., Salisbury, K.: A collaborative virtual environment for the simulation of temporal bone surgery. In: Barillot, C., Haynor, D.R., Hellier, P. (eds.) MICCAI 2004. LNCS, vol. 3217, pp. 319–327. Springer, Heidelberg (2004). https://doi.org/10.1007/978-3-540-30136-3_40
30. Murcia-Lopez, M., Steed, A.: A comparison of virtual and physical training transfer of bimanual assembly tasks. IEEE Trans. Visual Comput. Graphics **24**(4), 1574–1583 (2018)
31. O'leary, S.J., et al.: Validation of a networked virtual reality simulation of temporal bone surgery. Laryngoscope **118**(6), 1040–1046 (2008)
32. Ost, D., De Rosiers, A., Britt, E.J., Fein, A.M., Lesser, M.L., Mehta, A.C.: Assessment of a bronchoscopy simulator. Am. J. Respir. Crit. Care Med. **164**(12), 2248–2255 (2001)
33. Reddy-Kolanu, G., Alderson, D.: Evaluating the effectiveness of the voxel-man temposurg virtual reality simulator in facilitating learning mastoid surgery. Ann. R. Coll. Surg. Engl. **93**(3), 205–208 (2011)
34. Rowe, R., Cohen, R.A.: An evaluation of a virtual reality airway simulator. Anesth. Analg. **95**(1), 62–66 (2002)
35. Sacks, R., Perlman, A., Barak, R.: Construction safety training using immersive virtual reality. Constr. Manag. Econ. **31**(9), 1005–1017 (2013)

36. Schmidt, R., Lee, T.: Motor Control and Learning: A Behavioral Emphasis. Human Kinetics (2011)
37. Sedlack, R.E., Kolars, J.C.: Computer simulator training enhances the competency of gastroenterology fellows at colonoscopy: results of a pilot study. Am. J. Gastroenterol. **99**(1), 33 (2004)
38. Sedlack, R.E., Kolars, J.C., Alexander, J.A.: Computer simulation training enhances patient comfort during endoscopy. Clin. Gastroenterol. Hepatol. **2**(4), 348–352 (2004)
39. Sewell, C., et al.: Providing metrics and performance feedback in a surgical simulator. Comput. Aided Surg. **13**(2), 63–81 (2008)
40. Seymour, N.E., et al.: Virtual reality training improves operating room performance: results of a randomized, double-blinded study. Ann. Surg. **236**(4), 458 (2002)
41. Sonny, C., Peter, L., Dong Hoon, L., et al.: A virtual surgical environment for rehearsal of tympanomastoidectomy. Stud. Health Technol. Inform. **163**, 112 (2011)
42. Sørensen, M.S., Dobrzeniecki, A.B., Larsen, P., Frisch, T., Sporring, J., Darvann, T.A.: The visible ear: a digital image library of the temporal bone. ORL **64**(6), 378–381 (2002)
43. Sorensen, M.S., Mosegaard, J., Trier, P.: The visible ear simulator: a public PC application for GPU-accelerated haptic 3D simulation of ear surgery based on the visible ear data. Otol. Neurotol. **30**(4), 484–487 (2009)
44. Stefanidis, D.: Optimal acquisition and assessment of proficiency on simulators in surgery. Surg. Clin. North Am. **90**(3), 475–489 (2010)
45. Stefanidis, D., Heniford, B.T.: The formula for a successful laparoscopic skills curriculum. Arch. Surg. **144**(1), 77–82 (2009)
46. Van Merrienboer, J.J., Sweller, J.: Cognitive load theory and complex learning: recent developments and future directions. Educ. Psychol. Rev. **17**(2), 147–177 (2005)
47. Wiet, G.J., et al.: Virtual temporal bone dissection system: OSU virtual temporal bone system: development and testing. Laryngoscope **122**(S1), S1–S12 (2012)
48. Wiet, G.J., Stredney, D., Sessanna, D., Bryan, J.A., Welling, D.B., Schmalbrock, P.: Virtual temporal bone dissection: an interactive surgical simulator. Otolaryngol. Head Neck Surg. **127**(1), 79–83 (2002)
49. Wijewickrema, S., et al.: Development and validation of a virtual reality tutor to teach clinically oriented surgical anatomy of the ear. In: IEEE 31st International Symposium on Computer-Based Medical Systems (CBMS), pp. 12–17, June 2018. https://doi.org/10.1109/CBMS.2018.00010
50. Wijewickrema, S., et al.: Presentation of automated procedural guidance in surgical simulation: results of two randomised controlled trials. J. Laryngol. Otol. **132**(3), 257–263 (2018)
51. Wijewickrema, S., et al.: Region-specific automated feedback in temporal bone surgery simulation. In: IEEE 28th International Symposium on Computer-Based Medical Systems (CBMS), pp. 310–315. IEEE (2015)
52. Wijewickrema, S., et al.: Providing Automated Real-Time Technical Feedback for Virtual Reality Based Surgical Training: Is the Simpler the Better? In: Penstein Rosé, C., Martínez-Maldonado, R., Hoppe, H.U., Luckin, R., Mavrikis, M., Porayska-Pomsta, K., McLaren, B., du Boulay, B. (eds.) AIED 2018. LNCS (LNAI), vol. 10947, pp. 584–598. Springer, Cham (2018). https://doi.org/10.1007/978-3-319-93843-1_43

53. Wijewickrema, S., et al.: Developing effective automated feedback in temporal bone surgery simulation. Otolaryngol. Head Neck Surg. **152**(6), 1082–1088 (2015)

54. Wijewickrema, S., Zhou, Y., Bailey, J., Kennedy, G., O'Leary, S.: Provision of automated step-by-step procedural guidance in virtual reality surgery simulation. In: Proceedings of the 22nd ACM Conference on Virtual Reality Software and Technology, pp. 69–72. ACM (2016)

55. Xeroulis, G.J., Park, J., Moulton, C.A., Reznick, R.K., LeBlanc, V., Dubrowski, A.: Teaching suturing and knot-tying skills to medical students: a randomized controlled study comparing computer-based video instruction and (concurrent and summary) expert feedback. Surgery **141**(4), 442–449 (2007)

56. Zhao, Y.C., Kennedy, G., Yukawa, K., Pyman, B., O'Leary, S.: Improving temporal bone dissection using self-directed virtual reality simulation: results of a randomized blinded control trial. Otolaryngol. Head Neck Surg. **144**(3), 357–364 (2011)

57. Zhou, Y., Bailey, J., Ioannou, I., Wijewickrema, S., Kennedy, G., O'Leary, S.: Constructive real time feedback for a temporal bone simulator. In: Mori, K., Sakuma, I., Sato, Y., Barillot, C., Navab, N. (eds.) MICCAI 2013. LNCS, vol. 8151, pp. 315–322. Springer, Heidelberg (2013). https://doi.org/10.1007/978-3-642-40760-4_40

58. Zhou, Y., Bailey, J., Ioannou, I., Wijewickrema, S., O'Leary, S., Kennedy, G.: Pattern-based real-time feedback for a temporal bone simulator. In: Proceedings of the 19th ACM Symposium on Virtual Reality Software and Technology, pp. 7–16. ACM (2013)

Autonomy and Types of Informational Text Presentations in Game-Based Learning Environments

Daryn A. Dever[✉] and Roger Azevedo

University of Central Florida, Orlando, FL 32816, USA
ddever@knights.ucf.edu, roger.azevedo@ucf.edu

Abstract. Game-based learning environments (GBLEs) are being increasingly utilized in education and training to enhance and encourage engagement and learning. This study investigated how students, who were afforded varying levels of autonomy, interacted with two types of informational text presentations (e.g., non-player character (NPC) instances, traditional informational text) while problem solving with CRYSTAL ISLAND (CI), a GBLE, and their effect on overall learning by examining eye-tracking and performance data. Ninety undergraduate students were randomly assigned to two conditions, full and partial agency, which varied in the amount of autonomy students were granted to explore CI and interactive game elements (i.e., reading informational text, scanning food items). Within CI, informational text is presented in a traditional format, where there are large chunks of text presented at a single time represented as books and research articles, as well as in the form of participant conversation with NPCs throughout the environment. Results indicated significantly greater proportional learning gain (PLG) for participants in the partial agency condition than in the full agency condition. Additionally, longer participant fixations on traditionally presented informational text positively predicted participant PLG. Fixation durations were significantly longer in the partial agency condition than the full agency condition. However, the combination of visual and verbal text represented by NPCs were not significant predictors of PLGs and do not differ across conditions.

Keywords: Autonomy · Proportional learning gain ·
Game-based learning environment · Eye tracking

1 Introduction

1.1 Autonomy in Game-Based Learning Environments

Autonomy assumes people, or agents, actively interact with elements in their environment instead of being passive bystanders [1]. There is a need for autonomy within learning environments to promote understanding of content knowledge and skills critical for learning [2]. It is assumed learners who are active within a learning environment can reflect on their progress, whether it be while learning or regulating motivation and emotions, leading to effective planning and the execution of plans to achieve sub-goals [1]. In the context of game-based learning environments (GBLEs)

S. Isotani et al. (Eds.): AIED 2019, LNAI 11625, pp. 110–120, 2019.
https://doi.org/10.1007/978-3-030-23204-7_10

such as CRYSTAL ISLAND (CI) [3], learners are given autonomy to explore and interact with several game elements (e.g., choosing which text and science posters to read, generating hypotheses about potential pathogens, etc.), while also monitoring and regulating their cognitive, affective, metacognitive, and motivational (CAMM) self-regulatory processes, critical for effective learning with GBLEs [4].

As such, self-regulated learning (SRL) involves actively monitoring all thoughts, behaviors, and feelings to then activate and integrate prior knowledge with new information for future planning, monitoring, and achievement of learning goals [5]. Plan development occurs when a goal is made explicit and challenges the learner which increases their motivation to achieve the goal with efficiency [1]. If a goal is not specific, learners with effective SRL skills will identify and modify the plan and strategies used towards achieving the goal [6]. This may include redefining the goals to understand the task demands and steps needed to accomplish the task. In sum, SRL is extremely challenging for most learners, and it is even more challenging in GBLEs where the full agency afforded by these environments can further hinder effective SRL.

The amount of agency afforded to a learner can influence their ability and opportunity to use SRL effectively [2, 7]. GBLEs allow learners to choose how they interact with the environment, specifically while engaging in learning activities, such as reading about microbiology, collecting evidence, engaging in hypothesis testing, learning from biology experts, interviewing patients about their symptoms, etc. [8]. GBLEs are engaging environments for learners to practice SRL skills, accumulate content knowledge, and develop problem solving and reasoning through learning activities [9]. Learners exposed to these environments must monitor their CAMM SRL processes and adapt to the changing demands of the tasks within the environment to ensure successful goal achievement (e.g., identifying the disease causing the illness outbreak in CI). GBLEs are often criticized for their lack of scaffolding provided to the learner, where extraneous details within the game often distract learners from their role and the overall goal of the game [10]. Thus, the level of autonomy afforded to a learner within a GBLE should balance with the scaffolding provided to a learner within the environment [8, 11]. Scaffolding within GBLEs influences developing SRL competencies, where the components of the environment that introduce novel information, such as texts and diagrams, must be selected, organized, and evaluated for relevancy. If relevant, then the novel information is integrated with learners' prior knowledge to achieve their goal.

1.2   Application of the Cognitive Theory of Multimedia Learning to GBLEs

Multimedia learning occurs when the learner constructs a mental representation from the content provided through the combination of words and images presented within an advanced learning technology [15]. Multimedia is typically used to describe learning environments which are enhanced through the use of combining pictures (e.g., photographs, illustrations, and animations) and words (e.g., audio and text) [14]. GBLEs facilitate learners' construction of concepts and knowledge through navigating the environment (e.g., CI) and incorporating information that is received by the learner either through traditionally presented text via large blocks of information or through interactive elements in the environment such as non-player characters (NPCs).

The Cognitive Theory of Multimedia Learning (CTML) [12] can be presented within multimedia environments and their effect on learning processes. This theory is based on three assumptions: (1) visual and verbal/auditory processes have different channels; (2) these channels have a limit on the amount of information that can be processed at once; and, (3) learners actively process information in the environment [13]. In addition to the three assumptions, there is a set of five specified cognitive processes present during multimedia learning: (1) selecting relevant words from text, (2) selecting relevant visuals, (3) developing a mental model for selected relevant words, (4) developing a mental model for selected relevant visuals, and (5) integrating relevant text and visuals into conjoined representations [12]. These cognitive processes are important to note in this model as they require learner utilization of SRL skills (e.g., retention and transfer of learned information) and learner agency for cognitive development [15]. It is important to note that in deeper processing of multimedia presentations, information represented by words can be processed through either the visual (e.g., text) channel along with diagrams and graphs or auditory channel (e.g., spoken language) where they may then cross channels to be organized into either a verbal or pictorial model [10].

The multimedia principle specifically focuses on CTML's first basic assumption, visual and verbal information is processed through separate channels, and third basic assumption which asserts that learning with both channels simultaneously is more effective for deeper understanding than learning with information from a singular channel [14]. The interaction between the learner, more specifically the learner's ability to apply SRL strategies, and the presentation of information should be understood in order to optimally use multiple modes of presentations. This understanding will lead to the examination of the impact that these different modes can have on learning [14]. With both verbal and visual information being presented in conjunction with each other, the learner has a greater chance of recall with the information processed with two separate channels [14].

These channels of information can be presented in multiple ways, including computers and face-to-face interactions with artificial intelligent agents [12]. However, in GBLEs, which offer a unique learning opportunity through direct interaction and exploration of the environment, these presentations can occur through slightly different means. Instructional materials are integrated with the environment so that the learner can interact with the information, which should be regulated to control for the influence the environment can have on the learner and their ability to select and organize critical information for the goal [14]. Traditionally presented informational texts in GBLEs mimic books with blocks of written text appearing on the screen, whereas NPCs, serving as intelligent agents, offer a variation of face-to-face conversations through real-world interactions and character design. Dynamic content (e.g., animation) has found to be beneficial to overall learning outcomes compared to static content (e.g., graphs) when the dynamic content is realistic to the learner [14]. This has been supported through studies [16] to increase support to low-knowledge learners [14]. It has also been applied to GBLEs as the NPCs are typical within the design of GBLEs (refer to CRYSTAL ISLAND Environment section) and can appear to be realistic and provide information crucial to achieving goals. Within CI, participants were also presented with audio as the NPCs interact and answer the prompted questions.

1.3   Eye Tracking in GBLEs

Using eye tracking technology allows researchers to infer cognitive processes, specifically attention and implicit strategies, of a learner through observable behavior [17–19]. Understanding the relationship between cognitive processes and eye movements has become increasingly popular over the past decade, especially in education and science domains [19]. Using eye movements to measure cognitive processes, researchers use two types of measures: saccades and fixation durations [17, 19]. Saccades are rapid eye movements between fixations which can be represented by regressions [17, 19]. Fixation durations result from a relatively still eye motion lasting approximately 250 ms and may produce several variables such as the number of fixations, average duration of fixations, and total time fixating on an area of interest (AOI) [19]. For example, in a GBLE a learner could fixate on specific content within the environment and eye-tracking data captures how many times they fixate on an object, the proportion of time fixating on said object relative to other objects, as well as total amount of time the learner fixates on that object, providing inferences on what the learner may be thinking, the strategies they are using, and whether they are experiencing difficulties [18].

Eye tracking allows researchers to understand relationships between SRL strategy use and learner performance to increase understanding of learner problem-solving processes that occur in GBLEs [17] which introduces a large gap in current literature due to the limited study on these relationships. Problem-solving processes are described by transforming what occurs at the original state provided to the learner to the goal state when there is no evidence of the solution [17]. Past studies have indicated longer fixation durations within cognitive tasks perceived as difficult [19]. This includes problem solving within STEM education. Past research has also concluded improved problem-solving abilities in environments that highlight and emphasize critical components to the goal state [14, 19]. Eye tracking can support inferences about cognitive processes that are used while reading [19]. This can be combined with text structure and content within multimedia theories, such as CTML, to further understand the relationship between SRL strategy use, learning, and the acquisition of content knowledge within these environments [19]. Generally, understanding and integrating content is influenced by perceived difficulty of text and learners' reading ability, which affects eye movements where fixations increase as difficulty of text increases and saccades become shorter [19]. As such, learners' eye movements should allow for inferences in understanding learners' cognitive processes, progress throughout a GBLE, engagement, and SRL strategy use [18].

1.4   CRYSTAL ISLAND Environment

CRYSTAL ISLAND [3], a game-based learning environment, provides an opportunity for students to develop scientific reasoning skills through a microbiology-centered environment where students investigate an illness infecting an island of researchers. Participants are to identify the mysterious illness by interacting with NPCs and reading informational text (see Fig. 1), collecting and scanning food items that may be transmitting the disease, and organizing evidence by completing a diagnosis worksheet.

Once evidence has been gathered, participants make hypotheses about the illness and the source of the pathogen and then test their hypotheses. Once a hypothesis has been tested correctly, the game will end.

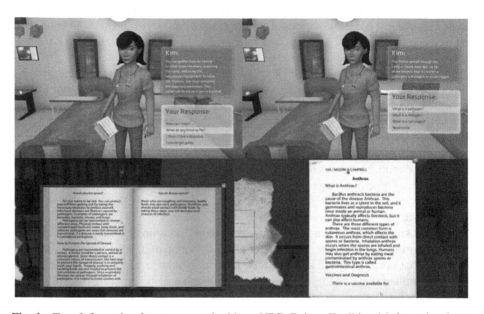

Fig. 1. Top: Informational text presented with an NPC; Below: Traditional informational text presentation

2   Current Study

To assess the role of autonomy and the types of presentation of informational text on PLGs within GBLEs, this study addresses the following research questions: (1) Do PLGs differ between the full and partial agency conditions?; (2) Do fixation durations on different types of informational text presentations in the environment predict PLGs?; and (3) Do fixation durations on different types of informational text presentations differ between the full and partial agency conditions? To address these questions, the hypotheses are as follows:

Hypothesis 1: Participants in the partial agency condition will demonstrate higher PLGs.

Hypothesis 2: The fixation durations of the different types of presentation of informational text in the environment will predict PLGs.

Hypothesis 3: Participants in the partial agency condition will have significantly greater fixation durations of both types of presentation of informational text.

3 Method

3.1 Participants

A total of 106[1] participants recruited from a large public North American university participated in the current study. Fifteen participants were removed due to eye tracking data inconsistencies while one participant was removed for not completing the post-task questionnaires. However, 90 (66% female) undergraduate students recruited from a large public North American university participated in the current study. Ages ranged from 18 to 26 years ($M = 20.01$, $SD = 1.66$). Participants were randomly assigned to one of three conditions: (1) full agency ($n = 53$), (2) partial agency ($n = 37$), or (3) no agency condition; we did not analyze data from the no agency condition, so details are excluded from this study. These conditions reflected the level of autonomy given to participants to navigate and problem solve with CRYSTAL ISLAND. Participants were compensated \$10/h and up to \$30 for completing the study.

3.2 Experimental Conditions

Participants were randomly assigned into one of three groups which allowed for varied control of gameplay: full agency, partial agency, and no agency. *Full agency* concedes full control to the participant where they can interact with the game at their own pace and discretion. Participants were free to move from building to building in whichever order they decided as well as choose whether or not to interact with certain game features such as opening a book or collecting a food item to later scan. *Partial agency* contains a "golden path" where participants are required to follow a set path through the game dictating which building to continue to next, requiring participants to interact with non-player characters, and having the participants look at each informational text to complete the concept matrices. For example, once past the tutorial portion of the game, participants in the partial agency condition were directed to the infirmary building first while the full agency participants could go to whichever building they desired. Once in the infirmary, participants in the partial agency condition were required to talk to both NPCs until the conversation options were exhausted, open all posters, books, and research articles, and then accurately complete the concept matrices for all books and research articles. Only after all of these actions were completed, were the participants able to leave the infirmary and directed to go to the next building. The *no agency* condition does not allow control to the participants as the participants will follow a video of an expert run-through of gameplay. This condition was not used in the study as the participants were not able to control for how long they fixated on informational text or NPC dialog.

3.3 Materials

Pre-task measures consisted of a demographic questionnaire and a microbiology pretest. The pretest quiz contained 21, four-option, multiple choice questions

---

[1] Our dataset derives from a larger study which was modified based on the quality of the data.

developed by an expert in the field. Post-task measures consisted of a microbiology posttest similar to the content knowledge pre-test. The SIM EYERED 250 eye tracker, using a 9-point calibration, recorded fixation duration and gaze movements of participants throughout the task. Log-file data was collected containing participant actions and timestamps.

3.4   Experimental Procedure

Participants read and completed the informed consent. Participants then completed the demographics questionnaire and the microbiology content knowledge quiz. After completion, the research assistant calibrated the eye-tracking device individualized to each participant. The research assistant then explained the scenario of CRYSTAL ISLAND, the role of the participant in the game, the goal of the game, and the actions available to the participant throughout the game, such as reading informational text, talking to NPCs, gathering possible sources of disease transmission, and completing the virtual worksheet. After the participants finished playing, they completed the post-task measures. This consisted of the microbiology content knowledge quiz which was similar to the pre-task version. Participants were then compensated, debriefed, and thanked for their time.

3.5   Coding and Scoring

A data pipeline that temporally aligned the multimodal, multichannel data was used to aggregate data during the experiment. Fixation durations were calculated by predefined areas of interest (AOIs) which included books, research articles, and NPCs. To calculate content knowledge of an individual after gameplay, differences in prior knowledge were accounted for in measuring the learning gains from the post-test score. PLGs are calculated using the pre- and post-test content knowledge scores using a formula accounting for prior knowledge [20].

4   Results

4.1   *Research Question 1*: Do PLGs Differ Between the Full and Partial Agency Conditions?

An independent samples t-test was conducted to compare the means of the PLGs between the full ($M = .218$, $SD = .231$) and partial ($M = .328$, $SD = .245$) agency conditions. There were significant differences in PLG ($t(88) = -2.18$, $p < .05$; $d = 0.46$) where participants in the partial agency condition had significantly higher PLGs than participants with full agency, suggesting those in the partial agency learned more about microbiology compared to those in the full agency.

4.2   Research Question 2: Do Fixation Durations on Different Types Of Informational Text Presentations in the Environment Predict PLG?

A linear regression was conducted to examine whether proportion of time fixating on NPCs over total game time (M = .124, SD = .035) predict PLG. There was no significant regression equation between the proportion of time fixating on NPCs and PLG (p > .05). An additional linear regression was calculated to assess whether the proportion of time fixating on traditional informational text over time (e.g., books and articles; M = .319, SD = .130) predict PLG. There was a

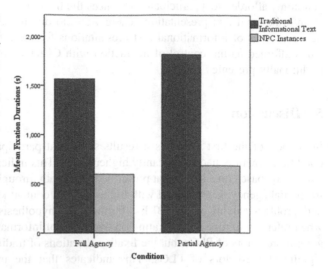

Fig. 2. Mean fixation durations of types of informational text presentation between conditions

significant positive correlation between the fixation duration of books and articles and PLGs (r = .233, p < .05), meeting the assumptions for our regression equation, and our results revealed a significant regression equation where the time spent fixating on informational texts was a significant positive predictor of PLG, $F(1,88) = 5.03, p < .05$ with an R^2 of .233, indicating that the longer participants fixated on traditionally presented informational texts, the higher their PLG ($ß$ = .233, p < .05). In sum, these findings showed that the fixation duration on traditionally presented text is a positive predictor of participants' PLG than the fixation duration of NPC instances, challenging the CTML model where text alone, not the integration of text and diagram, predicts PLGs.

4.3   Research Question 3: Do Fixation Durations on Different Types of Informational Text Presentations Differ Between the Full and Partial Agency Conditions?

A MANCOVA was conducted to examine differences in time spent fixating on different types of informational text between the two conditions with total game duration as a covariate (see Fig. 2). There were no significant differences in fixation duration of NPC instances between the full (M = 593.84, SD = 154.17) and partial (M = 682.20, SD = 170.52) agency conditions (p > .05). However, there were significant differences in time spent fixating on books and research articles, ($F(2,87) = 16.05, p < .0005$) between full ($M$ = 1565.97, SD = 755.05) and partial (M = 1851.13, SD = 915.84) agency conditions, where the partial agency condition has significantly higher fixation durations than the full agency condition. Overall, these results indicate that the autonomy afforded to a participant influences the fixation duration of the different types of informational text presentation. There were no differences in the fixation duration between types of informational text presentations for the full agency whereas participants afforded partial control of interaction with CI have greater fixation durations of traditionally presented text.

5 Discussion

In support of the first hypothesis, results show that participants in the partial agency condition generally had significantly higher PLGs. This indicates that learners with less autonomy, based on a somewhat prescribed ideal path through game elements allowing for partial agency, is associated with higher overall content knowledge during learning and problem solving with GBLEs. Further, the hypothesis was partially supported when referring to time spent fixating on two types of informational text. NPC instances are not predictors of PLGs, but the fixation durations of traditional information text are significant predictors of PLGs. This indicates that the traditional presentation of information through large amounts of text are better indicators and significantly correlate with higher content knowledge than interacting with NPCs who provide microbiology content knowledge through a more conversationalist approach. This finding runs counter to CTML in which the NPC instances, demonstrating a visual (e.g., the character itself) in conjunction with verbal (e.g., audio and text) information does not predict higher content knowledge whereas just the presentation of text does without the aid of an NPC or audio. This could be explained as the NPC presents verbal information when prompted by the participant that is not as representationally rich as a relevant diagram, and then participants are given small bits of information, but through predetermined prompts the participants may or may not have asked otherwise without room for adjustment of questions. Results partially supported the hypothesis where the partial agency condition had a higher fixation duration when referring to books and articles than the full agency condition, but no difference between conditions when calculating the fixation durations in NPC instances. This indicates that participants who have a set path fixate more on traditionally presentation of informational text over NPC instances. The partial agency condition required the participant to ask every prompt for NPCs as well as open every book and article to complete the concept matrices. From this, participants in the partial agency condition may identify the traditional presentation of text to hold a greater value in the information that is provided.

5.1 Future Directions: More AI in GBLEs?

This study supports the need for integrating more AI in GBLEs to support reading activities that are critical to learning about complex topics such as science. In general, GBLEs should support the development of learners' SRL strategies where the learner is guided by the environment in the completion of the goal, especially critical in GBLEs that afford full agency that may not be beneficial for all learning lacking CAMM SRL skills. As supported by our results, GBLEs that intelligently and actively guide the learner through the environment are needed to optimize proportional learning of complex instructional content. For example, GBLEs are often preferred over traditional learning technologies (e.g., hypermedia) due to perceived affordance to agency, autonomy, and engagement based on constructivist learning models, but our results show that full autonomy is not ideal since most learners do not have the cognitive and metacognitive self-regulatory skills need to make accurate instructional decisions such as when, how, and why instructional text embedded in GBLEs is critical for learning. In addition, our study also demonstrates that the NPCs (acting as intelligent agents interacting with learners) did not provide the information-rich instructional material that was needed and were disregarded, or not engaged with by the learners. The contrast between the roles on informational text and NPS highlights the careful attention that is needed in providing adaptive scaffolding during learning with GBLEs that should be based on time spent on different representations and sequences within and between representations and other related GBLEs activities. For example, information presented through large chunks of text are large components of learner interaction and theses affordances are influenced by the amount of autonomy afforded to a learner when interacting with a GBLE. The study further supports the need for appropriate direction towards the overall goal of the GBLE in order to obtain optimum learning from the learner exposed to the environment. In future versions of Crystal Island, or any text-dependent GBLE, limited, but present support should be given to the learner through the environment to increase the expected content knowledge gain. We envision intelligent agents embedded in GBLEs can play a more active role (a) in assisting learners to select, organize, and integrate instructional content; (b) providing adaptive scaffolding and feedback based on multimodal multichannel trace data from log-files, eye tracking, screen recording, facial expression of emotions, and natural language understanding, and (c) modeling specific self-regulatory processes by prompting and scaffolding students' planning, cognitive strategy use, metacognitive monitoring processes, etc.

Acknowledgements. This research was supported by funding from the Social Sciences and Humanities Research Council of Canada (SSHRC 895-2011-1006). Any opinions, findings, conclusions, or recommendations expressed in this material are those of the author(s) and do not necessarily reflect the views of the Social Sciences and Humanities Research Council of Canada. The authors would also like to thank members of the SMART Lab and the intelliMEDIA group at NCSU for their assistance and contributions.

References

1. Bandura, A.: Social cognitive theory: an agentic perspective. Ann. Rev. Psychol. **52**, 1–26 (2001)
2. Bradbury, A., Taub, M., Azevedo, R.: The effects of autonomy on emotions and learning in game-based learning environments. In: Proceedings of the 39th Annual Meeting of the Cognitive Science Society, pp. 1666–1671 (2017)

3. Rowe, J., Shores, L., Mott, B., Lester, J.: Integrating learning, problem solving, and engagement in narrative-centered learning environments. Int. J. Artif. Intell. Educ. **21**, 115–133 (2011)
4. Azevedo, R., Taub, M., Mudrick, N.: Understanding and reasoning about real-time cognitive, affective, and metacognitive processes to foster self-regulation with advanced learning technologies. In: The Handbook of Self-regulation of Learning and Performance, pp. 254–270. Routledge, New York (2018)
5. Schunk, D., Greene, J.: Handbook of Self-regulation of Learning and Performance, 2nd edn. Routledge, New York (2018)
6. Greene, J., Bolick, C., Robertson, J.: Fostering historical knowledge and thinking skills using hypermedia learning environments: The role of self-regulated learning. Comput. Educ. **54**(1), 23–243 (2010)
7. Azevedo, R., Mudrick, N., Taub, M., Bradbury, A.: Self-regulation in computer-assisted learning systems. In: Dunlosky, J., Rawson, K. (eds.) Handbook of cognition and education. Cambridge University Press, Cambridge (in press)
8. Sabourin, J., Shores, L., Mott, B., Lester, J.: Understanding and predicting student self-regulated learning strategies in game-based learning environments. Int. J. Artif. Intell. Educ. **23**, 94–114 (2013)
9. Taub, M., Mudrick, N., Azevedo, R., Millar, G., Rowe, J., Lester, J.: Using multi-level modeling with eye-tracking data to predict metacognitive monitoring and self-regulated learning with Crystal Island. In: ITS 2016 Proceedings of the 13th International Conference on Intelligent Tutoring Systems, pp. 240–246. Springer, New York (2016). https://doi.org/10.1007/978-3-319-39583-8_24
10. Mayer, R., Johnson, C.: Adding instructional features that promote learning in a game-like environment. J. Educ. Comput. Res. **42**(3), 241–265 (2010)
11. Burkett, C., Azevedo, R.: The effect of multimedia discrepancies on metacognitive judgments. Comput. Hum. Behav. **28**, 1276–1285 (2012)
12. Butcher, K.: The multimedia principle. In: The Cambridge Handbook of Multimedia Learning, pp. 174–205. Cambridge University Press, New York (2014)
13. Azevedo, R.: Multimedia learning of metacognitive strategies. In: The Cambridge Handbook of Multimedia Learning, pp. 647–672. Cambridge University Press, New York (2014)
14. Kalyuga, S., Chandler, P., Sweller, J.: Incorporating learner experience into the design of multimedia instruction. J. Educ. Psychol. **92**(1), 126–136 (2000)
15. Hu, Y., Wu, B., Gu, X.: An eye tracking study of high- and low-performing students in solving interactive and analytical problems. Educ. Technol. Soc. **20**(4), 300–311 (2017)
16. Tsai, M., Hou, H., Lai, M., Liu, W., Yang, F.: Visual attention for solving multiple-choice science problem: An eye-tracking analysis. Comput. Educ. **58**, 375–385 (2016)
17. Dogusoy-Taylan, B., Cagiltay, K.: Cognitive analysis of experts' and novices' concept mapping processes: An eye tracking study. Comput. Hum. Behav. **36**, 82–93 (2014)
18. Grant, E., Spivey, M.: Eye movements and problem solving: Guiding attention guides thought. Psychol. Sci. **14**(5), 462–466 (2003)
19. Rayner, K.: The 35th Sir Frederick Bartlett Lecture Eye movements and attention in reading, scene perception, and visual search. Q. J. Exp. Psychol. **62**(8), 1457–1506 (2009)
20. Witherspoon, A., Azevedo, R., D'Mello, S.: The dynamics of self-regulatory processes within self- and externally regulated learning episodes during complex science learning with hypermedia. In: Wolf, B.P., Nkambou, R., Lajoie, S. (eds.) Intelligent Tutoring Systems, pp. 260–269. Springer, Heidelberg (2008). https://doi.org/10.1007/978-3-540-69132-7_30

Examining Gaze Behaviors and Metacognitive Judgments of Informational Text Within Game-Based Learning Environments

Daryn A. Dever[(⊠)] and Roger Azevedo

University of Central Florida, Orlando, FL 32816, USA
ddever@knights.ucf.edu, roger.azevedo@ucf.edu

Abstract. Game-based learning environments (GBLEs) are often criticized for not offering adequate support for students when learning and problem solving within these environments. A key aspect of GBLEs is the verbal representation of information such as text. This study examined learners' metacognitive judgments of informational text (e.g., books and articles) through eye gaze behaviors within CRYSTAL ISLAND (CI). Ninety-one undergraduate students interacted with game elements during problem-solving in CI, a GBLE focused on facilitating the development of self-regulated learning (SRL) skills and domain-specific knowledge in microbiology. The results suggest engaging with informational text along with other goal-directed actions (actions needed to achieve the end goal) are large components of time spent within CI. Our findings revealed goal-directed actions, specifically reading informational texts, were significant predictors of participants' proportional learning gains (PLGs) after problem solving with CI. Additionally, we found significant differences in PLGs where participants who spent a greater time fixating and reengaging with goal-relevant text within the environment demonstrated greater proportional learning after problem solving in CI.

Keywords: Metacognitive judgments · Content evaluation ·
Game-Based Learning Environments

1 Introduction

1.1 Self-regulated Learning and Metacognitive Monitoring

Self-regulation, the modulation of behavior and internal cognitive processes due to experience and stimuli in the environment, involves the integration of prior knowledge and learning strategies to reach a goal [1]. Learners with self-regulated learning (SRL) skills discern and apply effective strategies needed to accomplish a set goal, commonly the attainment of knowledge [2, 3]. SRL models highlight the importance of planning, strategizing, and monitoring [4] to demonstrate an improved academic performance through the utilization of these SRL strategies when engaging, responding, and adapting to Game-Based Learning Environments (GBLEs) [5].

A significant component of SRL is monitoring and controlling progress of learning by modifying strategies and goals [2, 6]. In comprehending information, learners

© Springer Nature Switzerland AG 2019
S. Isotani et al. (Eds.): AIED 2019, LNAI 11625, pp. 121–132, 2019.
https://doi.org/10.1007/978-3-030-23204-7_11

interpret and integrate the meaning of information as it is presented [5]. To do so, learners may reread and reevaluate textual information they may not initially understand and judge their own learning through metacognition to evaluate their progress toward reaching the overall goal [5, 7]. Through metacognitive monitoring, learners identify discrepancies between their current state of learning and their desired state by modifying plans and goals to mitigate the discrepancy until the desired goal is reached [6, 8]. Learning outcomes are dependent upon the metacognitive monitoring strategies applied by learners [9, 10]. The implementation of these strategies has suggested an increase in the acquisition of deeper declarative knowledge through the integration of verbal and visual information provided by advanced learning technologies (ALTs; [11]). Further, a learner's ability to apply SRL strategies during learning significantly influences their performance and future learning [12]. Students using SRL strategies accurately apply monitoring judgments, such as identifying relevant information when encountering previously known information or when specified instructions and goals are provided [5, 13].

1.2 Metacognitive Judgments in Game-Based Learning Environments

Azevedo and colleagues [2] quantified student actions of metacognitive processes containing 35 micro-level metacognitive judgments under macro-processes (e.g., planning, monitoring, strategy use) to identify when students were using effective SRL processes and strategies. One of the micro-level metacognitive processes includes evaluating instructional content (e.g., textual information, diagram) known as content evaluations (CEs), which is described as the ability to monitor relevant information to attain goals [4]. For example, if a learner, while using a GBLE such as CRYSTAL ISLAND (CI), has a goal of learning about the Ebola virus, the learner should be able to discern relevant information related to the virus and disregard irrelevant information that is extraneous to the current goal (e.g., learning about smallpox). The actions taken towards achieving an overall goal within a GBLE are referred to as *goal-directed actions* within this paper (e.g., reading informational text, talking with non-player characters, scanning items, solving concept matrices, and consulting the scientific worksheet). Learners, with accurate CEs, should evaluate the relevancy of information and, once determined to be goal-relevant, expend more time and effort to studying and understanding that information which should then increase knowledge acquisition and improve learning outcomes [9].

ALTs, such as intelligent tutoring systems and GBLEs, are used to engage learners in educational tasks, such as problem solving, scientific inquiry, and reasoning to foster SRL processes such as selecting, organizing, and integrating novel or relevant information [14, 15]. GBLEs are designed to encourage learners to set and achieve goals [16] by providing tools that scaffold SRL processes. In GBLEs with narrative-focused goals, such as CRYSTAL ISLAND, a learner must be able to engage in SRL skills to properly interact and learn from the environment [12]. GBLEs provide an environment for learning through multiple modalities such as virtual text and interactive scenarios, where learning is supported through exposure to information accompanied by visual and verbal interactions, both enhancing the learning environment in conjunction [17, 18].

The Cognitive Theory of Multimedia Learning (CTML; [18]) assumes that processing visual and verbal information (text and diagrams) occurs separately, there is a limited amount of information that can be processed at once for visual and verbal information, and that learners actively process these types of multimedia information [7, 14]. These processes require the learner to successfully identify relevant information both from text and diagrams; as these processes occur separately, the identification of the level of relevancy may differ [7]. This model occurs differently within GBLEs where most information is presented with dynamic models such as videos, simulations, or as in CRYSTAL ISLAND, as interactions with the environment. This study focuses on the absence of dynamic models of information in CI when reading from a traditionally presented text.

Criticisms of GBLEs arise when several modalities with irrelevant content are presented to the learner, negating the support of the learner's self-regulatory development [19–21]. Learners within the environment may become distracted from the original plan or goal with content within the environment that does not directly support the overall goal. In GBLEs, learners' CEs result from self-monitoring actions throughout the game that helps distinguish relevant content from irrelevant content [9, 22]. In order to continue to develop and encourage the use of learners' metacognitive abilities, there needs to be monitoring of real-time cognitive processes and progress within an intelligent system which can lead to effective feedback [14].

1.3   Eye Tracking in Game-Based Learning Environments

Physiologically-based measures are becoming an increasingly utilized method to help infer the cognitive processes in conjunction with explicit behavior. Brain activity, trace data, log files, and eye movements have been used in order to supplement the traditional self-report measures of cognition [14, 22]. Recording eye behaviors can help track the cognitive processes of learners throughout the duration of a task, which may be reading, problem solving, or other actions available within a GBLE [23–25]. In order to record eye movements, there are two core measurements - eye fixations and saccades [24–26]. Eye fixations are relatively still positioning of the eye where researchers can measure how many times a learner fixates on an object, the average fixation duration, and the total time fixating on an object. Saccades are the rapid movements of the eye between fixations [23–25]. Information learners fixate on can be categorized by importance, by subject, or by the object within the environment. This type of information grouping is called the area of interest (AOI) [26].

These specific measures in eye-tracking technology allow for researchers to infer internal cognitive processes between domains of knowledge, expertise, and performance [23, 25–27]. Within text comprehension, eye movements differ with text difficulty where with more difficult texts, the fixation duration increases, saccades become shorter, and there is an increase in regressions [24]. Experts fixated on content less, had increased fixations on relevant areas, and was able to find the task-relevant information quicker than non-experts [27]. Eye tracking in GBLEs allow for researchers to understand learner engagement and SRL strategies in interactive environments by providing a better support system for the learner such as reorienting learner attention and highlighting task-relevant areas [25, 28].

1.4   CRYSTAL ISLAND: A Game-Based Learning Environment

CRYSTAL ISLAND [29], a GBLE which promotes the use of problem solving, scientific inquiry, reasoning focused on the development of knowledge in microbiology, begins on an island where its inhabitants have been infected with a mysterious illness. Participants are tasked with completing the game to identify the disease by engaging in goal-directed actions such as interacting with NPCs, consulting informational texts (e.g., books and research articles; Fig. 1), filling out concept matrices, collecting items (e.g., food) to later scan for diseases, gathering information via a worksheet, and creating and testing hypotheses to find what disease has infected the inhabitants. There is no difference between the two types of informational text where the books and articles do not provide varying quality of information to the participant in comparison to each other. The concept matrices measure the retainment of content knowledge To complete the game, the participant produces a final diagnosis that includes the type of illness (i.e., viral or bacterial), name of the illness (i.e., influenza or salmonellosis), and the transmission source (i.e., eggs, bread, or milk). With this, the learning gains can be used to investigate learners' metacognitive judgments as they evaluate the relevancy of content with GBLEs.

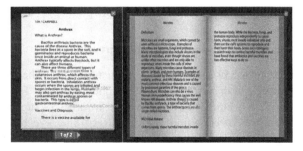

Fig. 1. Informational text (left: research article, right: book) with gaze behaviors indicated by the green markers. (Color figure online)

1.5   Related Works

Past studies investigating literacy and reading behaviors in CRYSTAL ISLAND have mostly examined participant performance throughout the game, measured by concept matrix attempts [15, 30]. These studies utilized eye gaze behaviors on books as well as the combination of books and articles to understand the metacognitive processes of participants. One study used CRYSTAL ISLAND to enhance student modeling through gaze behaviors in order to better predict the performance of a participant throughout the duration of the game [31]. An additional study incorporated relevancy of food item scanning and worksheet submission attempts to assess efficiency [32]. This current study aims to directly identify the importance of informational text within GBLEs as well as investigate the amount of time spent reading and the ability to make accurate CEs to predict proportional learning gains (PLGs).

2   Current Study, Research Questions, and Hypotheses

The objective of this study was to understand learners' metacognitive judgments of informational text by examining gaze behaviors and the PLGs within CI. The current study aimed to answer four research questions: (1) Are there differences in the proportions of fixation duration during goal-directed actions (e.g., reading informational text, talking with NPCs, scanning items, solving concept matrices, and consulting the scientific worksheet) over the duration of the game while problem solving in CI?; (2) Do the fixation duration proportions of goal-directed actions available to participants predict PLGs while problem-solving in CI?; (3) Do fixation durations of relevant informational text significantly predict PLGs?; and (4) Do the PLGs differ between groups of participants who revisit relevant texts more often and those who revisit relevant texts less often? To address the research questions, we hypothesize the following:

> *Hypothesis 1*: There are differences in the proportions of goal-directed action (e.g., reading informational text, talking with NPCs, scanning items, solving concept matrices, and consulting the scientific worksheet) fixation durations while problem-solving in CI.
> *Hypothesis 2*: The fixation duration proportions of goal-directed action available to participants throughout the game, specifically fixations on informational text, significantly predict PLGs while problem solving in CI.
> *Hypothesis 3*: The fixation durations of relevant informational texts significantly predict PLGs.
> *Hypothesis 4*: PLGs differ between groups of participants who revisit relevant texts more often and those who revisit relevant texts less often.

3   Method

3.1   Participants

107 undergraduate students from a public North American university participated in the current study. Fifteen participants were removed due to missing eye tracking data and one participant was removed as they did not complete the post-test. A total of 91 (66% females) participants' data were considered for these analyses. Participants were randomly assigned to either the full ($n = 54$) or partial ($n = 37$) agency conditions. Data loss resulted in the unequal number of participants assigned to each condition. A third condition (i.e., no agency condition) was not included in the current study. Participants' mean age was 20.01 ($SD = 1.66$). Participants were compensated $10/hour for a maximum of $30.

3.2   CRYSTAL ISLAND **Conditions**

Participants were randomly assigned to three groups: full agency, partial agency, and no agency. The conditions differed based on how students could freely navigate the environment. More specifically, *Full agency* allowed the most control, where participants can move freely throughout gameplay, while *Partial agency* provided a "golden path" to participants where they were directed to complete specific sequences of actions such as the order of building visitations. Participants in this condition were required to read all informational text (e.g., research articles and books) and complete all concept matrices from each book and research article. However, once the golden path was completed, participants were able to freely interact with the environment before they submitted their final diagnosis. The *no agency* condition did not allow any control as participants watched in third person as an expert solved the mystery illness. For the purposes of this study, only the full and partial agency conditions were used because the no agency condition did not allow for autonomy.

3.3   Materials

Participants were given a demographics questionnaire and a 21 item four-choice multiple choice microbiology content knowledge pre and posttest constructed by an expert in microbiology. The content knowledge pre and posttests contained questions that were randomized for both tests to diminish practice effects. The demographics questionnaire was distributed at pretest, asking about age, gender, and race along with video gaming habits of participants (e.g., frequency of play, self-perceived skill in video games, time spent playing games on a weekly basis, and the names of video games that participants play). Other self-report questionnaires investigating emotions and motivations were administered to participants, we do not provide more details on these measure as they were not used in our analyses. For purposes of this study, we only used demographics questionnaire and the content knowledge measured by the pre and posttests. A SMI EYERED 250 eye tracker was calibrated using a 9-point calibration to capture fixation duration and gaze movements during gameplay. Log-file data were also collected to track activity during game. However, for the purposes of this study, we only used eye-tracking and log-file data in our analyses.

3.4   Experimental Procedure

Participants were first asked to review and complete informed consent. Next, they completed pre-task measures including the demographics questionnaire and the content knowledge quiz. After completion, participants were given information about the study and were instrumented and calibrated to the SMI EYERED 250 eye-tracker by a researcher. All features in CRYSTAL ISLAND (e.g., informational text, NPCs, food item scanning, and the worksheet) were explained to the participant prior to gameplay. Multimodal multichannel data were collected on each participant throughout the duration of the experiment. After participants finished playing the game, they were instructed to immediately complete the content knowledge posttest. Participants then completed post-task self-report measures. After the completion of the post-task measures, participants were monetarily compensated for their participation, debriefed, and thanked for their time.

3.5 Coding and Scoring

Each participant's fixation duration for goal-directed actions were calculated by summing the fixation durations of all instances that action had occurred which was identified through AOIs which specified which action was occurring and for how long. We calculated the proportion of time fixating to control for differences in overall game time between participants by calculating the total time fixating on each goal-directed action (e.g., reading informational text, talking with NPCs, scanning items, solving concept matrices, and consulting the scientific worksheet) and dividing that time by total time in the game. PLG was calculated using pre- and post-test content knowledge scores using the following formula to control for differences in prior knowledge of microbiology [33]:

PLG = ((# correct post-test/total) − (# correct pre-test/total)) / (1 − (# correct pre-test/total)).

The fixation duration for informational text is the summation of the fixation durations of all instances for books, research articles, and posters. Relevant informational text was determined based on the correct diagnosis of the pathogen source for each participant. For example, if a participant's correct diagnosis was influenza, then the book on E. coli would be considered irrelevant, whereas the book on viruses would be relevant as it contains information crucial to concluding the correct diagnosis. The fixation duration on relevant informational text were then added for each participant. For the purposes of research question four in this study, participant data were split between two groups to identify the participants who engaged in more task-relevant informational texts (High; $n = 48$) and participants who did not (Low; $n = 43$). This was determined by the identification of relevant text revisits, where if a participant came back to a relevant text after an initial visit, it would be counted as one revisit. If a participant revisited a book that was relevant, it was also counted as one relevant revisit. The groups were determined by splitting the percentage, or value, so that the participants who revisited relevant texts over 50% of total revisits were placed in the high group and the others, who spent 50% of their time or greater revisiting irrelevant texts, were placed in the low group.

4 Results

4.1. *Research Question 1:* Are there differences in the proportions of fixation duration during goal-directed actions (e.g., reading informational text) over the duration of the game while problem solving in CI?;

A repeated measures ANCOVA was calculated to examine the differences of the proportion of time spent fixating on different goal-directed actions over the duration of gameplay with condition as a covariate. There was a significant difference between the fixation durations of the components over the duration of the game (F $(5,450) = 289.955$ s, $p < .0005$) where there are significant differences between the means for the proportion of informational text fixation duration over game time ($M = 0.323$ s, $SD = 0.145$ s) and the other components of the game with the exception of the proportion of concept matrix fixation durations over game time (see Table 1).

In sum, the time spent engaging with informational text within the game was significantly more than the time spent with other goal-directed actions except for concept matrices.

Table 1. Pairwise comparison of time spent on informational text to other goal-directed actions

Goal-directed action	N	Mean (s)	SD (s)	P-value
Talking to NPC	91	0.041	0.029	p < .0005
NPC dialog	91	0.084	0.036	p < .0005
Scanning food items	91	0.022	0.014	p < .0005
Concept matrices	91	0.325	0.145	p > .05
Worksheet	91	0.093	0.040	p < .0005

4.2. *Research Question 2:* Do the fixation duration proportions of goal-directed actions available to participants predict PLGs while problem-solving in CI?

A linear regression was run to examine whether goal-directed actions while problem solving in CI predict PLG. We found a significant correlation between fixation duration on concept matrices ($M = 1707.41$ s, $SD = 859.01$ s) and PLG ($M = 0.269$, $S = 0.246$; $r = .269$, $p < .01$) as well as the fixation duration on informational text ($M = 1694.13$ $SD = 8859.08$) and PLG ($r = .285$, $p < .01$). There was no relationship between PLG and fixation duration on other goal-directed actions (e.g., talking to NPCs, NPC dialog, scanning items, and worksheet instances). However, all components of the game used to achieve the goal of the game positively predicted PLG, ($F(6,84) = 2.653$, $p < .05$) with an R^2 of .159. An additional linear regression was run supporting the reading of informational text independently as a significant predictor of PLG, ($F(1,89) = 7.884$, $p < .01$) with an R^2 of .081, where as the fixation durations of informational text increased, so did participants' PLGs ($\beta = .285$, $p < .01$). Overall, time spent engaging with informational text positively predicted participant PLGs.

4.3. *Research Question 3:* Do fixation durations of relevant informational text significantly predict PLGs?

A linear regression was run to see if the amount of time spent fixating on relevant informational texts can predict PLGs. A significant positive correlation was found between the total fixation duration of relevant informational texts ($M = 1110.61$ s, $SD = 526.00$ s) and PLGs ($r = .299$, $p < .01$). The total fixation duration of relevant informational texts significantly predicts PLGs ($F(1,89) = 8.733$, $p < .01$) with an R^2 of .089 where as the fixation duration of information text increased, so did PLGs ($\beta = .285$, $p < .01$). Specifically within all engagement of informational text, the fixation on goal-relevant informational texts positively predicts participant PLGs. These results indicate that participants who engaged with relevant informational text for a greater period of time, demonstrated increased PLGs, supporting the presence of metacognitive monitoring.

4.4.   *Research Question 4:* Do the PLGs differ between groups of participants who revisit relevant texts more often and those who revisit relevant texts less often?

Participants were split into two separate groups: those who revisit goal-relevant texts for more than 50% of the total revisits (Low) and those who revisit goal-irrelevant texts for 50% or more of the total revisits (High). An ANCOVA was run to examine differences in the number of relevant revisits between groups and PLGs using the full and partial agency conditions as a covariates. The results revealed a significant difference between groups, $(F(2,88) = 3.226, p < .05)$, where participants who focused on relevant texts while revisiting text more than 50% of the time had significantly higher PLGs than those who revisited relevant texts 50% or less of the time. In sum, the fixation durations of participants in the High group have significantly higher PLGs than participants in the Low group. This supports the evidence for metacognitive judgments where participants discerned the relevancy of the text, how this may be relevant to their goal, understood their lack of knowledge in a subject, and adjusted their reading of informational text to optimize learning.

5 Discussion

The objective of this study was to examine learners' metacognitive judgments within CI, a GBLE. In support of the first hypothesis, results indicate that the proportion of time fixating on goal-directed actions differ from each other where the fixation duration of concept matrices and informational text are significant contributors to the overall game time. Further results support all goal-directed actions, including informational text independently, as predictors of PLGs. GBLEs contain activities that are crucial for the progress towards the goal of the game, but often lack in their ability to scaffold [19–21], especially informational text. In knowing the time distribution between actions as well as the ability for informational text within GBLEs to predict PLGs, more support can be directed toward these components of the game to increase overall content knowledge. Hypotheses were also supported where results indicated that participants who were able to make accurate metacognitive judgements as to the relevancy of informational text had higher PLGs. Further examination into evidence of metacognitive judgments yielded results in support of the hypothesis where participants who displayed a greater number of instances of content evaluation had higher PLGs. These results show that with informational text in GBLEs, without the aid of diagrams, are able to encourage the use of SRL skills while still having and positive impact on PLGs. This contradicts the CTML model and supporting studies in that within a game-based learning environment, text-only information presentations increase participants' proportional learning.

5.1 Implications for Adaptive Game-Based Learning Environments

This study investigates the importance of text within GBLEs as well as the need for increased scaffolding within these environments directed towards selecting relevant text-only information. As supported by the study, some students are not as adept at accurately monitoring and selecting relevant chunks of information from a large body of text and understand its relevancy towards the overall goal or their current learning goal. Time spent in informational text within these environments help to direct learners' improvements upon their SRL skills as well as having immediate impacts on their learning gains within the domain being studied. This encourages the need for intelligent feedback within CRYSTAL ISLAND and future adaptive GBLEs to provide needed real-time intelligent scaffolding to students in order to efficiently complete the game. These results propose that one way to identify the need for increased scaffolding on an individual level is to identify the patterns in reading informational text within GBLEs. In a narrative-based, text-centered GBLE like CRYSTAL ISLAND, participants are constantly engaging in informational text that may be relevant, irrelevant, or redundant to their overall goal that necessitates accurate metacognitive monitoring and regulation. A way to provide real-time scaffolding and improve metacognitive monitoring and regulation will be the use of real-time analysis of gaze behaviors supplemented with other trace data of self-regulatory processes (e.g., concurrent verbalizations, log-files, etc.).

Acknowledgements. This research was supported by funding from the Social Sciences and Humanities Research Council of Canada (SSHRC 895-2011-1006). The authors would also like to thank members of the SMART Lab and the intelliMEDIA group at NCSU for their assistance and contributions.

References

1. Schunk, D., Greene, J.: Handbook of Self-regulation of Learning and Performance, 2nd edn. Routledge, New York (2018)
2. Azevedo, R., Greene, J., Moos, D.: The effect of a human agent's external regulation upon college students' hypermedia learning. Metacogn. Learn. **2**, 67–87 (2007)
3. Greene, J., Azevedo, R.: A theoretical review of Winne and Hadwin's model of self-regulated learning: new perspectives and directions. Rev. Educ. Res. **77**(3), 334–372 (2007)
4. Greene, J., Azevedo, R.: A macro-level analysis of SRL processes and their relations to the acquisition of a sophisticated mental model of a complex system. Contemp. Educ. Psychol. **34**, 18–29 (2009)
5. de Bruin, A., van Got, T.: Improving self-monitoring and self-regulation: from cognitive psychology to the classroom. Learn. Instruct. **22**, 245–252 (2012)
6. Winne, P.: Theorizing and researching levels of processing in self-regulated learning. Br. J. Educ. Psychol. **88**, 9–20 (2018)
7. Burkett, C., Azevedo, R.: The effect of multimedia discrepancies on metacognitive judgments. Comput. Hum. Behav. **28**, 1276–1285 (2012)
8. Winne, P., Azevedo, R.: Metacognition. In: The Cambridge Handbook of the Learning Sciences, pp. 63–87. Cambridge University Press, New York (2014)
9. Mudrick, N., Taub, M., Azevedo, R.: Do accurate metacognitive judgments predict successful multimedia learning? In: Proceedings of the 39th Annual Conference of the Cognitive Science Society, Austin, Texas, pp. 2766–2771 (2017)
10. Azevedo, R.: Multimedia learning of metacognitive strategies. In: The Cambridge Handbook of Multimedia Learning, pp. 647–672. Cambridge University Press, New York (2014)

11. Aleven, V., Koedinger, K.: An effective metacognitive strategy: learning by doing and explaining with a computer-based cognitive tutor. Cogn. Sci. **26**, 147–179 (2002)

12. Lester, J., Rowe, J., Mott, B.: Narrative-centered learning environments: a story-centric approach to educational games. In: Mouza, C., Lavigne, N. (eds.) Emerging Technologies for the Classroom, pp. 223–237. Springer, New York (2013). https://doi.org/10.1007/978-1-4614-4696-5_15

13. de Bruin, A., Thiede, K., Camp, G., Redford, J.: Generating keywords improves metacomprehension and self-regulation in elementary and middle school children. J. Exp. Child Psychol. **109**, 294–310 (2011)

14. Mayer, R.: Educational psychology's past and future contributions to the science of learning, science of instruction, and science of assessment. J. Educ. Psychol. **110**(2), 174–179 (2018)

15. Taub, M., Mudrick, N.V., Azevedo, R., Millar, G.C., Rowe, J., Lester, J.: Using multi-level modeling with eye-tracking data to predict metacognitive monitoring and self-regulated learning with Crystal Island. In: Micarelli, A., Stamper, J., Panourgia, K. (eds.) ITS 2016. LNCS, vol. 9684, pp. 240–246. Springer, Cham (2016). https://doi.org/10.1007/978-3-319-39583-8_24

16. Conati, C., Vanlehn, K.: Toward computer-based support of meta-cognitive skills: a computational framework to coach self-explanation. Int. J. Artif. Intell. Educ. **11**, 398–415 (2000)

17. Butcher, K.: The multimedia principle. In: The Cambridge Handbook of Multimedia Learning, pp. 174–205. Cambridge University Press, New York (2014)

18. Mayer, R.: Cognitive theory of multimedia learning. In: The Cambridge Handbook of Multimedia Learning, pp. 43–71. Cambridge University Press, New York (2014)

19. Sabourin, J., Shores, L., Mott, B., Lester, J.: Understanding and predicting student self-regulated learning strategies in game-based learning environments. Int. J. Artif. Intell. Educ. **23**, 94–114 (2013)

20. Kirschner, P., Sweller, J., Clark, R.: Why minimal guidance during instruction does not work: an analysis of the failure of constructivist, discovery, problem-based, experiential, and inquiry-based teaching. Educ. Psychol. **41**(2), 75–86 (2006)

21. Mayer, R., Johnson, C.: Adding instructional features that promote learning in a game-like environment. J. Educ. Comput. Res. **42**(3), 241–265 (2010)

22. Azevedo, R., Taub, M., Mudrick, N.: Understanding and reasoning about real-time cognitive, affective, and metacognitive processes to foster self-regulation with advanced learning technologies. In: The Handbook of Self-Regulation of Learning and Performance, pp. 254–270. Routledge, New York (2018)

23. Dogusoy-Taylan, B., Cagiltay, K.: Cognitive analysis of experts' and novices' concept mapping processes: an eye tracking study. Comput. Hum. Behav. **36**, 82–93 (2014)

24. Rayner, K.: The 35th Sir Frederick Bartlett Lecture Eye movements and attention in reading, scene perception, and visual search. Q. J. Exp. Psychol. **62**(8), 1457–1506 (2009)

25. Tsai, M., Hou, H., Lai, M., Liu, W., Yang, F.: Visual attention for solving multiple-choice science problem: an eye-tracking analysis. Comput. Educ. **58**, 375–385 (2016)

26. Hu, Y., Wu, B., Gu, X.: An eye tracking study of high- and low-performing students in solving interactive and analytical problems. Educ. Technol. Soc. **20**(4), 300–311 (2017)

27. Gagenfurtner, A., Lehtinen, E., Säljö, R.: Expertise differences in the comprehension of visualizations: a meta-analysis of eye-tracking research in professional domains. Educ. Psychol. Rev. **23**, 523–552 (2011)

28. D'Mello, S., Olney, A., Williams, C., Hays, P.: Gaze tutor: a gaze-reactive intelligent tutoring system. Int. J. Hum. Comput. Stud. **70**, 377–398 (2012)

29. Rowe, J.P., Shores, L.R., Mott, B.W., Lester, J.C.: Integrating learning, problem solving, and engagement in narrative-centered learning environments. Int. J. Artif. Intell. Educ. **21**, 115–133 (2011)
30. Taub, M., Mudrick, N.V., Azevedo, R., Millar, G.C., Rowe, J., Lester, J.: Using multi-channel data with multi-level modeling to assess in-game performance during gameplay with Crystal Island. Comput. Hum. Behav. **76**, 641–655 (2017)
31. Emerson, A., Sawyer, R., Azevedo, R., Lester, J.: Gaze-enhanced student modeling for game-based learning. In: Proceedings of the Twenty-Sixth ACM Conference on User Modeling, Adaptation and Personalization, Singapore, pp. 63–72 (2018)
32. Taub, M., Azevedo, R., Bradbury, A., Millar, G., Lester, J.: Using sequence mining to reveal the efficiency in scientific reasoning during STEM learning with a game-based learning environment. Learn. Instr. **54**, 93–103 (2018)
33. Witherspoon, A., Azevedo, R., D'Mello, S.: The dynamics of self-regulatory processes within self- and externally regulated learning episodes during complex science learning with hypermedia. In: Wolf, B.P., Nkambou, R., Lajoie, S. (eds.) Intelligent Tutoring Systems, pp. 260–269. Springer, Heidelberg (2008). https://doi.org/10.1007/978-3-540-69132-7_30

Using "Idealized Peers" for Automated Evaluation of Student Understanding in an Introductory Psychology Course

Tricia A. Guerrero[✉] and Jennifer Wiley

University of Illinois at Chicago, Chicago, IL 60647, USA
{tguerr9, jwiley}@uic.edu

Abstract. Teachers may wish to use open-ended learning activities and tests, but they are burdensome to assess compared to forced-choice instruments. At the same time, forced-choice assessments suffer from issues of guessing (when used as tests) and may not encourage valuable behaviors of construction and generation of understanding (when used as learning activities). Previous work demonstrates that automated scoring of constructed responses such as summaries and essays using latent semantic analysis (LSA) can successfully predict human scoring. The goal for this study was to test whether LSA can be used to generate predictive indices when students are learning from social science texts that describe theories and provide evidence for them. The corpus consisted of written responses generated while reading textbook excerpts about a psychological theory. Automated scoring indices based in response length, lexical diversity of the response, the LSA match of the response to the original text, and LSA match to an idealized peer were all predictive of human scoring. In addition, student understanding (as measured by a posttest) was predicted uniquely by the LSA match to an idealized peer.

Keywords: Automated assessment · Natural language processing · Latent semantic analysis · Write-aloud methodology

1 Introduction

1.1 Generative Activities

Teachers may wish to use open-ended learning activities and tests, but they are burdensome to assess compared to forced-choice instruments. At the same time, forced-choice assessments suffer from issues of guessing (when used as tests) and may not encourage valuable behaviors of construction and generation of understanding (when used as learning activities). The use of generative learning activities such as prompting students to write explanations has been shown to be beneficial to improving understanding when learning in science [1–4]. Generating explanations can prompt students to engage in the construction of a mental model of the concepts in the text. The process of writing explanations may be effective because it prompts students to generate inferences and make connections across the text and to their own prior knowledge.

© Springer Nature Switzerland AG 2019
S. Isotani et al. (Eds.): AIED 2019, LNAI 11625, pp. 133–143, 2019.
https://doi.org/10.1007/978-3-030-23204-7_12

Prior work has shown that engaging in constructive learning activities, such as generating explanations, increases student understanding compared to other more passive activities such as re-reading [3]. However, other work suggests that the quality of the explanations that are generated may matter [2, 5]. This means that students may need feedback on the quality of their explanations in order to gain the benefits of engaging in this learning activity. In turn, this then places a large burden on teachers. However, if evaluation of student responses such as explanations could be accomplished using automated natural language processing indices, then teachers could utilize open-ended learning activities with increased frequency. And, the same methods could also be used to score open-ended test questions.

1.2   Using Latent Semantic Analysis in Automated Evaluation of Responses

Latent semantic analysis (LSA) has been useful in automated evaluation of constructed student responses as it can be used to generate an index representing the overlap in semantic space between two texts [6]. Foltz et al. [7] used multiple approaches with LSA to assess short-answer essays written about a cognitive science topic: how a particular connectionist model accounts for a psycholinguistic phenomenon (the word superiority effect). Measures of semantic overlap were obtained by comparing student essays to the original text in two ways: one using the whole text and one using selected portions that were deemed most important. Both approaches were found to be highly correlated with scores obtained from human graders who coded for content and quality of writing. Similarly, Wolfe et al. [8] derived LSA scores by comparing short student essays about heart functioning to a standard textbook chapter, and found these LSA scores predicted the grades assigned by professional graders (using a 5-point holistic measure of quality) as well as the scores that students received on a short-answer test of their knowledge of the topic.

In addition to comparing student responses to the original text or a standard text, another approach has compared student responses to an expert summary. León et al. [9] had students read either a narrative excerpt from a novel (The Carob Tree Legend) or an encyclopedia entry (The Strangler Tree) and write a short summary. The LSA comparison to the "gold standard" expert response was more predictive of human scoring than the LSA comparison to the original text. Similar results have been obtained in studies with students writing about ancient civilizations, energy sources and the circulatory system [10], and in response to conceptual physics problems [11].

Prior research has used LSA to make comparisons between student responses and expert responses; however, when experts write responses they tend to use more academic language and make different connections and elaborations than students based on their prior knowledge [12]. Thus, researchers have also explored making comparisons to peer responses. Both Foltz et al. [7] and León et al. [9] used exact responses written by peers to compute an average LSA score from comparisons of each student response with all other student responses. These average scores were predictive of human scoring. Other studies have used LSA to contrast student responses against "best peer" responses. Ventura et al. [12] had students write responses to conceptual physics problems within an intelligent tutoring system. Student responses were compared to both an expert response and a best peer response. The best peer response was taken randomly from all responses given the grade of an A. When comparing the LSA match to the expert response and the best peer response, the LSA match to the best peer more accurately predicted the letter grade assigned by a human grader.

Other work has used LSA measures based in "idealized" peer comparisons to predict not just human coding, but also student understanding. In Wiley et al. [13], students read texts as part of a multiple document unit on global warming, and were asked to generate an explanation about how global warming occurs. An idealized-peer response was constructed to include the key features from the best student essays. The LSA scores obtained by comparing the student responses to the idealized-peer response were predictive of both holistic human scoring, as well as student understanding as measured by an inference verification test given at the end of the unit.

The main goal for the present research was to further explore the effectiveness of automated scoring using peer-based LSA measures to predict understanding from a social science text in which a theory was presented along with supporting empirical research and examples to explain the theory. This text structure is representative of the style of many social science textbooks, including those in introductory psychology. With such texts, it is the responsibility of the reader to understand how and why the cited studies and examples support the theory as described. The present study tested whether the LSA match between student comments generated while reading and an experimenter-constructed idealized peer could serve not only as a predictor of holistic human coding, but also serve as a measure of student understanding.

2 Corpus and Human Scoring of Responses

2.1 Corpus

The corpus consisted of short written responses generated by 297 undergraduates while reading a text about cognitive dissonance, a key topic that is generally covered in most courses in introductory psychology. The comments were written by undergraduate students in an introductory psychology course (188 females; Age: $M = 18.93$, $SD = 1.16$) as a part of a homework assignment administered through the Qualtrics survey platform. All responses were edited to correct any typographical errors as well as to expand contractions and abbreviations. The textbook excerpt that was assigned for this topic had a Flesch-Kincaid reading level of 12.5 and contained 863 words in 5 paragraphs. The excerpt began with a real-world example followed by a description of the theoretical concept. The passage then described two research studies which provided empirical support for cognitive dissonance theory. Students were given an initial opportunity to read this textbook excerpt in an earlier homework assignment. During the target activity for this study, students were given a brief instructional lesson on how to generate explanations to support their learning from text:

As you read the texts again today, you should try to explain to yourself the meaning and relevance of each sentence and paragraph to the overall purpose of the text. At the end of each sentence and paragraph, ask yourself questions like:

- *What does this mean?*
- *What new information does this add?*
- *How does this information relate to the title?*
- *How does this information relate to previous sentences or paragraphs?*
- *Does this information provide important insights into the major theme of the text?*
- *Does this sentence or paragraph raise new questions in your mind?*

Students then saw an example text with associated example responses to these questions that could be written at various points in the text.

After the lesson, students reread the textbook excerpt on cognitive dissonance. At the end of each of the 5 paragraphs, they were prompted to "*write your thoughts*" for the current section of the text similar to a "type-aloud" or "write-aloud" procedure [14]. In addition, they were asked to write their thoughts at the end of the entire text. They were reminded to think about the questions given in the instructions which were present in a bulleted list on the screen as a reference while they wrote their thoughts. The 6 thought statements were concatenated into a single response for each student with an average length of 190 words ($SD = 114$, range: 6–728) and an average lexical diversity of 58.05 ($SD = 34.71$, range: .01–125.50).

Several additional measures were available for each student. Student understanding of the topic following the homework activity was measured by performance on a 5-question multiple–choice comprehension test ($M = 2.44$, $SD = 1.21$). As seen in Table 1, these questions were designed to test the ability to reason from information in the text, and to construct inferences about information left implicit in the text, not just verbatim memory for facts and details. Students did not have access to the text while completing the test. This was collected during the next week's homework activity which served as a practice test for the upcoming exam. The data set also included measures of reading ability (ACT scores, $M = 23.72$, $SD = 3.62$) and prior knowledge (performance on a 5-item multiple choice pretest on the topic given during the first week of the course, $M = 1.87$, $SD = 1.14$). Prior studies [except 13] have generally not included reading ability as a predictor when using automated evaluation systems. This leaves open the question of whether automated evaluation systems are solely useful in predicting general reading ability (and detecting features of essays written by better readers) rather than predicting the quality of features in specific responses.

2.2   Human Scoring of Responses

Student responses were scored by two human coders using a rubric adapted from McNamara et al. [15] and Hinze et al. [2], similar to what a teacher might use to quickly assess their quality. A score of 0 was assigned to responses that represented little to no effort: consisting of only non-word gibberish ("dfkashj"), two or fewer words per paragraph, or only verbatim phrases that were copied and pasted from the original text. Responses that included paraphrased ideas from the text (but no addi-tional elaborations) were assigned a 1 (e.g., "Possible ways to reduce cognitive dissonance include changing one's behavior," "Two scientist managed an experiment cognitive dissonance with children and their toys"). Responses that showed evidence of constructive processing, such as when students identified connections not explicit in the text, were assigned a score of 2. This could occur through identifying the relations between theories and evidence, or making connections to relevant prior knowledge (e.g., "Whenever people have conflicting beliefs and actions, some sort of resolution must occur. The conflict causes psychological distress and must be removed. In order to reduce cognitive dissonance, they must alter their beliefs to match the action or altering behaviors to match the belief"). Interrater agreement between two coders resulted in Cohen's kappa of .92.

Table 1. Paragraph 3 of cognitive dissonance text, idealized-peer response from concepts appearing in highest scoring student responses, and example test question.

Text Excerpt
In 1959, Festinger and Carlsmith conducted an experiment which tested cognitive dissonance theory. Participants were asked to spend an hour performing a very boring task.... These participants were asked to recommend the experiment they had just completed to other potential participants who were waiting to complete the experiment. They were instructed to tell these potential participants that the experiment was fun and enjoyable. Half of the participants in this group were paid $1 to recommend the experiment and the other half were paid $20. These participants were then taken to the interview room and asked the same questions as the participants in the control group, who were not paid and were not asked to talk to other participants. The participants in the $20 group responded similarly to the participants in the control group, namely that they did not find the experiment to be enjoyable and that they would not sign up to participate in a similar experiment. In contrast, participants in the $1 group rated the experiment as more enjoyable than participants in the other two groups, and indicated that they would be more willing to participate in another similar experiment.

Most frequent concepts in best responses	Idealized-peer response
- Identify groups performing similarly (18%) - Question the reasoning for results of study (72%)	The control group and the $20 group both told the truth that they did not enjoy the experiment. The $1 group rated the experiment as more enjoyable. This does not make sense. Why would the $1 group say it was fun?

Test Question
Imagine that the theory in the text was incorrect and that people do not experience cognitive dissonance. Which result of the Festinger experiment (about getting paid to do a boring task) would you expect? a. The control group who got paid nothing would have said they found the task very interesting b. The group paid $1 would have said they found the task to be boring c. The group paid $20 would have said they found the task to be very interesting d. How much people got paid would not have had a bigger effect on what they said about the task

2.3   Idealized-Peer Response

The idealized-peer response was constructed by selecting concepts and phrases that appeared most frequently in responses to each of the 5 paragraphs across the best student comments (i.e., scored as "2" by human raters). An example of the idealized-peer response for one paragraph is shown in Table 1. The idealized response, written at the 8th grade level, included a paraphrase of the main point and 1–2 of the most frequent elaborations for each paragraph. The elaborations were often written in the first and second person. Elaborations also included explicit connections between the theories presented and the experiments that were left implicit in the original text, and metacognitive comments (e.g., I am not sure why they would do that?).

3   Results Using Automated Scoring Indices

3.1   Automated Scoring Indices

Four automated measures were computed. Two measures were calculated using LSA. The first compared the student response to the actual text excerpt that was read (LSAORIG). The second compared the student response to the idealized-peer response (LSAIDEAL). In addition, the total response comment length (LENGTH) was computed using Linguistic Inquiry and Word Count (LIWC) [16] and the lexical diversity (LEXDIV) of all words in each student response was measured using Coh-Metrix index LDVOCDa [17]. The length of a response is often predictive of human scoring, accounting for over 35% of the variance in human-scored responses [18–20]. The variety of words used can also predict human scoring. In essays where students were asked to describe the popularity of comic books or wearing name-brand fashions, or to write letters responding to a complaint or welcoming an exchange student, the lexical diversity of the response was a positive predictor of essay grades assigned by human raters [21]. While features such as the number and diversity of words within a student response may influence human scoring, other work has found that length may not predict student understanding, and the relation between lexical diversity and understanding may became negative once the LSA match with the idealized-peer essay is taken into account [13]. To further explore these relations, two additional automated measures (LENGTH, LEXDIV) were included in the present analyses.

Table 2. Correlations among measures for student responses.

	HUMAN	LENGTH	LEXDIV	LSAORIG	LSAIDEAL
LENGTH	.46**	–			
LEXDIV	.55**	.54**	–		
LSAORIG	.68**	.55**	.55**	–	
LSAIDEAL	.79**	.50**	.54**	.83**	–
POSTTEST	.15**	.10	.08	.19**	.23**

**Correlations are significant at the 0.01 level.

As shown in Table 2, human scoring (HUMAN) predicted posttest performance (POSTTEST). LSA measures predicted human scoring, and were at least as strong of predictors of posttest performance as human scoring. Descriptively, the strongest single predictor of posttest performance was the match with LSAIDEAL (although this correlation was not significantly stronger than the correlation with HUMAN scoring, $z = 1.01$, $p = .16$). Despite the significant correlations among measures, variance inflation factors in all reported analyses remained below 1.8 indicating that multicollinearity was not an issue for analyzing the measures together in regressions.

3.2 Relation of Automated Scoring to Human Scoring

As shown in Table 2, the simple correlations between human scores and all four automated measures were significant. However, as shown in Table 3, when they were all entered simultaneously into a regression model, LSAORIG was no longer a significant predictor of human scoring. LSAIDEAL and LEXDIV both remained as positive unique predictors of the human scores, with the full model accounting for 58% of the variance in human scores, $F(4, 292) = 130.53$, $p < .001$.

Table 3. Human-scored quality as predicted by automated measures.

Variable	Unstandardized beta (B)	Std. error	Standardized beta (β)	t-value	p-value
(Constant)	0.27	0.08		3.17	.002
LENGTH	0.00	0.00	.04	0.81	.42
LEXDIV	0.00	0.00	.16	3.56	<.001
LSAORIG	−0.03	0.22	−.01	−0.15	.88
LSAIDEAL	2.68	0.25	.69	10.72	<.001

3.3 Relation of Automated Scoring to Student Understanding

As shown in Table 2, the simple correlations between student understanding (assessed by posttest scores) and automated measures were only significant for the two LSA measures (LSAORIG and LSAIDEAL). Posttest scores were not significantly predicted by response length (LENGTH) or lexical diversity (LEXDIV). Further, as shown in Table 4, only LSAIDEAL remained as a significant predictor, $R^2 = .04$ $F(4, 292) = 4.13$, $p = .003$, when all 4 automated measures were entered simultaneously.

Table 4. Student understanding as predicted by automated measures.

Variable	Unstandardized beta (B)	Std. error	Standardized beta (β)	t-value	p-value
(Constant)	1.23	0.31		4.01	<.001
LENGTH	0.00	0.00	.00	0.02	.99
LEXDIV	0.00	0.00	−.06	−0.79	.43
LSAORIG	0.13	0.82	.02	0.16	.88
LSAIDEAL	2.16	0.93	.24	2.33	.02

3.4   Unique Contribution of LSAIDEAL Over and Above Reader Characteristics

It is typically the case that students who are better readers or who have prior knowledge of a topic will develop better understanding when learning from text. Indeed, both ACT scores ($r = .25$) and prior knowledge measures (PRETEST, $r = . 29$) were significant predictors of posttest scores. However, as shown in Table 5, LSAIDEAL remained as a significant predictor even when both ACT scores and prior knowledge were included in the model, $R^2 = .17$, $F(3, 249) = 16.79$, $p < .001$.

Table 5. Student understanding as predicted by LSAIDEAL and reader characteristics.

Variable	Unstandardized beta (B)	Std. error	Standardized beta (β)	t-value	p-value
(Constant)	−0.80	0.54		−1.49	.14
ACT	0.07	0.02	.22	3.77	<.001
PRETEST	0.22	0.06	.21	3.63	<.001
LSAIDEAL	1.97	0.49	.23	4.01	<.001

3.5   Comparison of LSAIDEAL to Other LSA Alternatives

There are several possible reasons why idealized-peer responses were more predictive of understanding than the original text. One may be that sections in introductory textbooks contain a large number of ideas about each topic. The idealized-peer response may gain its power by selecting out the most relevant ideas from the section. Thus, when a student's response overlaps heavily with the content of the idealized-peer response, this may reflect that student's ability to identify, select, and attend to the most relevant features of the text. This may be similar to the predictive value of just the most important sentences within the text [7]. A second possible reason may be because idealized-peer comments are written in more colloquial language that other students may be more likely to use [12, 13]. A third possible reason is that idealized-peer responses may explicitly mention key inferences and connections that are left implicit in the text [12]. And finally, constructing an idealized-peer response from multiple high-quality student responses may be better than using only one randomly selected "best student" because comments vary and contain many idiosyncrasies that may be relevant based on the prior knowledge of one individual more so than another.

To better understand what may be responsible for the predictive power of the idealized-peer response, several alternative LSA comparisons were computed: the match of each student's comments to the same concepts in the LSAIDEAL but written in academic language at a 12th grade level (ACADEMIC), to an automated selection (selected by R package LSAfun [22]) of the important sentences in each section of the text (LSAFUN), to important sentences as selected by expert (SELECTED), to sentences written by an expert to represent the explicit connections that need to be made to comprehend the text (EXPLICIT), and to a randomly chosen single best peer response (BESTPEER). The partial correlations after controlling for the unique contributions from reading ability and prior knowledge are similar as shown in Table 6.

Table 6. Partial correlations among LSA measures and student understanding.

	Posttest
LSAORIG	.20**
LSAIDEAL	.25**
ACADEMIC	.24**
LSAFUN	.22**
SELECTED	.25**
BESTPEER	.23**
EXPLICIT	.23**

**Partial correlations are significant at the 0.01 level.
Note. Controlling for reading ability (ACT) and prior knowledge (pretest).

4 Discussion

This study tested multiple automated measures that may be useful for assessing student understanding. Students wrote responses while reading a textbook excerpt on cognitive dissonance, a commonly taught subject in introductory psychology courses. All responses were scored for quality by both humans and using automated measures.

Although lexical diversity of the comments was a significant positive predictor of human scoring, it was not predictive of student understanding as measured by the posttest. When the intended purpose of a learning activity is to promote student understanding, and when the goal for using automated measures is to predict student understanding (rather than to match holistic impressions of human scorers), then features such as length and lexical diversity may be less useful.

In contrast, the LSA match with the idealized-peer response provided a better fit for both human scoring and for student understanding than did the LSA match to the original text. Although this predictive model accounted for a relatively small proportion of the variance in test scores, it provides a first step in exploring how learning activities that prompt students to record their thoughts online as they are attempting to comprehend a text might be able to utilize automated evaluation techniques.

This study represents an advance beyond prior work by the inclusion of reading ability and prior knowledge in the prediction models, as well as by testing across a wide range of LSA metrics. Similar results were seen between idealized responses written in academic and more colloquial language indicating that the use of peer language may not be as important as hypothesized. Further, the use of idealized-peer responses that included multiple elements from several of the best students seemed to produce a better standard than a single randomly chosen best response (although this finding may be highly variable based on the single response chosen). Additionally, an expert may choose slightly better sentences than an automated system (LSAfun), but the advantage of automation may be important for broader implementation.

Another limitation of the current implementation was that the student responses needed to be edited to correct misspellings and abbreviations prior to processing to achieve these results. However, simply requiring students to use a spelling and grammar check tool prior to submission has been successful in properly editing responses for processing [10]. Adding that feature could also aid automation in this case.

5   Conclusion and Future Directions

The main goal for the present research was to further explore the effectiveness of automated scoring using LSA to predict understanding from a social science text in which a theory was presented along with supporting empirical research and examples to explain the theory. The results of the present study demonstrated that the LSA match between student comments and an idealized peer could serve not only as a predictor of holistic human coding, but also as a measure of student understanding.

Ultimately, the motivation behind developing and testing for effective means of automated coding of student responses is to enable the development of automated evaluation and feedback systems that support better student comprehension when attempting to learn from complex social science texts. Generative activities can be beneficial for learning, but they may be especially effective when feedback is provided to students. Moving forward, the next step in this research program is exploring how this automated scoring approach can be used to provide intelligent feedback to students as they engage in these learning activities.

Though the predictive power of this approach is limited, the results of the present study are promising as they suggest that evaluations of response quality derived from an LSA index based in the match between students' comments and an idealized-peer might be just as helpful as having a teacher quickly assess the quality of student comments made during reading. Utilizing these automated measures may make it more feasible for teachers to assign learning activities that contain open-ended responses, and for students to learn effectively from them.

Acknowledgements. This research was supported by grants from the Institute for Education Sciences (R305A160008) and the National Science Foundations (GRFP to first author). The authors thank Grace Li for her support in scoring the student responses, and Thomas D. Griffin and Marta K. Mielicki for their contributions as part of the larger project from which these data were derived.

References

1. Chi, M.T.H.: Self-explaining expository texts: the dual process of generating inferences and repairing mental models. In: Glaser, R. (ed.) Advances in Instructional Psychology, pp. 161–237. Erlbaum, Mahwah (2000)
2. Hinze, S.R., Wiley, J., Pellegrino, J.W.: The importance of constructive comprehension processes in learning from tests. J. Mem. Lang. **69**, 151–164 (2013)
3. Chi, M.T.H., de Leeuw, N., Chiu, M.H., LaVancher, C.: Eliciting self-explanation improves understanding. Cogn. Sci. **18**, 439–477 (1994)
4. McNamara, D.S.: SERT: self-explanation reading training. Discourse Process. **38**, 1–30 (2004)

5. Guerrero, T.A., Wiley, J.: Effects of text availability and reasoning processes on test performance. In: Proceedings of the 40th Annual Conference of the Cognitive Science Society, pp. 1745–1750. Cognitive Science Society, Madison (2018)
6. Landauer, T.K., Foltz, P.W., Laham, D.: An introduction to latent semantic analysis. Discourse Process. **25**, 259–284 (1998)
7. Foltz, P.W., Gilliam, S., Kendall, S.: Supporting content-based feedback in on-line writing evaluation with LSA. Interact. Learn. Environ. **8**, 111–127 (2000)
8. Wolfe, M.B., Schreiner, M.E., Rehder, B., Laham, D., Foltz, P.W., Kintsch, W., Landauer, T.K.: Learning from text: matching readers and text by latent semantic analysis. Discourse Process. **25**, 309–336 (1998)
9. León, J.A., Olmos, R., Escudero, I., Cañas, J.J., Salmerón, L.: Assessing short summaries with human judgments procedure and latent semantic analysis in narrative and expository texts. Behav. Res. Methods **38**, 616–627 (2006)
10. Kintsch, E., Steinhart, D., Stahl, G., LSA research group: Developing summarization skills through the use of LSA-based feedback. Interact. Learn. Environ. **8**, 87–109 (2000)
11. Graesser, A.C., Penumatsa, P., Ventura, M., Cai, Z., Hu, X.: Using LSA in AutoTutor: learning through mized initiative dialogue in natural language. In: Handbook of Latent Semantic Analysis, pp. 243–262 (2007)
12. Ventura, M.J., Franchescetti, D.R., Pennumatsa, P., Graesser, A.C., Hu, G.T.J.X., Cai, Z.: Combining computational models of short essay grading for conceptual physics problems. In: Lester, J.C., Vicari, R.M., Paraguaçu, F. (eds.) ITS 2004. LNCS, vol. 3220, pp. 423–431. Springer, Heidelberg (2004). https://doi.org/10.1007/978-3-540-30139-4_40
13. Wiley, J., et al.: Different approaches to assessing the quality of explanations following a multiple-document inquiry activity in science. Int. J. Artif. Intell. Educ. **27**, 758–790 (2017)
14. Muñoz, B., Magliano, J.P., Sheridan, R., McNamara, D.S.: Typing versus thinking aloud when reading: implications for computer-based assessment and training tools. Behav. Res. Methods **38**, 211–217 (2006)
15. McNamara, D.S., Boonthum, C., Levinstein, I.B., Millis, K.: Evaluating self-explanations in iSTART: comparing word-based and LSA algorithms. In: Handbook of Latent Semantic Analysis, pp. 227–241 (2007)
16. Pennebaker, J. W., Booth, R.J., Boyd, R.L., Francis, M.E.: Linguistic inquiry and word count: LIWC 2015. LIWC.net, Austin, TX (2015)
17. Graesser, A.C., McNamara, D.S., Louwerse, M.M., Cai, Z.: Coh-Metrix: analysis of text on cohesion and language. Behav. Res. Methods Instrum. Comput. **36**, 193–202 (2004)
18. Dikli, S.: An overview of automated scoring of essays. J. Technol. Learn. Assess. **5**, 1–35 (2006)
19. Kobrin, J.L., Deng, H., Shaw, E.J.: The association between SAT prompt characteristics, response features, and essay scores. Assessing Writ. **16**, 154–169 (2011)
20. Ferris, D.R.: Lexical and syntactic features of ESL writing by students at different levels of L2 proficiency. TESOL Q. **28**, 414–420 (1994)
21. Crossley, S.A., McNamara, D.S.: Predicting second language writing proficiency: the role of cohesion, readability, and lexical difficulty. J. Res. Read. **35**, 115–135 (2012)
22. Guenther, F., Dudschig, C., Kaup, B.: LSAfun: an R package for computations based on latent semantic analysis. Behav. Res. Methods **47**, 930–944 (2015)

4D Affect Detection: Improving Frustration Detection in Game-Based Learning with Posture-Based Temporal Data Fusion

Nathan L. Henderson[1]([⊠]), Jonathan P. Rowe[1]([⊠]),
Bradford W. Mott[1]([⊠]), Keith Brawner[2]([⊠]), Ryan Baker[3]([⊠]),
and James C. Lester[1]([⊠])

[1] North Carolina State University, Raleigh, NC 27695, USA
{nlhender, jprowe, bwmott, lester}@ncsu.edu
[2] U.S. Army Combat Capabilities Development Command,
Orlando, FL 32826, USA
keith.w.brawner.civ@mail.mil
[3] University of Pennsylvania, Philadelphia, PA 19104, USA
rybaker@upenn.edu

Abstract. Recent years have seen growing interest in utilizing sensors to detect learner affect. Modeling frustration has particular significance because of its central role in learning. However, sensor-based affect detection poses important challenges. Motion-tracking cameras produce vast streams of spatial and temporal data, but relatively few systems have harnessed this data successfully to produce accurate run-time detectors of learner frustration outside of the laboratory. In this paper, we introduce a data-driven framework that leverages spatial and temporal posture data to detect learner frustration using deep neural network-based data fusion techniques. To train and validate the detectors, we utilize posture data collected with Microsoft Kinect sensors from students interacting with a game-based learning environment for emergency medical training. Ground-truth labels of learner frustration were obtained using the BROMP quantitative observation protocol. Results show that deep neural network-based late fusion techniques that combine spatial and temporal data yield significant improvements to frustration detection relative to baseline models.

Keywords: Affect detection · Data fusion · Posture · Frustration ·
Deep learning

1 Introduction

Affect has a key role in shaping student learning outcomes [1]. Affective states such as *flow* tend to promote learning, while states such as *boredom* are not as conducive to learning. The affective state of *frustration* has a complex relationship with learning [2–5]. On the one hand, frustration often coincides with student efforts to overcome impasses, and it signifies situations in which students are grappling with a concept that is challenging [6]. On the other hand, frustration can lead to student disengagement, and it has been correlated with negative learning outcomes [7]. The ability to accurately detect

© Springer Nature Switzerland AG 2019
S. Isotani et al. (Eds.): AIED 2019, LNAI 11625, pp. 144–156, 2019.
https://doi.org/10.1007/978-3-030-23204-7_13

student affect at run-time is critical to the development of affect-sensitive learning technologies that dynamically intervene to support engagement and emotion regulation [3, 8]. The complex relationship between frustration and learning underscores the importance of reliable frustration detection to inform how affect-sensitive pedagogical interventions are delivered within intelligent tutoring systems [3].

Several methods for detecting student frustration have been investigated in recent years. These include both *sensor-free* methods and *sensor-based* methods. Sensor-free methods leverage trace log data from student interactions with a learning environment to train machine learning-based models of affect [9, 10]. Results have shown that sensor-free affect detection, in combination with deep recurrent neural networks, can yield accurate models across several affective states [9]. Alternatively, sensor-based methods utilize physical sensors to capture trace-level data on learner behavior and physiology, including facial expression, eye gaze, electrodermal activity (EDA), electroencephalography (EEG), and posture [3, 4, 11]. Sensor-based methods show promise for enabling *generalized affect detection*, which eschews domain-specific input feature representations, instead leveraging sensor data that can be gathered across a range of educational domains and learning environments. Notably, sensor-based approaches to affect detection do not necessarily require specialized hardware because a growing number of sensors are built directly into computers and tablets, including webcams, motion-tracking cameras, and increasingly, eye trackers.

Sensor-based frustration detection has shown good results when targeting self-reported affect data [12] or deploying sensors in laboratory settings [4]. Specific data channels, such as facial expression, have also shown promise using student data from classrooms [13], but other data channels, such as posture, have received less attention. Sensor-based frustration detection outside of the lab raises significant challenges [3]. Physical sensor data can be affected by reliability issues, background noise, poor calibration, subject mistracking, data storage constraints, and inconsistent sensor configurations. Further, trace-level data generated by sensors is intrinsically temporal, yet the input feature representations that are distilled from these data streams often contain limited temporal information [3]. Spatiotemporal data has been demonstrated to significantly improve the performance of sensor-based classifiers for action recognition [14] and engagement intensity [15], and it is likely to benefit affect-sensitive learning technologies as well.

In this paper, we investigate sensor-based frustration detection using deep neural network-based data fusion techniques integrating spatial and temporal data on student posture captured by Microsoft Kinect cameras. The dataset was gathered from a study involving students using a game-based learning environment for emergency medical training, TC3Sim. Ground-truth labels for learner frustration were obtained using the BROMP quantitative observation protocol [16]. We compare the effectiveness of deep neural network-based early- and late-fusion techniques across several evaluation metrics. Results show that deep neural network-based late-fusion yields significant improvements to frustration detection compared to several baseline techniques.

2   Related Work

There is growing interest in sensor-based affective modeling in advanced learning technologies. Bosch et al. [13] utilized webcam recordings of students engaged in a physics-based learning game to construct feature vectors extracted from observed head positions and movement, brow position, and gross body movement. Ground truth data was obtained through the BROMP protocol, using trained observers to mark instances of certain affective states at set time intervals. Utilizing BROMP observations as a target label, a multitude of classifiers were trained, including Bayesian classifiers and C4.5 decision trees, to detect affective states such as frustration, boredom, confusion, delight, and engagement. Motion-tracking cameras, such as the Microsoft Kinect, have also been utilized in sensor-based affect detection [17]. Grafsgaard et al. utilized learner posture and gesture data gathered by a Microsoft Kinect as learners engaged in computer-mediated tutoring sessions for introductory programming [17]. Posture estimation vectors were distilled from the Kinect's depth-channel data, and the vectors were used to determine correlations between specific postures and self-reported frustration, engagement, and learning gains. DeFalco et al. [3] utilized posture data from a Kinect sensor to detect learner affect in a game-based learning environment for emergency medical training. Separate classifiers were induced for each of five affective states: boredom, confusion, concentration, frustration, and surprise. The affect detectors performed only slightly better than chance, yielding Kappa values between 0 and 0.11.

As an alternative to sensor-based affect detection, Jiang et al. [10] utilized interaction trace log data in an investigation of deep neural network-based representation learning versus expert feature-engineering for sensor-free affect detection using BROMP data. Time, frequency, and ratio-based features were calculated for each student based on his/her individual interaction with a game-based learning environment for physics education. Overall, deep neural network-based models achieved equal or better performance compared to feature engineering-based models, with a lone exception being frustration (i.e., the feature-engineering approach was slightly more accurate). Subsequent work showed that recurrent neural networks (RNNs) outperformed the previous classification algorithms in the same affect detection task [9].

Recent efforts in affect detection have started to explore usage of temporal data channels as an input modality. Yang et al. [15] used several feature extraction approaches on spatiotemporal face and posture data to train long short-term memory (LSTM) networks alongside regression fusion to approximate engagement intensity in individuals watching an educational video. Temporal information has also been used to develop rule-based models to classify affect through recognition of sequences of joint movement and repetition of certain motions [18]. The frustration detection framework presented in this paper builds on recent advances in deep neural network-based data fusion and introduces an artificial temporal data stream (i.e., a "fourth dimension") derived from spatial 3D posture data to enhance run-time detector accuracy during student interactions with a game-based learning environment.

3 TC3Sim Game-Based Learning Environment

We investigate automatic detection of student frustration in the context of a game-based learning environment for training military medical personnel, the Tactical Combat Casualty Care Simulation (TC3Sim). Developed by Engineering and Computer Simulations (ECS), TC3Sim (Fig. 1) is widely used by the U.S. Army to train soldiers in the essential procedures required of an Army Combat Medic or Combat Life Saver. In TC3Sim, trainees complete a series of 3D simulated combat missions alongside a group of computer-controlled teammates. Each story-driven training scenario includes a series of simulated combat events that lead to the eventual injury of one or more teammates. Trainees must administer tactical combat casualty care in real-time, which includes securing the area, assessing casualties, performing triage, administering treatment, and preparing for medical evacuation. Trainees encounter opportunities to handle a wide range of injuries, including cuts, puncture wounds, blocked airways, amputations, and burns. In the present work, we focus on learner interactions with four training scenarios from TC3Sim, including a tutorial scenario, a leg injury scenario, a narrative scenario involving a squad of soldiers on patrol, and a final scenario that is impossible to complete successfully—the patient expires regardless of treatment. Prior work with TC3Sim has found evidence of a negative relationship between frustration and learning, and further, motivational feedback interventions that target frustration can positively impact learning outcomes [3]. We seek to improve the effectiveness of generalizable frustration detectors to enable affect-sensitive pedagogical support with enhanced effectiveness and reliability.

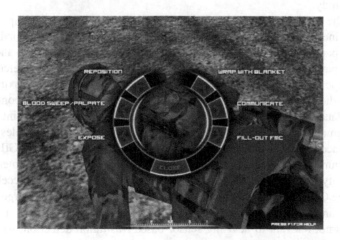

Fig. 1. Screenshot of injured soldier in TC3Sim.

4 Detecting Frustration with Posture-Based Temporal Data Fusion

The primary goal of this work was to induce machine learning-based classifiers for run-time frustration detection using student posture data collected by a Microsoft Kinect sensor. The detector's objective was to classify whether a student was frustrated or not given an input feature vector consisting of spatial and/or temporal posture data.

4.1 Dataset

We utilized a previously published dataset containing data from 119 students (83% male, 17% female) at the United States Military Academy. The training materials were administered using the Generalized Intelligent Framework for Tutoring (GIFT), an open-source software framework for building and deploying adaptive training systems [19]. All participants worked individually at laptops and received the same materials; there were no experimental conditions. Study sessions lasted approximately 1 h.

The study procedure was as follows. First, learners completed a brief demographic questionnaire and content pre-test. Next, they viewed a PowerPoint presentation about tactical combat casualty care. Afterward, participants completed a series of training scenarios in TC3Sim, each working at their own pace. The session concluded with a brief post-test, which included the same knowledge assessment items that were presented on the pre-test. Utilizing identical items on both the pre- and post-tests reduced the challenge of identifying items with matching difficulty for counterbalancing the assessments. Further, no feedback was given about student performance on the pre-test during the study.

During the study sessions, each participant was instrumented with a tripod-mounted Microsoft Kinect for Windows 1.0 sensor. The Kinect sensor was positioned in front of each participant to capture all head movements, body movements, and gestures throughout participants' interactions with TC3Sim using built-in skeletal-tracking features supported by GIFT. Kinect sensor data was recorded at approximately 10–12 Hz. The data consisted of a series of timestamped feature vectors containing 3D coordinate data for 91 vertices, each corresponding to a facial or body joint tracked by the Kinect. In addition to the Kinect, learners were equipped with a wireless Affectiva Q-Sensor bracelet, and their interaction trace log data was recorded by GIFT. The Q-Sensors captured timestamped data on learners' skin temperature, learners' electrodermal activity, and sensor 3D coordinates as measured by built-in accelerometers. However, the Q-Sensor data contained significant recording gaps for a large number of participants, and therefore it was not utilized in the current work. The interaction trace log data was not relevant to devising sensor-based frustration detection, so it was also not utilized.

To obtain ground-truth labels of learner affect, two field observers recorded learners' affect and behavior using the BROMP quantitative field observation protocol throughout the study [16]. The field observers, who were both BROMP-certified coders, walked around the perimeter of the classroom and used a hand-held Android device running the HART field-observation software to discreetly record each learner's affect and behavior at 20-s intervals in round robin sequence. The following emotional states were recorded: *Concentrating, Confused, Boredom, Surprised, Frustrated, Contempt,* and *Other*.

In total, the study yielded 3,066 BROMP observations by the two field observers. For the purpose of the current analysis, we utilize a subset of 755 observations coinciding with the time period during which participants interacted with the TC3Sim game-based learning environment and on which there was no disagreement between BROMP coders about the occurrence of a target affective state. The distribution of affective states across these observations were the following: 435 (57.6%) were coded as *Concentrating*, 174 (23.1%) as *Confused*, 73 (9.7%) as *Boredom*, 32 (4.2%) as *Frustrated*, 29 (3.8%) as *Surprised*, and 12 (1.6%) as *Contempt*.

To prepare the data for training posture-based frustration detectors, we re-coded the data into binary categories, yielding 32 instances of *Frustrated* and 723 instances of *Not-Frustrated*. The Kinect data was cleaned to remove instances of tracking anomalies and extraneous vertex data. Sessions containing fewer than 3 BROMP observations were also removed. Of the 91 vertices tracked by the Kinect, 3 were utilized for posture-based frustration detection: top_skull, head, and center_shoulder. These vertices were selected based on prior efforts investigating affect detection from Kinect data [17]. Next, Kinect and BROMP data were integrated and temporally aligned. A set of 73 posture-related features were computed for each BROMP observation after the initial data collection, serving as input features for frustration detection. These features captured spatial information about student posture, and they included summary statistics (e.g., median, variance, min, max) calculated over time windows of 5, 10, and 20 s preceding the BROMP coding event. These time window sizes are similar to prior work on affect detection, including a maximum window size that corresponds to the targeted maximum time between BROMP observations [3, 13]. In addition, features capturing aggregate changes in learner posture, as well as forward/backward lean behaviors, were computed. In aggregate, these features provided a detailed view of the spatial orientation of learners' posture.

4.2   Temporal Feature Engineering

The spatial features that were distilled from the Kinect posture data had ranges that varied widely, so feature scaling was performed. Each student's data was normalized using Z-score standardization: for each session, the difference between a single data point and session mean was divided by the session standard deviation. Temporal posture features were computed from the spatial posture feature vectors using the first derivative of each observation's posture coordinates [20]. Using the head vertex, for each set of (x, y, z) posture coordinates, the coordinate deltas across two consecutive sensor readings were calculated. The deltas were used to calculate *velocity features* averaged across time windows of 3, 5, 10, and 20 s. For each posture coordinate, the mean, median, max, and variance of the average corresponding velocity were calculated. This process provided an additional 48 temporally-related posture features. Due to the large number of additional features calculated per vertex, velocity information was not calculated for center_shoulder and top_skull. The temporal data was normalized using the same Z-score normalization described previously.

4.3   Feature Selection

Given the large number of available posture features, automated feature selection was utilized to reduce the size of the final feature sets for training frustration detectors. Forward feature selection was performed to investigate alternate configurations of feature vectors up to length 10. Greedy feature selection was performed using RapidMiner 9.0, and it was guided by classification performance with the sequential minimal optimization (SMO) variation of a polynomial-kernel support vector machine (SVM) [21]. We utilize RapidMiner because it is a convenient toolkit for processing and modeling data using a range of supervised learning algorithms, and it has been used widely in prior work on affect detection [3]. Forward feature selection is a common approach in prior work on affect detection, and SVMs trained with SMO have previously been found to outperform competing algorithms for frustration detection with learner data from the TC3Sim environment [3, 13].

4.4   Deep Neural Network Architecture

Each dataset produced by the feature-selection algorithm was used to train a multi-layer perceptron neural network. Each network was comprised of feed-forward layers containing 800, 800, 500, 100, 50, and 2 nodes, respectively. Each hidden layer utilized a Rectified Linear Unit (ReLU) activation function. The networks were trained for 10 epochs and used an ADADELTA [22] adaptive learning rate to help prevent overfitting. All deep neural network models were implemented using RapidMiner 9.0 [23].

4.5   Data Fusion

To investigate alternate approaches for integrating spatial and temporal posture features, we compared several classifiers induced with both early- and late-fusion techniques. Early fusion is based on the concept of "feature-level" fusion, or concatenation of multiple feature vectors to form a single vector prior to supervised learning [24]. To determine the best sequence of feature selection and feature-level fusion, we implemented two variants of early fusion. The first method, EarlyFusion1, performs feature selection after concatenating the spatial and temporal feature vectors (Fig. 2A). The second method, EarlyFusion2, performs feature selection on spatial and temporal features separately. After feature selection, feature-level fusion is performed on the top-selected features from each modality (Fig. 2B). LateFusion involves training a model on each modality separately and integrating the results of each classifier to produce a single prediction (Fig. 2C). This prediction can be determined using several different methods, such as majority voting, averaging, or weighting [25]. In this work, we compare the results of late fusion using match-score fusion [26] and the highest confidence level of the late-fusion output.

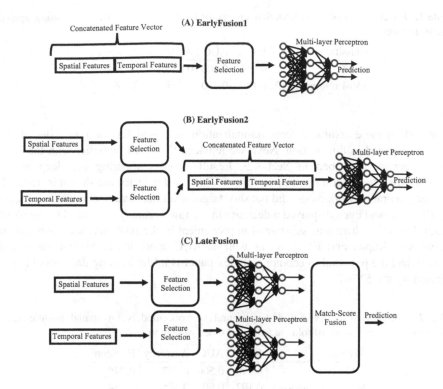

Fig. 2. Three data fusion techniques for integrating spatial and temporal posture-based frustration detection methods.

5 Results

Frustration detectors were trained using 10-fold student-level cross validation. Data splits were maintained across all modeling approaches to ensure fair comparisons. To ensure adequate training coverage for both target classes (i.e., *frustrated* and *not-frustrated*), the training data was oversampled using cloning of minority class instances. Feature selection and early fusion techniques were implemented in RapidMiner 9.0 [23]. RapidMiner does not support decision-level fusion, as required by our LateFusion method. Therefore, feature selection and deep neural network models were created using RapidMiner, the raw outputs of the models were recorded, and then decision-level fusion was performed outside of RapidMiner using Python.

We observed that z-score feature normalization has a sizable impact on the predictive accuracy of posture-based frustration detectors. As a baseline, we reproduced a machine learning pipeline for training SVM-based frustration detectors using spatial posture data, which had been previously reported in [3], and we investigated how the resulting models compared to an equivalent machine learning pipeline with z-score feature normalization added. Evaluation metrics included Cohen's kappa [27], area under the curve (AUC), total accuracy, and F1 score. Results are shown in Table 1.

Table 1. Effect of z-score normalization on sensor-based frustration detection using spatial posture features.

Classifier	Kappa	AUC	Accuracy	F1 Score
SVM	0.056	0.600	0.687	0.113
SVM (Normalized)	0.190	0.500	0.737	0.249

Based on these results, z-score normalization was used for the remainder of the analyses reported in this section. Next, we replaced the SVM classifier with the deep neural network described in Sect. 4.5. Results from comparing the deep neural network-based frustration detector with the SVM-based detector are shown in Table 2. The deep neural network model did not show significant improvement regarding Kappa and F1 score, and even displayed a decrease in the raw accuracy compared to the SVM model. However, there was substantial improvement in the AUC measurement. Slight increases in Kappa and F1 score, as well as AUC score indicated that the neural network had the potential to capture complex patterns in the training data possibly not detected by the SVM.

Table 2. Comparison of SVM and deep neural network models for spatial posture-based frustration detection under 10-fold student level cross validation.

Classifier	Kappa	AUC	Accuracy	F1 Score
SVM	0.190	0.500	0.737	0.249
Deep neural network	0.192	0.808	0.685	0.254

Next, temporal posture features were computed from the Kinect data, normalized, and used as input for the three data fusion methods. For the LateFusion model, two different selection schemes were tested. The first selection scheme used the model prediction with the highest confidence level. The second selection scheme took the average of all confidence levels for a predicted class and used the highest average, similar to match-score fusion [26]. However, detector accuracy did not change when the two selection methods were interchanged. This may be due to the high confidence levels of the classifiers for this particular data set, as well as the relatively small amount of test data available.

Results from a comparison of early- and late-fusion methods combining spatial and temporal posture data are shown in Table 3 alongside results from the deep neural network trained with spatial posture data only as a baseline. Best results for each evaluation metric are shown in bold. It is apparent that the addition of temporal feature information improved the quality of frustration detection, particularly for the LateFusion model. Due to the high proportion of non-frustration observations versus frustration observations in the test data, additional emphasis is placed on the Cohen's kappa metric, as it accounts for the potential of obtaining true-positives by chance.

Table 3. Results of early fusion and late fusion on posture and temporal feature data.

Classifier	Kappa	AUC	Accuracy	F1 Score
Baseline network	0.192	0.808	0.685	0.254
EarlyFusion1 network	0.178	0.780	0.845	0.213
EarlyFusion2 network	0.281	**0.854**	0.900	0.321
LateFusion network	**0.355**	0.809	**0.906**	**0.396**

EarlyFusion2 outperformed EarlyFusion1 across all evaluation metrics. This may be attributable to the dimensionality of the datasets used to train the respective models. Because feature selection operated on a single data stream for EarlyFusion1, the main difference between this model and the baseline deep neural network was the set of candidate features subjected to SVM-based feature selection, as EarlyFusion1 concatenated temporal velocity features with spatial posture features prior to feature selection. The temporal posture features added 48 additional attributes to the existing 73 spatial posture features, but feature selection only returned up to 10 features in each scenario. Alternatively, in EarlyFusion2, two separate feature selection processes are employed in parallel, yielding a maximum of 20 features as input to the neural network. This increase in number of attributes is a possible explanation for the improved accuracy of EarlyFusion2 over EarlyFusion1.

Late fusion offers a different approach due to its capacity to "correct" a single model's prediction during circumstances where the model's confidence level is relatively low. Upon closer examination, several instances were observed when the spatial posture-based detector made an incorrect prediction with a low confidence level, and the temporal posture-based model made a correct prediction with a high confidence level, and the latter was chosen as the representative prediction during match-score fusion. Several instances of the inverse scenario—the spatial posture-based model corrected a prediction by the temporal posture-based model—were also observed. This interaction contributed to the increased accuracy of LateFusion frustration detection over baseline SVM and deep neural network models, as well as the early fusion methods.

6 Conclusion

Detection of learner frustration is critical to the creation of affective-sensitive learning technologies. However, devising sensor-based run-time models of learner frustration using posture data poses significant challenges. We have introduced a data-driven framework that combines deep neural network-based data fusion and spatiotemporal representations of posture data to improve run-time models of frustration detection. Posture features were distilled using sensor data collected from participants engaging with a game-based learning environment for emergency medical training. We found that late fusion methods combining deep neural network-based frustration detectors trained with spatial and temporal posture feature data outperform several baseline techniques, including early fusion-based models and spatial posture-based models.

The results suggest several promising directions for future research. First, it will be important to investigate whether posture-based temporal data fusion techniques are transferable to other learning emotions (e.g., boredom, confusion, engaged concentration, surprise) as well as other learning environments. A key promise of sensor-based affect detection is the potential for creating computational models of learner affect that generalize across different educational subjects and settings. Second, alternate deep neural network architectures should be investigated, particularly those that are explicitly designed for modeling sequential data, such as recurrent neural networks, to better capture the temporal dynamics of affect as expressed through posture. Recent work has shown that recurrent neural network architectures, such as long short-term memory networks, yield significant improvements to sensor-free affect detection, but it remains to be seen how these methods are best utilized in sensor-based models of affect. Finally, there is significant promise in integrating posture-based temporal data fusion techniques for affect detection into run-time learning environments, enabling delivery of dynamic interventions designed to support student engagement and foster improved learning.

Acknowledgements. We wish to thank Dr. Jeanine DeFalco and Dr. Benjamin Goldberg at the U.S. Army Combat Capabilities Development Command – Simulation and Training Technology Center (CCDC-STTC), Dr. Mike Matthews and COL James Ness at the United States Military Academy, and Dr. Robert Sottilare at SoarTech for their assistance in facilitating this research. The research was supported by the U.S. Army Research Laboratory under cooperative agreement #W911NF-13-2-0008. Any opinions, findings, and conclusions expressed in this paper are those of the authors and do not necessarily reflect the views of the U.S. Army.

References

1. D'Mello, S.: A selective meta-analysis on the relative incidence of discrete affective states during learning with technology. J. Educ. Psychol. **105**, 1082–1099 (2013)
2. Grafsgaard, J.F., Wiggins, J.B., Vail, A.K., Boyer, K.E., Wiebe, E.N., Lester, J.C.: The additive value of multimodal features for predicting engagement, frustration, and learning during tutoring. In: Proceedings of the Sixteenth ACM International Conference on Multimodal Interaction, pp. 42–49. ACM (2014)
3. DeFalco, J.A., et al.: Detecting and addressing frustration in a serious game for military training. Int. J. Artif. Intell. Educ. **28**, 152–193 (2018)
4. Grafsgaard, J.F., Wiggins, J.B., Boyer, K.E., Wiebe, E.N., Lester, J.C.: Predicting learning and affect from multimodal data streams in task-oriented tutorial dialogue. In: Proceedings of the Seventh International Conference on Educational Data Mining, pp. 122–129. International Educational Data Mining Society, London, UK (2014)
5. Harley, J.M., Bouchet, F., Azevedo, R.: Aligning and comparing data on emotions experienced during learning with MetaTutor. In: Lane, H.C., Yacef, K., Mostow, J., Pavlik, P. (eds.) AIED 2013. LNCS (LNAI), vol. 7926, pp. 61–70. Springer, Heidelberg (2013). https://doi.org/10.1007/978-3-642-39112-5_7
6. Pardos, Z., Baker, R., Pedro, M.S., Gowda, S.M., Gowda, S.M.: Affective states and state tests: investigating how affect and engagement during the school year predict end-of-year learning outcomes. J. Learn. Anal. **1**, 107–128 (2014)

7. D'Mello, S., Graesser, A.: The half-life of cognitive-affective states during complex learning. Cogn. Emot. **25**, 1299–1308 (2011)
8. Cooper, D.G., Arroyo, I., Woolf, B.P.: Actionable affective processing for automatic tutor interventions. In: Calvo, R., D'Mello, S. (eds.) New Perspectives on Affect and Learning Technologies. Explorations in the Learning Sciences, Instructional Systems and Performance Technologies, vol. 3, pp. 127–140. Springer, New York (2011). https://doi.org/10.1007/978-1-4419-9625-1_10
9. Botelho, A.F., Baker, R.S., Heffernan, N.T.: Improving sensor-free affect detection using deep learning. In: André, E., Baker, R., Hu, X., Rodrigo, Ma.M.T., du Boulay, B. (eds.) AIED 2017. LNCS (LNAI), vol. 10331, pp. 40–51. Springer, Cham (2017). https://doi.org/10.1007/978-3-319-61425-0_4
10. Jiang, Y., et al.: Expert feature-engineering vs. deep neural networks: which is better for sensor-free affect detection? In: Proceedings of the International Conference on Artificial Intelligence in Education, pp. 198–211. Springer, Cham (2018)
11. Bosch, N., D'mello, S.K., Ocumpaugh, J., Baker, R.S., Shute, V.: Using video to automatically detect learner affect in computer-enabled classrooms. ACM Trans. Interact. Intell. Syst. **6**, 1–26 (2016)
12. Arroyo, I., Cooper, D.G., Burleson, W., Woolf, B.P., Muldner, K., Christopherson, R.: Emotion sensors go to school. In: Artificial Intelligence in Education, pp. 17–24 (2009)
13. Bosch, N., et al.: Detecting student emotions in computer-enabled classrooms. In: Proceedings of the 25th International Joint Conference on Artificial Intelligence, pp. 4125–4129 (2016)
14. Henderson, N., Aygun, R.: Human action classification using temporal slicing for deep convolutional neural networks. In: 2017 IEEE International Symposium on Multimedia (2017)
15. Yang, J., Wang, K.: Deep recurrent multi-instance learning with spatio-temporal features for engagement intensity prediction. In: Proceedings of the 2018 on International Conference on Multimodal Interaction, pp. 594–598. ACM (2018)
16. Ocumpaugh, J., Baker, R.S., Rodrigo, M.T.: Baker Rodrigo Ocumpaugh Monitoring Protocol (BROMP) 2.0 Technical and Training Manual (2015)
17. Grafsgaard, J., Boyer, K., Wiebe, E., Lester, J.: Analyzing posture and affect in task-oriented tutoring. In: FLAIRS Conference, pp. 438–443 (2012)
18. Patwardhan, A., Knapp, G.: Multimodal affect recognition using Kinect. arXiv preprint arXiv:1607.02652 (2016)
19. Sottilare, R.A., Baker, R.S., Graesser, A.C., Lester, J.C.: Special issue on the generalized intelligent framework for tutoring (GIFT): creating a stable and flexible platform for innovations in AIED research. Int. J. Artif. Intell. Educ. **28**, 139–151 (2018)
20. Sanghvi, J., Castellano, G., Leite, I., Pereira, A., McOwan, P.W., Paiva, A.: Automatic analysis of affective postures and body motion to detect engagement with a game companion. In: Proceedings of the 6th International Conference on Human-Robot Interaction, pp. 305–312. ACM (2011)
21. Platt, J.C.: Sequential minimal optimization: a fast algorithm for training support vector machines, pp. 1–21 (1998)
22. Zeiler, M.D.: ADADELTA: An adaptive learning rate method (2012)
23. Mierswa, I., Wurst, M., Klinkenberg, R., Scholz, M.: Yale: rapid prototyping for complex data mining tasks. In: Proceedings of the 12th ACM SIGKDD International Conference on Knowledge Discovery and Data Mining, pp. 935–940 (2006)
24. Soleymani, M., Pantic, M., Pun, T.: Multimodal emotion recognition in response to videos. IEEE Trans. Affect. Comput. **3**, 211–223 (2012)

25. Baltrušaitis, T., Ahuja, C., Morency, L.-P.: Multimodal machine learning: a survey and taxonomy. IEEE Trans. Pattern Anal. Mach. Intell. **41**, 423–443 (2018)
26. Rahman, W., Gavrilova, M.L.: Emerging EEG and kinect face fusion for biometric identification. In: Proceedings of the IEEE Symposium Series on Computational Intelligence (SSCI), pp. 1–8. IEEE (2017)
27. Cohen, J.: A coefficient of agreement for nominal scales. Educ. Psychol. Measur. **20**, 37–46 (1960)

Designing for Complementarity: Teacher and Student Needs for Orchestration Support in AI-Enhanced Classrooms

Kenneth Holstein[(⊠)], Bruce M. McLaren, and Vincent Aleven

Carnegie Mellon University, Pittsburgh, PA 15213, USA
{kjholste,bmclaren,aleven}@cs.cmu.edu

Abstract. As artificial intelligence (AI) increasingly enters K-12 classrooms, what do teachers and students see as the roles of human versus AI instruction, and how might educational AI (AIED) systems best be designed to support these complementary roles? We explore these questions through participatory design and needs validation studies with K-12 teachers and students. Using human-centered design methods rarely employed in AIED research, this work builds on prior findings to contribute: (1) an analysis of teacher and student feedback on 24 design concepts for systems that integrate human and AI instruction; and (2) participatory speed dating (PSD): a new variant of the speed dating design method, involving iterative concept generation and evaluation with multiple stakeholders. Using PSD, we found that teachers desire greater real-time support from AI tutors in identifying when students need human help, in evaluating the impacts of their own help-giving, and in managing student motivation. Meanwhile, students desire better mechanisms to signal help-need during class without losing face to peers, to receive emotional support from human rather than AI tutors, and to have greater agency over how their personal analytics are used. This work provides tools and insights to guide the design of more effective human–AI partnerships for K-12 education.

Keywords: Design · Classroom orchestration · Human-AI interaction

1 Introduction

When used in K-12 classrooms, AI tutoring systems (ITSs) can be highly effective in helping students learn (e.g., [32, 37]). However, in many situations, human teachers may be better suited to support students than automated systems alone (e.g., by providing socio-emotional support or flexibly providing conceptual support when continued problem-solving practice may be insufficient) [29, 44, 49, 53]. ITSs might be even more effective if they were designed not only to support students directly, but also to take advantage of teachers' complementary strengths and amplify their abilities to help their students [6, 27, 49, 65]. Yet the question

© Springer Nature Switzerland AG 2019
S. Isotani et al. (Eds.): AIED 2019, LNAI 11625, pp. 157–171, 2019.
https://doi.org/10.1007/978-3-030-23204-7_14

of how best to combine strengths of human and AI instruction has received relatively little attention in the AIED literature thus far [29,49,60].

Recent work has proposed the notion of human–AI "co-orchestration" systems that help teachers and AI agents work together to make complex yet powerful learning scenarios feasible [27,29,43,46,54,60]. For example, Olsen et al. explored how ITSs might best be designed to share control with teachers in orchestrating transitions between individual and collaborative activities during a class session [15,43]. Similarly, in our prior work [26,28,29], we designed a set of mixed-reality smart glasses that direct teachers' attention in real-time, during ITS class sessions, towards situations the software may be ill-suited to handle on its own (e.g., wheel spinning [7,31], gaming the system [5,58], or hint avoidance [2,51]). An in-vivo classroom experiment demonstrated that this form of real-time teacher/AI co-orchestration could enhance student learning, compared with an ITS classroom in which the teacher did not have such support [29].

While this work has begun to explore ways to combine strengths of human and AI instruction, many open questions remain regarding the design of classroom co-orchestration systems. If these tools are to be used in actual classrooms, beyond the context of research studies, it is critical that they are well-designed to respect the needs and boundaries of both teachers and students [3,14,42,52,66]. For example, prior design research with K-12 teachers has found that there is a delicate balance between automation and respecting teachers' autonomy [25,27,34,43]. Over-automation may take over classroom roles that teachers would prefer to perform and threaten their flexibility to set their own instructional goals. Yet under automation may burden teachers with tasks they would rather not perform, and may limit the degree of personalization they can feasibly achieve in the classroom [27,43]. Furthermore, this balance may depend heavily on the specific teacher tasks under consideration [26,55]. Yet prior work on co-orchestration systems has investigated the design of support for a relatively limited range of teacher tasks (e.g., monitoring student activities during class [45,50]). Furthermore, this research has generally focused on the needs of K-12 teachers, but not students' perspectives, in AI-enhanced classrooms [27,34,43].

The present work builds on prior findings to contribute: (1) an analysis of teacher and student feedback regarding 24 design concepts for human–AI co-orchestration systems, to understand key needs and social boundaries that such systems should be designed to address [13,21,66] and (2) "participatory speed dating": a new variant of the speed dating design method [12] that involves multiple stakeholders in the generation and evaluation of novel technology concepts.

2   Methods

To better understand and validate needs uncovered in prior ethnographic and design research with K-12 students and teachers (e.g., [20,27,43,52,53]), we adopted a participatory speed dating approach. Speed dating is an HCI method for rapidly exploring a wide range of possible futures with users, intended to help researchers/designers elicit unmet needs and probe the boundaries of what particular user populations will find acceptable (which otherwise often remain undiscovered until after a technology prototype has been developed and deployed) [12,42,67]. In speed dating sessions, participants are presented with a number of hypothetical scenarios in rapid succession (e.g., via storyboards) while researchers observe and aim to understand participants' immediate reactions.

Speed dating can lead to the discovery of unexpected design opportunities, when unanticipated needs are uncovered or when anticipated boundaries are discovered not to exist. Importantly, speed dating can often reveal needs and opportunities that may not be observed through field observations or other design activities [12,13,42,67]. For example, Davidoff et al. found that, whereas field observations and interview studies with parents had suggested they might appreciate smart home technologies that automate daily household tasks, a speed dating study revealed that parents strongly rejected the idea of automating certain tasks, such as waking or dressing their children in the morning. These findings led the researchers to dramatically reframe their project—away from creating smart homes that "do people's chores," towards homes that facilitate moments of bonding and connection between busy family members [12,67].

As described in the next subsection, we adapted the speed dating method to enable participants from multiple stakeholder groups (K-12 teachers and students) to reflect on other stakeholders' needs and boundaries, and contribute ideas for new scenarios and technology concepts. We refer to this adaptation as multi–stakeholder "participatory speed dating" (PSD). Like other speed dating approaches, PSD can help to bridge between broad, exploratory design phases and more focused prototyping phases (where associated costs may discourage testing a wide range of ideas) [12,18,67]. However, drawing from Value Sensitive Design [21,66], PSD emphasizes a systematic approach to balancing multiple stakeholder needs and values [38]. Drawing from Participatory Design [36,40,56], in addition to having stakeholders evaluate what is wrong with a proposed concept (which may address other stakeholders' needs), PSD also involves them in generating alternative designs, to address conflicts among stakeholder groups.

2.1 Needs Validation Through Participatory Speed Dating

We conducted PSD sessions one-on-one with 24 middle school teachers and students. To recruit participants, we emailed contacts at eight middle schools and advertised the study on Nextdoor, Craigslist, and through physical fliers. A total of 10 teachers and 14 students, from two large US cities, participated in the study. Sixteen sessions were conducted face-to-face at our institution, and eight were conducted via video conferencing. All participants had experience using some form of adaptive learning software in their classrooms, and 21 participants had used AI tutoring software such as ALEKS [23] or Cognitive Tutor [48].

We first conducted a series of four 30-minute study sessions focused on concept generation, with two teachers and two students. In each session, participants were first introduced to the context for which they would be designing: classes in which students work with AI tutoring software while their teacher uses a real-time co-orchestration tool that helps them help their students (specifically, a set of teacher smart glasses, following [29]). Participants were then shown an initial set of 11 storyboards, each created to illustrate specific classroom challenges uncovered in prior research (e.g., [20,27,47,53]), with multiple challenges hybridized [12,42] into a single storyboard in some cases.[1] For example, prior work suggests that teachers often struggle to balance their desire to implement personalized, mastery-based curricula with their need to keep the class relatively synchronized and "on schedule" [27]. Given this conflict, teachers often opt to manually push students forward in the curriculum if they have failed to master current skills in the ITS by a certain date, despite awareness that this practice may be harmful to students' learning [27,47]. As such, one storyboard (Fig. 1) presented a system that helps teachers make more informed decisions about when to move students ahead (based on the predicted learning benefits of waiting a few more class periods), but without strongly suggesting a particular course of action [27]. Each participant in these initial studies was then encouraged to generate at least one new idea for a storyboard, addressing challenges they personally face in AI-enhanced classrooms as opposed to imagined challenges of others (cf. [13]). To inform ideation, participants also reviewed storyboards generated by other teachers and students in prior study sessions. Participants were provided with editable storyboard templates, in Google Slides [22], and were given the options to generate entirely new concepts for orchestration tool functionality (starting from a blank template) or to generate a variation on an existing concept (starting from a copy of an existing storyboard). In either case, participants generated captions for storyboard panels during the study session, using existing storyboards for reference. Immediately following each session, a researcher then created simple illustrations to accompany each caption.

Following this concept generation phase, we conducted a series of PSD studies with an additional twelve students and eight teachers. Study sessions lasted approximately 60 min. In each session, storyboards were presented in randomized order. Participants were asked to read each storyboard and to describe their initial reactions immediately after reading each one. An interviewer asked follow-up and clarification questions as needed. Participants were then asked to provide an overall summary rating of the depicted technology concept as "mostly positive (I would probably want this feature in my classroom)", "mostly negative (I would probably not want this ...)", or "neutral" [13]. After participants rated each concept, they were asked to elaborate on their reasons for this rating. Before moving on to the next concept, participants were shown notes on reactions to a given concept, thus far, from other stakeholders. Participants were prompted to share their thoughts on perspectives in conflict with their own.

In addition, participants were encouraged to pause the speed dating process at any point, if they felt inspired to write down an idea for a new storyboard. Each time a participant generated a new idea for a storyboard, this storyboard was included in the set shown to the next participant. However, if a participant

---

[1] Please refer to https://tinyurl.com/Complementarity-Supplement for the full set of storyboards and more detailed participant demographics.

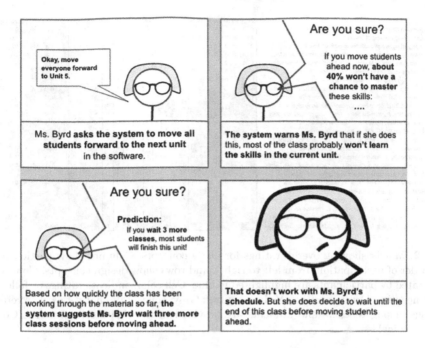

Fig. 1. Example of a storyboard addressing challenges raised in prior research.

saw an existing storyboard that they felt captured the same concept as one they had generated, the new, "duplicate" storyboard was not shown to subsequent participants (cf. [27]). In cases of disagreement between stakeholder groups, generating new storyboard ideas provided an opportunity for students and teachers to try to resolve these disagreements. For example, as shown in Fig. 2, the generation of concepts E.3 through E.6 over time represents a kind of "negotiation" between teachers and students, around issues of student privacy, transparency, and control. This phase of the study yielded a total of seven new storyboards.

3 Results

In the following subsections, we discuss teachers' and students' top five most and least preferred design concepts, according to the average overall ratings among those who saw a given concept [13]. To analyze participant feedback regarding each concept, we worked through transcriptions of approximately 19 h of audio to synthesize findings using two standard methods from Contextual Design: interpretation sessions and affinity diagramming [8,24]. High-level themes that emerged are briefly summarized below, organized by design concept. The most preferred concepts are presented in Sect. 3.1, and the least preferred are in Sect. 3.2. Within each subsection, preferences among teachers are presented first, followed by student preferences and those shared between teachers and students. Teacher participants are identified with a "T," and students are identified with an "S."

Concept	Teacher avg.	Student avg.
[A.1] Ranking Students by Need for Teacher Help	0.88	0.50
[A.2] Explaining Ranking of Students	0.13	0.08
[B] Suggesting Which Students to Help and How to Help	0.75	0.25
[C] Helping Teachers Mediate between Stu. and Student Models	-0.38	0.33
[D] Predicting Time to Mastery to Support Teacher Scheduling	0.63	0.33
[E.1] Alerting Teachers to Student Frustration, Misbehavior, ...	0.88	0.08
[E.2] Providing Automated Motivational Prompts ...	-0.13	0.00
[E.3] Allowing Stu. to Hide (All) of their Analytics from Teachers	-0.75	-0.30
[E.4] Notifying Stu. When the System has Alerted their Teacher	0.33	-0.13
[E.5] Allowing Students to Hide Emotion-related Analytics ...	0.00	0.60
[E.6] Asking Stu. Permission before Revealing (Some) Analytics ...	0.00	1.00
[F.1] "Invisible Hand Raises" and Teacher Reminders	0.88	0.75
[F.2] Suggesting Peer Tutors to Support Teachers ...	0.75	0.25
[G] Providing Teacher with Suggested "Conversation Starters" ...	0.50	0.33
[H.1] Enabling Students to Request Not to be Helped	0.38	0.83
[H.2] Enabling Stu. to Ask the Whole Class Anonymous Questions	0.13	0.67
[H.3] Student–System Joint Control Over Selection of Peer Tutors	0.63	0.89
[H.4] Showing Students Potential Peer Tutors' Skill Mastery	-0.25	-0.50
[I.1] Real-time Positive Feedback on Teacher Explanations.	0.75	0.58
[I.2] Real-time Negative Feedback on Teacher Explanations.	1.00	0.58
[J] Notifying Teachers about Stu. they Have Not Visited Recently	0.38	0.83
[K] Listening in on Teacher Help-giving to Improve AI Tutor's Hints	0.75	0.25
[L] Teacher-controlled Shared Displays to Foster Competition	0.88	0.67
[M] Allowing Parents to Monitor their Child's Behavior During Class	-0.17	-0.50

Fig. 2. Matrix showing overall ratings for all 24 concepts. Columns show participants (in order of participation, from left to right), and rows show design concepts. Concepts generated by participants are highlighted in blue. Cell colors indicate ratings as follows: Red: negative; Green: positive; Yellow: neutral; Grey: concept did not yet exist. Average ratings among teachers and students are provided in the rightmost columns. (Color figure online)

3.1 Most Preferred Design Concepts

Most Preferred Among Teachers

[I.2] Real-time Feedback on Teacher Explanations. Consistent with findings from prior design research [26,27], the most popular concept among teachers was a system that would provide them with constructive feedback, after helping a student, on the effectiveness of their own explanations. As one teacher explained, *"Usually our only chance to get [fast] feedback is, you ask [...] the kids [and] they just say, 'Oh, yeah, I get it,' when they don't really get it"* (T7).

[A.1] Ranking Students by their Need for Teacher Help. Another popular concept among teachers was a system that would allow them to see, at a glance, a visual ranking of which students most need the teacher's help at a given moment [27,49]. One teacher commented, *"Yeah. Welcome to teaching every day [...] trying to go to those kids that are [struggling] most"* (T5). However, several other teachers emphasized that such a ranking would be much more useful if it took into account the kind and extent of teacher help that would likely be needed to address a particular student issue. For example, T1 noted, *"If I could see how much time it would take [to help] I would start with the kids who I could get [moving again quickly] and then I'd spend more time with the other kids. [But] if it's a kid that I know is gonna get completely frustrated [...then I] wanna [go to] that kid first no matter what."* This concept was also generally well received by students. As one student put it, *"sometimes you just can't ask [for help] because you don't even know what [you're struggling with], and so it would just [be] hard to explain it to the teacher"* (S7). At the same time, as discussed

below, multiple students expressed preferences for systems that can support *students* in recognizing when (and with what) they need to ask the teacher for help, rather than always having the system alert the teacher on their behalf (cf. [51]).

[E.1] Alerting Teachers to Student Frustration, Misbehavior, or "Streaks". Consistent with [27], teachers were enthusiastic about a concept that would allow them to see real-time analytics about student frustration, misbehavior (e.g,. off-task behavior or gaming the system [5,58]), or high recent performance in the software. They felt that having access to this information could help them make more informed decisions about whom to help first and how best to help particular students (e.g., comforting a student or offering praise). Yet students reported finding aspects of this concept upsetting. While students generally liked the idea that the system would inform the teacher when they needed help, students often perceived real-time alerts about emotions like frustration as *"really creepy"* (S9) and alerts about misbehavior as *"basically the AI ratting out the child"* (S3).

[L] Teacher-controlled Shared Displays to Foster Competition. Finally, a popular concept among teachers was a system that would allow them to transition the classroom between different "modes," to help regulate students' motivation (cf. [1,43]). This system would allow teachers to switch the class into a "competitive mode," in which students would be shown a leaderboard of comparable classrooms in their school district and challenged to move their class to the top. Teachers expected that such a feature could work extremely well with some groups of students, while backfiring and potentially serving to demotivate others. As such, teachers emphasized the importance of teacher control and discretion.

Most Preferred Among Students

[E.6] Asking Students' Permission before Revealing (Some) Analytics to Teachers. In response to one of teachers' most preferred design concepts (*[E.1]*), students generated multiple new storyboards that preserved the idea of real-time teacher alerts, but provided students with greater control over alert policies. One of these emerged as the most popular design concept among students: a system that asks students' permission, on a case-by-case basis, before presenting certain kinds of information to the teacher on a student's behalf. Students and teachers were generally in agreement that an AI system should ask students' permission before alerting teachers about affective states, such as frustration. In this scenario, if a student opted not to share affective analytics with their teacher, the system might privately suggest other ways for students to regulate their own emotions. Interestingly, one student suggested that if a student opted to share their affect with the teacher, the system should also ask the student to specify *"How do you want the teacher to react? [...] Help you [in person]? Help you on the computer?"* (S12). This student noted that sometimes, they just want their teacher to *"know how I'm feeling,"* but do not actually want them to take action.

[H.3] Student–System Joint Control Over Selection of Peer Tutors. Whereas teachers often expressed that they know which groups of their students will not work well together, this did not align with students' perceptions of their own teachers. In contrast to teacher-generated concepts where teachers and AI worked together to match peer tutors and tutees (cf. [43]), the second most popular

concept among students was a student-generated storyboard that gave students the final say over peer matching decisions. In this storyboard, the system sends struggling students a list of suggested peer tutors, based on these students' estimated tutoring abilities (cf. [57]) and knowledge of relevant skills. Students could then send help requests to a subset of peers from this list who they would feel comfortable working with. Those invited would then have the option to reject a certain number of requests. Some students suggested that it would also be useful to have the option to accept but delay another student's invitation if they want to help but do not want to disrupt their current flow.

[H.1] Enabling Students to Request Not to be Helped. Another of the most popular concepts among students was a system that, upon detecting that a student seems to be wheel-spinning [7,31], would notify the student to suggest that they try asking their teacher or classmates for help. The system would then only notify the teacher that the student is struggling if the student both ignored this suggestion and remained stuck after a few minutes. By contrast, some teachers expressed that they would want the system to inform them *immediately* if a student was wheel-spinning: *"They shouldn't just get the option to keep working on their own, because honestly it hasn't been working"* (T5). Some students and teachers suggested a compromise: *"the AI should inform the teacher right away [...] that it suggested [asking for help] but the kid did something else"* (T7).

[J] Notifying Teachers of Students they Have Not Visited Recently. Finally, a popular concept among students was a system that would track a teachers' movement during class and occasionally highlight students they may be neglecting (cf. [4,19]). Several students noted that even when they are doing well on their own, they feel motivated when their teacher remembers to check in with them. Most teachers responded positively to this concept as, *"sometimes you forget about the kids that work well on their own, but sometimes those kids actually need help and don't raise their hands"* (T6). However, a few teachers perceived this system as overstepping bounds and inappropriately judging them: *"It's just too much in my business now. You better be quiet and give me a break"* (T4).

Most Preferred Among Both Teachers and Students

[F.1] "Invisible Hand Raises" and Teacher Reminders. A concept popular with both teachers and students was a system that would allow students to privately request help from their teacher by triggering an "invisible hand raise" that only the teacher could see. To preserve privacy, this system would also allow teachers to silently acknowledge receipt of a help request. After a few minutes, the teacher would receive a light reminder if they had not yet helped a student in their queue. S7 noted, *"I don't actually like asking questions since I'm supposed to be, like, 'the smart one' ...which I'm not. So I like the idea of being able to ask a question without [letting] others know."* Similarly, teachers suspected that students would request help more often if they had access to such a feature [26,53].

3.2   Least Preferred Design Concepts

Least Preferred Among Teachers
[C] Helping Teachers Mediate between Students and their Student Models. To our surprise, although prior field research [30] had suggested teachers might find it desirable to serve as "final judges" in cases where students wished to contest their student models (e.g., skill mastery estimates) [11], this was one of the least popular design concepts among teachers. Students generally viewed teacher-in-the-loop mediation desirable, since *"I feel like the teacher knows the student better, not the software"* (S9). However, teachers generally did not view this as an efficient use of their time: *"I would just trust the tutor on this one"* (T3). Furthermore, some teachers expressed concerns that from a student's perspective this concept *"pit[s] one teacher against the other, if you consider the AI as a kind of teacher"* (T1), and instead suggested having the system assign a targeted quiz if a student wants to demonstrate knowledge of particular skills (cf. [11]).

Least Preferred Among Students
[E.4] Notifying Students When the System has Automatically Alerted their Teacher. A teacher-generated concept intended to provide students with greater transparency into the analytics being shared about them was among those least popular with students overall. Interestingly, while students valued having more control over the information visible to their teachers, they generally did not want greater transparency into aspects of the system that were outside of their control (cf. [33]): *"That would make me really anxious [...] If it's not asking students' [permission], I don't think they should know about it"* (S10).

Least Preferred Among Both Teachers and Students
[E.3] Allowing Students to Hide (All) of their Analytics from Teachers. The least popular concept among teachers, and the third least popular among students, was a privacy feature that would enable individual students to prevent their AI tutor from sharing real-time analytics with their teacher. This was a student-generated concept intended to mitigate the "creepiness" of having their teacher "surveil" students' activities in real-time. Yet as discussed in Sect. 3.1, overall students felt that it should only be possible for students to hide certain kinds of analytics (e.g., inferred emotional states), *"but if the AI sees a student is really, really struggling [...] I don't think there should be that blanket option"* (S4).

[H.4] Showing Students Potential Peer Tutors' Skill Mastery. Consistent with prior research (e.g., [26]), teachers and students responded negatively to a student-generated concept that made individual students' skill mastery visible to peers. While this concept was intended to help students make informed choices about whom to request as a peer tutor, most teachers and students perceived that the risk of teasing among students outweighed the potential benefits.

[M] Allowing Parents to Monitor their Child's Behavior During Class. Somewhat surprisingly, T3 generated the concept of a remote monitoring system that would allow parents to *"see exactly what [their child is] doing at any moment in time."*, so that *"if a kid's misbehaving, their parent can see the teacher's trying [their] best"* (cf. [9,62]). While this concept resonated with one other teacher,

student and teacher feedback on this concept generally revealed an attitude that to create a safe classroom environment, *"we have to [be able to] trust that data from the classroom stays in the classroom"* (S11). Teachers shared concerns that data from their classrooms might be interpreted out of context by administrators: *"I don't ever want to be judged as a teacher [because] I couldn't make it to every student, if every kid's stuck that day. [But] using that data [as a teacher] is very useful"* (T5). Students shared fears that, depending on the data shared, parents or even future employers might use classroom data against them.

[E.2] Providing Automated Motivational Prompts to Frustrated Students. Finally, among the concepts least popular with both teachers and students was a system that automatically provides students with motivational prompts when it detects they are getting frustrated [16,64]. While teachers liked the idea of incorporating gamification elements to motivate students (cf. [35,62]), providing motivational messages was perceived as *"trying to [do] the teacher's job"* (T1). Similarly, several students indicated strongly that they would prefer these kinds of messages to come from an actual person, if at all. S8 said, *"I would just get more annoyed if the AI tried something like that"*, and S11 suggested *"No emotional responses, please. That feels just [...] not genuine. If it's from the AI it should be more analytical, like just [stick to] facts."*

4   Discussion, Conclusions, and Future Work

If new AI systems are to be well-received in K-12 classrooms, it is critical that they support the needs and respect the boundaries of both teachers and students. We have introduced "participatory speed dating" (PSD): a variant of the speed dating design method that involves multiple stakeholders in the iterative generation and evaluation of new technology concepts. Using PSD, we sampled student and teacher feedback on 24 design concepts for systems that integrate human and AI instruction—an important but underexplored area of AIED research.

Overall, we found that teachers and students aligned on needs for "hidden" student–teacher communication channels during class, which enable students to signal help-need or other sensitive information without losing face to their peers. More broadly, both teachers and students expressed nuanced needs for student privacy in the classroom, where it is possible to have "too little," "too much," or the wrong forms of privacy (cf. [41]). However, students and teachers did not always perceive the same needs. As discussed in Sect. 3.1, some of students' highest-rated concepts related to privacy and control were unpopular among teachers. Additional disagreements arose when teachers and students had different expectations of the roles of teachers versus AI agents and peer tutors in the classroom.

Interestingly, while students' expressed desires for transparency, privacy, and control over classroom AI systems extend beyond what is provided by existing systems [9,11,29,60], these desires are also more nuanced than commonly captured in theoretical work [10,59,61]. For example, we found that while students were uncomfortable with AI systems sharing certain kinds of personal analytics with their teacher without permission, they rejected design concepts that grant students full control over these systems' sharing policies. These findings indicate an important role for empirical, design research approaches to complement critical and policy-oriented research on AI in education (cf. [33,41,63]).

In sum, the present work provides tools and and early insights to guide the design of more effective and desirable human–AI partnerships for K-12 education. Future AIED research should further investigate teacher and student needs uncovered in the present work via rapid prototyping in live K-12 classrooms. While design methods such as PSD are critical in guiding the initial development of novel prototypes, many important insights surface only through deployment of functional systems in actual, social classroom contexts [30,42,53]. An exciting challenge for future research is to develop methods that extend the advantages of participatory and value-sensitive design approaches (e.g., [39,56,66]) to later stages of the AIED design cycle. Given the complexity of data-driven AI systems [17,26,66], fundamentally new kinds of design and prototyping methods may be needed to enable non-technical stakeholders to remain meaningfully involved in shaping such systems, even as prototypes achieve higher fidelity.

Acknowledgements. This work was supported by IES Grants R305A180301 and R305B150008. The opinions expressed are those of the authors and do not represent the views of IES or the U.S. ED. Special thanks to all participating teachers and students.

References

1. Alavi, H.S., Dillenbourg, P.: An ambient awareness tool for supporting supervised collaborative problem solving. IEEE Trans. Learn. Technol. **5**(3), 264–274 (2012)
2. Aleven, V., Roll, I., McLaren, B.M., Koedinger, K.R.: Help helps, but only so much: research on help seeking with intelligent tutoring systems. Int. J. Artif. Intell. Educ. **26**(1), 205–223 (2016)
3. Amershi, S., et al.: Guidelines for human-AI interaction. ACM, May 2019. https://www.microsoft.com/en-us/research/publication/guidelines-for-human-ai-interaction/
4. An, P., Bakker, S., Ordanovski, S., Taconis, R., Eggen, B.: Classbeacons: designing distributed visualization of teachers-physical proximity in the classroom. In: Proceedings of the Twelfth International Conference on Tangible, Embedded, and Embodied Interaction, pp. 357–367. ACM (2018)
5. Baker, R., Walonoski, J., Heffernan, N., Roll, I., Corbett, A., Koedinger, K.: Why students engage in "gaming the system" behavior in interactive learning environments. J. Interact. Learn. Res. **19**(2), 185–224 (2008)
6. Baker, R.S.: Stupid tutoring systems, intelligent humans. Int. J. Artif. Intell. Educ. **26**(2), 600–614 (2016)
7. Beck, J.E., Gong, Y.: Wheel-spinning: students who fail to master a skill. In: Lane, H.C., Yacef, K., Mostow, J., Pavlik, P. (eds.) AIED 2013. LNCS (LNAI), vol. 7926, pp. 431–440. Springer, Heidelberg (2013). https://doi.org/10.1007/978-3-642-39112-5_44
8. Beyer, H., Holtzblatt, K.: Contextual Design: Defining Customer-Centered Systems. Elsevier, San Francisco (1997)

9. Broderick, Z., O'Connor, C., Mulcahy, C., Heffernan, N., Heffernan, C.: Increasing parent engagement in student learning using an intelligent tutoring system. J. Interact. Learn. Res. **22**(4), 523–550 (2011)

10. Bulger, M.: Personalized learning: the conversations we're not having. Data Soc. **22** (2016)

11. Bull, S., Kay, J.: Smili: a framework for interfaces to learning data in open learner models, learning analytics and related fields. Int. J. Artif. Intell. Educ. **26**(1), 293–331 (2016)

12. Davidoff, S., Lee, M.K., Dey, A.K., Zimmerman, J.: Rapidly exploring application design through speed dating. In: Krumm, J., Abowd, G.D., Seneviratne, A., Strang, T. (eds.) UbiComp 2007. LNCS, vol. 4717, pp. 429–446. Springer, Heidelberg (2007). https://doi.org/10.1007/978-3-540-74853-3_25

13. Dillahunt, T.R., Lam, J., Lu, A., Wheeler, E.: Designing future employment applications for underserved job seekers: a speed dating study. In: Proceedings of the 2018 Designing Interactive Systems Conference, pp. 33–44. ACM (2018)

14. Dillenbourg, P.: Design for classroom orchestration. Comput. Educ. **69**, 485–492 (2013)

15. Dillenbourg, P., Prieto, L.P., Olsen, J.K.: Classroom orchestration (2018)

16. D'Mello, S., Picard, R.W., Graesser, A.: Toward an affect-sensitive autotutor. IEEE Intell. Syst. **22**(4), 53–61 (2007)

17. Dove, G., Halskov, K., Forlizzi, J., Zimmerman, J.: UX design innovation: challenges for working with machine learning as a design material. In: Proceedings of the 2017 CHI Conference on Human Factors in Computing Systems, pp. 278–288. ACM (2017)

18. Dow, S.P., Glassco, A., Kass, J., Schwarz, M., Schwartz, D.L., Klemmer, S.R.: Parallel prototyping leads to better design results, more divergence, and increased self-efficacy. ACM Trans. Comput. Hum. Interact. (TOCHI) **17**(4), 18 (2010)

19. Echeverria, V., Martinez-Maldonado, R., Power, T., Hayes, C., Shum, S.B.: Where is the nurse? towards automatically visualising meaningful team movement in healthcare education. In: Penstein Rosé, C., et al. (eds.) AIED 2018. LNCS (LNAI), vol. 10948, pp. 74–78. Springer, Cham (2018). https://doi.org/10.1007/978-3-319-93846-2_14

20. Feng, M., Heffernan, N.T.: Informing teachers live about student learning: reporting in the assistment system. Technol. Instr. Cogn. Learn. **3**(1/2), 63 (2006)

21. Friedman, B.: Value-sensitive design. Interactions **3**(6), 16–23 (1996)

22. Google: Google slides (2019). http://slides.google.com

23. Hagerty, G., Smith, S.: Using the web-based interactive software aleks to enhance college algebra. Math. Comput. Educ. **39**(3) (2005)

24. Hanington, B., Martin, B.: Universal Methods of Design: 100 Ways to Research Complex Problems, Develop Innovative Ideas, and Design Effective Solutions. Rockport Publishers, Beverly (2012)

25. Heer, J.: Agency plus automation: designing artificial intelligence into interactive systems. Proc. Nat. Acad. Sci. **116**(6), 1844–1850 (2019). https://doi.org/10.1073/pnas.1807184115, https://www.pnas.org/content/early/2019/01/29/1807184115

26. Holstein, K., Hong, G., Tegene, M., McLaren, B.M., Aleven, V.: The classroom as a dashboard: co-designing wearable cognitive augmentation for k-12 teachers. In: Proceedings of the 8th International Conference on Learning Analytics and Knowledge, pp. 79–88. ACM (2018)

27. Holstein, K., McLaren, B.M., Aleven, V.: Intelligent tutors as teachers' aides: exploring teacher needs for real-time analytics in blended classrooms. In: Proceedings of the Seventh International Learning Analytics & Knowledge Conference, pp. 257–266. ACM (2017)

28. Holstein, K., McLaren, B.M., Aleven, V.: Informing the design of teacher awareness tools through causal alignment analysis. In: International Conference of the Learning Sciences, pp. 104–111 (2018)

29. Holstein, K., McLaren, B.M., Aleven, V.: Student learning benefits of a mixed-reality teacher awareness tool in AI-enhanced classrooms. In: Penstein Rosé, C., et al. (eds.) AIED 2018. LNCS (LNAI), vol. 10947, pp. 154–168. Springer, Cham (2018). https://doi.org/10.1007/978-3-319-93843-1_12

30. Holstein, K., McLaren, B.M., Aleven, V.: Co-designing a real-time classroom orchestration tool to support teacher-AI complementarity. J. Learn. Anal. (Under review)

31. Kai, S., Almeda, M.V., Baker, R.S., Heffernan, C., Heffernan, N.: Decision tree modeling of wheel-spinning and productive persistence in skill builders. JEDM—J. Educ. Data Min. **10**(1), 36–71 (2018)

32. Kulik, J.A., Fletcher, J.: Effectiveness of intelligent tutoring systems: a meta-analytic review. Rev. Educ. Res. **86**(1), 42–78 (2016)

33. Lee, M.K., Baykal, S.: Algorithmic mediation in group decisions: fairness perceptions of algorithmically mediated vs. discussion-based social division. In: CSCW, pp. 1035–1048 (2017)

34. van Leeuwen, A., et al.: Orchestration tools for teachers in the context of individual and collaborative learning: what information do teachers need and what do they do with it? In: International Society of the Learning Sciences, Inc.[ISLS] (2018)

35. Long, Y., Aman, Z., Aleven, V.: Motivational design in an intelligent tutoring system that helps students make good task selection decisions. In: Conati, C., Heffernan, N., Mitrovic, A., Verdejo, M.F. (eds.) AIED 2015. LNCS (LNAI), vol. 9112, pp. 226–236. Springer, Cham (2015). https://doi.org/10.1007/978-3-319-19773-9_23

36. Luckin, R., Clark, W.: More than a game: the participatory design of contextualised technology-rich learning experiences with the ecology of resources. J. e-Learn. Knowl. Soc. **7**(3), 33–50 (2011)

37. Ma, W., Adesope, O.O., Nesbit, J.C., Liu, Q.: Intelligent tutoring systems and learning outcomes: a meta-analysis. J. Educ. Psychol. **106**(4), 901 (2014)

38. Miller, J.K., Friedman, B., Jancke, G., Gill, B.: Value tensions in design: the value sensitive design, development, and appropriation of a corporation's groupware system. In: Proceedings of the 2007 International ACM Conference on Supporting Group Work, pp. 281–290. ACM (2007)

39. Mitchell, V., Ross, T., May, A., Sims, R., Parker, C.: Empirical investigation of the impact of using co-design methods when generating proposals for sustainable travel solutions. CoDesign **12**(4), 205–220 (2016)

40. Muller, M.J., Kuhn, S.: Participatory design. Commun. ACM **36**(6), 24–28 (1993)

41. Mulligan, D.K., King, J.: Bridging the gap between privacy and design. U. Pa. J. Const. L. **14**, 989 (2011)

42. Odom, W., Zimmerman, J., Davidoff, S., Forlizzi, J., Dey, A.K., Lee, M.K.: A fieldwork of the future with user enactments. In: Proceedings of the Designing Interactive Systems Conference, pp. 338–347. ACM (2012)

43. Olsen, J.: Orchestrating Combined Collaborative and Individual Learning in the Classroom. Ph.D. thesis, Carnegie Mellon University (2017)

44. Pane, J.F.: Informing progress: insights on personalized learning implementation and effects. RAND (2017)
45. Prieto, L.P., Holenko Dlab, M., Gutiérrez, I., Abdulwahed, M., Balid, W.: Orchestrating technology enhanced learning: a literature review and a conceptual framework. Int. J. Technol. Enhanced Learn. **3**(6), 583–598 (2011)
46. Prieto Santos, L.P., et al.: Supporting orchestration of blended CSCL scenarios in distributed learning environments (2012)
47. Ritter, S., Yudelson, M., Fancsali, S.E., Berman, S.R.: How mastery learning works at scale. In: Proceedings of the Third (2016) ACM Conference on Learning@ Scale, pp. 71–79. ACM (2016)
48. Ritter, S., Anderson, J.R., Koedinger, K.R., Corbett, A.: Cognitive tutor: applied research in mathematics education. Psychon. Bull. Rev. **14**(2), 249–255 (2007)
49. Ritter, S., Yudelson, M., Fancsali, S., Berman, S.R.: Towards integrating human and automated tutoring systems. In: EDM, pp. 626–627 (2016)
50. Rodriguez Triana, M.J., et al.: Monitoring, awareness and reflection in blended technology enhanced learning: a systematic review. Int. J. Technol. Enhanced Learn. **9**, 126–150 (2017)
51. Roll, I., Aleven, V., McLaren, B.M., Koedinger, K.R.: Improving students' help-seeking skills using metacognitive feedback in an intelligent tutoring system. Learn. Instr. **21**(2), 267–280 (2011)
52. Schofield, J.W.: Psychology: computers and classroom social processes-a review of the literature. Soc. Sci. Comput. Rev. **15**(1), 27–39 (1997)
53. Schofield, J.W., Eurich-Fulcer, R., Britt, C.L.: Teachers, computer tutors, and teaching: the artificially intelligent tutor as an agent for classroom change. Am. Educ. Res. J. **31**(3), 579–607 (1994)
54. Sharplco, M.: Shared orchestration within and beyond the classroom. Comput. Educ. **69**, 504–506 (2013)
55. Sheridan, T.B.: Function allocation: algorithm, alchemy or apostasy? Int. J. Hum.-Comput. Stud. **52**(2), 203–216 (2000)
56. Trischler, J., Pervan, S.J., Kelly, S.J., Scott, D.R.: The value of codesign: the effect of customer involvement in service design teams. J. Serv. Res. **21**(1), 75–100 (2018)
57. Walker, E., Rummel, N., Koedinger, K.R.: Adaptive intelligent support to improve peer tutoring in algebra. Int. J. Artif. Intell. Educ. **24**(1), 33–61 (2014)
58. Walonoski, J.A., Heffernan, N.T.: Prevention of off-task gaming behavior in intelligent tutoring systems. In: Ikeda, M., Ashley, K.D., Chan, T.-W. (eds.) ITS 2006. LNCS, vol. 4053, pp. 722–724. Springer, Heidelberg (2006). https://doi.org/10.1007/11774303_80
59. Watters, A.: The Monsters of Education Technology. Smashwords Edition (2014)
60. Wetzel, J., et al.: A preliminary evaluation of the usability of an AI-infused orchestration system. In: Penstein Rosé, C., et al. (eds.) AIED 2018. LNCS (LNAI), vol. 10948, pp. 379–383. Springer, Cham (2018). https://doi.org/10.1007/978-3-319-93846-2_71
61. Williamson, B.: Calculating children in the dataveillance school: personal and learning analytics. In: Surveillance Futures, pp. 62–90. Routledge, London (2016)
62. Williamson, B.: Decoding classdojo: psycho-policy, social-emotional learning and persuasive educational technologies. Learn. Media Technol. **42**(4), 440–453 (2017)
63. Wong, R.Y., Mulligan, D.K., Van Wyk, E., Pierce, J., Chuang, J.: Eliciting values reflections by engaging privacy futures using design workbooks (2017)
64. Woolf, B., Burleson, W., Arroyo, I., Dragon, T., Cooper, D., Picard, R.: Affect-aware tutors: recognising and responding to student affect. Int. J. Learn. Technol. **4**(3–4), 129–164 (2009)

65. Yacef, K.: Intelligent teaching assistant systems. In: 2002 Proceedings of International Conference on Computers in Education, pp. 136–140. IEEE (2002)
66. Zhu, H., Yu, B., Halfaker, A., Terveen, L.: Value-sensitive algorithm design: method, case study, and lessons. Proc. ACM Hum. Comput. Interact. **2**(CSCW), 194:1–194:23 (2018). https://doi.org/10.1145/3274463, http://doi.acm.org/10.1145/3274463
67. Zimmerman, J., Forlizzi, J.: Speed dating: providing a menu of possible futures. She Ji: J. Des. Econ. Innov. **3**(1), 30–50 (2017)

The Case of Self-transitions
in Affective Dynamics

Shamya Karumbaiah[(✉)], Ryan S. Baker, and Jaclyn Ocumpaugh

University of Pennsylvania, Philadelphia, USA
shamya@upenn.edu, ryanshaunbaker@gmail.com,
jlocumpaugh@gmail.com

Abstract. Affect dynamics, the study of how affect develops and manifests over the course of learning, has become a popular area of research in learning analytics. Despite some shared metrics and research questions, researchers in this area have some differences in how they pre-process the data for analysis [17]. Specifically, researchers differ in how they treat cases where a student remains in the same affective state in two successive observations, referred to as *self-transitions*. While most researchers include these cases in their data, D'Mello and others have argued over the last few years that these cases should be removed prior to analysis. While this choice reflects the intended focus in their research paradigm on the transitions out of an affective state, this difference in data preprocessing changes the meaning of the metric used. For around a decade, the community has used the metric L to evaluate the probability of transitions in affect. L is largely believed to have a value of 0 when a transition is at chance, and this is true for the original use of the metric. However, this paper provides mathematical evidence that this metric does not have a value of 0 at chance if self-transitions are removed. This shift is problematic because previously published statistical analyses comparing L values to the value at chance have used the wrong value, incorrectly producing lowered p values and in many cases reporting transitions as significantly more likely than chance when they are actually less frequent.

Keywords: Affect dynamics · L statistics · Student affect · Engagement · Self-transitions · Data preprocessing

1 Introduction

In a data mining pipeline, data preprocessing is often considered a step separate from analysis. Data preprocessing steps like cleaning, sampling, normalization/ standardization, and imputation are performed to clean and consolidate the collected data into a format ready for input into an analytical technique. The choices made during pre-processing may, in many cases, have relatively minor implications on the analysis to follow. But in some cases, a seemingly small, theoretically-justified preprocessing step can change the meaning of the metric used in the analysis. In this paper, we present one such example of a misinterpreted metric, D'Mello, Taylor, and Graesser's [2007] L, that was used in affect dynamics research for over ten years, in over a dozen published studies [1–5, 8–11, 13, 14, 17, 19–21, 23–25]. However, a closer look at the

S. Isotani et al. (Eds.): AIED 2019, LNAI 11625, pp. 172–181, 2019.
https://doi.org/10.1007/978-3-030-23204-7_15

way data is pre-processed in some of these studies reveals how it changes the meaning of the metric. We discuss the implication of this new finding on the results of these studies, some of which appear to have reported results in the wrong direction due to this shift.

As mentioned, this work occurs in the domain of affect dynamics [18] - an area of research that studies how students transition between different emotional states, in this case in a learning setting. Based on increasing evidence that student affect is associated with learning and long-term outcomes [7, 22], affect has been used to understand the design of a learning environment [16] and affect-sensitive interventions have been designed and tested in some systems [12, 15]. Understanding how affect manifests over time is useful when designing real-time educational interventions that work with natural patterns and transitions in affect.

Perhaps the mostly widely used metric in research on affect dynamics is D'Mello, Taylor, and Graesser's [2007] L statistic. It measures whether a transition from one affective state to another is more likely than the second state's base rate. Approximately 20 studies have used this statistic to study the transitions between different emotional states of interest [17].

During data preprocessing, one key methodological question is whether self-transitions (when a student remains in the same affective state both before and after) should be considered or excluded from calculations, with most of the studies by D'Mello and his colleagues excluding self-transitions [3–5, 8–11, 17] and most of the work by other research groups including them [1, 2, 13, 14, 19–21, 23–25]. A recent review found that the exclusion of self-transitions leads to a higher proportion of transitions being found to be more likely than chance [17]. If valid, this result would suggest that it is beneficial to exclude self-transitions to increase statistical power. However, in one recent paper that excluded self-transitions, the researchers reported all the transitions into engaged concentration were more likely than chance [5], a mathematically impossible result. Further investigation with the original authors of this paper indicated that this result was not a typing error, raising questions about the validity and interpretation of this widely-used metric. In this paper, we extend prior work by explicitly investigating the mathematical basis of the L statistic, both when self-transitions are included and when they are excluded, to see how this impossible result was obtained and what its implications are for the use of this statistic.

2 L Statistics and Affect Dynamics Analysis

Given an affect sequence, the L statistic [10] calculates the likelihood that an affective state (*prev*) will transition to a subsequent (*next*) state, given the base rate of the next state occurring.

$$L(prev \rightarrow next) = \frac{P(next \mid prev) - P(next)}{1 - P(next)} \tag{1}$$

The expected probability, $P(next)$ for an affective state is the percentage of times that the state occurred as a next state. Thus, the first affective state in the sequence of a student will be excluded from this calculation since this state cannot take the role of a next state. Similarly, the calculation of the *prev* state excludes the last state in the sequence. The conditional probability, *P(next | prev)* is given by:

$$P(next \mid prev) = \frac{Count(prev \rightarrow next)}{Count(prev)} \tag{2}$$

where *Count(prev → next)* is the number of times the *prev* state transitioned to the next state, and *Count(prev)* is the number of times the state in *prev* occurred as the previous state.

There are several special cases in the calculation of *L* where there is no consensus in the literature on how to perform the calculation, and [17] has recommended the following treatment:

1. When any affective state (A_n) being considered in a given study is not present for a given student's observation period:
   a. Transitions to A_n do not occur for that student. In this case, *P(next)* = 0 and *P (next | prev)* = 0, and thus, *L* = 0.
   b. Transitions from A_n also do not occur. In this case, we do not know what affective state would have followed A_n, and thus, *L* = undefined.
2. Following from case 1, if a student remains in a single affective state (A_s) throughout an observation period, all other affective states being considered in the study behave as A_n. However, the calculations differ based on whether or not the self-transitions are included.
   a. If self-transitions are included in the analyses:
      (1) Transitions from A_s to any other affective state (e.g., A_n) do not occur, and therefore, as in 1a, *L* = 0 for any transition out of A_s.
      (2) Transitions to A_s from any other affective state (e.g., A_n) do not occur, and therefore, as in 1b, *L* = undefined.
   b. If self-transitions are discarded in the analyses, an affect sequence consisting of repeated observations of the same affective category is reduced to a single observation of that affective state. In this case, no transitions occur, and therefore *L* = undefined for all possible sequences being studied.

It is not always clear how these special cases are treated in past research. In this study, we follow [17] 's definition of *L* as outlined above.

The value of *L* varies from $-\infty$ to 1. D'Mello and Graesser [8] state in page 7 that "the sign and the magnitude of *L* is intuitively understandable as the direction and size of the association". As has been expanded in subsequent papers [1, 3–5, 8, 9, 11, 13, 14, 17, 19–21, 23–25], *L* = 0 is treated as chance, while *L* > 0 and *L* < 0 are treated as transitions that are more likely or less likely (respectively) than chance. To perform affect dynamics analysis across all students in an experiment, first the *L* value for each affect combination is calculated individually per student. Next, as [8, pg. 7] recommends, the researcher runs "one-sample [two-tailed] t-tests to test whether likelihoods were significantly greater than or equivalent to zero (no relationship between immediate and next state)", on the sample of individual student L values for each transition. Lastly, a Benjamini-Hochberg post-hoc correction procedure is often used [1, 5, 17, 21, 23–25] to control for false positive results since the set of hypotheses involves multiple comparisons.

3 Analysis

This straightforward procedure seems quite logical, but the result seen in [5], where, after removing self-transitions, all transitions into the affective state of engaged concentration were more likely than chance, suggests that something may be wrong. As such, it may be worth examining the mathematical assumptions of this procedure. Specifically, while calculating the transition likelihood from the affective state of M_t (*prev*) to M_{t+1} (*next*), D'Mello explains that, "...if M_{t+1} and M_t are *independent* [emphasis added], then $Pr(M_{t+1}|M_t) = Pr(M_{t+1})$" [8]. However, removing self-transitions breaks the independence between M_{t+1} and M_t as M_{t+1} can now only take values other than Mt. Hence, when self-transitions are excluded, $Pr(M_{t+1}|M_t) \neq Pr(M_{t+1})$.

Another sign of potential problems is found in [8], when that paper draws an analogy between L statistics and Cohen's kappa, saying, "The reader may note significant similarity to Cohen's kappa for agreement between raters and indeed the likelihood metric can be justified in a similar fashion." Although this analogy seems compelling, it is worth noting that there is a striking difference between the range of values the two statistics can take. While the value of L varies from $-\infty$ to 1 [2], the value of Cohen's kappa varies from -1 to 1 [6].

These raise the question: if a transition occurs at chance, and self-transitions are excluded, is the value of L still 0?

3.1 Understanding How Removing Self-transitions Affect L Values

Differences between a calculation based on a transition pattern (L) and a calculation based on a confusion matrix (e.g., Cohen's k) mean that the chance value takes a different value for L than for Cohen's k when transitions are altered. To illustrate, let's take an example with three states, A, B, and C, which allows for a total of nine unique transitions (AA, AB, AC, BB, BA, BC, CC, CA, and CB). We will consider the hypothetical sequence, ABBCAACCBA.

First, let us consider the case where we keep self-transitions within our calculations. Our hypothetical sequences contain all the 9 possible transitions occurring each occurring exactly once. As Table 1 shows, this makes all the possible transition types equally likely (as each occurs at the frequency expected given the base rate of the next state).

Now, consider the transition AB, where A is the *prev* state and B is the next state. The expected probability, $P(next)$, is $P(B\_next)$ i.e., the probability of occurrence of B in the next state.

$$P(next) = P(B\_next) = \frac{2}{6} = 0.33$$

Table 1. *L* statistics calculation for an example sequence of *ABBCAACCBA* when *self-transitions* are included

Transition	Count	P(next | prev)	P(next)	L
A -> A	1	0.33	0.33	0
A -> B	1	0.33	0.33	0
A -> C	1	0.33	0.33	0
B -> A	1	0.33	0.33	0
B -> B	1	0.33	0.33	0
B -> C	1	0.33	0.33	0
C -> A	1	0.33	0.33	0
C -> B	1	0.33	0.33	0
C -> C	1	0.33	0.33	0

The conditional probability, *P(next | prev)*, is *P(B_next | A_prev)*. Note that we are not including the last instance of A as it cannot take the *prev* state in any transition. Using Eq. (2), we have

$$P(next \mid prev) = P(B\_next \mid A\_prev) = \frac{1}{3} = 0.33$$

Substituting in Eq. (1), we get,

$$L(A \rightarrow B) - \frac{0.33 - 0.33}{1 - 0.33} = 0$$

This holds true for all the transitions. Recall that in Table 1, the conditional probability, *P(next | prev)*, is equal to the expected probability, *P(next)*. Thus, when self-transitions are included, all the transition likelihoods in this example take a value of zero, in line with the claim made in [D'Mello, p. 7].

Next, we consider what happens to the *L* value at chance when we omit self-transitions. If we consider the same hypothetical sequence (ABBCAACCBA), only six unique transitions remain ABCACBA. Though this sequence is different, each affective state is equally followed by all affective states. Again, consider the transition AB, where A is the *prev* state and B is the next state. The probability that B is the next state remains the same as it did when self-transitions were included.

$$P(next) = P(B\_next) = \frac{2}{6} = 0.33$$

However, the removal of A -> A sequences results in value of *P(next | prev)* that is different than in the original sequence.

$$P(next \mid prev) = P(B\_next \mid A\_prev) = \frac{1}{2} = 0.5$$

Table 2. L statistics calculation for an example sequence of *ABBCAACCBA* when self-transitions are excluded

Transition	Count	P(next | prev)	P(next)	L
A -> B	1	0.5	0.33	0.25
A -> C	1	0.5	0.33	0.25
B -> A	1	0.5	0.33	0.25
B -> C	1	0.5	0.33	0.25
C -> A	1	0.5	0.33	0.25
C -> B	1	0.5	0.33	0.25

Finally, we obtain.

$$L(A \rightarrow B) = \frac{0.5 - 0.33}{1 - 0.33} = 0.25$$

This value is obtained for all six possible transitions. As we see in Table 2, when all affective states allowed are equally likely as the next state, $L = 0.25$, not 0. Since self-transitions are excluded, a given state can only transition to the other two states as opposed to the three states in total. This contrasts with the claim that $P(next \mid prev) = P(next)$ [D'Mello, p. 7] and increases the conditional probability (i.e., $P(next \mid prev)$) to one out of two while the expected probability (i.e., $P(next)$) remains at two out of three. Thus, for a state space with three states, the chance value of L is at 0.25 instead of 0.

3.2   Redefining Chance L Value

We now generalize our observations above for a state space with n affective states (n > 2) and determine what L value would be expected at chance. Such a state space would have n^2 unique transitions if we include self-transitions, but only has $n^2 - n$ unique transitions if we exclude self-transitions. Thus, at chance, the expected probability is

$$P(next) = \frac{n}{n^2} = \frac{1}{n} \qquad \text{if self-transitions are included}$$
$$P(next) = \frac{n-1}{n^2-n} = \frac{1}{n} \qquad \text{if self-transitions are excluded}$$

However, at chance, the conditional probability is

$$P(next \mid prev) = \frac{1}{n} \qquad \text{if self-transitions are included}$$
$$P(next \mid prev) = \frac{1}{n-1} \qquad \text{if self-transitions are excluded}$$

Plugging these into the original equation of L (Eq. 1), the value of L at chance is

$$L = 0 \qquad \text{if self-transitions are included}$$
$$L = \frac{1}{(n-1)^2} \qquad \text{if self-transitions are excluded}$$

Generally, affect dynamics is studied in terms of the four academic emotions of confusion, frustration, boredom and engaged concentration (emotions like delight and surprise are also sometimes considered, somewhat more rarely). The otherwise unlabeled data segment in the timeline, which occurs when the primary states being investigated are not found, are sometimes given the label NA and considered in the analyses. In such a setup (n = 5), the L value at chance is $L = 0.0625$. For the smallest reasonable state space with n = 3, the L value at chance is at its maximum, 0.25. As the number of affective states observed increases, the impact of the difference between including and excluding self-transitions decreases (Table 3).

Table 3. The value of L that represents chance, for varying state space

n	3	4	5	6	7	8
Chance L	0.25	0.11	0.0625	0.04	0.0277	0.0204

4 Implications

The primary implication of this new finding is on the interpretation of the L value. If an affective dynamics study excludes self-transitions, the threshold to understand the direction of the transition must be set based on the number of affective states studied (see Table 3). For instance, for a study with four affective states, the transitions with L value less than 0.11 should be interpreted as being less likely than chance. Importantly, the test for significance of these transitions must set the null hypothesis at the appropriate chance levels and not zero.

This finding, thus, has implications on past published studies as well. In past studies which excluded self-transitions [3–5, 8–11, 17], we need to reconsider the results in terms of what the correct chance value was. Since these papers conducted hypothesis tests with $L = 0$ as the null hypothesis, they are likely to have overstated their possible effects, possibly finding positive results where negative results would have been more accurate. As such, these results need to be reanalyzed with the appropriate chance values for L (given in Table 3) to get the new significance values. For instance, in [5], the transition from boredom to frustration is reported to have an $L = 0.036$ and is significant with p < 0.001 – indicating that the transition from boredom to frustration is more likely than chance. But, with n = 5, the reported L value actually denotes a negative transition as the reported L value is less than the L value at chance (0.0625, as shown in Table 3). As such, it becomes essential to rerun the t-test on the original data with the null hypothesis of $L = 0.0625$ to confirm if this transition is actually significantly less likely than chance.

It is important to once again note that not all past publications using L are affected by this finding. Over half of the past studies using this metric included self-transitions [1, 2, 13, 14, 19–21, 23–25] and are therefore unchanged by this finding. The choice of whether or not one ought to include self-transitions in an affect dynamics analysis depends on the research goals and questions of the study. As [17] suggests, excluding self-transitions reveals a larger number of affective patterns that might otherwise be suppressed by the presence of persistent affective states. Including self-transitions in analysis helps us to better understand each state's persistence, but dilutes the transitions between different affective states. Better understanding transitions is likely important in theoretical models, but understanding true persistence might be particularly useful for algorithms being used to trigger interventions, for example.

5 Conclusion

In this paper, we demonstrate that commonly-used metric in affect dynamics research has been incorrectly interpreted when a common pre-processing step is also taken. The past 18 studies in this area can be divided into two groups - 10 studies that includes self-transitions [1, 2, 13, 14, 19–21, 23–25] and 8 that excludes self-transitions [3–5, 8–11, 17]. The studies that excluded self-transitions did so in order to concentrate on the transitions between states rather than on the persistence of each state [4]. While this focus can be justified, this paper demonstrates that doing so changes the interpretation of a key metric, and that the previous papers that excluded self-transitions did not account for this, invalidating many of their results.

Specifically, we find that when self-transitions are excluded, the value for L that represents chance shifts from 0 to $1/(n-1)^2$, where n is the number of affective states studied. This is because the exclusion of self-transitions leads to a violation of the assumption of independence in the equations used to calculate L. This new finding has a direct impact on the validity of the claims made by the 8 studies that excluded self-transitions as all the t-tests conducted in these studies have used $L = 0$ in their null hypothesis. As illustrated in Sect. 4, the t-tests in these studies should be re-run and re-examined for effects that switch from significantly more likely than chance to null effects or even effects that are significantly less likely than chance.

In conclusion, this paper illustrates the impact of a seemingly subtle data preprocessing step in the interpretation of the results of an analysis. As the use of data mining and automation becomes widespread in areas like education, we need to be more cautious about the impact of all the changes we do to the data processing pipeline - however independent the stages of the pipeline may look like. In some cases, as illustrated in this paper, a simple preprocessing step could potentially imply that you are attempting to answer a different research question. It is also necessary to be mindful of the underlying reasons and assumptions behind each step in the data mining pipeline. Only by carefully considering the validity of our complete processes can we ensure that our findings are valid, and that the adaptive systems we develop using those findings are optimally effective for learners.

Acknowledgements. We would like to thank Penn Center for Learning Analytics and Nigel Bosch and Luc Paquette for comments and discussions that helped to motivate this research and our thinking about the issues in this paper.

References

1. Andres, J.M.L., Rodrigo, M.M.T.: The incidence and persistence of affective states while playing Newton's playground. In: 7th IEEE International Conference on Humanoid, Nanotechnology, Information Technology, Communication and Control, Environment, and Management (2014)
2. Baker, R.S.J.d., Rodrigo, M.M.T., Xolocotzin, U.E.: The dynamics of affective transitions in simulation problem-solving environments. In: Paiva, Ana C.R., Prada, R., Picard, Rosalind W. (eds.) ACII 2007. LNCS, vol. 4738, pp. 666–677. Springer, Heidelberg (2007). https://doi.org/10.1007/978-3-540-74889-2_58
3. Bosch, N., D'Mello, S.: Sequential patterns of affective states of novice programmers. In The 1st Workshop on AI-supported Education for Computer Science (AIEDCS 2013), pp. 1–10 (2013)
4. Bosch, N., D'Mello, S.: The affective experience of novice computer programmers. Int. J. Artif. Intell. Educ. 1–26 (2017)
5. Botelho, A.F., Baker, R., Ocumpaugh, J., Heffernan, N. Studying affect dynamics and chronometry using sensor-free detectors. In: Proceedings of the 11th International Conference on Educational Data Mining, pp. 157–166 (2018)
6. Cohen, J.: A coefficient of agreement for nominal scales. Educ. Psychol. Measur. **20**(1), 37–46 (1960)
7. Craig, S., Graesser, A., Sullins, J., Gholson, B.: Affect and learning: an exploratory look into the role of affect in learning with AutoTutor. J. Educ. Media **29**(3), 241–250 (2004)
8. D'Mello, S., Graesser, A.: Dynamics of affective states during complex learning. Learn. Instr. **22**, 145–157 (2012)
9. D'Mello, S., Person, N., Lehman, B.: Antecedent-consequent relationships and cyclical patterns between affective states and problem solving outcomes. In: AIED, pp. 57–64 (2009)
10. D'Mello, S., Taylor, R., Graesser, A.: Monitoring affective trajectories during complex learning. In: McNamara, D., Trafton, J. (eds.) Proceedings of 29th Annual Cognitive Science Society, pp. 203–208 (2007)
11. D'Mello, S., Graesser, A.: Modeling cognitive-affective dynamics with Hidden Markov models. In: Proceedings of the 32nd Annual Cognitive Science Society, pp. 2721–2726 (2010)
12. DeFalco, J.A., et al.: Detecting and addressing frustration in a serious game for military training. Int. J. Artif. Intell. Educ. **28**(2), 152–193 (2018)
13. Guia, T.F.G., Rodrigo, M.M.T., Dagami, M., Sugay, J., Macam, F., Mitrovic, A.: An exploratory study of factors indicative of affective states of students using SQL-Tutor. Res. Pract. Technol. Enhanc. Learn. **8**(3), 411–430 (2013)
14. Guia, T.F.G., Sugay, J., Rodrigo, M.M.T., Macam, F., Dagami, M., Mitrovic, A.: Transitions of affective states in an intelligent tutoring system. In: Proceedings of Philippine Computing Society, pp. 31–35 (2011)
15. Karumbaiah, S., Lizarralde, R., Allessio, D., Woolf, B.P., Arroyo, I., Wixon, N.: Addressing student behavior and affect with empathy and growth mindset. In: EDM (2017)
16. Karumbaiah, S., Rahimi, S., Baker, R.S., Shute, V.J., D'Mello, S.: Is student frustration in learning games more associated with game mechanics or conceptual understanding. In: International Conference of Learning Sciences (2018)
17. Karumbaiah, S., Andres, J.M.A.L., Botelho, A.F., Baker, R.S., Ocumpaugh, J.S.: The implications of a subtle difference in the calculation of affect dynamics. In: 26th International Conference for Computers in Education (2018)

18. Kuppens, P.: It's about time: a special section on affect dynamics. Emot. Rev. 7(4), 297–300 (2015)
19. McQuiggan, S.W., Robison, J.L., Lester, J.C.: Affective transitions in narrative-centered learning environments. Educ. Technol. Soc. 13(1), 40–53 (2010)
20. McQuiggan, S.W., Robison, J.L., Lester, J.C.: Affective transitions in narrative-centered learning environments. In: Woolf, B.P., Aïmeur, E., Nkambou, R., Lajoie, S. (eds.) ITS 2008. LNCS, vol. 5091, pp. 490–499. Springer, Heidelberg (2008). https://doi.org/10.1007/978-3-540-69132-7_52
21. Ocumpaugh, J., et al.: Affect dynamics in military trainees using vMedic: from engaged concentration to boredom to confusion. In: André, E., Baker, R., Hu, X., Rodrigo, M. Mercedes T., du Boulay, B. (eds.) AIED 2017. LNCS (LNAI), vol. 10331, pp. 238–249. Springer, Cham (2017). https://doi.org/10.1007/978-3-319-61425-0_20
22. Pardos, Z.A., Baker, R.S., San Pedro, M.O., Gowda, S.M., Gowda, S.M.: Affective states and state tests: Investigating how affect throughout the school year predicts end of year learning outcomes. In: Proceedings of the Third International Conference on Learning Analytics and Knowledge, pp. 117–124. ACM, April 2013
23. Rodrigo, M.M.T., et al.: The effects of an embodied conversational agent on student affective dynamics while using an intelligent tutoring system. IEEE Trans. Affect. Comput. 2(4), 18–37 (2011)
24. Rodrigo, M.M.T., et al.: The effects of an interactive software agent on student affective dynamics while using; an intelligent tutoring system. IEEE Trans. Affect. Comput. 3(2), 224–236 (2012)
25. Rodrigo, M.M.T., et al.: The effects of motivational modeling on affect in an intelligent tutoring system. In: Proceedings of International Conference on Computers in Education, vol. 57, p. 64 (2008)

How Many Times Should a Pedagogical Agent Simulation Model Be Run?

David Edgar Kiprop Lelei[(✉)] and Gordon McCalla

ARIES Lab., Department of Computer Science, University of Saskatchewan,
Saskatoon, Canada
davidedgar.lelei@usask.ca, mccalla@cs.usask.ca

Abstract. When using simulation modeling to explore pedagogical phenomena, there are several issues a designer/practitioner should consider. One of the most important decisions has to do with determining how many runs of a simulation to perform in order to be confident in the results produced by the simulation [1]. With a deterministic model, a single simulation run is adequate. This issue becomes more challenging when part of the simulation model is based on stochastic elements. One of the solutions that has been used to address this challenge in other research communities is the use of Monte Carlo simulation [2]. Within the AIED research community, however, this question of how many times should a pedagogical simulation model be run to produce predictions in which the designer can have confidence has received surprisingly little attention. The aim of this paper is to explore this issue using a pedagogical simulation model, SimDoc, designed to explore longer term mentoring issues [3]. In particular, we demonstrate how to run this simulation model over many iterations until the accumulated results of the iteration runs reach a statistically stable level that matches real world performance but also has appropriate variability among the runs. We believe this approach generalizes beyond our simulation environment and could be applied to other pedagogical simulations and would be especially useful for medium and high fidelity simulations where each run may take a long time.

Keywords: Simulation · Simulated learners · Longer-term mentoring ·
Lifelong learning

1 Introduction

In this paper, we will explore an important aspect of pedagogical system evaluation when using simulation to help in the creation of environments to support longer-term learning. More specifically, we are interested in shedding light on the following question: how many runs of a simulation does it take for a practitioner to have confidence in the simulation model's output? As AIED researchers explore longer-term learning and mentoring contexts [4], the use of simulation to evaluate the functionality of pedagogical systems designed and built to support longer-term learning is important, given the high time and financial cost it would otherwise take to perform human studies [5]. Simulation offers a cheaper and quicker alternative to human studies and can be used to quickly accept and reject various hypotheses concerning the pedagogical

© Springer Nature Switzerland AG 2019
S. Isotani et al. (Eds.): AIED 2019, LNAI 11625, pp. 182–193, 2019.
https://doi.org/10.1007/978-3-030-23204-7_16

system being developed. Using simulation also enables practitioners to evaluate different aspects of a pedagogical system without exposing learners and their mentors to unfavorable learning conditions.

Simulated pedagogical agents have been part of AIED research for more than a quarter century [6]. Recently, there has been a growing trend towards the use of simulation to support several aspects of the design, development and evaluation lifecycles of a pedagogical system especially when considering longer-term learning [3], testing and validating adaptive educational systems [7], and using simulated students to overcome experimental complexity and unreliable student availability [8]. The simulations used come with varying degrees of model fidelity to the real world, including low fidelity [9], medium fidelity [10], and high fidelity [11].

Stochastic modeling is often an important part of these simulations, thus necessitating many iterations of a simulation in order to produce meaningful predictions. However, for system practitioners/designers to fully gain the benefits that simulation offers, they need to know how many times it is necessary to run a simulation model. Knowing this would allow practitioners to not only effectively explore and test various hypotheses but to also accept or reject them with confidence.

The main question in this paper is thus how many runs of a simulation are enough? We will propose several statistical measures to be used to determine the answer to this question. We will illustrate our discussion by drawing on the results of simulation runs we performed while evaluating a simulation model, SimDoc, that we have built of a longer-term mentoring environment, a doctoral program [5].

2 Related Work

2.1 Simulation Runs Within Simulation Research

Generally, the answer to the question of the number runs to make in a simulation depends on the question at hand and project-specific constraints. The number may range from 25 to 800 when using Monte Carlo approximation [2, 12]. Monte Carlo methods are used to explore behavior of statistical measures under controlled situations. Usually in any simulation study, a summary statistic is calculated after a finite number of iterations of a simulation have been performed. Often there is a between-run variability within the simulation results that depends on experimental settings and the number of iterations performed. Thus, determining the number of iterations required is critical.

One approach uses standard deviation and the confidence interval convergence rate to determine the stopping point as described in [1]. This approach has the advantage of minimizing the waste of simulation runs that would otherwise have been performed if too many iterations were specified a priori. A similar approach that recalculates sample standard deviation and mean when a new iteration is added until a stopping condition is achieved is proposed by Truong, Sarvi, Currie, and Garoni in [13]. Yet other methods may consider confidence intervals of measures of performance [13]. The domains for which these methods have been explored tend to be simpler and more predictable than in AIED, where simulation often involves many more variables, a range of statistical sub-models, and pedagogical agents.

2.2   Simulation Runs in AIED Research

In AIED there is no clear guidance on how to determine the number of runs a practitioner should use to evaluate their simulation model output. Unfortunately, the number of runs practitioners have used is rarely reported, with more general descriptions of the model and/or results being the focus of discussion (as in [14, 15] and [16] for example). In the few papers that have reported on the number of iteration runs used, the number varies greatly, ranging from as low as 2 (see [17]) to as high as 1000 (see [18]).

Sometimes the number of simulation runs is justified based on pedagogical or theoretical grounds. So, in an experiment to determine how students learn composite concepts, Liu in [17] used Bayesian Networks to represent student models since they are a popular way of capturing the relationship between students' competence and their performance. Liu indicates that the simulation needed at least two runs given that the number of concepts being explored is also two. Desmarais and Pu in [19] used Bayesian methods to model a new approach to Computer Adaptive Testing (CAT) based on a theory of knowledge spaces and item graphs with no hidden nodes called POKS (Partial Order Knowledge Structure). CAT systems are used to administer adaptive tests that are used to determine if the examinee is a master or a non-master using the least number of test items. In evaluating the performance of POKS, an average of 9 simulation runs were used.

Most often, though, the number of runs seem to have been arbitrarily chosen. A simulation based physics tutor, BEETLE II [18], was developed to encourage effective self-explanation using adaptive feedback. The BEETLE II tutor expected students to provide explanations for experiments using natural language in the form of sentences as input. An important statistical significance test that can be done is the F-Score [20]. The F-Score for BEETLE II was evaluated using the approximate randomization significance test with 1000 simulation runs. The evaluation was used to determine whether the system made a correct decision on either accepting or rejecting a student answer. In a proof of concept study exploring a medium fidelity simulation of a multi-agent pedagogical environment, Erickson, Frost, Bateman, and McCalla [21] chose 100 iterations to determine which learning condition is most desirable between unstructured, semi-structured, and structured approaches to assigning learning objects. In a study to explore the impact of an instructional planner that employed collaborative filtering based on learning sequences, Frost and McCalla [9] used 25 simulation runs to show how different groups of learners would perform. In another study, StudyWise [22], researchers used simulated learners to test an application meant to help students memorize collections of basic techniques required for an effective scheduling algorithm. The researchers performed 100 simulation runs to evaluate the pedagogical effectiveness of their system.

The goal of this paper is to explore this issue of how many simulation runs are adequate so that the designer of a pedagogical system can be confident in the results. We draw from a simulation of a doctoral program to illustrate our discussion.

3 SimDoc: Simulated Doctoral Program

In this section we will briefly introduce a simulation model of a doctoral program [3], SimDoc, complete with simulated students, simulated supervisors, and a simulated doctoral environment. Its conceptual model is illustrated in Fig. 1, below. SimDoc is based on the University of Saskatchewan doctoral program. We will use SimDoc to explore the issue of determining how many simulation runs are adequate for exploring issues with pedagogical simulations.

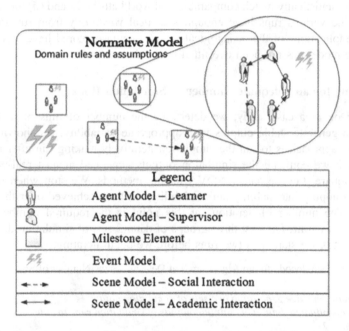

Fig. 1. SimDoc conceptual framework - element types and their interaction patterns

The normative model captures important features that affect doctoral students' progress towards their degree. We model two doctoral stakeholders in the form of agents: supervisors and learners (i.e. doctoral students). We use the milestone element to represent important goals a doctoral student must accomplish to complete their doctoral program. To trigger expected and other events that occur in doctoral studies we use the event model.

SimDoc is a medium fidelity simulation model, drawing on data from a real-world doctoral program (at the University of Saskatchewan) and, where such data is not available, from data derived from relevant studies on doctoral student-supervisor relationships [23] and supervisory styles [24]. In an initial phase, we ran the SimDoc model, tuning various parameters run by run with the goal that its outputs match a real-world doctoral program. 500 simulation runs were used in this phase. The best version of SimDoc resulted in a 93% similarity between the simulation's outputs and comparable data from the real world program. In the next phase we used this best version of the simulation to explore various pedagogical issues. It is in this experimental phase that we wish to determine when we have made enough simulation runs to be confident in the results.

4   Examining the Simulation Runs

We argue in this section that we can have confidence in the number of iterations when two conditions are met: (i) the simulation is *stable,* that is, the average of the simulation runs' outputs statistically match comparable real world attributes and (ii) the simulation outputs of the various runs have enough statistical *variability* from run to run that important implications of the pedagogical issue(s) being explored have come to light and that the model has not been overfitted to the data.

4.1   Testing for an Adequate Number of Simulation Runs

Using SimDoc as a case study, we determine the number of runs of a simulation necessary to generate stable outputs with appropriate variability. As shown in **Algorithm 1**, the approach is to run the simulation iteratively making run after run. After each iteration we compare the simulation outputs generated against real-world data using Chi-Square, Levene, and ANOVA testing methods. We stop when these tests indicate that appropriate stability and variability has been achieved over the accumulated runs. The number of iterations at this point is the required number of runs. Since ANOVA requires at least three groups of data (2 sets of simulation data and the real world data), we start with two runs before we begin iterating.

Algorithm 1 - Pseudocode for an Algorithm that Determines the Number of Simulation Runs

> *run simulation twice generating two sets of simulation data*
> *create consolidated dataset containing outputs of simulation runs to date*
> *iteration = 2*
> *compare consolidated dataset against the real-world dataset*
> *until p-values of Chi-Square & Levene's Test are>.05, & p-value of ANOVA is < .05*
> *iteration = iteration + 1*
> *run simulation generating the next set of simulation data*
> *consolidate simulation runs outputs*
> *compare consolidated dataset against the real-world dataset*
> *end until*
> *output iteration*

We implemented and ran this algorithm with the best version of SimDoc described above, and the iteration stopping conditions were met on iteration 100. The results from runs 91–100 are summarized in Table 1. We can observe that the Chi-Square p-value for each run is greater than 0.05. Similarly, the Levene Test's p-value for each run is also greater than 0.05. However, the ANOVA p-value does not become less than 0.05 until the 100th iteration. Therefore, the number of runs required for experiments with this best version of SimDoc is 100. More runs are not necessary, since we have the appropriate statistical significance on the relevant measures at 100 runs, and fewer runs won't be enough to give us this significance. Next, we explain in detail our measures of stability and variability, and show how and why we used the Chi-Square, Levene, and Anova methods that are central to this algorithm.

Table 1. P-values for Levene, Chi-Square, and ANOVA tests for the simulation runs (runs 91 to 100)

Run	Levene	Chi-Square	ANOVA	Run	Levene	Chi-Square	ANOVA
100	0.17	0.28	0.04	95	0.15	0.28	0.08
99	0.2	0.28	0.06	94	0.17	0.28	0.07
98	0.18	0.28	0.06	93	0.16	0.28	0.08
97	0.18	0.28	0.05	92	0.16	0.28	0.12
96	0.16	0.28	0.06	91	0.15	0.28	0.11

4.2 Test of Stability in Simulation Output

In this section we go into more detail to explain how we confirm the *stability* of the model using the output from 100 runs of our simulation. For simplicity we will consider only one output from the simulation: the pedagogically important output of a student's "time-in-program". We compare the time-in-program for simulated students in these simulation runs to the observed time-in-program for students in the real world dataset. A stable model should have simulation results that are statistically similar to the real world scenario. Table 2 shows the time-in-program frequency counts of real world students and the time-in-program frequency counts of simulated students averaged over 100 runs of the simulation (divided into students who graduated, those who withdrew, and the cumulative total of these two figures).

Table 2. Frequency counts of time-in-program between simulated and real world students

Time	Graduated		Withdrew		Total per year	
(years)	UofS	SimDoc	UofS	SimDoc	UofS	SimDoc
0	0	0	9	10	9	10
1	0	0	6	2	6	2
2	0	0	1	2	1	2
3	6	6	4	6	10	12
4	16	21	3	3	19	24
5	35	29	1	5	36	34
6	27	27	3	8	30	35
7	22	19	4	4	26	23
8	11	7	1	0	12	7
9	1	0	4	5	5	5

To determine whether these frequencies are distributed in a statistically similar manner, we conducted a Chi-Square test of homogeneity to check for consistency among the yearly distributions, separately for each column: Total per Year, Graduated, and Withdrew. Since the SimDoc model was tuned based on the real world dataset, we expect that SimDoc results are statistically like the real world data. Therefore, our null hypothesis is that the frequency counts of real world dataset and simulated dataset are equally distributed. Thus, the alternative hypothesis is that there is a difference between the distributions of the frequency counts. For this analysis, the significance level we use is 0.05. We then apply the Chi-Square test of homogeneity to the cumulative contingency table and compute the degree of freedom, the Chi-Square test statistic, and p-value. The results of the Chi-Square test show that the p-value is more than the significance level (0.05); therefore, we accept the null hypothesis that the frequency counts are statistically consistent between the real world dataset and simulated dataset, χ^2 (df = 9) = 5.0904, p = 0.8264.

The second analysis is to determine if the distribution of frequency counts in time-in-program among the graduated learners were similar between the simulated and real world datasets. Given that the distributions of the cumulative frequencies were statistically similar, we expect that the real world graduated frequency counts are similar to the simulated graduated frequency counts per year. Thus, our null hypothesis is that the frequency counts of real world (graduated) dataset and simulated (graduated) dataset are equally distributed. As such, the alternative hypothesis is that there is a difference between the frequency counts between these distributions. As in the first analysis, we choose a significance level at 0.05. We then conduct the Chi-Square test for homogeneity and the results show that we can accept the null hypothesis since the p-value is greater than the significance level, frequency counts are statistically consistent between the real world (graduated) dataset and simulated (graduated) dataset, χ^2 (df = 6) = 2.9945, p = 0.8095.

The third analysis is to assess whether the distribution of frequency counts in time-in-program among the learners who withdrew were similar between the real world dataset and simulated dataset. Given that the distributions of the cumulative datasets were statistically similar, our null hypothesis is that the real world withdrawal frequency counts are similar to the simulation withdrawal frequency counts per year. As such, the alternative hypothesis is that there is a difference between the frequency counts between these distributions. As in the first analysis, we choose a significance level at 0.05. We then conduct the Chi-Square test for homogeneity and the results show that there is no significant difference in the distribution of frequency counts per year between the real world (withdrew) dataset and simulated (withdrew) dataset, since the p-value is greater than the significant level, χ^2 (df = 9) = 7.9344, p = 0.5408.

Table 3 shows the overall total frequency count per outcome for the real world and SimDoc datasets. Since the resulting contingency table is small (2 × 2), to test whether the proportions for one nominal variable are different from another nominal variable, the Chi-Square test of homogeneity is not recommended but instead it is advisable to use a Fisher's exact test. In this analysis we are exploring if the frequency counts per outcome between the real world and the simulation datasets differ. Our null hypothesis is that the proportions of the outcome variables are not the same between the real world and the simulation datasets. Therefore, the alternative hypothesis is that there is no difference in the proportion of the frequency counts in the outcome variables. We then conduct a Fisher's exact test which yields a result with p-value = 0.3005, indicating that we can accept the null hypothesis that there is no significant difference in frequency counts per outcome between the real world and the simulation datasets.

Table 3. Summary of frequency count by outcome and data source

	UofS	SimDoc baseline
Graduated	118	109
Withdrew	36	45

The overall outcome of this analysis is that best version of the SimDoc model that came out of the tuning process in the initial 500 runs of the simulation turned out to be stable over 100 runs in producing overall results similar to the real world dataset. Next, we would like to look at the variability within the 100 runs.

4.3   Test of Variability Among Simulation Runs

In this section we go into more detail as to how we tested for appropriate *variability* in the simulation output, important to ensure that the tuning process has not gone too far and overfitted the simulation to the particular dataset. As in Sect. 4.2, we examine the characteristics of the results for the variable time-in-program, produced by the 100 iteration runs of the simulation discussed in Sect. 4.1. First, we randomly select 12 out of the 100 runs to examine graphically for insight into the variance among them. Figure 2 (Top) depicts density plots of the results for the 12 randomly selected runs. This figure shows that there is evidence of variation in the graduation and withdrawal rates between the runs.

A box plot sheds more insight into the nature of the simulation results as shown in Fig. 2 (Bottom). This box plot shows that indeed there are variations among the different runs, and, in fact, a few outliers exist. However, are these variations statistically significant? Are there any significant differences among the 100 iterations of the simulation? To answer these questions, we use one-way analysis of variance (ANOVA). ANOVA is an extension of the independent two-samples t-test that is used to analyze data organized in groups. With ANOVA we explore the variance in means of each of the runs between the distribution of student counts in the program per year in the real world dataset and the simulation's output (averaged over the 100 runs). In this case, there are 101 groups: 100 groups representing the 100 runs of the simulation and 1 group representing the student graduation and withdrawal counts gleaned from the real world dataset.

We are interested in exploring if there are any significant differences in the average mean time students were in the program either leading to completion of their degree or withdrawal from the program among the simulation's 100 runs and the real world dataset. Before performing the ANOVA, we establish that the three main ANOVA assumptions are met: independence of observations, homogeneity of variance, and dependent variable is normally distributed. To do this we perform a Chi-Square test of

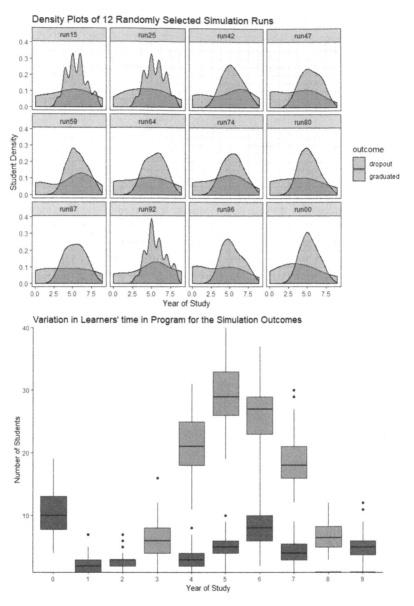

Fig. 2. Top: density plots of 12 randomly selected runs of the simulation. Bottom: variation in the graduation and attrition rates in the 12 simulation runs

independence (see previous section), a normality check and a Levene test. Given that the p-value (0.1707) > 0.05, equal variance can be assumed. With these three tests, the ANOVA assumptions are met. To show that there is variability among the results of the 100 simulation runs, we make the null hypothesis that there is no difference in the means among the 100 simulation iterations and the real world dataset.

We conduct a one-factor ANOVA to compare the difference in learners' time-in-program among the simulation's 100 runs and the real world dataset. The ANOVA results show that there is a statistically significant difference in the average time-in-program [$F(100,15453) = 1.272$, $p = 0.0352$] among the 100 iterations of the simulation and the real world dataset. Therefore, we reject the null hypothesis and thus accept the hypothesis that there is statistical evidence to suggest that there is a difference in the means among the 100 runs and the real world dataset. This result shows that there is a difference between at least one or more pairings. Whenever the null hypothesis is rejected in ANOVA, then all that is known is that at least 2 groups differ from each other. ANOVA cannot tell us which of these groups are different. Therefore, to explore how the mean for each of the 100 iterations compared to that of the real world dataset, we perform a post hoc test using the Tukey's Honest Significant Difference test at $p < .05$. The results show that there is no significant difference between the real world data when compared to each of the 100 iterations. The difference exists within the 100 iterations, thus ensuring appropriate variability among the simulation runs.

5   Conclusion

To the best of our knowledge, this paper is the first attempt to explore the appropriate number of runs there needs to be of a simulation model to get results about which the experimenter can be confident. As identified by Ritter et al. [25], many authors fail to report the number of runs used in testing a simulation model. Even when the number is included, the reason behind choosing a given number of runs is barely mentioned. Our approach is based on defining characteristics necessary of the simulation output, namely that the simulation runs, collectively, meet statistical standards of stability and variability when measured against comparable real world data. We provide a pseudo-algorithm that can be used to determine when the number of runs has reached this point, and therefore can determine that the simulation has an appropriate number of runs. We demonstrated this using data generated by our SimDoc simulation of a long term mentoring environment, a doctoral program, when compared with data from a real world doctoral program (at the University of Saskatchewan). The methods, though, are not specific to the particular doctoral program simulation, and should generalize to any simulation. Knowing when the simulation has been run an appropriate number of times should allow system designers to be confident in their results and should avoid them having to needlessly make extra simulation runs. This is especially important for medium and high fidelity simulations that can take a long time to run.

Acknowledgements. We would like to acknowledge the financial support of the Natural Sciences and Engineering Research Council of Canada for this research. We also would like to thank the University of Saskatchewan for providing (anonymized) data on its Ph.D. programs that we could use to inform the SimDoc simulation.

References

1. Bogdoll, J., Hartmanns, A., Hermanns, H.: Simulation and statistical model checking for modestly nondeterministic models. In: Schmitt, J.B. (ed.) International GI/ITG Conference on Measurement, Modelling, and Evaluation of Computing Systems and Dependability and Fault Tolerance, pp. 249–252. Springer, Heidelberg (2012). https://doi.org/10.1007/978-3-642-28540-0_20
2. Koehler, E., Brown, E., Haneuse, S.J.-P.A.: On the assessment of Monte Carlo error in simulation-based statistical analyses. Am. Stat. **63**(2), 155–162 (2009)
3. Lelei, D.E.K., McCalla, G.: How to use simulation in the design and evaluation of learning environments with self-directed longer-term learners. In: Penstein Rosé, C., et al. (eds.) AIED 2018. LNCS (LNAI), vol. 10947, pp. 253–266. Springer, Cham (2018). https://doi.org/10.1007/978-3-319-93843-1_19
4. Lane, H.C., McCalla, G.I., Looi, C.-K., Bull, S.: The next 25 years: how advanced interactive learning technologies will change the world. Int. J. Artif. Intell. Educ. **26**(1), 539–543 (2016)
5. Lelei, D.E.K., McCalla, G.I.: The role of simulation in the development of mentoring technology to support longer-term learning. In: The Proceedings of 3rd International Workshop on Intelligent Mentoring Systems Held in Conjunction with the 19th International Conference on Artificial Intelligence in Education (2018)
6. VanLehn, K., Ohlsson, S., Nason, R.: Applications of simulated students: an exploration. J. Artif. Intell. Educ. **5**(2), 1–42 (1994)
7. Dorça, F.: Implementation and use of simulated students for test and validation of new adaptive educational systems: a practical insight. Int. J. Artif. Intell. Educ. **25**(3), 319–345 (2015)
8. Laberge, S., Lin, F.: Simulated learners for testing agile teaming in social educational games. In: CEUR Workshop Proceedings, vol. 1432, pp. 65–77 (2015)
9. Frost, S., McCalla, G.: Exploring through simulation an instructional planner for dynamic open-ended learning environments. In: Conati, C., Heffernan, N., Mitrovic, A., Verdejo, M. Felisa (eds.) AIED 2015. LNCS (LNAI), vol. 9112, pp. 578–581. Springer, Cham (2015). https://doi.org/10.1007/978-3-319-19773-9_66
10. Lelei, D.E.K., McCalla, G.I.: Exploring the issues in simulating a semi-structured learning environment: the SimGrad doctoral program design. In: The Proceedings of the 2nd Workshop on Simulated Learners at the 17th International Conference on Artificial Intelligence in Education, vol. 5, pp. 11–20 (2015)
11. Carlson, R., Keiser, V., Matsuda, N., Koedinger, K.R., Penstein Rosé, C.: Building a conversational SimStudent. In: Cerri, S.A., Clancey, W.J., Papadourakis, G., Panourgia, K. (eds.) ITS 2012. LNCS, vol. 7315, pp. 563–569. Springer, Heidelberg (2012). https://doi.org/10.1007/978-3-642-30950-2_73
12. Booth, J.G., Sarkar, S.: Monte Carlo approximation of bootstrap variances. Am. Stat. **52**(4), 354–357 (1998)
13. Truong, L.T., Sarvi, M., Currie, G., Garoni, T.M.: How many simulation runs are required to achieve statistically confident results: a case study of simulation-based surrogate safety measures. In: IEEE 18th International Conference on Intelligent Transportation Systems, pp. 274–278 (2015)
14. Van Joolingen, W.: Design and implementation of simulation-based discovery environments: the SMISLE solution. J. Artif. Intell. Educ. **7**(4), 253–276 (1996)

15. Rosenberg-Kima, R.B., Pardos, Z.A.: Is this model for real? Simulating data to reveal the proximity of a model to reality. In: The Proceedings of the 17th International Conference on Artificial Intelligence in Education, pp. 78–87 (2015)
16. Weber, G.: Individual selection of examples in an intelligent learning environment. Int. J. Artif. Intell. Educ. 7(1), 3–31 (2015)
17. Liu, C.: A simulation-based experience in learning structures of bayesian networks to represent how students learn composite concepts. Int. J. Artif. Intell. Educ. 18(3), 237–285 (2008)
18. Dzikovska, M., Steinhauser, N., Farrow, E., Moore, J., Campbell, G.: BEETLE II: deep natural language understanding and automatic feedback generation for intelligent tutoring in basic electricity and electronics. Int. J. Artif. Intell. Educ. 24(3), 284–332 (2014)
19. Desmarais, M.C., Pu, X.: A Bayesian student model without hidden nodes and its comparison with item response theory. Int. J. Artif. Intell. Educ. 15(4), 291–323 (2005)
20. Yeh, A., Corp, M.: More accurate tests for the statistical significance of result differences. In: Proceedings of the 18th International Conference on Computational Linguistics, pp. 947–953 (2000)
21. Erickson, G., Frost, S., Bateman, S., McCalla, G.: Using the ecological approach to create simulations of learning environments. In: Lane, H.C., Yacef, K., Mostow, J., Pavlik, P. (eds.) AIED 2013. LNCS (LNAI), vol. 7926, pp. 411–420. Springer, Heidelberg (2013). https://doi.org/10.1007/978-3-642-39112-5_42
22. Riedesel, M.A., Zimmerman, N., Baker, R., Titchener, T., Cooper, J.: Using a model for learning and memory to simulate learner response in spaced practice. In: The Proceedings of the 18th International Conference on Artificial Intelligence in Education, pp. 644–649 (2017)
23. Heath, T.: A quantitative analysis of PhD students' views of supervision. High. Educ. Res. Dev. 21, 37–41 (2002)
24. Gatfield, T.: An investigation into PhD supervisory management styles: development of a dynamic conceptual model and its managerial implications. J. High. Educ. Policy Manage. 27(3), 311–325 (2005)
25. Ritter, F.E., Schoelles, M.J., Quigley, K.S., Klein, L.C.: Determining the number of simulation runs treating simulations as theories by not sampling their behavior. In: Rothrock, L., Narayanan, S. (eds.) Human-in-the-Loop Simulations: Methods and Practice, pp. 97–116. Springer, London (2011). https://doi.org/10.1007/978-0-85729-883-6_5

A Survey of the General Public's Views on the Ethics of Using AI in Education

Annabel Latham[1]([✉])(iD) and Sean Goltz[2](iD)

[1] Manchester Metropolitan University, Manchester M1 5GD, UK
a.latham@mmu.ac.uk
[2] Business and Law School, Edith Cowan University, Perth, Australia
n.goltz@gmail.com

Abstract. Recent scandals arising from the use of algorithms for user profiling to further political and marketing gain have popularized the debate over the ethical and legal implications of using such 'artificial intelligence' in social media. The need for a legal framework to protect the general public's data is not new, yet it is not clear whether recent changes in data protection law in Europe, with the introduction of the GDPR, have highlighted the importance of privacy and led to a healthy concern from the general public over online user tracking and use of data. Like search engines, social media and online shopping platforms, intelligent tutoring systems aim to personalize learning and thus also rely on algorithms that automatically profile individual learner traits. A number of studies have been published on user perceptions of trust in robots and computer agents. Unsurprisingly, studies of AI in education have focused on efficacy, so the extent of learner awareness, and acceptance, of tracking and profiling algorithms remains unexplored. This paper discusses the ethical and legal considerations for, and presents a case study examining the general public's views of, AI in education. A survey was recently taken of attendees at a national science festival event highlighting state-of-the-art AI technologies in education. Whilst most participants (77%) were worried about the use of their data, in learning systems fewer than 8% of adults were 'not happy' being tracked, as opposed to nearly two-thirds (63%) of children surveyed.

Keywords: Ethics · Trust · GDPR

1 Introduction

Although discussions of the ethics of Artificial Intelligence (AI) have been commonly found in popular writing and science fiction for decades, it is only relatively recently that the field of AI has become sufficiently advanced to bring the issues of an ethical and legal framework to the fore. The mainstream use of apps and search engines has led to the collection of large amounts of user interaction data, from which an increasing number of attributes can be inferred about the individual. Whilst for companies this has led to more efficient, highly targeted advertising campaigns, the question of whether this offers a benefit to the user, in terms of filtering information, or holds the risk of unwitting persuasion, hangs in the balance. Since 2011 Eli Pariser has campaigned to

© Springer Nature Switzerland AG 2019
S. Isotani et al. (Eds.): AIED 2019, LNAI 11625, pp. 194–206, 2019.
https://doi.org/10.1007/978-3-030-23204-7_17

raise awareness of the dangers of algorithmic personalization by search engines such as Google, warning that "the Internet is showing us what it thinks we want to see, but not necessarily what we need to see" [1]. Zuboff argues that knowledge and power are now asymmetrical in the business of 'surveillance capitalism', a major part of which is personalized communication, and that people's belief that they get something in return for their data is misled [2]. A major study of youth behavior online concluded that an important reason why most youths appear unconcerned about profiling by digital technology was a "lack of knowledge" rather than a "cavalier attitude toward privacy" [3]. Recent scandals involving the use of algorithms for user profiling to further political and marketing gain (e.g. Cambridge Analytica's alleged use of personal information to profile individual US voters for targeted political advertising [4]) have resulted in much publicity about the dangers of big data and algorithmic decision making in everyday lives. However, whether this additional publicity has translated into public awareness is still a subject for debate.

In education, research into the application of AI techniques to learning systems for the benefit of the learner has been an active field for several decades. The benefits of personalized, adaptive learning have long been argued and supported by results that show that learners can learn more efficiently and effectively with the inclusion of AI techniques as opposed to without [5, 6]. However, it is open to debate how many members of the general public have actually had access to learning systems that use AI techniques, or even if they have, whether they are aware of the use of AI profiling. With the popular use of apps that adapt to make our lives more convenient, AI techniques have moved into the mainstream and user expectations have shifted accordingly, so that many people would not categorize features such as predictive text or recommendation systems as using AI at all.

In order to benefit from the personalization of learning using AI techniques, learners must accept the trade-off of the system gathering personal data and tracking their learning experience. In fact, just like the Facebook/Cambridge Analytica scandal, learning systems use an individual's behavior within a system to infer information about personality, mood, learning styles and comprehension [7–11]. The question arises "How many learners are aware and understand that in order to personalize, learning systems gather user data in order to profile their personal traits?".

Whilst there have been a few studies investigating the public's perception of AI in everyday lives [12, 13], none have yet been published that specifically explore the issue of AI in education – and whether the perceived benefit of the educational context has any impact on views of AI generally. This pilot study aims to fill this gap by gathering views of AI in the Educational context. Its results will be of interest to AIED researchers, educators, and researchers with an interest in the legal and ethical aspects of AI.

This paper describes a survey of the general public's feelings on the use of AI in education. The survey involved collecting anonymous questionnaires completed voluntarily by some attendees at a free National Science Festival event held at Manchester Science Museum, called 'Me versus Machine'. The event included a number of activities designed to introduce people of different ages to Computer Science. One stand was dedicated to Artificial Intelligence in Education, where recent research in Conversational Intelligent Tutoring Systems was demonstrated and discussed. Interested attendees were asked to participate in a study of views on AI in Education, and completed a questionnaire.

The rest of the paper is organized as follows: Sect. 2 considers the legal and ethical context of AI and its use in education. Section 3 outlines profiling in education systems, with Sect. 4 presenting a case study of the general public's views on AI in education, followed by the conclusion in Sect. 5.

2   Legal and Ethical Considerations of AI in Education

2.1   Ethical Issues of AI in Education

The discussion of ethics in AIED is not new. In 2000, Aiken and Epstein published an article in the International Journal of Artificial Intelligence in Education titled 'Ethical Guidelines for AI in Education: Starting a Conversation' [14]. While the cited predictions for the future of AIED for 2010 are somewhat premature, most will agree that this is what we are expecting today, 20 years later, for 2025: "The teacher of 2010 will rarely spend a day lecturing…The artificial-intelligence tutor will become a valuable assistant, providing the individualized instruction that a teacher with 20 or more pupils does not have the time for. Learning can take place at the student's pace" [15].

Following Shneiderman's [16] quote from Mumford [17]: "The real question before us lies here: do these instruments further life and enhance its value, or not?", Aiken and Epstein propose two fundamental meta-principles as a basic philosophical underpinning for any discussion of AIED systems: (1) "The Negative Meta-Principle for AIED – AIED technology should not diminish the student along any of the fundamental dimensions of human being; and (2) The Positive Meta-Principle for AIED – AIED technology should augment the student along at least one of the fundamental dimensions of human being" [14].

Fast forward 20 years, and Nichols and Holmes propose eight principles constituting "an open ethical framework for implementing AI in educational setting in ways that empower students and provide transparency" [18]. These principles are required since while data is supposed to be applied in objective ways by AI, source data is not immune from bias and there is no such thing as "raw data" [19].

It is already established that algorithms designed by engineers to process data carry in them inherent bias with ethical consequences as was illustrated by sexist, racist and discriminatory consequences by AI systems [20]. The recent Cambridge Analytica scandal [21] further illustrates that even small amounts of personal data can be combined through AI algorithms with the potential to undermine democracy. The ethical issues faced by data analytics are shared with AI since they both draw on data [22].

While there is a consensus that ethical principles of AI are mandatory and urgent [23], and while multiple organizations are exploring this realm [21, 24–27], there is over representation to AI developers (e.g., DeepMind Ethics and Society) and the corporate perspective (e.g., OpenAI) raising questions whether they will be thinking broadly and critically enough [28].

With advances in AIED like the Ada [29] and Jill Watson [30] bots, the absence of a definite reference point for AI ethics is crucial to AIED ethics. Holmes argues that "around the world, virtually no research has been undertaken, no guidelines have been provided, no policies have been developed, and no regulations have been enacted to address the specific ethical issues raised by AIED" [31].

2.2   The Impact of GDPR

The General Data Protection Regulation (GDPR) was approved by the EU parliament on April 14, 2016 and came into force on May 25, 2018 (EUGDPR.org). According to the EUGDPR website, "The aim of the GDPR is to protect all EU citizens from privacy and data breaches in today's data-driven world" [32].

Certain aspects of the GDPR are particularly relevant to Artificial Intelligence. One of these is the principle of "accountability," which is an implicit requirement under the current law but has been explicitly introduced in the GDPR [33]. This principle requires organizations to demonstrate compliance with all the other principles in the GDPR, and several further provisions of the GDPR also promote accountability.

Another aspect relevant to AI is the tightened requirements for consent in the GDPR. The use of AI techniques by its nature (i.e., the collection and processing of massive amounts of data) stipulates that it would be challenging to obtain explicit consent from the individuals involved. This is especially relevant to AIED as the users are often minors, thus requiring both their own and their parents' explicit consent.

A further relevant challenge is the GDPR right to receive an explanation by a natural person of decisions based on automated processing. This right's scope and eligibility is not without doubts [34]. Nonetheless, even if we assume the right for explanation exists, AI decisions are made by complex and technical processes many times not even clear to their developers (e.g., neural networks). In addition, the algorithm structure and operation method may be proprietary information and considered a trade secret. Finally, it would be challenging to explain complex AI systems to a layperson, moreover to a minor.

To manage these challenges raised by AI and the GDPR regulations, the Ministers of the European Parliament (MEPs) asked the European Commission in February 2017, to propose EU-wide rules on robotics and AI. Following this request, a public consultation was held in October 2017. Interestingly, the consultation results showed that European public opinion appears to be much more positive towards automation technologies than U.S. public opinion, based on the results of a recently-release report by the Pew Research Centre [35].

Furthermore, in December 2018, the European Commission's High-Level Expert Group on Artificial Intelligence (AI HLEG) published its draft of the AI Ethics Guidelines for comments from the public with the aim to have a final version in March 2019 [36]. The group have adopted EU treaties and legislation on human rights as their ethical principles for AI. This has led to the following assertion: "It should also be noted that, in particular situations, tensions may arise between the principles when considered from the point of view of an individual compared with the point of view of society, and vice versa. There is no set way to deal with such trade-offs". In the context of AIED, the potential for such tension is high.

While the global (and mainly the Western) discussion around general ethical guidelines for AI is vibrant, there is yet to be a consensus around an established set of principles that would be easily applied to the different fields in which AI is applied. Moreover, as in many cases of challenging regulatory spheres in the past, the surplus of sources and mix of laws (e.g., privacy), regulation (e.g., GDPR), codes (The Asilomar AI Principles [37]) and standards (e.g., IEEE [38]) that apply to AI seems to complicate the problem and make compliance ever more challenging, rather than promoting its solution.

3 Profiling in Learning Systems

3.1 Intelligent Tutoring System Approaches

Intelligent Tutoring Systems (ITS) personalize learning based on traits of the individual learner held in a student model. Traditionally, student models were based on outcomes from ITS-designed assessment and self-reported affective information such as mood, motivation and learning style [39, 40]. More recently, construction of student models has been automated with ITS profiling learner behaviors, such as user interface interactions, to predict the affective state of a learner [8–11]. Most automated profiling techniques map tracked learner behaviors to typical behaviors described in psychological models (e.g. personality and learning styles [10]) to infer learner traits and preferences. Some ITS profile learner affect using physical indicators gathered from sensors worn by learners [41, 42].

Conversational Intelligent Tutoring Systems (CITS) are ITS with a conversational agent interface, enabling them to conduct tutoring via a mixed initiative conversation. Their advantage is that the learner does not have to self-motivate as the CITS leads the tutoring conversation, yet learners can ask questions and explore answers using natural language conversation. CITS capture rich interaction information from the conversation, that adds depth to the student model [6, 10, 11].

3.2 Oscar CITS

Oscar CITS aims to mimic a human tutor by delivering a personalized tutoring conversation based on an individual learner's knowledge and preferred learning style [10, 43]. Oscar CITS incorporates intelligent techniques to provide realtime problem solving support (hints), intelligent solution analysis (feedback) and curriculum sequencing. Learners are automatically profiled using 41 variables tracked based on behavior, preferences and language during the tutorial conversation [43]. Oscar CITS adapts to learner knowledge and learning style, by changing the style of conversation and support material presented, such as giving step-by-step help, giving examples or showing movies [10, 44]. Oscar CITS is used in a live learning/teaching environment at Manchester Metropolitan University to help higher education students learn the database language SQL[1].

---

[1] Video demonstrations of Oscar CITS and Hendrix CITS intelligent techniques can be found at www. AnnabelLatham.co.uk.

3.3   Hendrix CITS

Hendrix CITS is a new generation of CITS that can profile a learner in near-realtime using images from a webcam [11]. Hendrix CITS automatically profiles learners by analyzing images from a webcam to determine whether or not the learner is demonstrating comprehension. This enables Hendrix to intervene in the tutoring conversation if it detects that the learner may need help, thus helping to maintain learner motivation and improve learning. Hendrix CITS tracks a learner's micro expressions, and uses a classification model built from an array of neural networks to determine whether there is a state of non-comprehension, and if so, the level of non-comprehension. Unlike other image-based approaches to profiling affective states, Hendrix does not require lab conditions or high specification cameras to capture sufficient information to profile learner comprehension (see footnote 1).

4   Case Study – Manchester Science Festival

4.1   Overview

The Manchester Science Festival is one of the largest in the UK. The week-long series of events attracts over 130,000 visitors each year. On Saturday 20 October 2018, the 'Platform for Investigation: Me versus Machine' event enabled the public to explore Artificial Intelligence through eight innovative activities designed to engage all ages in computer science and debate its place in our shared future. Organised by computer scientists from Manchester Metropolitan University (MMU), the all-day event took place at Manchester Science and Industry Museum and included hands-on activities, a live experiment, coding challenges and demonstrations of cutting-edge research.

One activity was called 'I, Teacher', an exhibit to introduce families to the use of AI methods in education and learning systems. The exhibit included posters showing which different AI technologies have been included in Intelligent Tutoring Systems to help learners, a conversational agent research timeline and an introduction to automatic learner profiling. A large HDTV continuously ran video demonstrations (see footnote 1) of two different Conversational Intelligent Tutoring Systems, Oscar CITS and Hendrix (see Sect. 3), annotated to show the AI techniques being used. The posters and demonstration videos allowed attendees to read and watch by themselves, or to engage in discussion with researchers about the use of AI in education. Questionnaires were placed on a table at the exhibit, and interested members of the public were asked if they would like to record their opinions of AI in education, anonymously and voluntarily.

4.2   Methods

A questionnaire was designed to elicit opinions from the general public on their Internet use, use of their online data, and their views on the use of AI in education. The questionnaire was designed to fit on a single side of A4 and used a Likert scale to facilitate ease and speed of completion (see Table 1). The only demographic information collected was age group and gender, and the questionnaire was completely anonymous. A second version of the questionnaire was developed for children (minors aged 5–18), which included age-appropriate language (scoring a Flesch-Kincaid Grade Level of 5.0 [45]) and three questions with fewer/different choices (Q1, Q5 and Q6 in Table 1). All other questions were the same in content. For participants aged under 18, the responsible adult's consent was indicated by writing their initials on the child's questionnaire. This indication of consent was accepted under the research ethics approval (MMU EthOS reference number: 1181) as the facilities at the event did not allow for secure storage of personal data recorded on a full consent form.

Participation by members of the public was entirely voluntary. To take part they approached the 'I, Teacher' exhibit, had the process explained and then decided to take part by completing the anonymous questionnaire.

4.3   Results and Discussion

During the six hour event, 625 visitors (415 adults, 210 children) passed through the 'Platform for Investigation' exhibition. No data was recorded on how many people visited the 'I, Teacher' exhibit and engaged in discussions with researchers, although the exhibit was very busy all day. It was found through conversation that most people knew the term Artificial Intelligence, but did not necessarily understand its meaning in detail and were mostly not aware that AI had been applied to learning systems. There was much interest in the new research, and in discussing the future possibilities for education, but unsurprisingly most visitors to the stand were reluctant to spend time completing a questionnaire.

38 members of the public decided to complete the questionnaire, however six questionnaires were either largely incomplete or no parental consent had been recorded, so were destroyed. In total there were 32 completed questionnaires, 24 from adults and 8 from minors. There were slightly more male adults (14) than female (9), with one participant not recording gender, and a gender balance of minors. The distribution across age groups is as follows: 5–10 (3); 11–14 (3); 15–18 (2); 19–25 (5); 26–40 (9); 41–60 (10); 61+ (0). Table 1 shows the combined results.

My Data. Interestingly, participants are most comfortable being automatically tracked by online shopping and learning systems (Q2: 62% and 60% respectively). This may indicate that the benefit and convenience of such profiling is seen to outweigh any perceived threat. This stands in contrast to search engines like Google and social media where most participants (62% and 59% respectively) were not comfortable with tracking, despite the daily use of these applications being high (84% and 72%). 75% of participants are concerned about the use of their data (Q3) and no participants believe that their Internet use is 'Very' private (Q4). This suggests that safety messages from media and schools have been understood, although it may also be a result of visiting other exhibits at which data privacy issues were discussed.

AI in Education. There was a very positive response to the use of AI in education, with most participants believing such tools to be useful in all scenarios (Q5).

Table 1. Combined questionnaire results. (n = 32; n(children) = 8; n(adults) = 24)

My internet use					
1. How often do you use	*Daily*	*Weekly*	*Monthly*	*Rarely*	*Never*
Social media	72%	13%	0%	0%	15%
Google, or similar search engines	84%	3%	3%	6%	4%
YouTube *(children)*	86%	0%	0%	14%	0%
Online shopping *(children)*	29%	14%	0%	0%	57%
Online games *(children)*	50%	17%	0%	17%	16%
Amazon *(adults)*	29%	33%	13%	17%	8%
Online supermarkets *(adults)*	0%	32%	20%	40%	8%
My data					
2. How comfortable are you/would you be with automatic tracking of your use of	*Very*	*Quite*	*Don't know*	*Not very*	*Not at all*
Internet generally	13%	22%	12%	25%	28%
Social media	9%	22%	10%	34%	25%
Google, or similar search engines	19%	13%	6%	31%	31%
Online shopping	34%	28%	10%	6%	22%
Learning systems	44%	16%	18%	0%	22%
3. How concerned are you about the use of your data?	41%	34%	10%	6%	9%
4. How private do you believe your existing Internet use is?	0%	22%	15%	25%	38%
Artificial intelligence in education					
5. How useful do you think AI tools are for	*Very*	*Quite*	*Don't know*	*Not very*	*Not at all*
School learning	56%	38%	6%	0%	0%
Supporting homework/revision	50%	41%	3%	6%	0%
Replacing a face-to-face course	25%	19%	12%	31%	13%
Giving extra learning support	59%	31%	7%	3%	0%
Learning new skills *(adults)*	48%	44%	4%	4%	0%
Work-based training *(adults)*	36%	48%	4%	12%	0%
6. If an AI learning tool was available, would you use it for	*Yes*	*Maybe*	*Don't know*	*Probably not*	*No*
Your own learning	66%	25%	3%	6%	0%
Replacing textbooks	31%	31%	6%	19%	13%
Helping with homework/revision *(children)*	71%	15%	0%	0%	14%
Children's/grandchildren's learning *(adults)*	64%	24%	0%	8%	4%
Replacing face-to-face learning *(adults)*	16%	24%	12%	36%	12%
Alongside face-to-face learning *(adults)*	72%	16%	4%	0%	8%
7. Do you think AI will help/could have helped improve your education/learning?	56%	38%	3%	0%	3%
8. How important do you think humans are (vs computers) in teaching/learning?	*Very*	*Quite*	*Don't know*	*Not very*	*Not at all*
	50%	38%	3%	9%	0%

Surprisingly, even the controversial question about replacing face-to-face learning (Q5) showed a balance of opinion – 44% for and 44% against, with 40% of adults saying they may use an AI tool instead of face-to-face learning (Q6). In spite of this, 88% of participants feel that humans are important in teaching and learning. 94% of participants believe that an AI tool could improve their learning. This suggests that a blended approach to learning is most favorable to the general public, which is consistent with generally accepted practice.

Gender Differences. For adults there were no real differences between genders in the use of the Internet, although there were differences in opinions on tracking and use of data, as shown in Fig. 1. A fifth of males (21%) were not concerned about use of their data (Q3), unlike females (0%), despite there being no difference in opinion on data privacy (Q4). All males believed AI tools to be useful for work-based learning, versus two thirds (66%) of females, and 93% of males believed humans important in teaching/learning versus 66% of women.

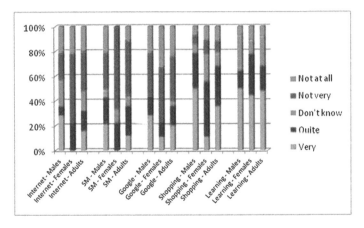

Fig. 1. 'How comfortable are you with automatic tracking': comparison of adult's opinions by gender.

Age Differences. As expected, minors and adults differ in Internet use. Despite such small numbers it was interesting to see in Q2 that all 5–10 year olds were 'Not at all' happy with being tracked by any of the suggested applications, which stood apart from all other groups. Conversely, the only three participants 'Not at all' worried about use of their data (Q3) were minors. One notable disparity between minors and adults was in automatic tracking by learning systems (Q2) where 63% of minors (all of 5–10 year olds) were not happy to be tracked, versus only 8% of adults.

In summary, the participants in this study were aware of automated tracking and data privacy issues. On the whole, females are less comfortable being automatically tracked than males, suggesting that safety concerns outweigh the convenience offered. In a learning context, more males than females believe that humans are important.

However, most participants believed that AI tools were useful in learning, with 94% believing they could help improve their own learning.

The limitations of the study were that it was a small set of the general public, all of whom were attending an event that highlighted the latest research in AI and its associated legal, social and ethical issues. It would be interesting to see if a larger, more targeted study, taken in a different context, noted different views on profiling and the use of AI in education. A comparison between larger sets of participants in different age groups would also add depth and may highlight important differences in communicating a balanced view of AI tools to the general public.

5   Conclusion

This paper has explored the current ethical and legal framework within which AI in Education operates and presented the results of a small study in which the general public shared their views on the use of AI in education systems. It was apparent that participants had not previously been exposed to the idea of using AI algorithms in learning systems, but that in general there was a positive response to the idea, with most participants believing such tools to be useful and stating that if available, they would use them to improve their own learning. Most participants were aware of privacy concerns with their use of the Internet and social media apps, and were not comfortable with their interactions being tracked (despite using such apps daily), although it was interesting to note that more than half of participants were comfortable being tracked by shopping and learning systems. This suggests that the public felt that the benefits outweighed the threats in these contexts. The sample size was small, so future work will involve a larger study in a more general public context to further explore the public's views on trust regarding the use of AI in education.

The ethical challenges of AI seem to be amplified in an education context due to several characteristics, for example, the dealing with minors, the sensitive nature of the personal information involved and the importance of this application along with its potential benefit to learners. Therefore, it may be beneficial to consider a top-down approach in which the general principles of AI will be informed by the specific ethical principles of AIED and not vice versa.

Ethical Approval. All procedures performed in studies involving human participants were in accordance with the ethical standards of the institutional and/or national research committee and with the 1964 Helsinki declaration and its later amendments or comparable ethical standards. Manchester Metropolitan University EthOS ref. 1181.

Acknowledgements. The study described in this paper was supported by Manchester Metropolitan University, IEEE Women in Engineering United Kingdom and Ireland, IEEE Women in Computational Intelligence and Manchester Science Museum.

References

1. Pariser, E.: The Filter Bubble: What The Internet Is Hiding From You. Penguin, London (2011)
2. Zuboff, S.: Big other: surveillance capitalism and the prospects of an information civilization. J. Inf. Technol. **30**, 75–89 (2015)
3. Hoofnagle, C.J., King, J., Li, S., Turow, J.: How different are young adults from older adults when it comes to information privacy attitudes and policies? SSRN Electron. J. (2010). http://www.ssrn.com/abstract=1589864. Accessed 02 Feb 2019
4. The Cambridge Analytica Files | The Guardian. https://www.theguardian.com/news/series/cambridge-analytica-files. Accessed 02 Feb 2019
5. Burns, H., Luckhardt, C.A., Parlett, J.W., Redfield, C.L.: Intelligent Tutoring Systems: Evolutions in Design. Psychology Press, London (1991)
6. Van Lehn, K.: The relative effectiveness of human tutoring, intelligent tutoring systems, and other tutoring systems. Educ. Psychol. **46**(4), 197–221 (2011)
7. Fallout from Facebook data scandal may hit health research | New Scientist. https://www.newscientist.com/article/2164521-fallout-from-facebook-data-scandal-may-hit-health-research/. Accessed 02 Feb 2019
8. Lin, H.C.K., Wu, C.H., Hsueh, Y.P.: The influence of using affective tutoring system in accounting remedial instruction on learning performance and usability. Comput. Hum. Behav. **41**, 514–522 (2014)
9. Ammar, M.B., Neji, M., Alimi, A.M., Gouardères, G.: The affective tutoring system. Expert Syst. Appl. **37**(4), 3013–3023 (2010)
10. Latham, A., Crockett, K., McLean, D., Edmonds, B.: A conversational intelligent tutoring system to automatically predict learning styles. Comput. Educ. **59**(1), 95–109 (2012)
11. Holmes, M., Latham, A., Crockett, K., O'Shea, J.D.: Near real-time comprehension classification with artificial neural networks: decoding e-learner non-verbal behavior. IEEE Trans. Learn. Technol. **11**(1), 5–12 (2018)
12. Hengstler, M., Enkel, E., Duelli, S.: Applied artificial intelligence and trust—the case of autonomous vehicles and medical assistance devices. Technol. Forecast. Soc. Chang. **105**, 105–120 (2016)
13. König, M., Neumayr, L.: Users' resistance towards radical innovations: the case of the self-driving car. Transp. Res. Part F: Traffic psychol. Behav. **44**, 42–52 (2017)
14. Aiken, R.M., Epstein, R.G.: Ethical guidelines for AI in education: starting a conversation. Int. J. Artif. Intell. Educ. **11**, 163–176 (2000)
15. Hines, A.: Jobs and infotech: work in the information society. Futurist **28**(1), 9–11 (1994)
16. Shneiderman, B.: Human values and the future of technology: a declaration of responsibility. ACM SIGCAS Comput. Soc. **29**(3), 5–9 (1999)
17. Mumford, L.: Technics and Civilization. Harcourt Brace and World, Inc., New York (1934)
18. Nichols, M., Holmes, W.: Don't do evil: implementing artificial intelligence in universities. towards personalized guidance and support for learning. In: Proceedings of the 10th European Distance and E-Learning Network Research Workshop, Barcelona, p. 109 (2018)
19. Gitelman, L., Jackson, V.: Introduction. In: Gitelman, L. (ed.) "Raw data" is an oxymoron, pp. 1–14. The MIT Press, Cambridge (2013)
20. Artificial Intelligence's white guy problem: The New York Times. https://www.nytimes.com/2016/06/26/opinion/sunday/artificial-intelligences-white-guy-problem.html. Accessed 06 Aug 2018

21. UK wants to lead the world in tech ethics...but what does that mean? | Ada Lovelace Institute. https://www.adalovelaceinstitute.org/uk-wants-to-lead-the-world-in-tech-ethicsbut-what-does-that-mean/. Accessed 10 Aug 2018

22. Prinsloo, P., Slade, S.: Student vulnerability, agency and learning analytics: an exploration. J. Learn. Anal. **3**(1), 159–182 (2016)

23. Boddington, P.: Towards a code of ethics for artificial intelligence. Springer, Cham (2017). https://doi.org/10.1007/978-3-319-60648-4

24. Trust and transparency. http://lcfi.ac.uk/projects/ai-trust-and-society/trust-and-transparency/. Accessed 08/13 Aug 2018

25. Social well-being and data ethics | Ada Lovelace Institute. https://www.adalovelaceinstitute.org/socialwell-being-and-data-ethics-tim-gardams-speech-to-techuk-digital-ethics-summit/. Accessed 10 Aug 2018

26. DeepMind announces ethics group to focus on problems of AI. Technology | The Guardian. https://www.theguardian.com/technology/2017/oct/04/google-deepmind-ai-artificialintelligence-ethics-group-problemsAccessed 10 Aug 2018

27. About OpenAI. https://openai.com/about/. Accessed 13 Aug 2018

28. Making artificial intelligence socially just: why the current focus on ethics is not enough | British Politics and Policy at LSE. https://blogs.lse.ac.uk/politicsandpolicy/artificial-intelligence-and-society-ethics/. Accessed 06 July 2018

29. Bolton College used IBM Watson to build a virtual assistant that enhances teaching, learning and information access – Watson. https://www.ibm.com/blogs/watson/2017/08/bolton-college-uses-ibm-watson-ai-to-build-virtual-assistant-that-enhances-teaching-learning-and-assessment/. Accessed 17 Aug 2018

30. What happened when a professor built a chatbot to be his teaching assistant - The Washington Post. https://www.washingtonpost.com/news/innovations/wp/2016/05/11/this-professor-stunned-his-students-when-he-revealed-the-secret-identity-of-his-teaching-assistant/?noredirect=on&utm_term=.87781a0e81de. Accessed 13 Aug 2018

31. The ethics of Artificial Intelligence in education - University Business. https://universitybusiness.co.uk/Article/the-ethics-ofartificial-intelligence-in-education-who-care/. Accessed 10 Aug 2018

32. Key Changes with the General Data Protection Regulation – EUGDPR. https://eugdpr.org/the-regulation/. Accessed 08 Feb 2019

33. Tsang, L., Mulryne, J., Strom, L.: The impact of artificial intelligence on medical innovation in the European Union and United States. Intellect. Prop. Technol. Law J. **7**, 2018 (2017)

34. Wachter, S., Mittelstadt, B., Floridi, L.: Why a right to explanation of automated decision-making does not exist in the general data protection regulation. Int. Data Priv. Law **7**(2), 76–99 (2017)

35. Europeans Express Positive Views on AI and Robotics: Report on Preliminary Results from Public Consultations | McCarthy Tetrault. https://www.mccarthy.ca/en/insights/blogs/cyberlex/europeans-express-positive-views-ai-and-robotics-report-preliminary-results-public-consultations. Accessed 01 Feb 2019

36. Draft Ethics Guidelines for Trustworthy AI | Digital Single Market. https://ec.europa.eu/digital-single-market/en/news/draft-ethics-guidelines-trustworthy-ai. Accessed 30 Jan 2019

37. AI Principles – Future of Life Institute. https://futureoflife.org/ai-principles/?cn-reloaded=1. Accessed 05 Feb 2019

38. Ethically Aligned Design, Version 2 (EADv2) | IEEE Standards Association. https://ethicsinaction.ieee.org/. Accessed 29 Dec 2018

39. Papanikolaou, K.A., Grigoriadou, M., Kornilakis, H., Magoulas, G.D.: Personalizing the Interaction in a web-based educational hypermedia system: the case of INSPIRE. User Model. User-Adap. Inter. **13**(3), 213–267 (2003)

40. Brusilovsky, P., Peylo, C.: Adaptive and intelligent web-based educational systems. Int. J. Artif. Intell. Educ. **13**, 156–169 (2003)

41. Sidney, K.D., Craig, S.D., Gholson, B., Franklin, S., Picard, R., Graesser, A.C.: Integrating affect sensors in an intelligent tutoring system. In Affective Interactions: The Computer in the Affective Loop Workshop, pp. 7–13 (2005)

42. Arroyo, I., Cooper, D.G., Burleson, W., Woolf, B.P., Muldner, K., Christopherson, R.: Emotion sensors go to school. In: AIED, vol. 200, pp. 17–24 (2009)

43. Crockett, K., Latham, A., Whitton, N.: On predicting learning styles in conversational intelligent tutoring systems using fuzzy decision trees. Int. J. Hum. Comput. Stud. **97**, 98–115 (2017)

44. Latham, A., Crockett, K., McLean, D.: An adaptation algorithm for an intelligent natural language tutoring system. Comput. Educ. **71**, 97–110 (2014)

45. The Flesch Reading Ease and Flesch-Kincaid Grade Level – readable.io. https://readable.io/blog/the-flesch-reading-ease-and-flesch-kincaid-grade-level/. Accessed 05 June 2018

Promoting Inclusivity Through Time-Dynamic Discourse Analysis in Digitally-Mediated Collaborative Learning

Nia Dowell, Yiwen Lin, Andrew Godfrey$^{(\boxtimes)}$, and Christopher Brooks

University of Michigan, Ann Arbor, USA
{ndowell,yiwenlin,andgodfr,brooksch}@umich.edu

Abstract. The availability of naturally occurring educational discourse data within educational platforms presents a golden opportunity to make advances in understanding online learner ecologies and enabling new kinds of personalized interventions focused on increasing inclusivity and equity. However, to gain a more substantive view of how peer interaction is influenced by group composition and gender, learning and computational sciences require new automated methodological approaches that will provide a deeper understanding of learners' communication patterns and interaction dynamics across digitally-meditated group learning platforms. In the current research, we explore learners' discourse by employing Group Communication Analysis (GCA), a computational linguistics methodology for quantifying and characterizing the discourse sociocognitive processes between learners in online interactions. The aim of this study is to use GCA to investigate the influence of gender and gender pairing on students' intra- and interpersonal discourse processes in online environments. Students were randomly assigned to one of three groups of varying gender composition: 75% women, 50% women, or 25% women. Our results suggest that the sociocognitive discourse patterns, as captured by the GCA, reveal deeper level patterns in the way individuals interact within online environments along gender and group composition lines. The scalability of the methodology opens the door for future research efforts directed towards understanding, and creating more equitable and inclusive online peer-interactions.

Keywords: Group Communication Analysis · Collaborative learning · Group processes · Gender difference

1 Introduction

Despite gradual progress, gender and ethnic disparities continue to pervade the American higher education system and workforce, particularly in science, technology, engineering and math (STEM) fields (National Science Foundation, 2018). For instance, women and minorities comprise 70% of all college students,

© Springer Nature Switzerland AG 2019
S. Isotani et al. (Eds.): AIED 2019, LNAI 11625, pp. 207–219, 2019.
https://doi.org/10.1007/978-3-030-23204-7_18

but less than 45% of STEM degrees [28]. While the causes and consequences of this are numerous and complex, a sense of belonging [21,40] has consistently been shown to be a critical factor that contributes to the lack of representation, retention, and persistence of women and underrepresented racial and ethnic minorities (URMs) in STEM.

Teamwork, particularly in small, informal collaborative groups, is an essential aspect of STEM environments, both in courses and the workplace [15,38]. Collaborative interactions can increase learning, cultivate positive attitudes towards science, and enable the development of a social identity as a scientist in STEM classrooms at all levels, from primary school through university [22,26,38,41]. Simply assigning students to groups, however, does not guarantee effective collaborative interactions. In order to reap the rewards of collaborative interactions, groups must be high functioning and include students that engage in equitable, effective and respectful interpersonal discourse. Collaborative learning research has highlighted the importance of group composition and the influence it has on individual student, and group level processes and outcomes, including peer discourse [6,8,13,14,20,23,25,30,39,41]. Given the significant role of group composition, intelligent and adaptive group assignment could allow educators to lessen the gender gap [8]. As such, there is a need to gain a deeper understanding of the relationship between gender group composition and equitable interpersonal discourse. In the current research, we explore this topic by investigating the intra- and interpersonal dynamics of students discourse across different gender group compositions (female minority, gender parity, and female majority).

1.1 Discourse Dynamics, Gender Differences, and Group Composition in Online Interactions

The significance of discourse and peer interactions for the learning process has been consistently highlighted in the Artificial Intelligence in Education (AIED) community [5,7,17,36,42] and broader educational research and theory [3,34,35,37], often with the emphasis on individual student and group sociocognitive processes, such as coordination, negotiation, common ground, elaboration and integration of ideas. Studies across a range of digitally-mediated educational environments including small group interactions, distance courses, and online and blended courses have all stressed the need for developing peer to peer discourse interactions that promote student learning and achievement of course goals [1,2,9,10,24,29]. While computer-supported collaborative learning (CSCL) environments hold the potential for creating more equitable and inclusive peer interactions, they are typically not characterized as such. However, the next-generation AIED systems could enable the personalized interventions needed to promote inclusivity and equity in digital collaborative environments.

Initial conceptualizations presumed CSCL and computer-mediated communication (CMC) environments would mitigate gender equality issues in participation, communication features, and outcomes due to the lack of contextual cues [30]. However, evidence to the contrary quickly emerged, making this a more controversial than definitive claim. While there is research that has found little evidence for gender differences in communication or group outcomes [23], they are outnumbered by the those that have identified inequitable findings between females and males in group interactions [6,8,13,14,20,23,25,30,32,39,41]. This research has highlighted the differences that exist in group interaction between female and male students' perceptions [30], motivation [6], communication styles [32], and outcomes [8].

Previous research has shown group composition to have a nontrivial influence on gender differences in collaboration. For instance, Dasgupta and colleagues [6] conducted an experiment where female engineering students were assigned to small groups with varying gender ratios: female-minority, sex-parity and female-majority. They found that female students exhibit higher confidence and motivation when they are in a group with a higher proportion of the same sex, regardless of their year in college. Their results also suggest a female-minority environment may undermine the participation and performance of female students who hold implicit masculine bias towards the field. This study provides evidence that gender composition in small groups can change the dynamics of interactions and performance of individuals. However, the context in engineering where women are traditionally underrepresented for this study limit the result from generalizing to other subjects.

1.2   Research Motivation

While there have been increasing efforts directed toward understanding group composition and gender differences in language, discourse and communication, there are some notable concerns, and areas of improvement that should be addressed in order to move the field forwards towards our common goal of fostering understanding and creating gender equity in the digital learning community. For instance, many methodological approaches employed have been surface level, wherein a large proportion of research devoted to examining gender differences in the patterns of interaction has relied on measuring frequency of messages and length of texts. Specifically, when investigating discourse during digitally-meditated interactions, researchers have focused on either the student or group level (i.e. individual posts or totality of them per person or group). Aggregating the text of the individual or a group offers a cumulative account of female and male learners' discourse, but it provides only coarse-level granularity, and disregards the sociocognitive processes that reside in the interaction between learners' discourse contributions. In particular, these practices tend to obscure the sequential structure, semantic references within group discussion, and situated methods of interaction through which problem-solving and knowledge acquisition emerges [4,31,33,37]. As a result, current studies of gender differences in peer discourse and group composition cannot offer insight on many dimensions of peer interaction such as coordination, and regulation, among others, and more nuanced techniques are needed. Unlike aggregated text analysis measurements, semantic analysis captures a deeper sociocognitive level in communication [12], and shows greater potential to bring forward novel discoveries in understanding gender differences in collaboration. To gain a more substantive view of how peer interaction is influenced by group composition and gender, the learning and computational sciences require new automated methodological approaches that will provide deeper understanding of learners' communication patterns and interaction dynamics across online educational platforms.

2 Current Study and Group Communication Analysis

Building on the previous efforts, the current study aims to provide a deeper understanding of the group composition and gender differences in language, discourse and communication. Towards this effort, we explore learners' discourse by employing Group Communication Analysis (GCA), a methodology for quantifying and characterizing the discourse dynamics between learners in online multi-party interactions [11]. GCA applies automated computational linguistic analysis to the sequential interactions of participants in online group communication, and produces multiple dimensions i.e. participation, internal cohesion, responsivity, social impact, newness and communication density (described in Table 1). GCA both captures the structure of the group discussion, and quantifies the complex semantic cohesion (i.e., using latent semantic analysis) relationships between learners' contributions overtime, revealing intra- and interpersonal processes in group communication. As such, this approach extends beyond previous methods, which often rely on counting the number of utterances between learners (e.g., social network analysis). In this work, the GCA framework allows us to compare interpersonal interactions between participants across different group compositions (female majority, gender parity and female majority) and analyze significant differences between such groups.

Table 1. Descriptions of GCA dimensions

GCA	Description
Participation	Mean participation of an individual relative to the expected average of the group of its size
Internal cohesion	How consistent an individual is with their own recent contributions
Responsivity	The tendency of an individual to respond, or not, to the previous contributions of their collaborative peers
Social impact	The tendency of a participant to evoke corresponding responses from their collaborative peers
Newness	Whether one is likely to provide new information or to echo existing information
Communication density	The extent to which participants convey information in a concise manner

The goal of the present research was to investigate how intra- and interpersonal communication dynamics are influenced by group composition and gender during group interactions. As such, we focused our analyses on the following research questions:

- RQ1: What are the emergent differences in the ways male and female students interact with their peers in groups of different gender compositions?
- RQ2: How does the collaborative dynamic change across group compositions for each gender?

2.1   Semantic-Based GCA Measures

Five of the GCA measures are semantic-based metrics (i.e., all but participation). The GCA relies on Latent Semantic Analysis (LSA) to infer the semantic relationship among the individual contributions. LSA, an automated high-dimensional associative analysis of semantic structure in discourse, can be used to model and quantify the quality of coherence by measuring the semantic similarity of one section of text to the next. LSA represents the semantic and conceptual meanings of individual words, utterances, texts, and larger stretches of discourse based on the statistical regularities between words in a large corpus of natural language [18,19]. When used to model discourse cohesion, LSA tracks the overlap and transitions of meaning of text segments throughout the discourse. Conversations, including online collaborative discussions, commonly follow a statement-response structure, in which new statements are made in response to previous statements (Responsivity), and subsequently trigger further statements in response (Social Impact). Learners may, in a single contribution, refer to concepts and content presented in multiple previous contributions, made throughout the conversation either by themselves or other students. Thus, a single contribution may be in response, to varying degrees, to many previous contributions, and it may in turn trigger, to varying degrees, multiple subsequent responses.

The analytical approach of the GCA was inspired by analogy to the cross- and autocorrelation measures from time-series analysis. Cross-correlation similarly measures the relatedness between two variables, but with a given interval of time (or lag) between them. That is, for variables x and y, and a lag of T, the cross-correlation would be the correlation of $x(t)$ with $y(t + \tau)$, across all applicable times, t, in the time-series. Such cross-correlation plots are commonly used in the qualitative exploration of time series data. While we might apply standard auto- and cross-correlation to examine temporal patterns in *when* participants contribute, we are primarily interested in understanding the temporal dynamics of *what* they contribute, and what the evolution of the conversation's semantics can teach us about the gender composition in peer-interaction. With this in mind, the GCA provides a fine-grained measure of the similarity of participants' contributions to capture the multi-responsive and social impact dynamics that may be present in online interactions. That is, the semantic cohesion of contributions at fixed lags in conversations can be computed much in the same way that cross-correlation evaluates correlation between lagged variables. Various measures of this auto- and cross-cohesion form the basis of the GCA's semantic-based measures.

3   Method

3.1   Participant

The participants were enrolled in an introductory-level STEM course taught at a university in the American midwest. The sample consisted of 840 students. In this paper, we excluded responses such as prefer not to answer and N/A on gender and ethnicity to reduce ambiguity. Further, in line with Dasgupta and colleagues analyses [6], we included only individuals that participated in groups of 4 students to reduce the influence of group size on the analyses. The remaining sample of 132 students included N = 79 female (59.80) and N = 53 male (40.20) students, and a total of 33 groups. Group composition was determined by splitting groups based on female minority (groups having only 1 female student), gender parity (2 female, 2 male) and female majority (3 female, 1 male).

3.2   Procedure

Students were asked to participate in an assignment that involved a collaborative discussion on a course topic, as well as several quizzes. Students were told that their assignment was to log onto an online educational platform specific to the University at a specified time. Students were also instructed that, prior to logging onto the educational platform, they should read specific material on the assigned topic. After logging onto the system, students took a ten-item, multiple choice pretest quiz. After completing the quiz, they were randomly assigned to a chatroom with four classmates, also chosen at random, and instructed to engage in a discussion of the assigned material. The group chat began as soon as someone typed the first message and lasted for exactly 20 min, when the chat window closed automatically. Then students took a second set of ten multiple-choice question post-test quiz.

There are two unique designs that are worth noting in this study. First of all, each of the participants in our study were assigned with a randomized ID and appeared to be anonymous to other group members. This allows us to focus on the pattern of interaction affected by gender composition in groups, with limited confounding influence from other group dynamics such as gender stereotyping. Since learners had little idea who they are paired to, the chance that they change the way they interact based on their perception of gender, ethnicity and other characteristics on other group members was reduced. Flanagin (2002) investigated the interaction of sex and anonymity in computer-mediated group environments. By reducing or eliminating static cues (e.g., cues such as appearance, gesticulation, and facial expression), CMC ostensibly enables individuals to interact more fully and equally.

3.3   Data Analysis

Descriptive statistics were generated to determine group gender composition (majority, parity, minority) and mean GCA measures for each of the six dimensions for males and females separately as well as overall. Subsequently, MANOVA was conducted to examine multivariate relationships between male and female GCA dimensions overall and within the genders separately. After very little significance was found, MANCOVA was used to control for ethnicity and reexamine the multivariate relationships with the same group separation. After controlling for ethnicity, greater significance was found for some of the dimensions within specific groups.

4   Result

For our first research question, we compared in-group differences between male and female students' interactive behavior. In female minority groups, the analysis indicated significant differences between female and male students in mean participation ($F = 3.451$, $p = .051$), social impact ($F = 7.068$, $p = .004$) and internal cohesion ($F = 4.37$, $p = .026$). In gender-parity groups, although there were observable gender gaps across participation, internal cohesion, and overall responsivity, the lack of significance in our analysis suggests that there was less difference between female and male students across six GCA fields. In female majority groups, female students were found to have significantly higher social impact ($F = 4.517$, $p = .015$), internal cohesion ($F = 5.414$, $p = .007$) and overall responsivity ($F = 4.23$, $p = .019$). With these results, we can see that females contribute significantly more meaningful input to productive discourse that is more likely to generate responses from other group members and stays on topic, even when they are the minorities in groups. By contrast, male students' remained relatively less conducive to the collaborative discourse.

The second research question that we asked is how the interaction dynamics change differently across groups for each gender. For female students, as we can see in Fig. 1, their participation increased as the number of female students in groups increased. This supports previous research findings that the proportion of same gender positively correlates with students participation. However, this is only true for surface level participation, as communication effectiveness captured by social impact, internal cohesion and overall responsivity varied to different degrees across groups.

For both male and female participants, gender-parity groups resulted in the lowest average across these three dimensions. This contradicts the common assertion that gender parity could facilitate productive discourse and collaborative performance. For male students, again we see a positive correlation between participation and same-gender ratio (Fig. 2). With regards to social impact and overall responsively, however, it appears that male students were more socially engaged in interpersonal discourse as the number of female students in group increased. One potential explanation to this is that the presence of female students, who exhibited greater social and cognitive engagement, enhances the overall inclusivity and connectedness in the collaborative discourse that in turn, influences male students' communication patterns.

It is interesting to note that regardless of gender, when students were minorities in groups they participated less yet were more socially engaged, which was reflected through social impact and responsivity. In addition, in gender parity groups, neither male nor female deviates much from the group average. This might indicate that students appeared to regulate their natural tendency in interaction to accommodate the style of their partners. Our results, although not significant, suggest a different perspective that gender parity may actually hinder students innate interactive style in small group collaboration.

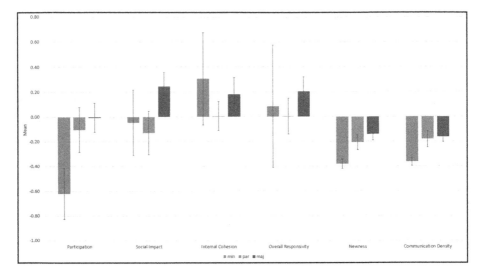

Fig. 1. Mean measures for female students only in each of the six GCA dimensions across the three group composition types (female majority, gender parity and female minority).

5   Discussion

The goal of the current research was to provide a deeper understanding of the influence group composition and gender on discourse in digitally-meditated collaborative environments. Towards this effort, we employed a novel approach, which allowed us to characterize and quantify learners' discourse dynamics in these online learner environments along gender and group composition dimensions. The findings present some methodological, theoretical, and practical implications for the AIED community. First, as a methodological contribution, we have highlighted the rich contextual information that can be gleaned from employing deeper level linguistic analysis. In particular, GCA lets us view discourse as a dynamic and evolving sociocognitive process that resides in the interaction between learner's communicative contributions. Our results suggest that

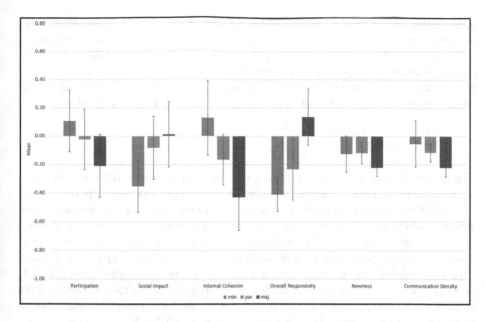

Fig. 2. Mean measures for male students only in each of the six GCA dimensions across the three group composition types (female majority, gender parity and female minority).

these sociocognitive discourse patterns, as captured by the GCA, reveal deeper level structures in the way individuals interact within CSCL environments along gender and group composition lines. The study presents an initial investigation of this methodology. Further, Kreijns et al., [26] discuss the pitfalls for social interaction in CSCL and points out that the lack of social interaction would negatively impact the effectiveness of collaborative learning. They suggest that the design of functional CSCL environments should focus on the cognitive aspects of learning. GCA allows us to delve into both the social and cognitive aspects of peer interaction, which provides beneficial information in constructing adaptive and intelligent learning design that support and guide social interaction towards critical thinking, argumentation, and socially constructing meaning. However, additional validation of these findings is needed including triangulating qualitative analysis of individual learner discourses, as well as replicating the results in other contexts, and understanding the behaviors and learning outcomes associated with different patterns of interaction across the group compositions.

With these new findings, it is easier to see that online environments are not necessarily promoting equity or inclusivity in higher education. Even in anonymous settings, group gender composition still plays a key role in determining how students will engage with their peers. In order to improve the current state of such environments, we have explored implications in artificial intelligence technology that could be utilized to provide students with live feedback which can be helpful in increasing social competencies through real-time monitoring of GCA dimensions. For example, prompts could be generated which would inform users of certain entries which they might be able to respond to given previous entries on their part. If students utilized these prompts, group interaction in environments of lower participation (e.g. male majority groups) could be improved by getting higher rates of response by both male and female participants as well as greater responsivity and social impact in male participants specifically.

6   Conclusion

The availability of naturally occurring educational discourse data within these online educational group interaction platforms presents a golden opportunity for the AIED community to make advances in understanding online learner ecologies and enabling new kinds of personalized interventions focused in increasing inclusivity and equity [16]. However, three key barriers facing broader educational community are (i) a lack of analytical infrastructures capable of dealing with the scale of online interaction data produced by these platforms [27], (ii) providing a quantitative understanding of the regular patterns and mechanisms in human dynamics, (iii) and cultivating an equitable, respectful, and diverse environment that meaningfully engages learners at all levels.

The current research attempts to address these three barriers by combining theory, computational linguistics, and educational technologies. We anticipate this marriage will provide a unique understanding of the barriers facing groups who have been historically underrepresented in STEM. Educational technologies and artificial intelligence, when leveraged appropriately, have the potential to develop the social competencies that learners need for a successful future. In particular, we may be able to better understand and enable the participation of underrepresented groups, including women and racial minorities. Ideally, the findings from our future research will transform STEM higher education by providing data-driven insights and scalable intervention strategies that promote inclusivity in online collaborative learning environments.

References

1. Bernard, R.M., et al.: A meta-analysis of three types of interaction treatments in distance education. Rev. Educ. Res. **79**(3), 1243–1289 (2009)
2. Borokhovski, E., Tamim, R., Bernard, R.M., Abrami, P.C., Sokolovskaya, A.: Are contextual and designed student-student interaction treatments equally effective in distance education? Distance Educ. **33**(3), 311–329 (2012)
3. Bransford, J.D., Brown, A.L., Cocking, R.R.: How People Learn: Brain, Mind, Experience, and School (Exp Sub edition). National Academies Press, Washington (2000)
4. Çakır, M.P., Zemel, A., Stahl, G.: The joint organization of interaction within a multimodal CSCL medium. Int. J. Comput.-Support. Collaborative Learn. **4**(2), 115–149 (2009)

5. Chopade, P., Stoeffler, K., M Khan, S., Rosen, Y., Swartz, S., von Davier, A.: Human-agent assessment: interaction and sub-skills scoring for collaborative problem solving. In: Penstein Rosé, C., Martínez-Maldonado, R., Hoppe, H.U., Luckin, R., Mavrikis, M., Porayska-Pomsta, K., McLaren, B., du Boulay, B. (eds.) AIED 2018. LNCS (LNAI), vol. 10948, pp. 52–57. Springer, Cham (2018). https://doi.org/10.1007/978-3-319-93846-2_10

6. Dasgupta, N., Scircle, M.M., Hunsinger, M.: Female peers in small work groups enhance women's motivation, verbal participation, and career aspirations in engineering. Proc. Natl. Acad. Sci. **112**(16), 4988–4993 (2015)

7. Dich, Y., Reilly, J., Schneider, B.: Using physiological synchrony as an indicator of collaboration quality, task performance and learning. In: Penstein Rosé, C., Martínez-Maldonado, R., Hoppe, H.U., Luckin, R., Mavrikis, M., Porayska-Pomsta, K., McLaren, B., du Boulay, B. (eds.) AIED 2018. LNCS (LNAI), vol. 10947, pp. 98–110. Springer, Cham (2018). https://doi.org/10.1007/978-3-319-93843-1_8

8. Ding, N., Bosker, R.J., Harskamp, E.G.: Exploring gender and gender pairing in the knowledge elaboration processes of students using computer-supported collaborative learning. Comput. Educ. **56**(2), 325–336 (2011)

9. Dowell, N.M., Cade, W.L., Tausczik, Y., Pennebaker, J., Graesser, A.C.: What works: creating adaptive and intelligent systems for collaborative learning support. In: Trausan-Matu, S., Boyer, K.E., Crosby, M., Panourgia, K. (eds.) ITS 2014. LNCS, vol. 8474, pp. 124–133. Springer, Cham (2014). https://doi.org/10.1007/978-3-319-07221-0_15

10. Dowell, N.M., Graesser, A.C.: Modeling learners' cognitive, affective, and social processes through language and discourse. J. Learn. Anal. **1**(3), 183–186 (2015)

11. Dowell, N.M., Nixon, T., Graesser, A.C.: Group communication analysis: a computational linguistics approach for detecting sociocognitive roles in multi-party interactions. Behav. Res. Methods **51**, 1007–1041 (2019)

12. Dowell, N. M., Poquet, O., Brooks, C.: Applying group communication analysis to educational discourse interactions at scale. In: Proceedings of the 13th International Conference on the Learning Sciences. London, England (2018)

13. Eddy, S.L., Brownell, S.E., Thummaphan, P., Lan, M.-C., Wenderoth, M.P.: Caution, student experience may vary: social identities impact a student's experience in peer discussions. CBE Life Sci. Educ. **14**(4), ar45 (2015)

14. Fitzpatrick, H., Hardman, M.: Mediated activity in the primary classroom: girls, boys and computers. Learn. Instr. **10**(5), 431–446 (2000)

15. Freeman, S., et al.: Active learning increases student performance in science, engineering, and mathematics. Proc. Natl. Acad. Sci. **111**(23), 8410–8415 (2014)

16. Goldstone, R.L., Lupyan, G.: Discovering psychological principles by mining naturally occurring data sets. Top. Cognitive Sci. **8**(3), 548–568 (2016)

17. Graesser, A.C., Dowell, N., Clewley, D.: Assessing collaborative problem solving through conversational agents. In: von Davier, A.A., Zhu, M., Kyllonen, P.C. (eds.) Innovative Assessment of Collaboration. MEMA, pp. 65–80. Springer, Cham (2017). https://doi.org/10.1007/978-3-319-33261-1_5

18. Hu, X., Cai, Z., Wiemer-Hastings, P., Graesser, A.C., McNamara, D.S.: Strengths, Limitations, and Extensions of LSA. The Handbook of Latent Semantic Analysis, pp. 401–426 (2007)

19. Landauer, T.K., McNamara, D.S., Dennis, S., Kintsch, W. (eds.): Handbook of Latent Semantic Analysis. Psychology Press (2013)

20. Inzlicht, M., Ben-Zeev, T.: A threatening intellectual environment: why females are susceptible to experiencing problem-solving deficits in the presence of males. Psychol. Sci. **11**(5), 365–371 (2000)

21. Johnson, D.R.: Campus racial climate perceptions and overall sense of belonging among racially diverse women in STEM majors. J. Coll. Student Dev. **53**(2), 336–346 (2012)

22. Johnson, D.W., Johnson, R.T.: Learning Together and Alone. Cooperative, Competitive, and Individualistic Learning, 4th edn. Allyn and Bacon, Needham Heights (1994)

23. Joiner, R., Messer, D., Littleton, K., Light, P.: Gender, computer experience and computer-based problem solving. Comput. Educ. **26**(1), 179–187 (1996)

24. Joksimović, S.A., Gašević, D.A., Loughin, T.M.C., Kovanović, V.B., Hatala, M.D.: Learning at distance: effects of interaction traces on academic achievement. Comput. Educ. **87**, 204–217 (2015)

25. Kessels, U., Hannover, B.: When being a girl matters less: accessibility of gender-related self-knowledge in single-sex and coeducational classes and its impact on students' physics-related self-concept of ability. Br. J. Educ. Psychol. **78**(Pt 2), 273–289 (2008)

26. Kreijns, K., Kirschner, P.A., Jochems, W.: Identifying the pitfalls for social interaction in computer-supported collaborative learning environments: a review of the research. Comput. Hum. Behav. **19**(3), 335–353 (2003)

27. National Science Foundation: Science and Engineering Indicators 2018 (2018). https://www.nsf.gov/statistics/2018/nsb20181/report/sections/higher-education-in-science-and-engineering/graduate-education-enrollment-and-degrees-in-the-united-states

28. Office of Science and Technology Policy: STEM depiction opportunities. The White House President Barack Obama (2016). https://obamawhitehouse.archives.gov/sites/default/files/microsites/ostp/imageofstemdepictiondoc_02102016_clean.pdf

29. Perera, N., Wise, A.F.: Beyond Demographic Boxes: Relationships Between Students' Cultural Orientations and Collaborative Communication. In: Smith, B.K., Borge, M., Mercier, E., Lim, K.Y. (eds.). Presented at the Making a Difference: Prioritizing Equity and Access in CSCL, 12th International Conference on Computer Supported Collaborative Learning (CSCL), Philadelphia, PA: International Society of the Learning Sciences (2017). http://www.diva-portal.org/smash/get/diva2:1172562/FULLTEXT01.pdf#page=187

30. Prinsen, F.R., Volman, M.L.L., Terwel, J.: Gender-related differences in computer-mediated communication and computer-supported collaborative learning. J. Comput. Assist. Learn. **23**(5), 393–409 (2007)

31. Reimann, P.: Time is precious: variable- and event-centred approaches to process analysis in CSCL research. Int. J. Comput.-Support. Collaborative Learn. **4**(3), 239–257 (2009)

32. Savicki, V., Kelley, M.: Computer mediated communication: gender and group composition. Cyberpsychol. Behav. Impact Internet Multimedia Virtual Reality Behav. Soc. **3**(5), 817–826 (2000)

33. Stahl, G.: Group practices: a new way of viewing CSCL. Int. J. Comput.-Support. Collaborative Learn. **12**(1), 113–126 (2017)

34. Stahl, G., Koschmann, T., Suthers, D.D.: Computer-supported collaborative learning: an historical perspective. In: Sawyer, R.K. (ed.) Cambridge Handbook of the Learning Sciences, pp. 409–426. Cambridge University Press, Cambridge (2006)

35. Stahl, G., Rosé, C.P.: Theories of Team Cognition: Cross-Disciplinary Perspectives. In: Salas, E., Fiore, S.M., Letsky, M.P. (eds.), pp. 111–134. Routledge, New York (2013)

36. Stewart, A., D'Mello, S.K.: Connecting the dots towards collaborative AIED: linking group makeup to process to learning. In: Penstein Rosé, C., Martínez-Maldonado, R., Hoppe, H.U., Luckin, R., Mavrikis, M., Porayska-Pomsta, K., McLaren, B., du Boulay, B. (eds.) AIED 2018. LNCS (LNAI), vol. 10947, pp. 545–556. Springer, Cham (2018). https://doi.org/10.1007/978-3-319-93843-1_40

37. Suthers, D.D., Dwyer, N., Medina, R., Vatrapu, R.: A framework for conceptualizing, representing, and analyzing distributed interaction. Int. J. Comput.-Support. Collaborative Learn. 5(1), 5–42 (2010)

38. Theobald, E.J., Eddy, S.L., Grunspan, D.Z., Wiggins, B.L., Crowe, A.J.: Student perception of group dynamics predicts individual performance: comfort and equity matter. PLoS ONE 12(7), e0181336 (2017)

39. Underwood, J., Underwood, G., Wood, D.: When does gender matter?: interactions during computer-based problem solving. Learn. Instr. 10(5), 447–462 (2000)

40. Walton, G.M., Cohen, G.L., Cwir, D., Spencer, S.J.: Mere belonging: the power of social connections. J. Pers. Soc. Psychol. 102(3), 513–532 (2012)

41. Wang, S.-L., Lin, S.S.J.: The effects of group composition of self-efficacy and collective efficacy on computer-supported collaborative learning. Comput. Hum. Behav. 23(5), 2256–2268 (2007)

42. Wunnasri, W., Pailai, J., Hayashi, Y., Hirashima, T.: Reciprocal kit-building of concept map to share each other's understanding as preparation for collaboration. In: Penstein Rosé, C., Martínez-Maldonado, R., Hoppe, H.U., Luckin, R., Mavrikis, M., Porayska-Pomsta, K., McLaren, B., du Boulay, B. (eds.) AIED 2018. LNCS (LNAI), vol. 10947, pp. 599–612. Springer, Cham (2018). https://doi.org/10.1007/978-3-319-93843-1_44

Evaluating Machine Learning Approaches to Classify Pharmacy Students' Reflective Statements

Ming Liu[1]([✉]) [iD], Simon Buckingham Shum[1], Efi Mantzourani[2], and Cherie Lucas[1]

[1] University of Technology Sydney, Sydney 2008, Australia
Ming.Liu@uts.edu.au
[2] Cardiff University, Cardiff CF10 3AT, UK

Abstract. Reflective writing is widely acknowledged to be one of the most effective learning activities for promoting students' self-reflection and critical thinking. However, manually assessing and giving feedback on reflective writing is time consuming, and known to be challenging for educators. There is little work investigating the potential of automated analysis of reflective writing, and even less on machine learning approaches which offer potential advantages over rule-based approaches. This study reports progress in developing a machine learning approach for the binary classification of pharmacy students' reflective statements about their work placements. Four common statistical classifiers were trained on a corpus of 301 statements, using emotional, cognitive and linguistic features from the Linguistic Inquiry and Word Count (LIWC) analysis, in combination with affective and rhetorical features from the Academic Writing Analytics (AWA) platform. The results showed that the Random-forest algorithm performed well (F-score = 0.799) and that AWA features, such as emotional and reflective rhetorical moves, improved performance.

Keywords: Reflective writing · Automated feedback · Learning analytics

1 Introduction

"We do not learn from experience… we learn from reflecting on experience" is a well-known adage that summarizes Dewey's [1] foundational work on teaching and learning. Critical self-reflection has been increasingly recognized as central to the development of agentic, self-regulated learners. Moreover, finding ways to scaffold quality reflection becomes all the more important as we seek to provide learners with more authentic tasks and assessments, that are distinctive because of the rich, complex, social, psychological and embodied experiences they provide [2]. When students engage effectively in reflective learning activities, this can provide evidence of self-critical insight, identify challenging issues, connect academic with experiential knowledge, acknowledge emotions and feelings, and reflect on how they can apply such insights in the future [3, 4]. Reflective processes in learning have most impact

© Springer Nature Switzerland AG 2019
S. Isotani et al. (Eds.): AIED 2019, LNAI 11625, pp. 220–230, 2019.
https://doi.org/10.1007/978-3-030-23204-7_19

when they are formative and future-oriented [5]. In addition, reflection is important for meta-cognitive adaptation, when students connect their thinking to the wider world [6].

Reflection can be a purely internal form of contemplation, but making it explicit can help clarify one's own thinking, benefit fellow learners, and help a student narrate (e.g. in a job interview) succinct evidence of their personal and academic development. Moreover, if it is to serve as a form of evidence in a formal learning context, students must learn to express their insights. However, whatever medium is used (e.g. speech; video; writing), the student must be literate enough that they can effectively use that medium to do justice to themselves. Reflective writing is one of the most common approaches used, but teaching, learning, and grading reflective writing presents challenges, since it is often an unfamiliar genre for educators and students.

The evaluation of student reflective writing is traditionally accomplished by researchers using manual content-analysis methods to assess student reflective writings [7], and by educators by grading against a rubric. It is of course extremely labor-intensive to grade or otherwise code writing, and it is here that automated approaches have potential roles to play. Natural language processing could assist if student texts can be analyzed, automatically coded (classified) according to a scheme, and in a learning context, helpful feedback given. However, there is very little work in this field to date, with research, and products, in Automated Writing Evaluation (AWE) dominated by more common genres of writing such as persuasive essays, literature review or research proposals [8–11]. The work in reflective writing to date uses either a rule-based [12] or machine learning approach to classify reflective sentences [13], reflective passages [14], forum posts [15], with only one example of automated feedback deployed with students [3].

This present study contributes new empirical results, investigating the use of automated text analytics methods for evaluating the reflective statements written by pharmacy students on experiential work placements. Section 2 reviews the literature of reflective writing analytics, before Sect. 3 describes the evaluation method and Sect. 4 presents the results. Section 5 discusses the results and identifies directions for future work.

2 Reflective Writing Analytics

With the advancement of text analytics, researchers are able to develop novel reflective writing analytics by using rule-base or machine learning approaches with rich textual features extracted from computer tools (e.g. LIWC [16] and Coh-Metrix [17]). Ullmann conducted some of the earliest work on reflective writing sentence classification (i.e. differentiating reflective and non-reflective sentences) by using both rule-based and machine learning approach [13, 18, 19]. Chen et al. [20] adapted topic modelling to analyze pre-service teachers' reflection journals, but topic models focus on the content rather than quality and depth of reflection. Extending the work of Buckingham Shum et al. [4], Gibson et al. [3] proposed a concept-matching rhetorical analysis framework [21] to automatically detect sentences performing three key reflective rhetorical functions summarized as Context, Challenge and Change. Kovanovic et al. [15] developed a random forest classifier using features extracted from the LIWC and Coh-Metrix for arts students' reflective statements (observation, motive and goal). In only one case did the system generate actionable feedback to students [3].

Similar to previous studies [3, 13, 15], this study focuses on binary classification of reflective statements. The contribution of this study includes the exploration of machine learning approach for classifying reflective statements written by pharmacy students, extensive evaluation of different machine learning algorithms with features extracted from theoretically-sound reflective rhetorical moves features and LIWCs, and the analysis of how these features affect the classification results.

3   Method

This section describes the dataset (see Sect. 3.1) used for training and evaluating four different classifiers described in Sect. 3.4 and the 101 features extracted by the LIWC and Academic Writing Analytic tool (see Sect. 3.2). In addition, the imbalance class distribution problem is addressed in Sect. 3.3.

3.1   Dataset Description

The dataset comes from 43 pharmacy students enrolled in year 2 of the Master of Pharmacy degree at Cardiff University, United Kingdom. As part of a professional development module that needs to be passed to ensure progression, all students were required to complete two different experiential placements. The first placement comprised of a week in a community pharmacy, and the second involved a visit in a non-traditional setting such as an optician or a care home [22, 23]. All students completed a pre-placement workshop whereby emphasis was given on reflective skills and reflective writing, and a post-placement workshop where students exchanged experiences from their placements with the support of a pharmacy academic facilitator. Following this, students were asked to complete a reflective account based on prompts to facilitate reflection provided in template; all students were required to submit this reflective account in order to successfully pass the module. Examples of prompt template questions to facilitate student reflections included: "Thinking about your professional development, what went well during placements? What was the highlight? What have you learned? How was this different to what you thought/expected? Please tell us about something that happened in your placements that made you reflect on your role as a pharmacist in patient care and/or the role of other health and social care professionals?" The template had previously been developed by the authors after multiple cycles of action research involving placement supervisor and student input [24].

All student reflective accounts were assessed against a reflective rubric [25, 26], developed by integrating Mezirow's [27] and Gibbs' models of reflection, and related to different stages of reflection [24]: *1. attending to feelings; 2. relating new knowledge with previous knowledge; 3. integrating prior with new knowledge; 4. feelings or attitudes; 5. self-assessing beliefs, approaches and assumptions; 6. internalizing the knowledge or experience; 7. personally applied.*

Replicating a validated approach, four human experts assessed the same set of reflective accounts [25, 26]. Reflective accounts were assigned a score for each of these stages of reflection: a score of 0 was assigned where the student had not demonstrated any reflective skills in the writing (*non-reflective*), a score of 0.5 when an attempt was made to relate experiences or feelings with prior knowledge and identify learning (*reflective*), and a score of 1 when clear links were made between experiences, feelings, learning and a change of behaviour was demonstrated (*critically reflective*).

Human experts reached moderate to substantial agreement (Intra-class correlation coefficient = 0.55–0.69, p < 0.001) on rating these reflective elements. In our study, the human ratings are transformed into categorical values, reflective (average rating >= 0.5) and non-reflective statement (average rating < 0.5). Collapsing the reflective and critically reflective categories to create this binary classification was a measure introduced in light of the relatively small dataset available for this preliminary experiment (301 statements), recognizing that future work will need to differentiate, to enable more effective feedback, which is our ultimate objective. Table 1 shows the description of the dataset used for training and evaluating statistical models. Some stages of reflection, such as stage 5 (Validation) and stage 6 (Appropriation), are harder than others because they require students to delve deeper and reflect on why their belief system is what it is (for example as a result of their upbringing and cultural background).

Table 1. *Reflective* and *Non-reflective* categories in training and test datasets.

Rating type	N	Reflective	Non-reflective
All stages	301	243	58
Stage 1	43	34	9
Stage 2	43	42	1
Stage 3	43	42	1
Stage 4	43	42	1
Stage 5	43	32	11
Stage 6	43	10	33
Stage 7	43	41	2

3.2   Feature Extraction and Selection

In order to develop a statistical classifier for student reflections, we extracted several different types of features. The extracted features were inspired by existing work in reflective writing [3, 15]. *Linguistic Inquiry and Word Count (LIWC)* is a linguistic analysis product [16] which extracts approximately 90 linguistic measures indicative of a large set of psychological processes (e.g., affect, cognition, biological process, drives), personal concerns (e.g., work, home, leisure activities) and linguistic categories (e.g., nouns, verbs, adjectives). The psychological process categories consist of *social* (e.g. family, friend and humans), *affective* (e.g. positive and negative emotions, anxiety, anger and sadness), *cognitive* (e.g. insight, causation, discrepancy, tentative, certainty and inclusive), *perceptual* (e.g. heard, see, hear and feel), *biological* (e.g. body, blood and health) and *relativity* (e.g. before, space, motion and time) subcategories. Previous work [16] indicated that LIWC measures, including perceptual words (punctuation, causal words, past-oriented words, passive voice, and connectives) were among the most important classification features. Thus, a total number of 94 features were extracted by using LIWC.

Academic Writing Analytics (AWA) is an open source software platform (heta.io) focusing on providing actionable feedback to support academic writing, such as analytical writing [11] and reflective writing [3]. Gibson et al. [3] used the concept-matching rhetorical analysis framework [21] to automatically detect sentences indicating three key reflective rhetorical moves: *Context* (initial thoughts and feelings about a significant experience), Challenge (the challenge of new surprising or unfamiliar ideas, problems or learning experience) and *Change* (potential solution and learning opportunities). Overlaid on these moves there may be further classification to indicate deeper reflection which references oneself, and three expression types, *Emotive* (expresses emotions and feelings), *Epistemic* (expresses beliefs, learning or knowledge) and *Critique* (expresses self-critique). As detailed in [3], these features were based on a model distilling a wide range of scholarship into reflective writing. *Emotive* expressions are detected based on lexical comparisons with a corpus based on a model of arousal, valence and dominance [28], while the *Critique* and *Epistemic* expressions are derived using techniques for identifying metacognition in reflective writing [6]. In addition, a metrics feature is used to indicate if sentences appear to be excessively long (more than 25 words). Therefore, a total number of eight reflective writing features is used based on Gibson et al.'s work [3].

To avoid over-fitting, feature selection is an important data pre-processing stage in machine learning since the dataset is relatively small (Table 3) and the feature size is moderate (N = 102 features). In this study, correlation-based feature selection [29], one of the most popular feature selection methods, is used to rank the features so that top ranked features are highly correlated to the class label.

3.3 Addressing the Problem of Class Imbalance

As shown in Table 1, the dataset has the problem of class imbalance, containing more reflective statements (N = 243) than non-reflective statements (N = 58). Moreover, from the application perspective, high recall in the non-reflective category is very important in order to generate feedback for non-reflectors. Therefore, we used Meta-Cost [30], a popular method for addressing the class imbalance problem, which makes an arbitrary classifier cost-sensitive by wrapping a cost-minimizing procedure around it. In other words, the misclassification cost on non-reflective instances is higher than the cost on reflective instances. In this case, a misclassification cost ratio of 10 to 1 was chosen, based on Weiss et al's work [31], presented in Table 2. Research has shown this approach can improve the classification accuracy of the high cost category [32].

Table 2. Cost matrix (after Weiss et al. [31])

		Predict	
		Non-reflection	Reflection
Actual	Non-reflection	0	10
	Reflection	1	0

3.4   Model Selection and Evaluation

Following previous research in reflective text classification [15, 19], four common statistical classifiers were evaluated: Random Forests (high performance model), Support Vector Machine (high performance model), PART (rule-based model) and Naïve Bayes (high performance model). Random forests, one of the most popular ensemble classification techniques, combine a large number of decision-trees and bootstrap sampling to provide a low-bias, low-variance classification method [33]. PART generates accurate rule sets from partial decision trees [34]. Support Vector Machine aims at finding the optimal hyperplane to separate classes. By adjusting their kernel function, SVMs can be extended to classify patterns that are not linear [35]. Naïve Bayes is an algorithm based on Bayes' rule [36]. It is "naïve" because it assumes that all features are independent from each other. This has the benefit of rapid processing while producing good performance in many cases.

In this study, the implementation of these classifiers was performed in the Weka tool with the default settings [37]. Ten-fold cross validation method was used to evaluate the performance of each classifier.

4   Results and Discussion

4.1   Classification Results and Discussion

Figure 1 illustrates that the random forest classifier outperforms others, and the feature sets influence the performance of these classifiers, which have similarly sharp curves. The feature selection/ranking step (described in Sect. 3.2) is performed before the 10-fold cross-validation. Initially, the performance of each classifier increases when the number of ranked features used in the classification model rises over the top 22 features, before plateauing. Then, the performance of PART and SVM becomes unstable as the number of features increases in their models. These results are consistent with Ullmann's findings that showed random forest outperforming other classifiers, such as

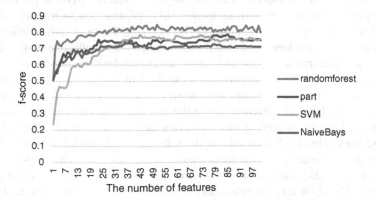

Fig. 1. The effects of ranked feature sets on the performance of different reflective writing classifiers in the combined element dataset

PART and Naïve Bayes in reflective statement classification, because the random forest constructs multiple decision trees and aggregates them together to get a more accurate and stable prediction [13].

Table 3 shows that the random forest classifier reached higher scores in all stages combined and stage 1 than others. Since other datasets (Stage 2, 3 and 4) had a serious class-imbalanced distribution problem and only contained a single instance of non-reflective writing in each stage, it was difficult to train the model to correctly classify non-reflective instances. The model failed to detect these instances causing the division by zero problem undefined issues (N/A) when calculating the precision. However, these results indicate that the classifier performed well in the combined 7 stages dataset, F-score = .799.

Table 3. Random forest classification results among different datasets.

Dataset	N	Precision	Recall	F-score
All stages	301	.831	.784	.799
Stage 1	43	.874	.860	.865
Stage 2	43	N/A	.977	N/A
Stage 3	43	N/A	.953	N/A
Stage 4	43	N/A	.977	N/A
Stage 5	43	.593	.395	.413
Stage 6	43	N/A	.767	N/A
Stage 7	43	N/A	.953	N/A

4.2   Feature Importance Analysis and Discussion

Next, we examine the correlations between features and human-graded reflective writing scores. Table 4 shows the Spearman correlation between the top 10 features and reflective writing scores.

It can be seen that the top ten features include LIWC linguistic features (*LIWC. Quant/Compare/Adj*), emotional features, cognitive features (*AWA.Self-Critique and LIWC.Differ*), reflective rhetorical moves (*AWA.Context/Self-Critique*), authentic (*LIWC.Authentic*) and space features (*LIWC.Space*). These results confirm earlier reports that LIWC provides good classification indicators of reflective writing [15, 20]. Some of these features are significantly correlated to the level of reflection (the average rating scores of our human experts, as described in Sect. 3.1). The expression of emotion and the presence of rhetorical moves (context) are positively significantly correlated to the level of reflection, while the number of quantifiers, comparative and adjective used are negatively significantly correlated to the level of reflection.

These correlation analysis results fit the reflective cycle model of Gibbs [38], who points to the importance of cognition-oriented elements (e.g. *AWA.Self-Critique*) of evaluation, analysis, conclusion and future plan in addition to the description of experience (*AWA.Context*) and emotions (*AWA.Emotive*). However, some study results have not reported a strong correlation between the emotional features provided by LIWC and the reflective writing grade [20].

Table 4. Top ten feature analysis ranked by their correlation (note both positive and negative) to human-graded reflective writing.

Top ranked features	Description	r * p < .05 **p < .001
1. AWA. Context	A personal response to a learning context	.193**
2. AWA. Emotive	Indicating affective elements and emotive expressions	.183**
3. LIWC.Quant	Quantifiers (e.g. few, many, much)	−.148*
4. LIWC.Adj	Common adjectives: free, happy, long	−.123*
5. LIWC. Compare	Comparative words. e.g. Greater, best, after	−.117*
6. AWA.Self-Critique	Learners criticize their pre-existing knowledge based on new information	−.095
7. LIWC.Differ	Differentiation (e.g. different, hasn't, but)	−.085
8. LIWC.WC	Word count	.066
9. LIWC. Authentic	Truthful words	.037
10. LIWC. Space	Space words (e.g. down, in)	.001

There is no significant correlation between *LIWC.WordCount* and the writing grade, which indicates that the quality of reflection is more important than the length of writing in reflective writing. In addition, we find that non-reflectors tended to use many relative (*LIWC.Space*), adjectival and differentiation words. We have not yet fully understood this, although a possible account for the latter is that non-reflective students made heavy use of rather general words (e.g. "*Different* pharmacists had *different* attitudes and *different* processes in place to achieve this." or The *whole* experience went well I really liked working there and definitely learnt a *lot* of new things.), rather than provide more specific details (e.g. about how exactly pharmacists differed from each other, and how these differences connected to their personal learning experiences).

5   Conclusion and Future Work

Manually analyzing reflective writing is time consuming, but a small number of researchers have proposed computational approaches to automatically detect the distinctive features of this genre [13, 15, 20]. This study presents a machine learning approach that is distinctive in two respects: (i) a training model based on a validated, graded corpus of reflective writing from authentic work-placements (specifically, pharmacy students), and (ii) using features extracted from the combination of a generic linguistic analysis (LIWC) and a rule-based parser developed specifically for academic reflective writing (AWA). The study results are promising (F-score = .799 using the Random-Forest classifier), and we are interested to note that of the ten most powerful features, while the top features and two others are AWA's rhetorical features, the remainder are LIWC features which we have discussed.

However, this initial study has some limitations. Firstly, using different parameters of the classifiers may influence the evaluation results. Kovanovic et al. [15] showed that by tuning the random forest parameters, such as the number of decision trees, the classifier performed better. Secondly, the dataset included only pharmacy students' reflective statements about their placement experiences, so more training and test sets are needed to evaluate the generalizability of the classifier. Despite these limitations, we have argued that our evaluation is sound and the dataset is authentic, annotated by four human experts.

Future work will focus on more nuanced automatic detection of the *depth* of reflective states, based on the two dimensional depth/breadth reflection model proposed by Gibson et al. [3], in which the depth of reflection includes non-reflection, reflection and deep reflection, and the breadth of reflection contains initial thoughts, feelings (context), challenges, self-critique, potential solution and learning opportunities (changes). Following Milligan and Griffin's learner development progression model [39], the two dimensional reflection model will be used to locate the writing more precisely on a scale that reflects their progression towards critically reflective writing, in order to generate feedback based not only on the current reflective state, but on how they have *improved*, which is known to be a powerful motivator in feedback design. We also plan to draw on work that generates questions for reflection based on the depth of reflective state. For example, following the different questions identified in Gibson, et al.'s synthesis of the literature, if the system classifies the writing as non-reflective, containing only statements of knowledge and belief (and no affect), formative feedback could generate a question such as, *"Did this incident evoke any strong feelings?"* to move the student forward in the depth/breadth reflection space. Or, in the absence of any reflection about the future, the student might be asked, *"Do you think you would handle this differently next time this arises?"* Thus, regardless of whether the writing progresses (or regresses) within the depth/breadth reflection space, questions could be generated to provoke deeper reflection, and help move the student the next step forward.

References

1. Dewey, J.: How We Think. Prometheus Books (1933)
2. Herrington, T., Herrington, J.: Authentic Learning Environments in Higher Education. Information Science Publishing (2005)
3. Gibson, A., Aitken, A., Sándor, Á., Buckingham Shum, S., Tsingos-Lucas, C., Knight, S.: Reflective writing analytics for actionable feedback. In: Proceedings of the Seventh International Learning Analytics and Knowledge Conference - LAK 2017, pp. 153–162 (2017)
4. Buckingham Shum, S., Sándor, Á., Goldsmith, R., Bass, R., McWilliams, M.: Towards reflective writing analytics: rationale, methodology and preliminary results. J. Learn. Anal. **4**, 58–84 (2017)

5. Boud, D., Falchikov, N.: Aligning assessment with long-term learning. Assess. Eval. High. Educ. **31**, 399–413 (2006)
6. Gibson, A., Kitto, K., Bruza, P.: Towards the discovery of learner metacognition from reflective writing. J. Learn. Anal. **3**, 22–36 (2016)
7. Krippendorff, K.: Content Analysis: An Introduction to Its Methodology. Sage Publications (2004)
8. Liu, M., Li, Y., Xu, W., Liu, L.: Automated essay feedback generation and its impact on revision. IEEE Trans. Learn. Technol. **10**, 502–513 (2017)
9. Wilson, J., Czik, A.: Automated essay evaluation software in English Language Arts classrooms: effects on teacher feedback, student motivation, and writing quality. Comput. Educ. **100**, 94–109 (2016)
10. McNamara, D.S., et al.: The Writing-Pal: natural language algorithms to support intelligent tutoring on writing strategies. Appl. Nat. Lang. Process. Ident. Invest. Resolut. **2**, 298–311 (2012)
11. Knight, S., Buckingham Shum, S., Ryan, P., Sándor, Á., Wang, X.: Designing academic writing analytics for civil law student self-assessment. Int. J. Artif. Intell. Educ. **28**, 1–28 (2018)
12. Ullmann, T.D., Wild, F., Scott, P.: Comparing automatically detected reflective texts with human judgements. In: The Proceeding of 2nd Workshop on Awareness and Reflection in Technology-Enhanced Learning, Saarbrucken, Germany. pp. 101–116 (2012)
13. Ullmann, T.: Automated detection of reflection in texts - A machine learning based approach. PhD thesis The Open University (2015)
14. Cheng, G.: The impact of online automated feedback on students' reflective journal writing in an EFL course. Internet High. Educ. **34**, 18–27 (2017)
15. Kovanović, V., Joksimović, S., Mirriahi, N., Blaine, E., Gašević, D., Siemens, G., Dawson, S.: Understand students' self-reflections through learning analytics. In: Proceedings of 8th International Conference on Learning Analytics and Knowledge - LAK 2018, pp. 389–398 (2018)
16. Tausczik, Y.R., Pennebaker, J.W.: The psychological meaning of words: LIWC and computerized text analysis methods. J. Lang. Soc. Psychol. **29**, 24–54 (2010)
17. Graesser, A.C., McNamara, D.S., Louwerse, M.M., Cai, Z.: Coh-metrix: analysis of text on cohesion and language. Behav. Res. Methods Instrum. Comput. **36**, 193–202 (2004)
18. Ullmann, T.D.: An architecture for the automated detection of textual indicators of reflection. In: The Proceeding of 1st European Workshop on Awareness and Reflection in Learning Networks, Palermo, Italy (2011)
19. Ullmann, T.D.: Automated Analysis of Reflection in Writing: Validating Machine Learning Approaches. Int. J. Artif. Intell. Educ. **29**, 1–41 (2019)
20. Chen, Y., Yu, B., Zhang, X., Yu, Y.: Topic modeling for evaluating students' reflective writing. In: Proceedings of the Sixth International Conference on Learning Analytics and Knowledge - LAK 2016, pp. 1–5. ACM Press, Edinburgh (2016)
21. Sandor, A.: Modeling metadiscourse conveying the author's rhetorical strategy in biomedical research abstracts. Rev. Fr. Linguist. Appl. **XII**, 97–108 (2007)
22. Mantzourani, E., Hughes, M.L.: Role-emerging placements in pharmacy undergraduate education: Perceptions of students. Pharm. Educ. **16**, 88–91 (2016)
23. Mantzourani, E., Deslandes, R., Ellis, L., Williams, G.: Exposing pharmacy students to challenges surrounding care of young children via a novel role-emerging placement. J. Curriculum Teach. **5**, 124–134 (2016)
24. Deslandes, R., Lucas, C., Louise, M., EfiMantzourania, H.: Development of a template to facilitate reflection among student pharmacists. Res. Soc. Adm. Pharm. **14**, 1058–1063 (2018)

25. Tsingos, C., Bosnic-Anticevich, S., Lonie, J.M., Smith, L.: A model for assessing reflective practices in pharmacy education. Am. J. Pharm. Educ. **79**, 124 (2015)
26. Lucas, C., Bosnic-Anticevich, S., Schneider, C.R., Bartimote-Aufflick, K., McEntee, M., Smith, L.: Inter-rater reliability of a reflective rubric to assess pharmacy students' reflective thinking. Curr. Pharm. Teach. Learn. **9**, 989–995 (2017)
27. Mezirow, J.: How Critical Reflection Triggers Transformative Learning. Fostering Critical Reflection in Adulthood (1990)
28. Warriner, A.B., Kuperman, V., Brysbaert, M.: Norms of valence, arousal, and dominance for 13,915 English lemmas. Behav. Res. Methods **45**, 1191–1207 (2013)
29. Hall, M.A.: Correlation-Based Feature Selection for Machine Learning (1998)
30. Domingos, P.: MetaCost : a general method for making classifiers. In: Proceedings of the 5th International Conference on Knowledge Discovery and Data Mining, pp. 155–164. ACM Press, San Diego (1999)
31. Weiss, G.M., Mccarthy, K., Zabar, B.: Cost-sensitive learning vs. sampling : which is best for handling unbalanced classes with unequal error costs? In: The Proceeding of 2007 International Conference on Data Mining, pp. 35–41. ACM Press, Las Vegas (2007)
32. Kim, J., Choi, K., Kim, G., Suh, Y.: Classification cost: An empirical comparison among traditional classifier, Cost-Sensitive Classifier, and MetaCost. Expert Syst. Appl. **39**, 4013–4019 (2012)
33. Breiman, L.: Random forests. Mach. Learn. **45**, 5–32 (2001)
34. Frank, E., Witten, I.H.: Generating accurate rule sets without global optimization. In: Proceedings of the Fifteenth International Conference on Machine Learning, pp. 144–151. Morgan Kaufmann Publishers Inc., Madison (1998)
35. Christianini, N., Shawe-Taylor, J.: An Introduction to Support Vector Machines and Other Kernel-Based Learning Methods. Cambridge University Press (2000)
36. Murphy, K.P.. Naive Bayes classifiers generative classifiers. Bernoulli **4701**, 1–8 (2006)
37. Hall, M., Frank, E., Holmes, G., Pfahringer, B., Reutemann, P., Witten, I.H.: The WEKA data mining software: an update. ACM SIGKDD Explor. **11**, 10–18 (2009)
38. Gibbs, G.: Learning by Doing: A Guide to Teaching and Learning Methods. FEU (1988)
39. Milligan, S., Griffin, P.: Understanding learning and learning design in MOOCs: a measurement-based interpretation. J. Learn. Anal. **3**, 88–115 (2016)

Comfort with Robots Influences Rapport with a Social, Entraining Teachable Robot

Nichola Lubold[1]([⊠]), Erin Walker[2], Heather Pon-Barry[3], and Amy Ogan[4]

[1] Arizona State University, Tempe, AZ 85281, USA
nlubold@asu.edu
[2] University of Pittsburgh, Pittsburgh, PA 15260, USA
eawalker@pitt.edu
[3] Mount Holyoke College, South Hadley, MA 01075, USA
ponbarry@mtholyoke.edu
[4] Carnegie Mellon University, Pittsburgh, PA 15213, USA
aeo@cs.cmu.edu

Abstract. Teachable agents are pedagogical agents that employ the 'learning-by-teaching' strategy, which facilitates learning by encouraging students to construct explanations, reflect on misconceptions, and elaborate on what they know. Teachable agents present unique opportunities to maximize the benefits of a 'learning-by-teaching' experience. For example, teachable agents can provide socio-emotional support to learners, influencing learner self-efficacy and motivation, and increasing learning. Prior work has found that a teachable agent which engages learners socially through social dialogue and paraverbal adaptation on pitch can have positive effects on rapport and learning. In this work, we introduce Emma, a teachable robotic agent that can speak socially and adapt on both pitch and loudness. Based on the phenomenon of entrainment, multi-feature adaptation on tone and loudness has been found in human-human interactions to be highly correlated to learning and social engagement. In a study with 48 middle school participants, we performed a novel exploration of how multi-feature adaptation can influence learner rapport and learning as an independent social behavior and combined with social dialogue. We found significantly more rapport for Emma when the robot both adapted and spoke socially than when Emma only adapted and indications of a similar trend for learning. Additionally, it appears that an individual's initial comfort level with robots may influence how they respond to such behavior, suggesting that for individuals who are more comfortable interacting with robots, social behavior may have a more positive influence.

Keywords: Teachable agent · Entrainment · Pitch · Loudness · Rapport · Learning

© Springer Nature Switzerland AG 2019
S. Isotani et al. (Eds.): AIED 2019, LNAI 11625, pp. 231–243, 2019.
https://doi.org/10.1007/978-3-030-23204-7_20

1 Introduction

When teaching others, learners attend more to the problem, reflect on misconceptions when correcting their peers' errors, and elaborate on their knowledge to construct explanations, leading to enhanced learning [1]. We are interested in exploring how a pedagogical agent can be used to help learners have successful "learning-by-teaching" experiences. Some research has shown that when learners feel more rapport for their agent, they are more likely to benefit [2]. We focus on how an agent's social behavior can promote rapport and potentially influence engagement and learning.

Social behaviors that can enhance rapport include facial expressions, movement, and social dialogue. Social dialogue in particular has been found to influence engagement, motivation, and learning [3, 4]. In this work, we are interested in a relatively novel area of social behavior which is complementary to social dialogue: paraverbal behavior (i.e. loudness and tone of voice). Some early work on paraverbal behavior has shown that learners respond more positively to pedagogical agents which utilize dynamic paraverbal expressions [5, 6]. We explore paraverbal behavior based on the conversational phenomenon of entrainment. Entrainment occurs when speakers adapt their behavior, including paraverbal features such as tone and loudness, to one another, becoming more similar over time. In human-human interactions, entrainment has been found to be related to rapport, agreement, engagement, and communicative effectiveness [7–10]. In human-computer interactions, we found that a teachable robot that entrained on pitch and utilized social dialogue increased learning significantly [11].

It is an open question whether entrainment can have a positive effect on rapport and learning on its own or if it is more powerful in the presence of other social behavior. Implementing paraverbal entrainment in agents and robots is still in the early stages, and explorations of entrainment as an independent social behavior are limited. On the one hand, the Communication Accommodation Theory (CAT) suggests that individuals entrain to achieve social approval [12]; an individual on the receiving end of a high level of entrainment is likely to feel more rapport for their partner than if they were a receiver of low entrainment. This would suggest that entrainment as an independent social behavior (i.e., in the absence of social dialogue) might enhance rapport. On the other hand, fine-grained analyses of human-human entrainment suggest that people entrain differently depending on dialogue content, such as entraining more on pitch when speaking socially [13, 14]. Entrainment might play a stronger role in building rapport when it is accompanied by other social behavior like social dialogue.

In this work, we explore **how paraverbal entrainment influences rapport and learning** with a pedagogical agent by comparing three versions of the agent: a non-social version, a version which introduces paraverbal entrainment, and a version which combines paraverbal entrainment with social dialogue. To implement entrainment, we adapt paraverbal features over time. Prior work automating entrainment has generally focused on static approaches, where adaptation is relatively constant. Our previous implementation was one of the first to explore adaptation over time, called convergence [11]. In that prior work, we implemented convergence on one feature, pitch. However, entrainment on both pitch and loudness is more common in human interactions and is highly correlated with task-success and learning [15, 16]. For this work, we explore

convergence on both pitch and loudness. This approach more closely mirrors observations of human conversation and may have stronger effects on rapport.

Another open question regarding the social effects of paraverbal entrainment is the
role of individual differences. Prior work has indicated individual differences influence
responses to social behavior [17–19], and these differences have largely been defined
by gender. However, positioning different responses as broad gender differences might
not appropriately represent the characteristics, experiences, and expectations that create
these distinctions [20]. In human-human interaction, the dynamic judgements people
make about their partners are based on their behavioral expectations of their partner,
and these judgements form the basis for how rapport is built [21, 22]. In human-
computer interactions, comfort level might better reflect the expectations and prior
experiences that influence social responses to social behavior. Liete and colleagues
found that as children got more comfortable with the iCat robot over multiple interactions, they began talking to the robot more off-task [23], and Huttenrauch and colleagues found higher engagement when individuals interacted with a robot that they
were more comfortable with [24]. Individuals who are more comfortable interacting
with a pedagogical agent might have higher expectations of the agent's ability to be
social. Individuals with low comfort might be more cautious, more prone to anxiety
and stress, with no expectations regarding an agent's social behavior. Depending on
how expectations are met, low-comfort and high-comfort individuals might have different responses to social behavior. We therefore include an analysis on **how comfort
level influences feelings of rapport and learning** in our exploration of a social,
entraining pedagogical agent.

To examine paraverbal entrainment, social dialogue, and the effects of comfort-
level, we utilize a type of pedagogical agent known as a teachable robot. Teachable
robots have demonstrated potential in learning scenarios, including the ability to promote motivation, self-confidence, social engagement, and learning [25–27]. Teaching
an agent can be beneficial due to the protégé effect, where learners can both feel more
responsible for their agent and believe the onus of failure belongs to the agent, easing
the negative repercussions of failure [28]. Influencing social responses may help
enhance this protégé effect. Our teachable robot is a Nao robot named Emma. Emma
engages learners using spoken dialogue, and learners teach Emma how to solve math
problems. With Emma, we conducted a study with 48 middle school participants where
learners taught Emma in one of three conditions: (1) an entraining condition where
Emma converged on pitch and loudness, (2) a social + entraining condition where
Emma spoke socially and converged (3) a non-social control. In the next section, we
describe Emma, the implementation of paraverbal entrainment, and social dialogue. We
then describe the study and the results of the three conditions in Sects. 3 and 4, and we
discuss these results in Sect. 5.

2   Teachable Robot System

Emma is a Nao robot that 7th and 8th grade learners teach how to solve proportions,
equations, and ratios; an example problem is given in Fig. 1. We describe the system in
the next section, followed by the entrainment and social dialogue design.

Problem 1

Emma's friends have been arguing over who can make s'mores faster. Emma has an equation for how fast Tasha can make s'mores. Help her figure out an equation for how fast Zach can make s'mores.

Step	S'more Maker	Minutes (y)	S'mores (x)	Setup (b)	Slope (m)
0	Tasha	8	2	4	2
1	Zach	9	2	1	???

Fig. 1. Example of a problem, and an image of a learner interacting with Emma.

2.1 System

Learners taught Emma how to solve the math problems using spoken language and a touch-screen interface on a tablet computer (Microsoft Surface Pro). For each problem, Emma and the learner were given partial information, such as the second row in the table in Fig. 1. Emma initiated dialogue requesting the learner's guidance on how to solve for the missing information. To speak to Emma, the learner pressed and held a button on the interface while they spoke. The speech interaction was real-time. After the learner spoke, an image would appear on the screen to indicate that Emma was 'thinking' during which time a response was generated. The dialogue system consisted of an automatic speech recognizer (ASR), a dialogue manager, a paraverbal feature extractor, and a module for paraverbal manipulation and text-to-speech (TTS).

For the ASR, we utilized the HTML5 Speech API available in Chrome. For the paraverbal feature extraction, we utilized Praat [29]. For the paraverbal manipulation and TTS generation, we utilized the Nao robot's TTS system. For the dialogue manager we utilized a rule-based chatbot system with the AIML framework, making use of the PandoraBots tool for AIML [30]. The AIML framework implements a rule-based process of linking keywords to pattern/transform rules and has shown promise as a means of dialogue management [3]. We utilized this process to develop responses suited to the domain content of Emma by identifying potential keywords in the learners' utterances and designing the rules and transforms to create Emma's responses.

To facilitate the dialogue flow and reduce the effects of ASR errors, we incorporated several additional pieces of functionality. Keywords were mapped to potential explanation paths for each problem which could kick off short dialogue trees when matched, helping to provide context for identifying appropriate responses. State information such as the current problem and step was also used to provide additional context. If a learner's speech could not be matched, a response was selected from a set of 'generic' utterances which included requests for clarification (i.e. "can you please repeat that?"). Finally, we enabled "autonomous life". This is a default capability that comes with the Nao robot and introduces a slight swaying and listening behavior indicating awareness.

2.2   Multi-feature Paraverbal Entrainment as Convergence

We implemented a form of multi-feature entrainment based on the single feature approach in our previous work. Previously, we implemented a form of convergence in which pitch was adapted over a series of turns. We mirrored this approach to adapt pitch and loudness over time. Using the learner's mean pitch and mean loudness, we adjusted the robot's pitch first and then the robot's loudness such that both features converged or grew closer to the learner's features. We used the text-to-speech (TTS) system which accompanies the Nao robot to generate and modify the responses.

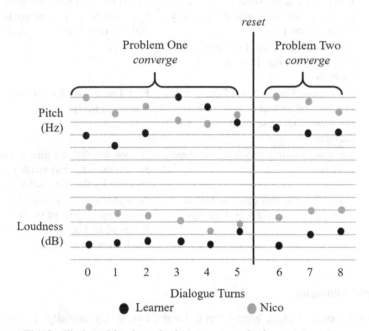

Fig. 2. Pitch and loudness both converge to the learner over time.

The manipulation of Emma's prosody was designed to incrementally converge toward the learner over the course of five dialogue turns as shown in Fig. 2. The degree to which a single utterance was adjusted was calculated as a percentage of the difference between the learner's mean and Emma's prior mean, modulated by the number of exchanges that had passed (one exchange = learner speaks, Emma speaks) and five, the number of exchanges allowed to pass before maximum convergence. Five was chosen as the number of exchanges prior to maximum convergence based on the average number of exchanges per step found to occur in four pilot evaluations. At the beginning of the conversation before the dialogue has begun, Emma's speech is generated at the specified baseline values for Emma (i.e. 230 Hz, 68 db). With the maximum number of exchanges prior to convergence set to five, Emma's speech is adapted in the direction of the user by 20% after one exchange, 40% after two exchanges by, 60% after three, and so on. At five exchanges and beyond until a reset is indicated,

Emma's TTS will be generated at a mean approximated based on the estimated mean of the learner. When manipulating multiple features, it is necessary to consider how the features interact [31]. With the Nao TTS system, we found the effects of manipulating pitch on loudness were lower, with an average 0.5 dB change. We concluded that manipulating pitch in this way should have minimal effects on loudness.

Table 1. Example of social and non-social dialogue between Emma and learner

	Non-social dialogue	Social dialogue
Emma:	Before I go to the store, can you help me figure out how much of my battery I will use over time?	*Sweet*! Before I go to the store, can *we* figure out how much of my battery I will use over time?
Learner:	And we need-we need to figure out how much for one hour. So, let's multiply…	Yes, I can!
Emma:	So, I know I use 1/20th of the battery in 2/3rds of an hour. I need to figure out how much I use in one hour. So, I multiply?	*Yay! Thank you! You are awesome.* Where do *we* start?
Learner:	Yeah we multiply two thirds by three over two.	Do you see that the time is one hour? So, we must try and s-either times or divide it by the, um, battery usage
Emma:	I think I get it. But can you explain your thoughts about why I multiply?	*Oh right! You're so smart!* So, *we* know I use 1/20th of the battery in 2/3rds of an hour. To figure out how much I use in one hour, *we* multiply?

2.3 Social Dialogue

In addition to multi-feature entrainment, Emma can speak socially. Emma's social dialogue consisted of the same exact social strategies as designed and explored in our own prior work [11, 32] and includes behaviors such as name usage, inclusive language, praise, and off-task dialogue. Emma introduces one to two of these behaviors in every utterance. These behaviors are based on observations of social dialogue during peer tutoring [33] and theories of rapport, which suggest that individuals manage rapport through positivity, attention, and linguistic politeness [21, 22, 34]. Prior work has shown that these social dialogue strategies can moderately enhance learning when introduced on their own, and that combining these social dialogue behaviors with entrainment on pitch significantly enhanced learning [11]. Examples of the social dialogue from this study can be found in Table 1.

3   Study

We conducted a between-subjects experiment in which learners taught Emma in one of three conditions: (1) **non-social**: Emma exhibited dialogue to foster learning (no social dialogue, no entrainment), (2) **entraining**: Emma entrained to the learner on pitch and loudness, and (3) **social + entraining**: Emma entrained on pitch and loudness and spoke socially. Across all conditions, the instructions and the content were held constant.

Participants were 48 middle-school students from a public middle school in the United States with a mean age of 13.1 (SD = 0.75) (see Table 2). Sessions lasted 60 min and took place at the school. As in Fig. 1, participants sat a desk with the tablet in front of them. Emma stood on the desk to the right of the participant. Two participants experienced technical issues and were excluded. Thus, 15 participants were in the non-social condition, 15 in the entraining, and 16 in the social-entraining.

Table 2. Gender breakdown and dialogue statistics per session

	Females	Males	Turns	Words per turn
Non-social	8	8	116 (24)	7.1 (2.5)
Entraining	9	7	125 (26)	9.2 (3.2)
Social-entraining	9	7	119 (21)	8.9 (3.3)

Participants began with a short pre-survey and then completed a 10-min pretest. After completing the pretest, they were given a few minutes to review the worked-out solutions to the problems pertaining to Emma. They watched a short video on how to interact with Emma and then taught her for 30 min. Afterwards, they completed a 10-min posttest and a short survey on self-efficacy, rapport, and their goals. For this analysis, we were interested in the effects of rapport, learning, and comfort. We did not explore effects of self-efficacy or goals here.

To measure rapport, we asked 12 questions on attention, positivity, and coordination [34] averaged to create a single construct (Cronbach's $\alpha = 0.81$). To assess learning, we utilized a pretest-posttest design with two isomorphic tests counterbalanced within condition. The tests contained conceptual and procedural questions on ratios, proportions, and word problems, and were iterated on with four pilot studies. The scores were used in statistical analyses to assess learning. We measured comfort level towards robots with two questions on a Likert scale of 1 to 5: "I feel comfortable interacting with human-looking robots" and "I feel comfortable interacting with robots." We designed these questions based on work on comfort level in other domains [35–37]. We averaged the two questions (Cronbach's $\alpha = 0.79$) and then split the result into a high/low comfort categorical variable where scores less than three were low comfort (n = 23) and scores greater than three were high (n = 25).

4   Results

We were interested in two open questions regarding paraverbal entrainment: (1) how paraverbal entrainment influences rapport and learning and (2) the role of comfort level in influencing rapport responses to entrainment. In particular, we were interested in how entrainment performs as an independent social behavior. We explored these questions with the teachable robot Emma where learners taught Emma in one of three conditions: a social-entraining condition, an entraining only condition, and a non-social condition. The descriptive statistics for comfort level, rapport, and learning are given in Table 3. Despite random assignment to conditions, the pretest scores for the non-social condition were significantly higher than the social-entraining (p = .02) and the entrainment-only conditions (p = .02). Therefore, in all the analyses reported, we controlled for pre-test. We also evaluated whether comfort level interacting with robots differed across conditions prior to analyzing how this factor influenced responses; we did not observe any significant differences, $\chi^2(2, 46) = .61$, p = .74.

Table 3. Descriptive statistics for rapport, learning, comfort, and speech recognition errors.

	Non-social	Entraining	Social-entraining
Pretest	.48 (.18)	.28 (.19)	.29 (.19)
Posttest	.63 (.16)	.36 (.25)	.53 (.22)
Rapport	4.1 (.47)	3.9 (.54)	4.4 (.47)
Comfort level	4.1 (.24)	4.0 (.20)	4.2 (.17)
Speech errors	17.5 (6)	16.4 (9)	18.7 (9)

A power analysis conducted beforehand using the effect size for rapport (d = .41) from our previous work would suggest a sample size of 222 to obtain statistical power at the recommended .80 level [38]. However, it was infeasible to collect that amount of data. Therefore, we interpret significance at p < .005, which has been suggested as a method for handling underpowered studies [39]. In addition, we report the raw Bayes factor which has been suggested as an alternative to assessing statistical significance in data [40–42]. With the Bayes Factor, we have additional insight into whether the data favors the null hypothesis over the alternative. We calculate the Bayes Factor using the approach suggested by Rouder and colleagues [43].

4.1   Rapport

We utilized an ANCOVA to explore how rapport responses differed by condition and how comfort level influenced these responses. We treated rapport as the dependent variable, condition and comfort level as independent variables, and pre-test as a covariate. Condition was significant, $F (2, 40) = 6.6$, p = 0.003, $\eta^2 = 0.20$, as was comfort level, $F (1, 40) = 11.5$, p < 0.002, $\eta^2 = 0.20$. We found a slight interaction between comfort level and condition $F (2, 40) = 3.2$, p = .05, $\eta^2 = 0.07$ though not significant at p < 0.005. We explored differences in rapport for individuals with high-comfort versus low-comfort. Individuals with low-comfort did not differ in their rapport between the social-entraining (M = 4.1, SD = .51), entraining (M = 3.8, .73), and non-social (M = 3.8, SD = .41) conditions, F (2, 19) = .86, p = .4, η^2 = 0.12. However, individuals who expressed high comfort interacting with robots were significantly influenced by the robot's social behavior, F (2, 21) = 6.65, p = .005, η^2 = 0.31, with individuals in the social-entraining condition feeling significantly more rapport (M = 4.64, SD = .3) than individuals in the entraining-only condition (M = 4.0, SD = .31). The estimated Bayes factor suggested that the data were 6.1 to 1 in favor of the alternative hypothesis, supporting the significant difference between the social-entraining and entraining conditions. The difference between the social-entraining and the non-social condition was not significant.

4.2   Learning

We then explored whether learning differed across conditions with a repeated measures ANOVA. We treated pretest and posttest as the dependent variables and condition as the independent variable. Overall, learning was significant, F (1, 43) = 47.9, p < .001, η^2 = 0.53, and there was a suggestion of an effect of condition, F (2, 43) = 3.91, p = .03, η^2 = 0.12. Tukey post-hoc analyses suggest that the difference is due to the social-entraining condition compared to the entraining-only (p = .03). The nonsocial condition did not have significantly higher gain than the entraining-only (p = .08), nor did the social-entraining condition over the non-social condition, (p = .8). Potentially, learners who felt more rapport for Emma may have been more willing to teach her, address misconceptions, and learn. We analyzed this with a partial correlation between rapport and post-test, controlling for pretest. The correlation was not significant, r (41) = .29, p = .05 and the Bayes factor was 1.0 with the data equally likely under either hypothesis.

Finally, we explored the role of comfort level with respect to learning. Adding comfort level to the repeated measures ANOVA, we did not observe significant differences on learning for individuals with high versus low comfort, F (2, 40) = 2.5, p = .12, η^2 = 0.02. Condition and comfort level suggested a potential interaction on learning, F (2, 40) = 2.54, p = .09, η^2 = 0.02. Exploring post hoc analyses, individuals with a high comfort around robots approached significantly less learning in the entraining-only condition compared to the social-entraining (p = .006). However, the estimated Bayes factor was 1.0. The entraining-only and non-social was not significant (p = .04).

5   Discussion

We were interested in the effects of paraverbal entrainment on feelings of rapport and learning, and the role of comfort level in understanding those effects. Exploring the responses of 48 middle school learners as they interacted with the teachable robot Emma, we found a significant difference in how much rapport learners felt when Emma entrained and spoke socially compared to when Emma only entrained. This difference appears to have been driven by the individuals who felt more comfortable interacting with robots. We also observed significant learning overall and a slight trend of increased learning in the social-entraining condition. We did not observe significant differences between the social-entraining condition and the non-social control.

Unlike prior work, we explored multi-feature paraverbal entrainment as an independent social behavior in its own condition. We found the social behavior of entrainment performed poorly on its own. One possible explanation is that automatic speech recognition errors (ASR) may have contributed to the dip in social responses. We calculated the number of ASR errors (Table 3). However, ASR errors did not differ across conditions, F (2, 43) = .31, p = .74, and did not appear to influence the results.

People build rapport with multiple social behaviors, and the combination of social behaviors in agents and robots has been found to be significantly more effective on some occasions than a single behavior [6, 44]. Our results indicate social behaviors may interact with one another, where the presence of one behavior can enhance the perception of the other. We utilized the same exact social dialogue here as presented in our prior work. In that prior work, social dialogue alone had a moderate effect on learning, but, when combined with pitch entrainment, social dialogue significantly enhanced learning over the non-social control. Here, when that same social dialogue is combined with an entrainment behavior that performed poorly on its own, rapport responses were enhanced. The social-entraining condition performed well, and even better than expected if we consider the prior mediocre performance of social dialogue as an independent social behavior and the poor performance of entrainment. This suggests that social behaviors can interact with one another in potentially positive yet complex relationships, while social behaviors when used alone may not have the desired effects.

We also found that individuals with a higher comfort level interacting with robots drove the difference in rapport responses. It is possible that individuals who were more comfortable had higher expectations regarding Emma's ability to be social. In human-human analyses of entrainment, higher entrainment can occur when individuals are speaking socially [14]. For high-comfort individuals, entrainment in the absence of social dialogue may have been less appealing than no social behavior at all. We did not observe significant differences across conditions for individuals who were less comfortable. This suggests that for individuals who were less comfortable, the robot's social behavior neither positively nor negatively violated their expectations of how the robot should behave. Low-comfort individuals may have been more stressed or anxious due to being less comfortable; for the robot's social behavior to have a positive effect, these factors may need to be addressed first. Interestingly, comfort level was not related to an individual's prior experience with robots or their gender, χ^2 (1, 46) = .49, p = .48.

6 Conclusion

We explored the potential of paraverbal entrainment for enhancing rapport and learning with the teachable robot Emma. We found that individuals felt more rapport for Emma when the robot both adapted and spoke socially than when Emma only adapted and indications of a similar trend for learning. This appeared to be driven by individuals who were more comfortable around robots. These findings suggest several directions for future work. First, in designing entrainment, there are alternative approaches based on how people adapt; exploring these additional patterns and combinations with social behavior is an important area of future work. Secondly, the social plus entraining condition was more appealing to individuals highly comfortable interacting with robots. Future work should explore whether increasing how comfortable individuals are around robots is needed before social behavior can have positive effects on rapport. Overall, paraverbal entrainment is a complex phenomenon and responses to it are influenced by individual differences; understanding these differences is vital for use of social behavior to enhance rapport and learning.

Acknowledgements. This work is supported by the National Robotics Initiative and the National Science Foundation, grant #CISE-IIS-1637809.

References

1. Roscoe, R.D., Chi, M.T.H.: Understanding tutor learning: knowledge-building and knowledge-telling in peer tutors' explanations and questions. Rev. Educ. Res. **77**(4), 534–574 (2007)
2. Ogan, A., Finkelstein, S., Mayfield, E., Adamo, C.D.: 'Oh, dear Stacy !' Social Interaction, Elaboration, and Learning with Teachable Agents (2012)
3. Gulz, A., Haake, M., Silvervarg, A., Sjödén, B., Veletsianos, G.: Building a social conversational pedagogical agent. In: Conversational Agents Natural Language Interaction: Techniques and Effective Practices, pp. 128–155 (2011)
4. Kumar, R., Ai, H., Beuth, J.L., Rosé, C.P.: Socially capable conversational tutors can be effective in collaborative learning situations. In: Aleven, V., Kay, J., Mostow, J. (eds.) ITS 2010. LNCS, vol. 6094, pp. 156–164. Springer, Heidelberg (2010). https://doi.org/10.1007/978-3-642-13388-6_20
5. Kory Westlund, J.M., et al.: Flat vs. expressive storytelling: young children's learning and retention of a social robot's narrative. Front. Hum. Neurosci. **11**, 295 (2017)
6. Lubold, N., Walker, E., Pon-Barry, H.: Effects of voice-adaptation and social dialogue on perceptions of a robotic learning companion. In: ACM/IEEE International Conference on Human-Robot Interaction, vol. 2016–April, pp. 255–262 (2016)
7. Vaughan, B.: Prosodic synchrony in co-operative task-based dialogues: a measure of agreement and disagreement. In: Interspeech, pp. 1865–1868 (2011)
8. Friedberg, H., Litman, D., Paletz, S.B.F.: Lexical entrainment and success in student engineering groups. In: Spoken Language Technology Workshop, pp. 404–409 (2012)
9. Borrie, S., Barrett, T., Willi, M., Berisha, V.: Syncing up for a good conversation: a clinically-meaningful methodology for capturing conversational entrainment in the speech domain. J. Speech, Lang. Hear. Res. (2018)
10. Lubold, N., Pon-Barry, H.: Acoustic-prosodic entrainment and rapport in collaborative learning dialogues categories and subject descriptors. In: Proceedings ACM Workshop on Multimodal Learning Analytics and Grand Challenge (2014)
11. Lubold, N., Walker, E., Pon-Barry, H., Ogan, A.: Automated pitch convergence improves learning in a social, teachable robot for middle school mathematics. In: Penstein Rosé, C., et al. (eds.) AIED 2018. LNCS (LNAI), vol. 10947, pp. 282–296. Springer, Cham (2018). https://doi.org/10.1007/978-3-319-93843-1_21

12. Giles, H., Coupland, N., Coupland, J.: Accommodation theory: communication, context, and consequence. In: Contexts of Accomodation: Developments in Applied Sociolinguistics, pp. 1–68 (1991)
13. Levitan, R., Gravano, A., Willson, L., Benus, S., Hirschberg, J., Nenkova, A.: Acoustic-prosodic entrainment and social behavior. In: Proceedings of the 2012 Conference of the North American Chapter of the Association for Computational Linguistics: Human Language Technologies, pp. 11–19 (2012)
14. Lubold, N., Walker, E., Pon-Barry, H.: Relating entrainment, grounding, and topic of discussion in collaborative learning dialogues. In: 12th International Conference on Computer Supported Collaborative Learning, pp. 0–1 (2015)
15. Ward, A., Litman, D.: Dialog convergence and learning. In: Proceedings 2007 Conference Artificial Intelligence in Education: Building Technology Rich Learning Contexts That Work, pp. 262–269 (2007)
16. Borrie, S.A., Lubold, N., Pon-Barry, H.: Disordered speech disrupts conversational entrainment: a study of acoustic-prosodic entrainment and communicative success in populations with communication challenges. Front. Psychol. **6**, 1187 (2015)
17. Arroyo, I., Burleson, W., Tai, M., Muldner, K., Woolf, B.P.: Gender differences in the use and benefit of advanced learning technologies for mathematics. J. Educ. Psychol. **105**(4), 957–969 (2013)
18. Siegel, M., Breazeal, C., Norton, M.I.: Persuasive robotics: the influence of robot gender on human behavior. In: 2009 IEEE/RSJ International Conference on Intelligent Robots and Systems, pp. 2563–2568 (2009)
19. Schermerhorn, P., Scheutz, M., Crowell, C.R.: Robot social presence and gender: do females view robots differently than males? In: ACM/IEEE International Conference on Human Robot Interaction, pp. 263–270 (2008)
20. Schwalbe, M.L., Staples, C.L., Demo, D., Gecas, V., Kleinman, S., Risman, B.: Gender differences in sources of self-esteem*. Soc. Psychol. Q. **54**(2), 158–168 (1991)
21. Spencer-Oatey, H.: Managing rapport in talk: using rapport sensitive incidents to explore the motivational concerns underlying the management of relations. J. Pragmat. **34**(5), 529–545 (2002)
22. Spencer-Oatey, H.: (Im)Politeness, face and perceptions of rapport: unpackaging their bases and interrelationships. Politeness Res. **1**(1), 95–119 (2005)
23. Leite, I., Martinho, C., Pereira, A., Paiva, A.: As time goes by: long-term evaluation of social presence in robotic companions. In: Proceedings - IEEE International Workshop on Robot and Human Interactive Communication, pp. 669–674 (2009)
24. Huttenrauch, H., Green, A., Norman, M., Oestreicher, L., Eklundh, K.S.: Involving users in the design of a mobile office robot. IEEE Trans. Syst. Man Cybern. Part C (Appl. Rev.) **34**(2), 113–124 (2005)
25. Jacq, A., Lemaignan, S., Garcia, F., Dillenbourg, P., Paiva, A.: Building successful long child-robot interactions in a learning context. In: ACM/IEEE International Conference on Human-Robot Interaction, vol. 2016–April, pp. 239–246 (2016)
26. Tanaka, F., Matsuzoe, S.: A self-competitive method for the development of an educational robot for children. In: 2016 IEEE/RSJ International Conference on Intelligent Robots and Systems (IROS), pp. 2927–2933 (2016)
27. Walker, E., Girotto, V., Kim, Y., Muldner, K.: The effects of physical form and embodied action in a teachable robot for geometry learning. In: IEEE 16th International Conference on Advanced Learning Technologies, pp. 381–385 (2016)
28. Chase, C.C., Chin, D.B., Oppezzo, M.A., Schwartz, D.L.: Teachable agents and the protege effect: increasing the effort towards learning. J. Sci. Educ. Technol. **18**(4), 334–352 (2009)
29. Boersma, P.: Praat, a system for doing phonetics by computer. Glot Int. **5** (2002)

30. Wallace, R.S.: The Elements of AIML Style. ALICE AI Foundation INC (2013)
31. Levitan, R., et al.: Implementing acoustic-prosodic entrainment in a conversational avatar. In: Annual Conference of the International Speech Communication Association, Interspeech, 08–12 September, pp. 1166–1170 (2016)
32. Lubold, N., Walker, E., Pon-Barry, H., Flores, Y., Ogan, A.: Using iterative design to create efficacy-building social experiences with a teachable robot. In: 13th International Conference of the Learning Sciences, pp. 737–744 (2018)
33. Bell, D.C., Arnold, H., Haddock, R.: Linguistic politeness and peer tutoring. Learn. Assist. Rev. **14**(1), 37–54 (2009)
34. Tickle-degnen, L., Rosenthal, R.: The nature of rapport and its nonverbal correlates. Pyschological Inq. **1**(4), 285–293 (1990)
35. Inoue, K., Nonaka, S., Ujiie, Y., Takubo, T., Arai, T.: Comparison of human psychology for real and virtual mobile manipulators. In: IEEE International Workshop on Robot and Human Interactive Communication, ROMAN 2005, pp. 73–78 (2005)
36. Kuthy, R.A., Mcquistan, M.R., Riniker, K.J., Heller, K.E., Qian, F.: Students' comfort level in treating vulnerable populations and future willingness to treat: results prior to extramural participation. J. Dent. Educ. **69**(12), 1307–1314 (2005)
37. Hicks, C.M., Gonzales, R., Morton, M.T., Gibbons, R.V., Wigton, R.S., Anderson, R.J.: Procedural experience and comfort level in internal medicine trainees. J. Gen. Intern. Med. **15**(10), 716–722 (2000)
38. Cohen, J.: A power primer. Pyschol. Bull. **112**(1), 155–159 (1992)
39. Boeck, D., et al.: Redefine statistical significance. Nat. Hum. Behav. **2** (2018)
40. Wagenmakers, E.-J.: A practical solution to the pervasive problems of p values. Psychon. Bull. Rev. **14**(5), 779–804 (2007)
41. Rouder, J.N., Speckman, P.L., Sun, D., Morey, R.D., Iverson, G.: Bayesian t tests for accepting and rejecting the null hypothesis. Psychon. Bull. Rev. **16**(2), 225–237 (2009)
42. Jarosz, A.F., Wiley, J.: What are the odds? A practical guide to computing and reporting bayes factors. J. Probl. Solving Spec. Issue **7** (2014)
43. Rouder, J.N., Morey, R.D., Speckman, P.L., Province, J.M.: Default Bayes factors for ANOVA designs. J. Math. Psychol. **56**, 356–374 (2012)
44. Kennedy, J., Baxter, P., Belpaeme, T.: The robot who tried too hard: social behaviour of a robot tutor can negatively affect child learning. In: Proceedings ACM/IEEE International Conference Human-Robot Interaction, no. 801, pp. 67–74 (2015)

A Concept Map Based Assessment of Free Student Answers in Tutorial Dialogues

Nabin Maharjan[(✉)] and Vasile Rus

Department of Computer Science, Institute for Intelligent Systems,
The University of Memphis, Memphis, TN 38152, USA
nabin247@gmail.com

Abstract. Typical standard Semantic Textual Similarity (STS) solutions assess free student answers without considering context. Furthermore, they do not provide an explanation for why student answers are similar, related or unrelated to a benchmark answer. We propose a concept map based approach that incorporates contextual information resulting in a solution that can both better assess and interpret student responses. The approach relies on a novel tuple extraction method to automatically map student responses to concept maps. Using tuples as the unit of learning (learning components) allows us to track students' knowledge at a finer grain level. We can thus better assess student answers beyond the binary decision of correct and incorrect as we can also identify partially-correct student answers. Moreover, our approach can easily detect missing learning components in student answers. We present experiments with data collected from dialogue-based intelligent tutoring systems and discuss the added benefit of the proposed method to adaptive interactive learning systems such as the capability of providing relevant targeted feedback to students which could significantly improve the effectiveness of such intelligent tutoring systems.

Keywords: Concept map · Student answer assessment ·
Tutorial dialogues · Intelligent tutoring system ·
Student knowledge representation · semantic textual similarity ·
Interpretable similarity

1 Introduction

In general, standard semantic textual similarity methods (STSs) [1,2,13,20,38] just assign a similarity score for measuring how similar a given pair of short texts are. Following the 0..5 similarity score scale proposed by Agirre and colleagues [1] in the recent STS shared task, a score of 5 indicates the text pairs are semantically equivalent while a score of 0 means that the two texts are completely unrelated whereas a score of 3 means that they differ in some important details or concepts. However, these STS systems do not explain why the two texts are similar, related or unrelated. For example, consider the student answer to a

© Springer Nature Switzerland AG 2019
S. Isotani et al. (Eds.): AIED 2019, LNAI 11625, pp. 244–257, 2019.
https://doi.org/10.1007/978-3-030-23204-7_21

question asked by DeepTutor [34,36], an Intelligent Tutoring System (ITS) for Newtonian Physics, and the corresponding ideal answer as shown in Table 1. A typical STS method would fail to explain that the student is missing information about *direction*.

Table 1. A student answer to a computer tutor question with the expectation i.e. ideal expected answer for the question.

Question: Because it is a vector, acceleration provides what two types of information?
Student Answer: Acceleration gives magnitude
Expectation: Acceleration provides magnitude and direction

One approach towards providing an explanatory layer is to align the chunks in a given pair of texts and label them with semantic relation types and scores. Such interpretable Semantic Textual Similarity tasks (iSTS) were proposed in 2015 and 2016 [1,3] in which the top performing systems were proposed by Banjade et al. [10,11]. However, iSTS capabilities are added layers on the top of standard STS methods for interpreting the similarity score.

Another drawback of an STS system is that they typically ignore contextual information when computing a similarity score. In dialogue-based ITSs, it has been shown, based on an analysis of conversational tutorial logs, that contextual information is important to assess student responses [28]. They reported that 68% of pronouns in student responses were referring to entities in the previous dialogue turn or the problem description. For instance, the student answer to the tutor question in the Table 1 might be elliptical: *"magnitude and direction"* or containing a pronoun referring to entities mentioned earlier in the previous dialogue turn: *"it gives magnitude and direction."* A typical STS system might fail to deem such student answers as correct.

To address these issues, we propose a novel concept map based approach to both better assess and interpret student free-response answers. A concept map is a graphical representation of knowledge where concepts are labeled nodes and relationships between the concepts are the directed labeled edges of the graph. A concept map can be hierarchical [30] where the most general concepts are at the top. A concept map can also be associative with no hierarchy, i.e., the concept map is a semantic network of concepts and their interrelations [14]. Since the concept map derived from student free responses in the domain of Newtonian Physics, where our experimental data comes from, is typically associative, we consider associative concept maps in our work.

Using concepts maps for assessment leads to an important shift in the granularity of assessment. That is, breaking down an expectation into one or more tuples (a triplet consisting of two concepts and their relationship) essentially means that the unit of analysis shifts from a full sentence, i.e., an expectation, to tuples. Ideally, a concept map with a single tuple, *(acceleration, gives, magnitude)*, is extracted from the student response in Table 1. For assessment purpose, we consider a tuple to be either *neutral* or *learning* depending upon its pedagogical value. For example, the *Expectation* in Table 1 is represented by two *learning* tuples *(acceleration, provides, magnitude)* and *(acceleration, provides, direction)*. We may also have a composite tuple *(acceleration, provides, magnitude and direction)* that covers both *learning* tuples. Similarly, the equivalent expectation: "*Because it is a vector, acceleration provides magnitude and direction*" can be represented by the above two *learning* tuples and one *neutral* tuple *(acceleration, is, a vector)*. The notion of *neutral* and *learning* tuple is useful when assessing correct student responses that have extra information/concepts relative to the expert provided ideal answers. In fact, a good portion of student responses may contain such extra information. Banjade and colleagues [9] found that 11% of student answers contain such additional, learning-neutral information.

The advantages of using finer grain learning components in the form of tuples are many. First, we can track students' knowledge at a finer grain level leading to more subtle differences among different knowledge states. Unlike in binary assessment, the concept map approach allows tracking how much of a given expectation is covered by student response. For example, if we assess the student answer against the ideal answer in Table 1, we can conclude that the student has mastered 50% of the expectation. Second, we can give proper credit to student answers based on the percentage of the *learning* tuples covered. Lastly, by comparing the two maps (student's vs. expert's), we can determine missing or incorrect tuples from student answers. This finer grain assessment enables adaptive interactive learning systems to provide better feedback and to better plan the next move, e.g., providing hints about the missing learning components. In summary, we propose a novel concept map-based approach which accounts for context and jointly functions as an STS and iSTS system.

2   Related Work

Concept maps were first proposed by Novak and Musonda [30] to represent knowledge of science for identifying learning specific changes in children. The concept maps were developed based on the learning psychology of Asubel [7] whose fundamental idea was that people learn new concepts and propositions by asking questions and getting clarification of relationships between old concepts and new concepts and between old propositions and new propositions.

Concept maps have been used for many purposes. They have been used for checking students' knowledge on a topic (CMap Tools; Canas et al. [12]) and for collaborative learning of a topic [24]. Also, they have been used as instructional tools for meaningful learning, i.e., linking new information with already known information [4,18,29,37,39]. Some ITSs use them as instructional tools to scaffold the learning process [31].

Concept maps have also been used for assessment. For example, students might be asked to fill in a skeleton map [5], to construct a concept map [32, 40], or to write an essay [19]. Recently, Wu et al. [40] developed a method that evaluates student concept maps on-the-fly and provides real-time feedback by comparing them with the expert's concept map. Also, the COMPASS [17] system provides individualized feedback based on the diagnostic assessment of the learner's concept map against an ideal concept map.

From a task perspective, our assessment approach is similar to the concept map-based assessment approach of Lomask and colleagues [19], with some differences. In their work, students wrote essays on two central topics in biology and then trained teachers derived concept maps from the essays. No hierarchical structure was assumed. Similarly, in our approach, we do not assume any hierarchy. However, we automatically extract concepts maps from student-generated responses during problem-solving tutorial interactions with a dialogue-based intelligent tutoring system. In our case, the target domain is conceptual Newtonian Physics. Finally, we assess their correctness by comparing against the corresponding ideal concept maps.

We also incorporate context when using concept maps for assessment. There have been recent attempts that consider contextual information for assessing two short text pairs. Bailey et al. [8] resolved pronouns implicitly by distinguishing between new and given concepts based on the context, i.e., a previous question. On the other hand, Banjade and colleagues [9] accounted for context by giving less weights to words already mentioned in the context. Maharjan et al. [21] handled context by using context-based count features and word-weighted similarity scores. Recently, an LSTM model has been proposed for assessing student answers in context [22]. We combine the approaches of both Bailey et al. [8] and Banjade et al. [9] for implicitly resolving pronouns and ellipsis at the tuple level.

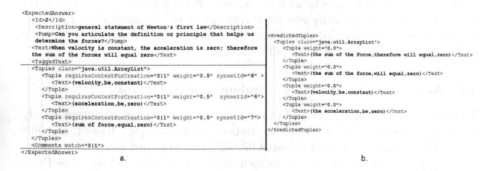

Fig. 1. A comparison of an ideal concept map (a) and computer generated concept map (b) for the ideal answer: *"When velocity is constant, the acceleration is zero; therefore the sum of the forces will equal zero"*.

3 Concept Map Based Approach

Using concept maps for knowledge representation is grounded on a key assumption in most cognitive theories: *"the knowledge within a content domain is well structured and organized around central concepts"*. Glaser and Bassok [16] defined competence in a domain as *"well-structured knowledge"*. Therefore, as students acquire expertise in a domain, their knowledge becomes increasingly interconnected and resembles the subject-matter expert's representation of the domain [16,33]. Our approach is based on those theories. In our approach to assessing students' natural language responses, we build an ideal concept map to represent an instructional task, say a Physics problem, based on the set of ideal steps (expectations) in the solution provided by domain experts. We then extract from the expectations the concepts and relationships between them. Second, we create a concept map for a student based on their responses to the task related questions. The concept map thus created can be considered a representation of the student's mental model. The student answers may be either right or wrong which implies that parts of their concept map or graph may be correct and parts of it may be incorrect. For example, in Table 1, the concept map for the expectation would be a part of an ideal concept map for the task whereas the concept map derived for the corresponding student answer would be part of a student concept map (knowledge graph) for the task. By comparing the two, we can determine which tuples are matched or unmatched. Our approach consists of three steps which we describe next.

3.1 Creation of Ideal Concept Maps

Currently, in dialogue-based ITSs, the subject-matter experts (SMEs) create ideal answers in the form of a set of logical steps or expectations for each instructional tasks. In our method, ideal concept maps must be created. To generate the ideal maps, we used a syntactic pattern based method [15] to map ideal solutions from sets of expectations to concepts maps and then asked the SMEs to curate those automatically generated concept maps. An annotation guidelines manual was created for this purpose which experts used to correct the concept maps. We describe the step in detail in Sect. 4.1.

3.2 Automated Extraction of Student Concept Maps

This step extracts concept maps from student-generated answers automatically. There are several existing open information extraction (IE) tools that could be used including the state-of-the-art Ollie [25] and Stanford systems (Stanford-OpenIE; [6]). However, these systems focus on solving the Knowledge Base Problem (KBP) and as such tuples produced by these systems are not suited for the task of student answer assessment.

The Stanford-OpenIE tool ignores shorter clauses not entailed from the original text. For example, given the text: *"If the acceleration of a system is zero, the net force is zero."*, no tuples are extracted. Another issue with the tool is that its natural logic inference system tends to over-produce tuples from the text. For example, for the text *"the frictional force cancels normal force"*, the desirable tuple output is *(frictional force, cancels, normal force)*; however, the tool also generates *(frictional force, cancels, force)*, *(force, cancels, normal force)* and *(force, cancels, force)* which are all misleading for assessment. On the other hand, the Ollie system might retrieve false tuples sometimes. For example, the system retrieves the incorrect tuple *(the desk, increase its speed as, the net force)* and misses *(net force, is anymore, not zero)* when processing the following text: *"The desk increases its speed as the net force is not zero anymore"*. Also, it fails to extract any tuple from the simpler text *"Mover's push equals friction"*.

To address these concerns, we developed a new extraction method which exploits the strengths of these two systems called DT-OpenIE [23]. Our extraction system first extracts shorter clauses from them and then, generates tuples from the shorter clauses by using Ollie extraction system [25] and a list of manually-crafted patterns. Figure 1(b) shows a concept map generated by our extraction system.

3.3 Assessment System

The Assessment System consists of two steps, namely, (i) *tuple filtering* and (ii) *tuple assessment*.

Tuple Filtering: Bailey and Meurers [8] implicitly resolved co-references by ignoring words that are already known, i.e., words already mentioned in the context. However, if a student answer repeats known words/concepts, it doesn't necessarily mean that they are repeating the same information because they may use the same concepts for making different propositions. However, if a tuple in a student concept map is repeated, we can assert that the student is simply repeating the given information with high confidence because a tuple is a higher level construct that itself represents a proposition showing the relationship between two or more concepts. We exploit this inherent characteristic of the tuple to filter out redundant or known tuples such that the resulting student concept map contains only tuples which are most likely to carry new information or propositions.

In our approach, we deem the tuples coming from the problem description (global context) and tutor question (local context) as externally known information. We consider *neutral* tuples in the ideal concept map to be redundant as they don't cover pedagogical aspects of the target ideal answer. Therefore, if a tuple in a student concept map matches with a tuple from any of these sources, we filter out the matching tuple. For filtering purpose, we match tuples using the M_1 and M_2 tuple matching methods described below without incorporating context.

Contradictory Tuple: We consider two tuples as contradictory if they are opposite or contrasting to each other. For example, we consider *(tension, is equal to, gravity)* and *(tension, is greater than, gravity)* as contradictory tuples. However, we don't consider *(Newton's first law, is, relevant)* and *(Newton's second law, is, relevant)* as contradictory. They are related, but in the context of answer evaluation, we consider such tuples as *disjoint* tuples. In our work, if two words are in an antonymy relation in WordNet [26] or they match any of the sixteen rules for antonym pairs [27], we consider them contradictory. We also created a domain specific antonym lookup table to address certain contrasting concepts. For example, we consider "equal", "less" and "greater" to be contradictory to each other. Similarly, we created a domain disjoint lookup list where we consider concepts such as "first" and "second" to be disjoint. We also filter out contradictory and disjoint tuples from the student concept maps.

Tuple Assessment: We assess the filtered student concept maps against corresponding ideal concept maps. We only assess if the *learning* tuples (corresponding to learning components) in the ideal concept map are covered/matched by tuples in the student concept map. We describe our tuple matching approach below.

Given a pair of tuples, (T_1, T_2), *where* $T_1 = (e_1, r_1, e_2)$, $T_2 = (e_{1'}, r_{1'}, e_{2'})$, T_1 is a tuple from ideal concept map and T_2 is a tuple from student concept map, we consider T_1 is semantically equivalent to T_2 using following three methods.

1. Matching by element-by-element (M_1): If e_1 matches $e_{1'}$, r_1 matches $r_{1'}$ and e_2 matches $e_{2'}$. We consider e_1 matches $e_{1'}$ if the set of words formed from e_1 is equal to the set of words formed from $e_{1'}$. Similarly, we define the *matches* function for element pairs $(r_1, r_{1'})$ and $(e_2, e_{2'})$.
2. Matching by bag-of-words (M_2): If the set of words formed from T_1 is equal to the set of words formed from T_2.
3. Matching by best similarity score (M_3): If similarity score $sim(T_1, T_2) > 0.55$, where T_2 is the tuple in the student concept map that has the maximum similarity score with T_1. 0.55 is empirically set threshold.

Incorporating Context: The filtering step discards tuples in the student answers which are deprived of new information. However, the remaining tuples with new information might also need context to match them properly. For example, we need context to accurately assess elliptical student response: "*magnitude and direction*" or student response containing pronoun: "*it gives magnitude and direction*" with the expectation in Table 1. In such cases, we attempt to match the tuples by incorporating context using the following two strategies. For the M_1 and M_2 methods, we account for context by considering only new concepts in student answers that were not mentioned in the previous context [8]. That is, if the student response repeats concepts mentioned in the tutor question, we ignore those concepts. We do the same for repeated concepts in the ideal answer. Specifically, we use the set of words Q from the tutor question as context for matching element pairs (E_1, E_2), where $(E_1, E_2) \in \{(e_1, e_{1'}), (r_1, r_{1'}), (e_2, e_{2'}), (T_1, T_2)\}$. We consider E_1 to be semantically equivalent to E_2 under the following two cases.

1. if the set of words formed from E_1 is a proper subset of the set of words formed from E_2 and the set of unmatched words from $E_2 - E_1$ is a subset of set Q.
2. if the set of words formed from E_2 is a proper subset of the set of words formed from E_1 and the set of unmatched words from $E_1 - E_2$ is a subset of set Q.

On the other hand, in the case of the M_3 matching method, we follow a word weighting approach based on context where we don't completely discard the words in the student tuple or ideal answer tuple that are also present in context. Instead, we give less weight to such words [9]. Specifically, we give full weight of 1 to new concepts/words and 0.4 (empirically set) to repeated concepts. Subsequently, we compute an alignment based similarity score between the two tuples as:

$$sim(T_1, T_2) = 2 * \frac{\sum_{(r,a)\in OA} w_r w_a sim(r, a)}{\sum_{r\in T_1} w_r + \sum_{a\in T_2} w_a}, \tag{1}$$

where OA is optimal alignment of words between T_1 and T_2 such that $a \in T_2$ and $r \in T_1$, using Hungarian algorithm as described in Rus and Lintean [35]. The $0 \le w_r \le 1$ and $0 \le w_a \le 1$ refer to weight of the words in T_1 and T_2 respectively.

The strictness of matching relaxes as we go from the M_1 to the M_3 method. Therefore, we preferentially match *learning* tuples in the ideal concept map against the student concept in the following order: M_1, M_2 and M_3.

4   Experiment and Results

4.1   Data

We used student answer data from logged interactions of 41 high school students with the DeepTutor ITS [36]. During the summer of 2014, students participated in an experiment on which they were given 9 different Physics problems to solve. The experiment produced 370 tutorial interactions in total (one student performed a task twice).

Ideal Concept Maps: Two subject-matter-experts (SMEs) manually created ideal concept maps for all 9 tasks. They were provided with a reference guide for creating the ideal concept map. For each instructional task, annotators were provided with an XML file containing a skeletal concept map that needed to be checked. The map was automatically generated by using syntactic patterns [15]. The annotators updated the concept maps by deleting invalid tuples, modifying tuples and by adding missing tuples.

While creating concept maps, the annotators also annotated the tuples with a rich set of attributes. For example, we assigned tuples covering identical concepts with same id (*synsetId*). We also differentiated between *learning* and *neutral* tuples depending on their pedagogical importance. We assign a weight 0 to *neutral* tuples. In case of *learning* tuples, if the expected answer has n unique *learning* tuples, we assign each of these tuples weight of $1/n$ (we consider only tuples with different ids as unique). These annotations are useful to handle variations in student answers as described in Sect. 3.3. Figure 1(a) shows a final human generated ideal concept map for one expectation. The annotators also set the *watch* attribute in their comments to flag their tuples for review later. After creating three concept maps, the SMEs met and compared their maps to resolve any discrepancies. A refined annotation guide was created which was then followed for the whole data annotation.

```
<Utterance>
   <Speaker>DEEPTUTOR</Speaker>
   <Text>Can you articulate the definition or principle that
   helps us determine the forces?</Text>
</Utterance>
<Utterance>
   <Speaker>STUDENT</Speaker>
   <Text>Newton's law says that an object remains at rest or at
   constant velocity unless acted on by a net force.</Text>
   <CoveredExpectation>
       <Gold>T:2_6,2_7,2 F:</Gold>
   </CoveredExpectation>
</Utterance>
```

Fig. 2. Annotating data for binary classification. TupleIds 2_6 and 2_7 consist of expectation id 2 concatenated with synsetIds 6 and 7 respectively.

4.2 Experiments

We evaluated the performance of our concept map-based approach against both binary and multi-level classification tasks.

Binary Classification Task: First, we assessed whether a particular student answer is correct or incorrect. The 41 student session data were randomly divided into training (21 sessions, 1296 instances) and test (20 sessions, 1296 instances) sets.

Figure 2 shows an annotation example for the binary classification task. The annotators judged which of the targeted *learning* tuples are covered/uncovered by the student answer. If a tuple is covered, its tupleId is recorded in the T group. On the other hand, if it is uncovered, its tupleId is recorded in the F group. Similarly, judges annotated full expectations as covered or uncovered by recording its expectationId either under T or F group, respectively. An expectation is covered if its all *learning* tuples are covered.

We ran our assessment method (Sect. 3.3) with and without context against both training and test data. For comparison, we also derived results by applying the word alignment based sentence similarity method with context [9] and without context [35]. Specifically, in the case of the context agnostic method, we first computed similarity score by giving full weight of 1 to each word in the student and reference answer using Eq. 1. For the context-based approach, we applied the same equation such that the words in the student answer or reference answer that are also present in the context, i.e., problem description and tutor question, are given lower weights.

Multi-level Classification Task: In this task, we classified the student answers into one of four correctness classes: *correct, correct-but-incomplete, contradictory* or *incorrect*. We used the DT-Grade data annotated with these four correctness classes for evaluating our approach. The DT-Grade [9] dataset consists of 900 student responses sampled from logged data recorded during the same experiment with the DeepTutor system as described above.

We adapted our approach for multi-level classification task by using a simple classification rule as follows. If the student concept map covers all *learning* tuples, we classify the student answer as *correct*. If at least one of the *learning* tuples is covered, then student answer is classified as *correct-but-incomplete*. If no *learning* tuples are covered, and a *contradictory* tuple is present in the student concept map, we classify the instance as *contradictory*. If none of the above conditions satisfy, we classify the student answer as *incorrect*.

Table 2. Results of different methods for binary classification. WA-Sim and WA-Sim-C are optimal word alignment based STS system without context [35] and with context [9] respectively. The value inside the bracket alongside the accuracy is Cohen's Kappa.

Training data			Test data	
System	F-score	Accuracy	F-score	Accuracy
WA-Sim	0.739	73.8(0.43)	0.742	74.6(0.39)
WA-Sim-C	0.756	77.2(0.45)	0.752	77.2(0.40)
Concept map (no context)	0.763	75.8(0.49)	0.783	77.7(0.52)
Concept map	**0.783**	**78(0.53)**	**0.802**	**79.9(0.55)**

Table 3. Performance on Multi-level classification task. Comparison of concept map approach against Logistic Model [9], GMM Model [21] and LSTM Model [22]. The value inside the bracket alongside the accuracy is Cohen's Kappa.

System	Logistic model	GMM model	LSTM model	Concept map
F-score	-	0.58	0.62	0.59
Accuracy	49.3(0.22)	58.2(0.40)	62.2(0.45)	59.3(0.41)

4.3   Results and Discussions

Table 2 shows the performance of our concept map approach for the binary classification task. Moreover, we found that our system does a better job at identifying whether a *learning* tuple is covered or not by the student concept map. At the tuple level, our system yielded an accuracy of 76.8% (F-score = 0.769) on the training set (2408 tuple instances), and an accuracy of 77.7% (F-score = 0.778) on the test set (2360 tuple instances). This result implies that we can detect most of the missing *learning* tuples from the student answer. Therefore, our assessment method, if incorporated in adaptive learning systems, can enable such systems to provide more targeted relevant feedback to students and also provide useful information for planning system's next move targeting the missing learning components. This targeted, adaptive feedback based on individual student performance could significantly improve the effectiveness of adaptive learning systems.

For the multi-level classification task, our approach performed slightly lower than the state-of-the-art LSTM model [22] but performs better than other models such as logistic model [9] and the GMM based method [21] as shown in Table 3. However, LSTM model lacks explanatory capability of our concept map approach.

It is evident that improving the quality of extracted student concept maps would improve performance our system. We note that the desirable student concept map was not extracted from complex elliptical student response such as *"The force of the man pushing the box and the force of friction acting on the box"* to the tutor question: *"What forces balance each other?"*. Exploring machine learning methods with concept maps can also improve student answer assessment.

Our concept map doesn't model temporal and conditional relationships. For example, given the statement *"If the net force is zero, the meteor moves with constant velocity"*, *(meteor, moves with, constant velocity)* is true only *if the net force is zero*. In future work, we can augment concept maps with temporal and conditional dependence information. Many bag of word approaches including LSA perform very well for semantic assessment. We believe that our simple tuple representation can capture the meaning of text more strongly since tuples are higher semantic structures than words.

5   Conclusion

We presented a novel concept map-based approach to assess student answers in tutorial dialogues. The approach takes context into account and implicitly handles linguistic phenomena such as ellipsis and pronouns for assessment. We combined the approaches of both Bailey and Meurers [8] and Bajande et al. [9] for implicitly resolving pronouns and ellipsis at the tuple level. We use tuples as the unit of learning to track students' knowledge at a finer grain level which enables us to better assess student answers rather than just classifying them as correct or incorrect. Moreover, our approach can easily detect missing learning components in student answers which can be used for dynamic and automatic generation of personalized of next moves such as hints. As such, adaptive tutoring systems can provide targeted adaptive feedback and scaffolding in the form of hints to the students based on their individual performance.

Our future work is to improve the concept map-based approach further. Second, we plan to study the impact of our approach to improving the effectiveness of adaptive intelligent tutoring systems.

Acknowledgments. This work was partially supported by The University of Memphis, the National Science Foundation (awards CISE-IIS-1822816 and CISE-ACI-1443068), and a contract from the Advanced Distributed Learning Initiative of the United States Department of Defense.

References

1. Agirre, E., et al.: Semeval-2015 task 2: semantic textual similarity, English, Spanish and pilot on interpretability. In: Proceedings of the 9th International Workshop on Semantic Evaluation SemEval 2015, pp. 252–263. Association for Computational Linguistics (2015). http://aclweb.org/anthology/S15-2045

2. Agirre, E., et al.: Semeval-2016 task 1: semantic textual similarity, monolingual and cross-lingual evaluation. In: Proceedings of the 10th International Workshop on Semantic Evaluation (SemEval-2016), pp. 497–511. Association for Computational Linguistics (2016). http://aclweb.org/anthology/S16-1081

3. Agirre, E., Gonzalez-Agirre, A., Lopez-Gazpio, I., Maritxalar, M., Rigau, G., Uria, L.: Semeval-2016 task 2: interpretable semantic textual similarity. In: Proceedings of the 10th International Workshop on Semantic Evaluation (SemEval-2016), pp. 512–524. Association for Computational Linguistics (2016). http://aclweb.org/anthology/S16-1082

4. All, A.C., Huycke, L.I., Fisher, M.J.: Instructional tools for nursing education: concept maps. Nurs. Educ. Perspect. **24**(6), 311–317 (2003)

5. Anderson, T.H., Huang, S.c.C.: On using concept maps to assess the comprehension effects of reading expository text. Center for the Study of Reading Technical report, no. 483 (1989)

6. Angeli, G., Johnson Premkumar, M.J., Manning, C.D.: Leveraging linguistic structure for open domain information extraction. In: Proceedings of the 53rd Annual Meeting of the Association for Computational Linguistics and the 7th International Joint Conference on Natural Language Processing (Volume 1: Long Papers), pp. 344–354. Association for Computational Linguistics (2015). http://aclweb.org/anthology/P15-1034

7. Ausubel, D.P.: The Psychology of Meaningful Verbal Learning (1963)

8. Bailey, S., Meurers, D.: Diagnosing meaning errors in short answers to reading comprehension questions. In: Proceedings of the Third Workshop on Innovative Use of NLP for Building Educational Applications, pp. 107–115. Association for Computational Linguistics (2008). http://aclweb.org/anthology/W08-0913

9. Banjade, R., Maharjan, N., Niraula, N.B., Gautam, D., Samei, B., Rus, V.: Evaluation dataset (DT-Grade) and word weighting approach towards constructed short answers assessment in tutorial dialogue context. In: Proceedings of the 11th Workshop on Innovative Use of NLP for Building Educational Applications, pp. 182–187. Association for Computational Linguistics (2016). http://aclweb.org/anthology/W16-0520

10. Banjade, R., Maharjan, N., Niraula, N.B., Rus, V.: DTSim at SemEval-2016 task 2: interpreting similarity of texts based on automated chunking, chunk alignment and semantic relation prediction. In: Proceedings of the 10th International Workshop on Semantic Evaluation (SemEval-2016), pp. 809–813. Association for Computational Linguistics (2016). http://aclweb.org/anthology/S16-1125

11. Banjade, R., et al.: NeRoSim: a system for measuring and interpreting semantic textual similarity. In: Proceedings of the 9th International Workshop on Semantic Evaluation (SemEval 2015), pp. 164–171. Association for Computational Linguistics (2015). http://aclweb.org/anthology/S15-2030

12. Cañas, A.J., et al.: CmapTools: a knowledge modeling and sharing environment (2004)

13. Cer, D., Diab, M., Agirre, E., Lopez-Gazpio, I., Specia, L.: SemEval-2017 task 1: semantic textual similarity multilingual and crosslingual focused evaluation. In: Proceedings of the 11th International Workshop on Semantic Evaluation (SemEval-2017), pp. 1–14. Association for Computational Linguistics (2017). http://aclweb.org/anthology/S17-2001

14. Deese, J.: The Structure of Associations in Language and Thought. Johns Hopkins University Press, Baltimore (1966)

15. Fader, A., Soderland, S., Etzioni, O.: Identifying relations for open information extraction. In: Proceedings of the Conference on Empirical Methods in Natural Language Processing, pp. 1535–1545. Association for Computational Linguistics (2011)

16. Glaser, R., Bassok, M.: Learning theory and the study of instruction. Ann. Rev. Psychol. **40**(1), 631–666 (1989)

17. Gouli, E., Gogoulou, A., Papanikolaou, K., Grigoriadou, M.: COMPASS: an adaptive web-based concept map assessment tool (2004)

18. Horton, P.B., McConney, A.A., Gallo, M., Woods, A.L., Senn, G.J., Hamelin, D.: An investigation of the effectiveness of concept mapping as an instructional tool. Sci. Educ. **77**(1), 95–111 (1993)

19. Lomask, M., Baron, J., Greig, J., Harrison, C.: ConnMap: connecticut's use of concept mapping to assess the structure of students' knowledge of science. In: Annual Meeting of the National Association of Research in Science Teaching, Cambridge, pp. 21–25 (1992)

20. Maharjan, N., Banjade, R., Gautam, D., Tamang, L.J., Rus, V.: Dt_team at SemEval-2017 task 1: Semantic similarity using alignments, sentence-level embeddings and Gaussian mixture model output. In: Proceedings of the 11th International Workshop on Semantic Evaluation (SemEval-2017), pp. 120–124. Association for Computational Linguistics (2017). http://aclweb.org/anthology/S17-2014

21. Maharjan, N., Banjade, R., Rus, V.: Automated assessment of open-ended student answers in tutorial dialogues using Gaussian mixture models. In: Proceedings of the Thirtieth International Florida Artificial Intelligence Research Society Conference, pp. 98–103 (2017). https://aaai.org/ocs/index.php/FLAIRS/FLAIRS17/paper/view/15489

22. Maharjan, N., Gautam, D., Rus, V.: Assessing free student answers in tutorial dialogues using LSTM models. In: Penstein Rosé, C., et al. (eds.) AIED 2018. LNCS (LNAI), vol. 10948, pp. 193–198. Springer, Cham (2018). https://doi.org/10.1007/978-3-319-93846-2_35

23. Maharjan, N., Rus, V.: Towards concept map based free student answer assessment. In: Proceedings of the Thirty-second International Florida Artificial Intelligence Research Society Conference (2019)

24. Martinez Maldonado, R., Kay, J., Yacef, K., Schwendimann, B.: An interactive teacher's dashboard for monitoring groups in a multi-tabletop learning environment. In: Cerri, S.A., Clancey, W.J., Papadourakis, G., Panourgia, K. (eds.) ITS 2012. LNCS, vol. 7315, pp. 482–492. Springer, Heidelberg (2012). https://doi.org/10.1007/978-3-642-30950-2_62

25. Mausam, Schmitz, M., Soderland, S., Bart, R., Etzioni, O.: Open language learning for information extraction. In: Proceedings of the 2012 Joint Conference on Empirical Methods in Natural Language Processing and Computational Natural Language Learning, pp. 523–534. Association for Computational Linguistics (2012). http://aclweb.org/anthology/D12-1048

26. Miller, G.A.: WordNet: a lexical database for english. Commun. ACM **38**(11), 39–41 (1995)

27. Mohammad, S., Dorr, B., Hirst, G.: Computing word-pair antonymy. In: Proceedings of the 2008 Conference on Empirical Methods in Natural Language Processing, pp. 982–991. Association for Computational Linguistics (2008). http://aclweb.org/anthology/D08-1103

28. Niraula, N.B., Rus, V., Banjade, R., Stefanescu, D., Baggett, W., Morgan, B.: The DARE corpus: a resource for anaphora resolution in dialogue based intelligent tutoring systems. In: LREC, pp. 3199–3203 (2014). http://www.lrec-conf. org/proceedings/lrec2014/pdf/372_Paper.pdf

29. Novak, J.D., Bob Gowin, D., Johansen, G.T.: The use of concept mapping and knowledge vee mapping with junior high school science students. Sci. Educ. **67**(5), 625–645 (1983)

30. Novak, J.D., Musonda, D.: A twelve-year longitudinal study of science concept learning. Am. Educ. Res. J. **28**(1), 117–153 (1991). https://www.jstor.org/stable/pdf/1162881.pdf

31. Olney, A.M., et al.: Guru: a computer tutor that models expert human tutors. In: Cerri, S.A., Clancey, W.J., Papadourakis, G., Panourgia, K. (eds.) ITS 2012. LNCS, vol. 7315, pp. 256–261. Springer, Heidelberg (2012). https://doi.org/10.1007/978-3-642-30950-2_32. http://www.academia.edu/download/46942204/Guru_A_Computer_tutor_that_models_expert20160701-24127-rkuna9.pdf

32. Roth, W.M., Roychoudhury, A.: The concept map as a tool for the collaborative construction of knowledge: a microanalysis of high school physics students. J. Res. Sci. Teach. **30**(5), 503–534 (1993)

33. Royer, J.M., Cisero, C.A., Carlo, M.S.: Techniques and procedures for assessing cognitive skills. Rev. Educ. Res. **63**(2), 201–243 (1993)

34. Rus, V., D'Mello, S., Hu, X., Graesser, A.: Recent advances in conversational intelligent tutoring systems. AI Mag. **34**(3), 42–54 (2013). http://www.aaai.org/ojs/index.php/aimagazine/article/viewFile/2485/2378

35. Rus, V., Lintean, M.: A comparison of greedy and optimal assessment of natural language student input using word-to-word similarity metrics. In: Proceedings of the Seventh Workshop on Building Educational Applications Using NLP, pp. 157–162. Association for Computational Linguistics (2012). http://aclweb.org/anthology/W12-2018

36. Rus, V., Niraula, N.B., Banjade, R.: DeepTutor: an effective, online intelligent tutoring system that promotes deep learning. In: AAAI, pp. 4294–4295 (2015). http://www.aaai.org/ocs/index.php/AAAI/AAAI15/paper/download/10019/9857

37. Schmid, R.F., Telaro, G.: Concept mapping as an instructional strategy for high school biology. J. Educ. Res. **84**(2), 78–85 (1990)

38. Tian, J., Zhou, Z., Lan, M., Wu, Y.: ECNU at SemEval-2017 task 1: leverage kernel-based traditional NLP features and neural networks to build a universal model for multilingual and cross-lingual semantic textual similarity. In: Proceedings of the 11th International Workshop on Semantic Evaluation (SemEval-2017), pp. 191–197. Association for Computational Linguistics (2017). http://aclweb.org/anthology/S17-2028

39. Wallace, J.D., Mintzes, J.J.: The concept map as a research tool: exploring conceptual change in biology. J. Res. Sci. Teach. **27**(10), 1033–1052 (1990)

40. Wu, P.H., Hwang, G.J., Milrad, M., Ke, H.R., Huang, Y.M.: An innovative concept map approach for improving students' learning performance with an instant feedback mechanism. Br. J. Educ. Technol. **43**(2), 217–232 (2012)

Deep (Un)Learning: Using Neural Networks to Model Retention and Forgetting in an Adaptive Learning System

Jeffrey Matayoshi[(✉)], Hasan Uzun, and Eric Cosyn

McGraw-Hill Education/ALEKS Corporation, Irvine, CA, USA
{jeffrey.matayoshi,hasan.uzun,eric.cosyn}@aleks.com

Abstract. ALEKS, which stands for "**A**ssessment and **LE**arning in **K**nowledge **S**paces", is a web-based, artificially intelligent, adaptive learning and assessment system. Previous work has shown that student knowledge retention within the ALEKS system exhibits the characteristics of the classic Ebbinghaus forgetting curve. In this study, we analyze in detail the factors affecting the retention and forgetting of knowledge within ALEKS. From a dataset composed of over 3.3 million ALEKS assessment questions, we first identify several informative variables for predicting the knowledge retention of ALEKS problem types (where each problem type covers a discrete unit of an academic course). Based on these variables, we use an artificial neural network to build a comprehensive model of the retention of knowledge within ALEKS. In order to interpret the results of this neural network model, we apply a technique called permutation feature importance to measure the relative importance of each feature to the model. We find that while the details of a student's learning activity are as important as the time that has passed from the initial learning event, the most important information for our model resides in the specific problem type under consideration.

Keywords: Forgetting curves · Neural networks ·
Knowledge space theory · Adaptive learning ·
Permutation feature importance

1 Introduction

ALEKS, which stands for "**A**ssessment and **LE**arning in **K**nowledge **S**paces", is a web-based, artificially intelligent, adaptive learning and assessment system [18]. The artificial intelligence of ALEKS is a practical implementation of knowledge space theory (KST) [5,7,8], a mathematical theory that employs combinatorial structures to model the knowledge of learners in various academic fields of study including math [14,22], chemistry [12,26] and even dance education [31].

Understanding the behavior of retention and forgetting within adaptive systems is an important area of research, as it has been shown that student models

© Springer Nature Switzerland AG 2019
S. Isotani et al. (Eds.): AIED 2019, LNAI 11625, pp. 258–269, 2019.
https://doi.org/10.1007/978-3-030-23204-7_22

can be significantly improved when these aspects of learning are accounted for [21,27]. Furthermore, some previous results have emphasized the importance of identifying the variables that affect forgetting [28,29], while others have shown that personalized interventions and review schedules can improve students' long-term retention of knowledge [16,30].

Motivated by these previous works, in this study we analyze in detail the factors that affect the forgetting and retention of knowledge within the ALEKS system. Given that the retention of knowledge within ALEKS exhibits the characteristics of the famous Ebbinghaus forgetting curve [1,6,17], we begin with some exploratory data analysis of the factors affecting this curve. Based on these results, we then build a comprehensive model of forgetting and retention within ALEKS using an artificial neural network. Finally, by combining our exploratory data analysis with an application of permutation feature importance [2,24,25], we are able to get a clearer understanding of the relative importance of each of the features to our final neural network model.

2   Background

In KST, an *item* is a problem type that covers a discrete unit of an academic course. Each item contains many examples called *instances*, and these examples are carefully chosen to be equal in difficulty and to cover the same content. A *knowledge state* in KST is a collection of items that, conceivably, a student at any one time could know how to do.

Another concept important to our study is the *inner fringe* of a knowledge state. An item is contained in the inner fringe of a knowledge state when the item can be removed from the state and the remaining set of items forms another knowledge state. An inner fringe item can be viewed as being at the edge of a student's knowledge, as complete knowledge of the item is not required to know any of the other items in the knowledge state.

Within the ALEKS system, the student is guided through a course via a cycle of learning and assessments. In an assessment, a student is presented an item for which they can attempt to answer, or they can respond "I don't know" if they, presumably, have little knowledge of how to solve the problem. If the student attempts to answer the item, the response is classified as either correct or incorrect. A course begins with an *initial assessment*, the goal of which is to accurately measure the starting knowledge of the student. Then, in the learning mode, the student is presented items based on her knowledge state, with the system tracking the student's performance and continually updating the student's knowledge state. Each subsequent *progress assessment* is given to a student after some time has been spent in the learning mode, and the process continues. The purpose of these progress assessments is to verify the student's recent learning, as well as to act as a mechanism for enforcing spaced practice and retrieval.

For the purposes of this study, we define *retention* as the act of answering an item correctly on a progress assessment at a point in time after the item is learned in ALEKS. We can then define the *retention rate* as the correct answer rate to these assessment questions. For our analyses, we gather data from over 3.3 million ALEKS progress assessment questions that are drawn from 10 different math and chemistry courses. The questions we use are restricted to items that are contained in the inner fringe of the student's knowledge state. In looking only at inner fringe items, we are attempting to reduce any bias from students reinforcing the core knowledge of an item by working on related content. We partition the data into a training set of 2, 989, 835 questions, along with validation and test sets each consisting of 166, 102 questions. In addition to being used to train our neural network models, we also use the training data to perform all our exploratory data analysis. The validation set is used to test different neural network architectures and tune the hyperparameters, while the test set is used for the final model evaluation.

3   Forgetting Curve

As shown in [17], the retention rate of an inner fringe item in ALEKS changes as a function of the time since the item was learned (with an item being "learned" after a certain amount of success on the item has been demonstrated in the learning mode). To see this, for each assessment question in our training data we compute the number of days from the time the student learned the item to the time the item appeared in the progress assessment, and then we group these questions based on the outcome (correct, incorrect, or "I don't know"). The results are shown in Fig. 1, where the solid curve (the proportion of corrects) can be considered a forgetting curve [1,6]. Note that this curve is analogous to the curve first shown in [17].

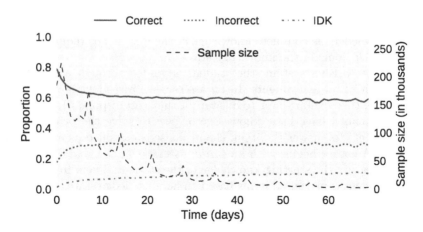

Fig. 1. Proportions of responses as a function of the time (in days) since the inner fringe item appearing as a progress assessment question was learned.

At this point it should be emphasized that inner fringe items are at a very specific, and critical, place in a student's knowledge state. The overall retention rate on these items is relatively modest, with the average correct rate for our dataset being 0.64. Since an inner fringe item has recently been learned by a student, without any learning that reinforces the skill(s) contained in the item, the relatively low correct rate is not unexpected. Thus, predicting the retention of inner fringe items is a difficult task, as the items that are most likely to be retained (i.e., the items with highest correct rates) would not generally be found in the inner fringe. (See Fig. 1 in [4] for an example of how the correct rate increases for items "deeper" in the knowledge state.) That being said, there are many factors that can affect the inner fringe correct rate, and it is important to identify these factors when building models of retention [28,29]. Thus, in the next section we take a look at these factors in more detail.

4   Exploratory Data Analysis

Now that we have established a baseline forgetting curve, we can look at what factors, or variables, affect this curve. The first variable we discuss is the knowledge of the student at the beginning of the course, which is measured by what we call the student's initial score; this is simply the proportion of the items in the course that are in the student's knowledge state at the end of the initial assessment. The results are shown in Fig. 2, which compares the forgetting curves for students in the first decile (in terms of the initial score) and in the tenth decile. We can see that there is a relatively large gap between the two correct answer curves. Additionally, the "I don't know" curves show an interesting difference, in that the students in the first decile show an increasing rate of "I don't know" answers over time, while the students in the tenth decile have a constant rate after about a week.

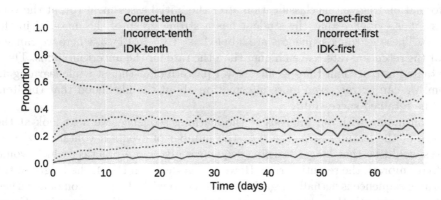

Fig. 2. Proportions of responses conditioned on the student being in the first decile (in terms of initial score) or the tenth decile. The top set of lines (blue) represents the correct responses, the middle set (green) represents the incorrect responses, and the bottom set (red) represents the "I don't know" responses. (Color figure online)

The next factor we consider is the classification of the learned item after the student's initial assessment. An ALEKS assessment finishes with each item classified into one of three distinct categories.

- Items that are most likely in the student's knowledge state (in-state)
- Items that are most likely not in the student's knowledge state (out-of-state)
- The remaining items (uncertain).

The learned items in our dataset are exclusively composed of items classified as either out-of-state or uncertain after the initial assessment. The out-of-state items are items that the ALEKS system, at the conclusion of the initial assessment and based on the student's responses to the assessment questions, strongly believes the student does not know. On the other hand, the uncertain items are those for which the system does not have enough information to make a confident classification of either in-state or out-of-state. Thus, it stands to reason that a good portion of these uncertain items are actually items that the student already knows, in which case the forgetting curve and retention may take a different form. The results in Fig. 3 support this conjecture, where there is a clear separation between the forgetting curves for the uncertain items and the out-of-state items, with the uncertain items being retained at a higher rate.

We next look at how retention is affected by a student's *learning sequence*, which is the sequence of events taken by the student when learning an item. The possible events in a learning sequence are (a) submitting a correct answer, (b) submitting an incorrect answer, and (c) viewing an explanation of the current instance. If an item is deemed uncertain, a student can demonstrate mastery of the item in the learning mode by correctly answering the first two given instances of the item. Intuitively, if the first two instances of an uncertain item are answered correctly, this would appear to be strong evidence that the student does actually know the item, and that the ALEKS system simply lacked the information to give this classification after the initial assessment (or, at the very least, it is evidence that the student has a strong grasp of the material in the item). These learning sequences are labeled as "CC" in Fig. 3, where we can see that the retention rate is even higher than the rate for the uncertain items. Thus, by taking into account the specific answer pattern of a student while learning an item, we can extract even more information about the likelihood that the item will be retained successfully.

Continuing with our analysis of the learning sequence, we next look at the length of the learning sequence (i.e., the number of events it contains). A first guess would be that longer learning sequences give more practice, which would help to improve the retention rate. However, as shown in Fig. 4, the length of the learning sequence is actually negatively correlated with the retention of a learned item. A moment's thought shows that this is not actually that surprising. Given that the learning sequence ends when the ALEKS system decides the student has shown mastery of an item, there is a selection effect when partitioning the items by the learning sequence length. More specifically, the shorter sequences tend to involve simpler items for which it is easier to demonstrate mastery (and for which it is also easier to retain the knowledge), or are from students who have a

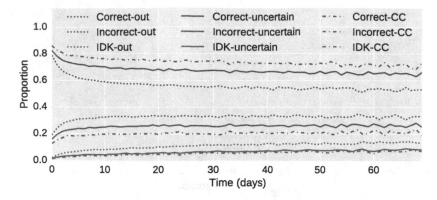

Fig. 3. Proportions of responses conditioned on whether the ALEKS initial assessment classified the item as out-of-state or uncertain (which are mutually exclusive categories), or if the item has a CC learning sequence (which is a subset of the uncertain items). The top three lines (blue) represent the correct responses, the middle three lines (green) represent the incorrect responses, and the bottom three lines (red) represent the "I don't know" responses. (Color figure online)

stronger grasp of the material (again leading to a higher retention rate). On the other hand, the longer learning sequences either involve noisy and difficult items (for which we would expect a lower retention rate) or students who are struggling with the material (in which case we might again expect a lower retention rate).

The last variable we consider is the item itself. While the majority of ALEKS items have a similar open-ended answer format, the actual content, as well as the intrinsic difficulty, can vary. To get a sense of these differences, Fig. 5 shows a histogram of the item correct rates in the training set, restricted to the 1664 items with at least 200 data points each. The mean and median correct rates are 0.64 and 0.65, respectively, with a standard deviation of 0.11. While the majority of the items cluster around the mean, there are certain items with somewhat extreme behavior. For example, the maximum and minimum values for retention are 0.92 and 0.17, respectively, with 88 items having a rate above 0.8 and 168 having a rate below 0.5. Thus, the specific characteristics of an item appear to have a significant effect on its retention rate.

5   Retention Models

To make use of the information discussed in the previous section, we next develop a model of retention using an artificial neural network. This neural network model takes the form of a classifier that attempts to predict whether or not a student will give a correct answer when presented an inner fringe item during a progress assessment. The following features are used to build this model.

– ALEKS course: categorical variable with 10 values
– Item: categorical variable with 2190 values

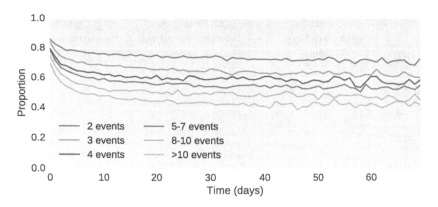

Fig. 4. Proportion of correct responses conditioned on the number of events in the associated learning sequence. Note that, as grouped, the correct rate is a decreasing function of the number of events.

- Initial score: continuous variable with values in $[0, 1]$
- Time in days since item was learned: discrete variable with values in $[0, 399]$
- Learning sequence: responses encoded as a sequence of categorical variables, each with three values corresponding to the student's action (correct answer, incorrect answer, reading the explanation)

The learning sequence variable is fed to a recurrent neural network (RNN), a type of neural network that is well-suited to handling sequential data [11]. The output from this RNN is then concatenated with the original set of features, and this combined set of features is then fed to a multilayer perceptron (MLP). For the hidden units of our RNN, we evaluate two different recurrent units on our validation set: gated recurrent units (GRU) [3] and long short-term memory (LSTM) units [13]. Additionally, the learning rate, number of hidden layers, and number of units in each hidden layer are also tuned on the validation set. In all cases we use batch normalization [15] while training, and we also apply early stopping [20] and dropout [9,23] to help prevent overfitting.

Our best performing model on the validation set, which we evaluate in detail in the next section, is comprised of an RNN containing four layers of LSTM units. The output from the last LSTM layer is then combined with the other features and fed to an MLP. The MLP consists of an initial hidden layer with 800 units and 2 subsequent hidden layers with 400 units each. Lastly, each hidden unit of the MLP uses a rectified linear unit (ReLU) as the activation function.

6   Model Evaluation and Feature Importance

One use of an accurate model of retention and forgetting would be to optimize the set of items that are chosen to be tested in an ALEKS progress assessment. That is, if it is very likely that the student will answer a learned item correctly

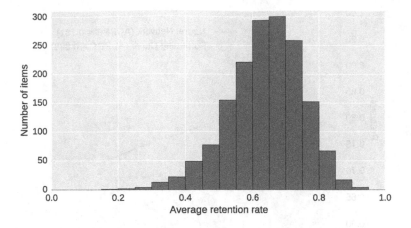

Fig. 5. Histogram of average inner fringe retention rate by item.

on a progress assessment, it may be more beneficial to the student's learning if a different item, one that the student is struggling to retain, be tested instead. In this case, the student would gain the benefits of retrieval and spaced practice focused on the more troublesome items [16]. Under this implementation, an effective model is one that can correctly identify items that are very likely to be retained; thus, a natural measure of this ability is precision. Additionally, the model must also identify a large enough subset of these items to be effective, which can be measured by the true positive rate or recall. To that end, Fig. 6 shows the precision-recall curve on the data in the test set. For comparison, we also give the results for a baseline forgetting curve that uses only time as a parameter and is fit to the correct rate data in Fig. 1 (specifically, we use the power function model that is discussed at length in [1]).

One common criticism of neural networks is that they are difficult to interpret, and that in some cases a simpler model such as a logistic regression may be preferable because of this. However, if the goal is to have an idea of the relative importance of each feature to the model (which is typically the argument for using a regression model where, in theory, the coefficients can be interpreted), this can be accomplished using a technique called permutation feature importance [2,24,25]. The idea behind permutation feature importance is the following. Given a metric to evaluate the performance of our classification model, we first compute the score for the classifier on our test set. Then, to determine the relative importance of a feature (or, set of features), we randomly shuffle the values for that feature (or, again, set of features) across all the data points in our test set; importantly, however, while doing this we leave the order of the rest of the features untouched. We then run this modified test set through our classifier, extract the predicted probabilities, and then recompute the score of our chosen metric. Comparing this score to the score on the unshuffled test set gives an idea of how "important" this feature is to the performance of the model; if the feature is very important, we can expect a large negative effect on the metric's

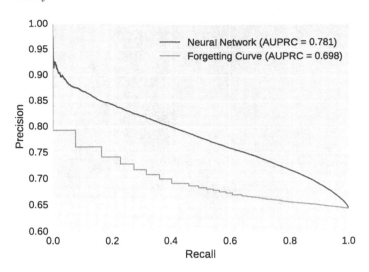

Fig. 6. Comparison of precision-recall curves for the neural network classifier and the single-parameter forgetting curve model.

score on the shuffled test set, while a minor change in the score indicates that the feature is not as crucial to the performance of the model. While some other measures of feature importance may exhibit a bias towards categorical variables with many values, permutation feature importance does not suffer from these same shortcomings [25], and thus is well-suited to our neural network model.

The results from applying permutation feature importance to our classifier are shown in Table 1, where we display the area under the curve for both the precision-recall (PR) and receiver operating characteristic (ROC) curves, averaged over 10 trials (i.e., each set of features is randomly shuffled 10 times, and the average scores over these 10 trials are reported). We can see that the variable with the greatest effect on the scores is the item categorical variable. Taking the histogram in Fig. 5 into account, this makes intuitive sense given that some of the items vary widely in their overall retention rates. Additionally, while not quite as impactful as the item variable, both the time since the item was learned, and the information from the learning sequence, are important to the model. In the latter case, this is supported by Figs. 3 and 4, which show large differences in retention based on the properties of the learning sequence. On the other hand, the course variable and the initial score are the least important variables. Regarding the initial score, while the differences shown in Fig. 2 are significant, these differences are from a comparison of the most extreme decile groups. The initial score has less of an effect when looking at other deciles, which most likely explains the smaller importance of this variable to the final model.

Table 1. Area under the precision-recall (PR) and receiver operating characteristic (ROC) curves using permutation feature importance.

Permuted feature	PR (change)	ROC (change)
None (optimal classifier)	0.781	0.680
ALEKS product	0.765 (−0.016)	0.658 (−0.022)
Item	0.713 (−0.068)	0.591 (−0.089)
Initial score	0.760 (−0.021)	0.653 (−0.027)
Time	0.753 (−0.028)	0.641 (−0.039)
Learning sequence	0.749 (−0.032)	0.634 (−0.046)

7   Discussion and Future Work

In this paper we give a detailed study of how the retention of knowledge works within the ALEKS system. By aggregating data from a large number of ALEKS assessments, we are able to look at the effects of several different variables on this retention. Based on these results, we then build a neural network model of retention within ALEKS. This neural network combines the sequential data from the student learning sequences with several other (non-sequential) variables to make predictions of the likelihood an item will be retained, improving upon the basic one-dimensional forgetting curve model. Furthermore, to help address a common criticism that neural network models are difficult to interpret, we show that an application of permutation feature importance to our neural network model, combined with our exploratory analysis of the data, gives a coherent picture of the relative importance of these variables to our model. Both the learning sequence of the student, and the time since an item was learned, are more informative to our model; on the other hand, the starting knowledge of the student and the specific ALEKS course being used have relatively smaller effects. However, the most influential information came from the categorical variable representing the items, an indication that being able to differentiate between the items is important when building an accurate model of retention. This last result is seemingly consistent with studies that have shown improvements in Bayesian knowledge tracing (BKT) models when item-specific information is taken into account [10,19,32].

Given the importance of the item variable when predicting retention, it would be of interest to explore this topic further. For example, are there certain skills and content that characterize, or are inherent to, hard or easy to retain items? Alternatively, it is possible that the outsized influence of the item variable is due to something specific to ALEKS. As an example, a low retention rate could be an indication that an item is placed at a suboptimal position within the ALEKS system, and in such a case a student would benefit from seeing additional prerequisite material before learning the item. Thus, it is not a stretch to think that the information contained within the item variable may be due to factors such as this. Answering these questions would give an even more complete picture of how retention works within ALEKS.

References

1. Averell, L., Heathcote, A.: The form of the forgetting curve and the fate of memories. J. Math. Psychol. **55**, 25–35 (2011)
2. Breiman, L.: Random forests. Mach. Learn. **45**(1), 5–32 (2001)
3. Cho, K., van Merrienboer, B., Gülçehre, Ç., Bougares, F., Schwenk, H., Bengio, Y.: Learning phrase representations using RNN encoder-decoder for statistical machine translation. CoRR abs/1406.1078 (2014). http://arxiv.org/abs/1406.1078
4. Doble, C., Matayoshi, J., Cosyn, E., Uzun, H., Karami, A.: A data-based simulation study of reliability for an adaptive assessment based on knowledge space theory. Int. J. Artif. Intell. Educ. (2019). https://doi.org/10.1007/s40593-019-00176-0
5. Doignon, J.P., Falmagne, J.C.: Spaces for the assessment of knowledge. Int. J. Man-Mach. Stud. **23**, 175–196 (1985)
6. Ebbinghaus, H.: Memory: A Contribution to Experimental Psychology. Originally published by Teachers College, Columbia University, New York (1885). Translated by Ruger, H.A., Bussenius, C.E (1913)
7. Falmagne, J.C., Albert, D., Doble, C., Eppstein, D., Hu, X. (eds.): Knowledge Spaces: Applications in Education. Springer, Heidelberg (2013). https://doi.org/10.1007/978-3-642-35329-1
8. Falmagne, J.C., Doignon, J.P.: Learning Spaces. Springer, Heidelberg (2011). https://doi.org/10.1007/978-3-642-01039-2
9. Gal, Y., Ghahramani, Z.: A theoretically grounded application of dropout in recurrent neural networks. In: Advances in Neural Information Processing Systems, vol. 29 (2016)
10. González-Brenes, J., Huang, Y., Brusilovsky, P.: General features in knowledge tracing to model multiple subskills, temporal item response theory, and expert knowledge. In: Proceedings of the 7th International Conference on Educational Data Mining, pp. 84–91 (2014)
11. Graves, A.: Supervised Sequence Labelling with Recurrent Neural Networks. Studies in Computational Intelligence. Springer, Heidelberg (2012). https://doi.org/10.1007/978-3-642-24797-2
12. Grayce, C.: A commercial implementation of knowledge space theory in college general chemistry. In: Falmagne, J.C., Albert, D., Doble, C., Eppstein, D., Hu, X. (eds.) Knowledge Spaces: Applications in Education, pp. 93–114. Springer, Heidelberg (2013)
13. Hochreiter, S., Schmidhuber, J.: Long short-term memory. Neural Comput. **9**, 1735–1780 (1997)
14. Huang, X., Craig, S., Xie, J., Graesser, A., Hu, X.: Intelligent tutoring systems work as a math gap reducer in 6th grade after-school program. Learn. Individ. Differ. **47**, 258–265 (2016)
15. Ioffe, S., Szegedy, C.: Batch normalization: accelerating deep network training by reducing internal covariate shift. In: International Conference on Machine Learning, pp. 448–456 (2015)
16. Lindsey, R.V., Shroyer, J.D., Pashler, H., Mozer, M.C.: Improving students long-term knowledge retention through personalized review. Psychol. Sci. **25**(3), 639–647 (2014)

17. Matayoshi, J., Granziol, U., Doble, C., Uzun, H., Cosyn, E.: Forgetting curves and testing effect in an adaptive learning and assessment system. In: Proceedings of the 11th International Conference on Educational Data Mining, pp. 607–612 (2018)
18. McGraw-Hill Education/ALEKS Corporation: What is ALEKS? https://www.aleks.com/about_aleks
19. Pardos, Z.A., Heffernan, N.T.: KT-IDEM: introducing item difficulty to the knowledge tracing model. In: Konstan, J.A., Conejo, R., Marzo, J.L., Oliver, N. (eds.) UMAP 2011. LNCS, vol. 6787, pp. 243–254. Springer, Heidelberg (2011). https://doi.org/10.1007/978-3-642-22362-4_21
20. Prechelt, L.: Early stopping—but when? In: Montavon, G., Orr, G.B., Müller, K.-R. (eds.) Neural Networks: Tricks of the Trade. LNCS, vol. 7700, pp. 53–67. Springer, Heidelberg (2012). https://doi.org/10.1007/978-3-642-35289-8_5
21. Qiu, Y., Qi, Y., Lu, H., Pardos, Z.A., Heffernan, N.T.: Does time matter? modeling the effect of time with Bayesian knowledge tracing. In: Proceedings of the 4th International Conference on Educational Data Mining, pp. 139–148 (2011)
22. Reddy, Λ., Harper, M.: Mathematics placement at the University of Illinois. PRIMUS 23, 683–702 (2013)
23. Srivastava, N., Hinton, G., Krizhevsky, A., Sutskever, I., Salakhutdinov, R.: Dropout: a simple way to prevent neural networks from overfitting. J. Mach. Learn. Res. 15, 1929–1968 (2014)
24. Strobl, C., Boulesteix, A.L., Kneib, T., Augustin, T., Zeileis, A.: Conditional variable importance for random forests. BMC Bioinform. 9(1), 307 (2008)
25. Strobl, C., Boulesteix, A.L., Zeileis, A., Hothorn, T.: Bias in random forest variable importance measures: illustrations, sources and a solution. BMC Bioinform. 8(1), 25 (2007)
26. Taagepera, M., Arasasingham, R.: Using knowledge space theory to assess student understanding of chemistry. In: Falmagne, J.C., Albert, D., Doble, C., Eppstein, D., Hu, X. (eds.) Knowledge Spaces: Applications in Education, pp. 115–128. Springer, Heidelberg (2013). https://doi.org/10.1007/978-3-642-35329-1_7
27. Wang, Y., Heffernan, N.: Towards modeling forgetting and relearning in ITS: preliminary analysis of ARRS data. In: Proceedings of the 4th International Conference on Educational Data Mining, pp. 351–352 (2011)
28. Wang, Y., Beck, J.E.: Incorporating factors influencing knowledge retention into a student model. In: Proceedings of the 5th International Conference on Educational Data Mining (2012)
29. Xiong, X., Li, S., Beck, J.E.: Will you get it right next week: Predict delayed performance in enhanced ITS mastery cycle. In: The Twenty-Sixth International FLAIRS Conference (2013)
30. Xiong, X., Wang, Y., Beck, J.B.: Improving students' long-term retention performance: a study on personalized retention schedules. In: Proceedings of the Fifth International Conference on Learning Analytics and Knowledge, pp. 325–329. ACM (2015)
31. Yang, Y., Leung, H., Yue, L., Deng, L.: Automatic dance lesson generation. IEEE Trans. Learn. Technol. 5, 191–198 (2012)
32. Yudelson, M.: Individualizing Bayesian knowledge tracing. Are skill parameters more important than student parameters? In: Proceedings of the 9th International Conference on Educational Data Mining (2016)

Checking It Twice: Does Adding Spelling and Grammar Checkers Improve Essay Quality in an Automated Writing Tutor?

Kathryn S. McCarthy[1]($\boxtimes$), Rod D. Roscoe[2], Aaron D. Likens[3],
and Danielle S. McNamara[2]

[1] Georgia State University, Atlanta, GA, USA
kmccarthy12@gsu.edu
[2] Arizona State University, Tempe, AZ, USA
{rod.roscoe,dsmcnamal}@asu.edu
[3] University of Nebraska at Omaha, Omaha, NE, USA
alikens@unomaha.edu

Abstract. This study investigated the effect of incorporating spelling and grammar checking tools within an automated writing tutoring system, Writing Pal. High school students ($n = 119$) wrote and revised six persuasive essays. After initial drafts, all students received formative feedback about writing strategies. Half of the participants were also given access to spelling and grammar checking tools during the writing and revision periods. Linear mixed effects models revealed that essay quality for students in both conditions improved from initial draft to revision in terms of all aspects except essay unity. The availability of spelling and grammar checking yielded added improvements from initial draft to revision for several aspects of essay quality (i.e., conclusion, organization, voice, grammar/mechanics, and word choice), but other aspects were unaffected (i.e., introduction, body, unity, and sentence structure). The availability of spelling and grammar checking tools had no effect on holistic essay scores. These results indicate that automated spelling and grammar feedback contribute to modest, incremental improvements in writing quality that may complement automated strategy feedback.

Keywords: Automated writing evaluation · Feedback ·
Natural language processing · Spelling and grammar checking ·
Writing strategies

1 Introduction

National standards report that a majority of students are not proficient writers [1]. One challenge is that the development of writing skills requires repeated, deliberate practice with formative feedback [2], yet providing opportunities for students to compose essays and receive extensive feedback is time-consuming for instructors. Although the U.S. Department of Education recommends at least one hour of writing each day, writing instruction is often overlooked compared to other skills [3, 4]. However, sophisticated natural language processing (NLP) tools have made it possible for automated writing

© Springer Nature Switzerland AG 2019
S. Isotani et al. (Eds.): AIED 2019, LNAI 11625, pp. 270–282, 2019.
https://doi.org/10.1007/978-3-030-23204-7_23

evaluation systems (AWEs) to partially address these challenges. AWEs can provide (a) rapid and accurate evaluations and (b) formative feedback that are otherwise unfeasible in classroom settings constrained by time and resources. Students who receive AWE support have also demonstrated improved writing skills [5–9].

As one example, the Writing Pal (W-Pal [10–13]) is an AWE and intelligent tutoring system (ITS) that targets persuasive writing. Learners are typically given 25 min to write an essay in response to an SAT-like persuasive writing prompt. W-Pal employs NLP-driven algorithms to evaluate the essays [14]. W-Pal algorithms deliver both summative feedback (i.e., a holistic score on a 6-point scale) and formative feedback about writing strategies and improving essay quality. This strategy feedback is related to eight different aspects of writing: freewriting, planning, introduction, body, conclusion, unity, paraphrasing, and revision (see [13]). W-Pal lessons and practice improve students' writing quality and strategy knowledge, and both students and teachers find W-Pal instruction to be valuable and informative [13, 15, 16].

1.1 What About Spelling and Grammar?

The development of technologies that deliver valid automated formative feedback has required innovations in NLP, modeling, and artificial intelligence. Nonetheless, a common question from educators is "Can it check my students' spelling and grammar?" Research suggests that classroom teachers devote extensive time and attention to grammar and mechanics, often at the expense of other writing instruction [17, 18]. Although spelling and grammar checkers have proven useful in ITSs for second language learning (e.g., [19]), the benefits of these tools for writing composition may be questionable. Research suggests that spelling and grammar instruction has little effect on writing quality after early elementary school [20], and spelling and grammar errors are only weakly correlated to expert judgments of essay quality [21]. More effective instruction focuses on writing strategies, such as drafting, organizing, and revising in combination with individualized feedback to improve these strategies (see [22]). Indeed, strategy instruction interventions have yielded large effect sizes [23].

Although spelling and grammar instruction alone may be insufficient for improving students' overall writing skill, this type of feedback may still be valuable. Struggling with spelling can exhaust cognitive resources that would be better put toward other aspects of the writing process [24, 25]. Moreover, in contrast to writing experts, non-expert readers may base their evaluations of quality on spelling and grammar [26–28]. For instance, non-experts use spelling and grammar mistakes to make judgments about a writers' intelligence and other personality traits [29, 30]. Johnson and colleagues [30] asked college students to read writing samples that contained no errors, low-level errors (spelling and grammar mistakes), high-level errors (structural and conceptual errors), or both types of errors. Participants rated the quality of the essays as well as their perceptions of the authors. Low-level errors resulted in more harsh judgments, both in terms of the essay quality and negative personality traits to the writers. In other words, spelling and grammar are still important considerations for writing quality, assessment, and audience perceptions.

1.2   The Current Study

The current study examines the effect of incorporating spelling and grammar checking (SGC) tools into an automated writing evaluation and tutoring system for writing (i.e., W-Pal). W-Pal already provides strategy training and formative feedback for more complex aspects of the writing process (e.g., planning, elaboration, and unity). Thus, this study poses the research question: *In an AWE system, what is the added value of spelling and grammar feedback when combined with higher-level strategy feedback?* In this experiment, all students write essays, receive strategy feedback, and then revise their essays. However, half of the students are also given access to (and reminders to use) spelling and grammar checking tools while writing and revising.

Spelling and grammar feedback could *benefit* students' overall essay quality by helping them focus on content and structure-based writing and revision [24, 25]. Alternatively, spelling and grammar information may have *no effect* on performance. This prediction is driven by the research that indicates essay quality is a function of deeper aspects of the content and style (e.g., [21]). Finally, spelling and grammar feedback could *reduce* essay quality. This outcome might arise because too much prompting and feedback can be distracting rather than helpful (e.g., [31]). Students also often default to "unproductive" superficial revisions [32, 33], and the inclusion of spelling and grammar feedback may misdirect students away from substantive revisions.

2   Method

2.1   Participants

High school students ($n = 143$) were recruited from a large metropolitan area in the southwestern United States and received financial compensation for their participation. Twenty-four participants were omitted due to incomplete data as a result of either technical or experimenter errors, resulting in a total of 119 participants.

One student completed the study but did not provide demographic information. Thus, the sample ($M_{age} = 17.19$, $SD = 1.28$, Range: 13–19) was 61.3% female and 37.5% male; 53.8% Caucasian, 21% Hispanic (Latin American), 10.1% Asian, 7.6% African American, and 6.7% reported as other. Finally, 85.7% of participants were native English speakers.

2.2   Design and Procedure

Students were randomly assigned one of two conditions in which they received either writing strategy feedback (Strategy Condition, $n = 60$) or writing strategy feedback along with spelling and grammar checking tools (Strategy + SGC Condition, $n = 59$).

The study included four sessions. In the first, session, students provided demographic information and then completed the Gates-MacGinitie Reading Test (GRMT; [34]) as a measure of reading skill. In each of the subsequent three sessions, the students wrote and revised two essays on different persuasive writing prompts

(Table 1). Each of the prompts had a brief introduction to frame the issue and ended in a question. For example, the prompt about images and impressions read:

> All around us appearances are mistaken for reality. Clever advertisements create favorable impressions but say little or nothing about the products they promote. In stores, colorful packages are often better than their contents. In the media, how certain entertainers, politicians, and other public figures appear is sometimes considered more important than their abilities. All too often, what we think we see becomes far more important than what really is. Do images and impressions have a positive or negative effect on people?

Table 1. Essay prompt questions in order of presentation

Prompt question
1. Do images and impressions have a positive or negative effect on people?
2. Do people achieve more success by cooperation or by competition?
3. Do people place too much emphasis on winning?
4. Should people always maintain their loyalties, or is it sometimes necessary to switch sides?
5. Is it better for people to act quickly and expect quick responses from others rather than to wait patiently for what they want?
6. Do personal memories hinder or help people in their effort to learn from their past and succeed in the present?

Participants were allotted 25 min to write an initial draft. After this time elapsed, W-Pal assigned a holistic score from "Poor" (1) to "Great" (6) along with individual strategy feedback aligned with W-Pal strategy lessons (e.g., planning, paragraph quality, unity, and paraphrasing). After viewing this feedback, participants were given 10 min to revise. Feedback messages provide actionable steps to help the participant improve their essay. For example, if the algorithm determined that the essay was too short, the participant might receive the following strategy feedback message:

> This essay may not have enough paragraphs to fully support the main argument. If you need help developing support for future drafts or essays, it may be helpful to freewrite.

- Write down possible arguments that may relate to your thesis
- Brainstorm as many relevant facts and examples as you can
- Try to think of details from school classes, news stories, and your own life that may relate to the arguments!

For participants in the Strategy + SGC Condition, "Check Spelling" and "Check Grammar" buttons appeared at the bottom of the interface during the writing and revision periods (Fig. 1). Participants could access either function at any time while writing or revising. Errors were detected using the open source API LanguageTool [35, 36]. When these tools were selected, errors were underlined similar to common word processors. Clicking on the error opened a small pop-up window with potential corrections. A reminder about the SGC tools appeared when there were 5 min remaining in the writing session, but students were not forced to use the tools.

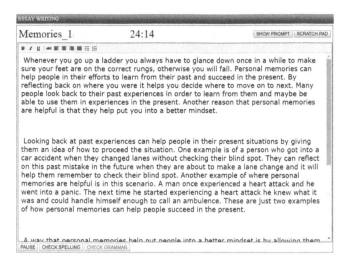

Fig. 1. W-Pal writing window with spelling and grammar checkers

2.3   Writing Assessment and Scoring

In the study, W-Pal algorithms determined immediate scores and feedback. For later analyses, human raters (following the same criteria underlying the algorithm) assigned holistic scores and nine subscores (i.e., specific writing traits) on both drafts. Raters were four graduate students of English. Rater pairs were trained to a high level of reliability (all kappas > .80) on all metrics. Holistic scores (1–6 scale) were based on the standard SAT rubric and subscores (1–6 scale) aligned with W-Pal lessons:

- **Grammar, Style, & Mechanics:** essay conveys strong control of the standard conventions of writing; avoiding errors in grammar, syntax, and mechanics
- **Introduction:** writer demonstrates mastery in meeting the goals of an introduction (e.g., presenting a topic, providing a purpose, clearly stating a thesis, and previewing arguments)
- **Body:** writer demonstrates mastery in meeting the goals of body arguments (e.g., transition between arguments, using topic sentences, supporting arguments with evidence, and maintaining a flow throughout the arguments)
- **Conclusion:** writer demonstrates mastery in meeting the goals of a conclusion (e.g., summarizing the essay, re-establishing the significance of discussion, capturing the reader's attention, and effectively closing the essay)
- **Organization:** essay follows a logical structure (e.g., introduction, body arguments and evidence, conclusion)
- **Unity:** details support the thesis and do not stray from the prompt and the main ideas and organizing principles presented in the introduction
- **Voice:** writer is expressive, engaging, and sincere; a strong sense of audience
- **Word Choice:** writer is precise and effective in word choice
- **Sentence Structure:** sentence patterns are varied effectively, enhancing the quality of the essay

Note that holistic scores was not calculated from the subscales (e.g., an average)—holistic scores and subscores were distinct judgments although likely to be correlated.

Each essay ($N = 1435$) was scored by two raters. Across raters, the reliability of ratings for scores ranged from ICC = .79 to .90. The final scores for each essay reflect the average score of the two raters.

3 Results

Table 2 presents average scores as a function of draft and feedback condition. Overall, participants exhibited modest yet significant increases in average essay scores from initial draft ($M = 3.63$, $SD = .55$) to revision ($M = 3.76$, $SD = .53$), $F(1, 117) = 25.44$, $p < .000$, $\eta^2 = .18$. Holistic writing scores were strongly correlated with all writing subscores ($r = .73$ to .93), supporting the concurrent validity of the assessment. Consistent with existing research [37], reading skill (i.e., GMRT scores) was also positively correlated with holistic score and subscores ($r = .47$ to .68).

Table 2. Means and standard deviations of essay scores as a function of experimental condition and essay draft

Score (1–6)	Strategy condition		Strategy + SGC condition	
	Initial	Revised	Initial	Revised
Holistic	3.49 (0.66)	3.64 (0.64)	3.38 (0.69)	3.52 (0.70)
Grammar, Mechanics*	3.55 (0.61)	3.57 (0.56)	3.51 (0.60)	3.69 (0.60)
Introduction	3.83 (0.69)	3.93 (0.62)	3.66 (0.65)	3.76 (0.55)
Body	3.77 (0.60)	3.82 (0.55)	3.64 (0.65)	3.75 (0.62)
Conclusion*	3.43 (0.74)	3.52 (0.72)	2.98 (0.83)	3.35 (0.77)
Organization*	3.83 (0.57)	3.93 (0.55)	3.53 (0.60)	3.77 (0.66)
Unity	4.13 (0.52)	4.10 (0.49)	3.94 (0.49)	4.01 (0.50)
Voice*	3.77 (0.54)	3.85 (0.54)	3.61 (0.58)	3.90 (0.53)
Word choice*	3.82 (0.56)	3.89 (0.47)	3.67 (0.54)	3.88 (0.57)
Sentence structure	3.59 (0.54)	3.73 (0.55)	3.53 (0.58)	3.59 (0.60)

Note. An asterisk (*) indicates that the Strategy + SGC condition demonstrated statistically significant gains from initial to revised whereas those in the Strategy Condition did not.

3.1 Analysis of Essay Improvement and Condition

Linear mixed effects (LME) models were conducted to detect the effects of the spelling and grammar checker tools on both initial draft and revision while also accounting for potential influences of the essay prompt or reading skill.

Holistic writing score and all nine writing subscores were fit with the same series of LME models using the lme4 package [38] in R [39]. Simple slopes were estimated using the *reghelper* package [40]. Table 3 shows the variables entered into each model.

For the *Baseline* model (M0), only GMRT scores and a prompt factor were included. Second, a *Draft* model (M1) included a draft factor (i.e., initial vs. revised draft) to investigate how scores changes as a function of revising. Finally, a *Condition* model (M2) added a condition factor (i.e., Strategy vs. Strategy + SG) along with two interaction terms to assess the effect of condition. The interaction terms included "Draft × Condition" and "GMRT × Condition". Likelihood ratio tests were used to compare model fit. Significant chi-square (χ^2) tests indicate that adding the additional variable(s) improved fit as compared to the previous model (Table 4). These analyses revealed mixed results for the added value of SGC tools in an AWE. Benefits were observed for several subscores but not holistic quality.

Table 3. Linear mixed effects model design

Model	Variables included
Baseline (M0)	GMRT + Prompt
Draft (M1)	GMRT + Prompt + Draft
Condition (M2)	GMRT + Prompt + Draft + Condition + (Draft × Condition) + (GMRT × Condition)

Table 4. Likelihood ratio tests (χ^2) comparing fit with additional variables

	Draft (M1) $\chi^2(1)$	Condition (M2) $\chi^2(3)$
Holistic	24.15***	ns
Grammar, Mechanics	13.87***	15.06***
Introduction	24.15***	ns
Body	6.56*	ns
Conclusion	26.68***	14.01***
Organization	20.34***	7.42+
Unity	ns	ns
Voice	27.02***	11.56**
Word choice	26.04***	9.08*
Sentence structure	9.59**	ns

Note. ns = not significant; ***$p < .001$; **$p < .01$; *$p < .05$; +$p < .06$.

3.2   Effects on Holistic Writing Quality and Grammar, Style, and Mechanics

One overarching question was whether the accessibility of SGC tools influenced overall writing quality. The LME models suggest that this was not the case. With regards to *Holistic Score*, the Draft model demonstrated improved model fit compared to the Baseline model. However, the Condition model did not further improve model fit. Although students improved their essays through revising, the availability of spelling and grammar checking tools did not appear to significantly contribute to holistic gains.

If students receive valid and useful feedback and writing mechanics, these benefits should influence *Grammar, Style, and Mechanics* subscores, even if holistic quality was not affected. Indeed, this was the observed pattern. The Draft model exhibited improved model fit compared to Baseline. Further, the Condition model demonstrated further improved model fit. Simple slopes revealed that Strategy + SGC Condition participants improved from initial draft to revision, Estimate = 0.26, SE = 0.04, t (1283) = 4.86, $p < 0.001$, whereas participants in the Strategy Condition did not, Estimate = 0.02, SE = 0.04, $t(1283) = 0.50$, $p = 0.62$. The availability of spelling and grammar feedback had little effect on initial drafts, but facilitated revising with respect to grammar, style, and mechanics, as should be expected. In broader terms, spelling and grammar tools contributed to incremental improvements in writing quality that were not necessarily reflected in holistic essay quality.

3.3   Additional Benefits of Spelling and Grammar Feedback

Several other subscores provided evidence that the availability of spelling and grammar tools facilitated incremental gains in specific aspects of writing. The quality of *Conclusion, Organization, Voice,* and *Word Choice* significantly improved from the initial draft to the revision for students in the Strategy + SGC Condition, whereas there was no improvement for students in the Strategy Condition. These findings are reflected within the LME by improved model fit compared to baseline in the Draft model, and improved fit of the Condition model, specifically driven by the Draft × Condition interaction term. Simple slopes—with draft as the focal predictor and condition as the moderator—revealed that participants who had access to spelling and grammar tools tended to increase in subscores from draft to revision (*Conclusion* Estimate = 0.38 SE = 0.06, $t(1283) = 6.07$, $p < 0.001$; *Organization* Estimate = 0.23, SE = 0.05, t (1283) = 4.86, $p < 0.001$; *Voice* Estimate = 0.29, SE = 0.05, $t(1283) = 5.97$, $p < 0.001$; *Word Choice* Estimate = 0.22, SE = 0.04, $t(1283) = 5.54$, p < 0.001), whereas there was no observed improvements for those in the Strategy Condition (all Estimates < .09; $t < 2$).

3.4   Writing Subscores Unaffected by Spelling and Grammar Feedback

As hinted by the lack of effects on holistic writing quality, the benefits of SGC tools were not universal—several writing subscores showed no effect of feedback condition. These included *Introduction, Body,* and *Sentence Structure*. Participants improved on these traits from initial to revised draft (i.e., revising improved the essay) but condition (i.e., adding spelling and grammar feedback) had no influence (see Table 4).

Finally, contrary to other subscores, participants did not show any change in *Essay Unity* scores across drafts; the Draft model did not improve model fit compared to Baseline. The Condition model also did not improve model fit.

4   Discussion and Future Work

Computer-based educational tools can provide a wealth of feedback to student writers, including summative scores on writing quality and individualized, formative feedback on strategies and ways to improve [41]. Similar to word processing programs, these tools might also incorporate spelling and grammar checking. However, one question is whether having access to SGC tools benefits students above and beyond strategy feedback. Prior research suggests that an overemphasis on writing mechanics can be useless or detrimental [21], whereas strategy instruction and feedback are beneficial [22, 23].

To explore these questions, the current study examined how the inclusion of spelling and grammar feedback in an AWE system affected the quality of essays and revisions. This study focused on high school students, a critical target population for writing instruction and intervention. Analyses examined gains in holistic essay scores and nine subscores (rated by writing experts) as a function of feedback condition.

4.1   Incremental Improvements in Essay Quality

Linear mixed effect models indicated that essay quality improved holistically and along all subscores (except unity) as a function of revising with feedback. When students wrote essays in W-Pal, received feedback, and revised, their essays improved. This finding replicates previous results showing that strategy feedback results in improvements from initial draft to the revision [42–45].

As one might expect, the availability of checking tools improved the grammar and mechanics in the essays. However, spelling and grammar feedback, in conjunction with strategy feedback, *also* improved essays from initial draft to revision on the dimensions of conclusion paragraphs, organization, voice, and word choice. This finding is important because it suggests that providing students with tools to check their spelling and grammar might (a) free up resources to consider other aspects of writing when writing and/or (b) inspire a greater willingness to revise. Grammar checkers have been shown to increase students' motivation and confidence in their writing [46, 47]. Thus, spelling and grammar checkers may have direct benefits on mechanics that then afford indirect benefits elsewhere. Future research should consider students' subjective reactions to these tools, such as whether they indeed perceive writing to be less burdensome or more engaging when they have a "safety net" of checking tools in an AWE system.

Importantly, these tools did *not* appear to benefit holistic essay scores. This is consistent with prior work showing that expert evaluations of essay quality rely on deeper features of text [21]. Likewise, there were no apparent benefits of spelling and grammar feedback on introduction quality, body quality, unity, or sentence structure. Thus, although the tools were moderately useful, they were not universally beneficial.

Finally, it is worth noting that this feedback did not *reduce* performance on any subscores. Overall, these findings indicate that adding spelling and grammar checkers *in conjunction with strategy feedback* is moderately beneficial for AWEs.

4.2   Directions for Future Research

In the current study, participants did not improve the unity of their essays, regardless of feedback or checking tools. One explanation may be that unity requires the writer to step back and evaluate the connectivity, or cohesion, of the essay as a whole. A 10-minute window for revision may not provide sufficient time to conceptualize and implement such revisions. An alternative explanation could be that participants received fewer messages about essay unity than other feedback topics. Further analyses of the types of feedback that participants received, and their viewing of that feedback, may shed further light on how we might help students to improve the unity of their essays.

In addition to assessing the specific feedback messages received and viewed, future work might also analyze how students approached the revising task (i.e., the frequency and type of edits made). One possibility is that students in the strategy feedback condition used their 10 min of revision time to address mechanical errors—a typical bias toward proofreading over substantive revising. In contrast, because students in the spelling and grammar condition had already completed this task (at least partially), they spent more time during revising on substantive edits. Given that spelling and grammar feedback did not impact the subscores equally, these findings may shed light on which aspects of writing, or which writing strategies, are prioritized by students. Eliciting students' self-reported rationale for revising or using AWE tools could help to understand how they navigate AWE functions and uptake feedback. For instance, students might focus on the "easiest" or "fastest" edits, or they may prioritize the most "critical" flaws.

Finally, having these tools available might change the dynamics of the writing process [48, 49]—students may write *more* or *faster* when they feel they can rely on SGC tools to make the task easier. To explore these plausible changes in writing production, the use of log-files, key strokes, and similar data may help to elucidate how students use the spelling and grammar functions, and the extent that tool use—as opposed to mere availability or accessibility—influences drafting and revising activities. The benefits of using spelling and grammar tools are likely to be more nuanced than a simple "more is better" assumption. Students might rely on the tools to mechanically and mindlessly fix typos instead of reflecting on the meaning of their writing or the possible reactions of their audience.

4.3   Conclusion

Questions about optimal feedback—including feedback content, timing, methods, and effects on performance—are among the most critical challenges facing educators, researchers, or others who develop and implement adaptive educational technologies. A number of nuanced factors influence feedback quality and students' feedback uptake.

This study explored feedback in the context of essay writing with AWE support. Research has previously established that formative feedback on writing strategies is effective whereas feedback that exclusively targets grammar and mechanics is not effective. Nonetheless, teachers, students, and writers intuitively crave feedback and automated corrections on these writing features. This study provides compelling evidence that students benefit from both types of feedback, and that guidance on writing mechanics does not inhibit writing quality or deter from revising.

Several questions remain concerning the source, loci, and dynamics of these effects. Do benefits stem from helping students manage their resources and focus on rhetorical aspects of writing? Do benefits stem from a sense of writer empowerment? How do varying feedback tools influence real-time writing and revising behaviors? Ultimately, the objective is to enhance students' ability to improve their writing, and automated feedback affords multiple resources for accomplishing that goal.

Acknowledgements. This research was supported by grants from the U.S. DoEd Institute of Education Sciences (R305A120707 and R305A180261) and the U.S. DoD Office of Naval Research (N000141712300). Opinions, findings, or recommendations expressed in this material are those of the authors and do not necessarily reflect the views of the funding sources.

References

1. National Center for Education Statistics: The Nation's report Card: Washington, D.C.: Institute of Education Sciences, U.S. Department of Education (2012)
2. Kellogg, R.T., Raulerson III, B.A.: Improving the writing skills of college students. Psychon. Bull. Rev. **14**(2), 237–242 (2007)
3. Graham, S., Capizzi, A., Harris, K.R., Hebert, M., Morphy, P.: Teaching writing to middle school students: a national survey. Read. Writ. **27**(6), 1015–1042 (2014)
4. Kiuhara, S.A., Graham, S., Hawken, L.S.: Teaching writing to high school students: a national survey. J. Educ. Psychol. **101**(1), 136–160 (2009)
5. Allen, L.K., Jacovina, M.E., McNamara, D.S.: Computer-based writing instruction. In: MacArthur, C.A., Graham, S., Fitzgerald, J. (eds.) Handbook of Writing Research, 2nd edn, pp. 316–329. The Guilford Press, New York (2016)
6. Palermo, C., Thomson, M.M.: Teacher implementation of self-regulated strategy development with an automated writing evaluation system: effects on the argumentative writing performance of middle school students. Contemp. Educ. Psychol. **54**, 255–270 (2018)
7. Shermis, M.D., Burstein, J. (eds.): Handbook of Automated Essay Evaluation: Current Applications and New Directions. Routledge, New York (2013)
8. Stevenson, M., Phakiti, A.: The effects of computer-generated feedback on the quality of writing. Assess. Writ. **19**, 51–56 (2014)
9. Wilson, J., Czik, A.: Automated essay evaluation software in English language arts classrooms: effects on teacher feedback, student motivation, and writing quality. Comput. Educ. **100**, 94–109 (2016)
10. Crossley, S.A., Allen, L.K., McNamara, D.S.: The Writing Pal: a writing strategy tutor. In: Crossley, S.A., McNamara, D.S. (eds.) Adaptive educational technologies for literacy instruction, pp. 204–224. Routledge, New York (2016)
11. Dai, J., Raine, R.B., Roscoe, R., Cai, Z., McNamara, D.S.: The writing-pal tutoring system: development and design. J. Eng. Comput. Innov. **2**, 1–11 (2011)
12. McNamara, D.S., et al.: The writing-pal: natural language algorithms to support intelligent tutoring on writing strategies. In: McCarthy, P.M., Boonthum-Denecke, C. (eds.) Applied Natural Language Processing and Content Analysis: Identification, Investigation, and Resolution, pp. 298–311. IGI Global, Hershey (2012)
13. Roscoe, R.D., McNamara, D.S.: Writing Pal: feasibility of an intelligent writing strategy tutor in the high school classroom. J. Educ. Psychol. **105**, 1010–1025 (2013)
14. McNamara, D.S., Crossley, S.A., Roscoe, R.D., Allen, L.K., Dai, J.: A hierarchical classification approach to automated essay scoring. Assess. Writ. **23**, 35–59 (2015)

15. Allen, L.K., Crossley, S.A., Snow, E.L., McNamara, D.S.: Game-based writing strategy tutoring for second language learners: game enjoyment as a key to engagement. Lang. Learn. Technol. **18**, 124–150 (2014)

16. Roscoe, R.D., Allen, L.K., Weston, J.L., Crossley, S.A., McNamara, D.S.: The Writing Pal intelligent tutoring system: usability testing and development. Comput. Compos. **34**, 39–59 (2014)

17. Cutler, L., Graham, S.: Primary grade writing instruction: a national survey. J. Educ. Psychol. **100**(4), 907–919 (2008)

18. Morris, D., Blanton, L., Blanton, W.E., Perney, J.: Spelling instruction and achievement in six classrooms. Elem. Sch. J. **96**(2), 145–162 (1995)

19. Heift, T., Rimrott, A.: Learner responses to corrective feedback for spelling errors in CALL. System **36**(2), 196–213 (2008)

20. Graham, S., Perin, D.: A meta-analysis of writing instruction for adolescent students. J. Educ. Psychol. **99**(3), 445–476 (2007)

21. Crossley, S.A., Kyle, K., Allen, L.K., McNamara, D.S.: The importance of grammar and mechanics in writing assessment and instruction: evidence from data mining. In: Stamper, J., Pardos, Z., Mavrikis, M., McLaren,B.M. (eds.) Proceedings of the 7th International Conference on Educational Data Mining, London, UK, pp. 300–303 (2014)

22. Graham, S., Harris, K.R., Chambers, A.B.: Evidence-based practice and writing instruction. Handb. Writ. Res. **2**, 211–226 (2016)

23. Graham, S., Harris, K.R.: Strategy instruction and the teaching of writing. Handb. Writ. Res. **5**, 187–207 (2006)

24. Graham, S., Santangelo, T.: Does spelling instruction make students better spellers, readers, and writers? A meta-analytic review. Read. Writ. **27**(9), 1703–1743 (2014)

25. Morphy, P., Graham, S.: Word processing programs and weaker writers/readers: a meta-analysis of research findings. Read. Writ. **25**(3), 641–678 (2012)

26. Marshall, J.C.: Composition errors and essay examination grades re-examined. Am. Educ. Res. Assoc. **4**, 375–385 (1967)

27. Marshall, J.C., Powers, J.M.: Writing neatness, composition errors, and essay grades. J. Educ. Meas. **6**, 97–101 (1969)

28. Figuredo, L., Varnhagen, C.K.: Didn't you run the spell checker? Effects of type of spelling error and use of a spell checker on perceptions of the authors. Read. Psychol. **26**, 441–458 (2005)

29. Boland, J.E., Queen, R.: If you're house is still available, send me an email: personality influences reactions to written errors in email messages. PLoS ONE **11**, e0149885 (2016)

30. Johnson, A.C., Wilson, J., Roscoe, R.D.: College student perceptions of writing errors, text quality, and author characteristics. Assess. Writ. **34**, 72–87 (2017)

31. McCarthy, K.S., Likens, A.D., Johnson, A.M., Guerrero, T.A., McNamara, D.S.: Metacognitive overload!: Positive and negative effects of metacognitive prompts in an intelligent tutoring system. Int. J. Artif. Intell. Educ. **28**(3), 1–19 (2018)

32. Crawford, L., Lloyd, S., Knoth, K.: Analysis of student revisions on a state writing test. Assess. Eff. Interv. **33**, 108–119 (2008)

33. Fitzgerald, J.: Research on revision in writing. Rev. Educ. Res. **57**, 481–506 (1987)

34. MacGinitie, W.H., MacGinitie, R.K., Cooter, R.B., Curry, S.: Assessment: Gates-MacGinitie reading tests. Read. Teach. **43**(3), 256–258 (1989)

35. Miłkowski, M.: Developing an open-source, rule-based proofreading tool. Softw. Pract. Exp. **40**(7), 543–566 (2010)

36. Naber, D.: A rule-based style and grammar checker. Master's thesis, Universität Bielefeld (2003). http://www.danielnaber.de/publications

37. Allen, L.K., Snow, E.L., Crossley, S.A., Jackson, G.T., McNamara, D.S.: Reading comprehension components and their relation to the writing process. L'année psychologique/Top. Cogn. Psychol. **114**, 663–691 (2014)
38. Bates, D., Maechler, M., Bolker, B., Walker, S.: lme4: linear mixed-effects models using Eigen and S4. R package version 1.1–7 (2015)
39. R Core Team: A language and environment for statistical computing. R Foundation for Statistical Computing, Vienna, Austria (2017)
40. Hughes, J., Team, R.C. Reghelper: helper functions for regression analysis. R package, version 0.3.3 (2017)
41. Strobl, C., et al.: Digital support for academic writing: a review of technologies and pedagogies. Comput. Educ. **131**, 33–48 (2018)
42. Attali, Y., Burstein, J.: Automated essay scoring with e-rater® v. 2.0. ETS Research Report Series 2004. 2, i-21(2004)
43. Crossley, S.A., Roscoe, R.D., McNamara, D.S.: Using automatic scoring models to detect changes in student writing in an intelligent tutoring system. In: Proceedings of the 26th International Florida Artificial Intelligence Research Society (FLAIRS) Conference, pp. 208–213. The AAAI Press, Menlo Park (2013)
44. Roscoe, R.D., Snow, E.L., McNamara, D.S.: Feedback and revising in an intelligent tutoring system for writing strategies. In: Lane, H.C., Yacef, K., Mostow, J., Pavlik, P. (eds.) AIED 2013. LNCS (LNAI), vol. 7926, pp. 259–268. Springer, Heidelberg (2013). https://doi.org/10.1007/978-3-642-39112-5_27
45. Roscoe, R.D., Snow, E.L., Allen, L.K., McNamara, D.S.: Automated detection of essay revising patterns: application for intelligent feedback in a writing tutor. Technol. Instr. Cogn. Learn. **10**(1), 59–79 (2015)
46. Cavaleri, M.R., Dianati, S.: You want me to check your grammar again? The usefulness of an online grammar checker as perceived by students. Journal of Academic Language and Learning **10**(1), A223–A236 (2016)
47. Potter, R., Fuller, D.: My new teaching partner? Using the grammar checker in writing instruction. Engl. J. **98**(1), 36–41 (2008)
48. Allen, L.K., et al.: {ENTER}ing the time series {SPACE}: Uncovering the writing process through keystroke analysis. In: Proceedings of the 9th International Conference on Educational Data Mining (EDM 2016), pp. 22–29. International Educational Data Mining Society, Raleigh, NC (2016)
49. Likens, A.D., Allen, L.K., McNamara, D. S.: Keystroke dynamics predict essay quality. In: Gunzelmann, G., Howes, A., Tenbrink, T., Davelaar, E. (eds.) Proceedings of the 39th Annual Meeting of the Cognitive Science Society (CogSci 2017), pp. 2573–2578. Cognitive Science Society, London, UK (2017)

What's Most Broken? Design and Evaluation of a Tool to Guide Improvement of an Intelligent Tutor

Shiven Mian[1(✉)], Mononito Goswami[2], and Jack Mostow[3]

[1] IIIT-Delhi, Delhi, India
shiven15094@iiitd.ac.in
[2] Delhi Technological University, Delhi, India
mononitog@hotmail.com
[3] Carnegie Mellon University, Pittsburgh, PA, USA
mostow@cs.cmu.edu

Abstract. Intelligent Tutoring Systems (ITS) have great potential to change the educational landscape by bringing scientifically tested one-to-one tutoring to remote and under-served areas. However, effective ITSs are too complex to perfect. Instead, a practical guiding principle for ITS development and improvement is to fix what's most broken. This paper presents SPOT (Statistical Probe of Tutoring), a tool that mines data logged by an ITS to identify 'hot spots' most detrimental to its efficiency and effectiveness in terms of its software reliability, usability, task difficulty, student engagement, and other criteria. SPOT uses heuristics and machine learning to discover, characterize, and prioritize such hot spots in order to focus ITS refinement on what matters most. We applied SPOT to data logged by RoboTutor, an ITS that teaches children basic reading, writing and arithmetic. A panel-of-experts experimental evaluation found SPOT's selected video clips of RoboTutor's hot spots as significantly more informative than video clips selected randomly.

1 Introduction

An Intelligent Tutoring System (ITS) is a computer system that enables learning by providing personalized instruction to learners. Intelligent Tutoring Systems are becoming increasingly popular for education across a wide variety of subjects from algebra and geometry to foreign languages. Prior research has demonstrated that ITSs take extensive time to author, with reported estimates of 200–300 h of development per hour of instruction [1]. As a result, various authoring tools have been built to make ITS development more efficient.

But, despite our growing understanding of human cognition and the tutor authoring process, developing effective tutoring systems with limited

S. Mian and M. Goswami—authors contributed equally. Work was partially done while the primary authors were Summer Scholars at the Robotics Institute, Carnegie Mellon University, PA.

© Springer Nature Switzerland AG 2019
S. Isotani et al. (Eds.): AIED 2019, LNAI 11625, pp. 283–295, 2019.
https://doi.org/10.1007/978-3-030-23204-7_24

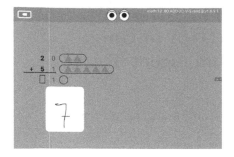

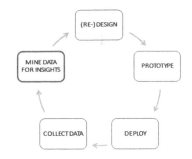

Fig. 1. Math activity in RoboTutor   **Fig. 2.** Data-driven iterative development

development resources remains hard, due to their massive design space [19]. In this situation, a practical guiding principle for ITS development and improvement is to fix what is most broken.

The design of RoboTutor [25] (see Fig. 1) was guided by this principle. Robo-Tutor is a finalist in the Global Learning XPRIZE Competition [27] to develop an open-source Android tablet application that teaches children aged 7–10 in developing countries basic reading, writing, and arithmetic. As Fig. 2 illustrates, RoboTutor's iterative development process involves designing prototype activities, deploying them for field testing, collecting data as the children do the activity, and mining this data to identify and modify what's most broken.

This paper addresses how to use data from RoboTutor to automate discovery of design issues pertaining to RoboTutor's:

– **Reliability:** How often and under what conditions does RoboTutor crash or hang? How fast does a child recover from the crash or hang?
– **Recognition:** How accurately does RoboTutor recognize written and spoken input?
– **Usability:** How easily and efficiently can children operate RoboTutor? Which situations do they find hard to navigate?
– **Engagement:** When are the children disengaged, and why?

To answer these questions we present Statistical Probe of Tutoring (SPOT), an educational data mining tool intended to help ITS developers identify "what's most broken" - i.e, "hot spots" with respect to design criteria such as software reliability, recognition accuracy, UI/UX usability, student engagement and task difficulty. SPOT uses quantitative metrics to evaluate such criteria; for example, metrics of reliability include the frequency of crashes and hangs.

SPOT uses heuristic metrics to identify undesirable events, and trains a decision tree to discover hot spots. A hot spot is a subtree with a high proportion of undesirable events.

The intuitions that inspired this approach are as follows:

– Within a decision tree, undesirable events in the same subtree are likely to have the same underlying cause.

- The feature combination associated with a subtree - that is, the sequence of tests from the root to the subtree - characterizes when the undesirable events tend to occur and may reflect this underlying cause.
- Screen capture videos undesirable events in the subtree may shed further light on a certain cause and inspire ideas for how to address it.

SPOT may be especially useful when user-testing in person is impractical, such as when the users are far away, when children may behave differently because adults are present, and when hot spots are important to fix but too rare to observe in person.

Earlier work [16] designed a UI/UX for SPOT and prototyped it using simulated data. In this paper, we describe an implementation that uses automated decision tree learning and runs on real data from RoboTutor.

Section 2 discusses prior related work. Section 3 describes how RoboTutor is organized, the data collection process and the SPOT workflow in detail. Section 4 summarizes SPOT's discoveries about RoboTutor, Sect. 5 reports an experimental evaluation of SPOT, and Sect. 6 summarises our contributions and avenues of future work.

2 Relation to Prior Work

Many Intelligent Tutoring Systems record student interactions in log files [21]. To the best of our knowledge SPOT is the first tool that specifically facilitates the discovery of hot spots in Reliability, Recognition Accuracy, Usability and Student Engagement.

2.1 Software Reliability

Intelligent Tutoring Systems must be reliable. Frequent crashes or hangs can disengage learners and negatively impact user experience. Prior studies of software reliability growth models relied on assumptions about the nature of faults and the non-deterministic behavior of failures [14]. Connectionist models have also been used to predict software reliability, as they can generate models automatically from the history of past failures [18]. The version of RoboTutor we analyzed did not explicitly log crashes, so we could not use reliability models based on data of past failures. Instead, we used simple heuristics to label likely instances of a crash or hang, for e.g. successive sessions on the same tablet separated by only a short hiatus.

2.2 Usability

Usability is concerned with making a system easy to learn and use. Realizing usable ITSs involves alternating between usability evaluation and re-design until a satisfactory usable design is achieved. Several approaches have been used to evaluate ITS usability, including objective performance measures of effectiveness and efficiency, users' subjective assessments, walkthrough usability tests, and memo tests to measure interface memorability [9,10,15].

2.3   Engagement, Off-Task Behavior and Gaming the System

Children must focus on solving problems and engage with a tutor to effectively utilize their time and learn. A child is disengaged when he or she is not actively thinking about the subject material, and/or attempts to "game the system" by systematically misusing the ITS's feedback and hints to advance through the curriculum. Prior work [11] has shown that off-task and gaming behaviors are associated with poor learning outcomes. There is also extensive literature on detecting instances of student disengagement and lack of motivation. Beck's Engagement Tracing [17] model showed that time on task is an important predictor of disengagement. Cocea et al. [12] used a rich bag of features from log files to train a decision tree to assess students' motivation. Baker et al. [3,4,6] developed latent response models to classify instances of gaming the system. Successful detectors of engagement and off-task behavior have often relied on either a limited set of activities (or questions), for example student performance on multiple-choice cloze questions [17], or field observations of student affect to train statistical or machine learning models [11]. In contrast, SPOT does not use data from observers to ascertain whether a child is off-task while using Robo-Tutor. Also, SPOT analyzes log data from many different kinds of activities, including story reading, multiple choice tasks and writing, and uses only on log files and screen video, unlike previous studies that utilize time intensive field observations and questionnaire evaluations.

Metrics like number of bailouts in activities, and average duration per attempt (see Table 2) were inspired by predictors of gaming behavior. Previous work [5,7,12,13,17] has shown that these predictors are accurate in detecting disengagement.

3   Methodology

3.1   Organization of RoboTutor

RoboTutor uses a variety of activities to teach basic literacy and numeracy to children. Figure 1 illustrates a math activity teaching addition of two numbers. RoboTutor's hierarchy of events is as follows:

1. **Session:** The sequence of zero or more activities between launching Robo-Tutor and exiting or crashing.
2. **Activity:** A tutor or game that teaches a certain skill, such as addition of two numbers, by giving assisted practice and feedback on a sequence of items.
3. **Item:** A stimulus presented visually and/or orally. Depending on the activity, the child taps, speaks or writes a response. In Fig. 1, the problem of adding 20 and 51 is an item of a math activity. The next item is presented after the child gets the current item right. An item may require multiple steps.
4. **Step:** A part of performing an item. For example Fig. 1 is a two-step item. Adding numbers in the ones' place is the first step, while adding numbers in the tens' place is the second step.

5. **Attempt:** A single input by the child within a step. An attempt can be either correct or incorrect. A step may take more than one attempt. Activities vary in the number of incorrect attempts allowed before RoboTutor automatically advances to the next step.

The RoboTutor log data consists of transactions. A transaction is a record of an attempt at a step of an item, either successful, unsuccessful, or unfinished if the child quits the activity. We derived three levels of features from logged transactions:

- **Step level:** Aggregation over all the attempts at a step, e.g. number of attempts
- **Item level:** Aggregation over all steps of an item, e.g. time spent on the item
- **Activity level:** Aggregation over all items of an activity, e.g. percent of items correct on the first attempt

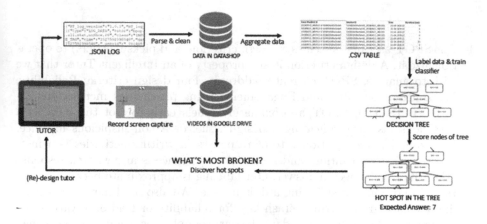

Fig. 3. Workflow of SPOT

3.2 Data Collection

The dataset used by SPOT to discover hot spots is derived from the performance logs from RoboTutor's beta field testing sites between April and July 2018, comprising 357,115 student transactions from a total of 198 user IDs (students aged 7–10), spanning approximately 212 student hours. After data cleaning, we get the fields listed in Table 1. The dataset also includes screen capture videos of RoboTutor recorded using AZ Screen Recorder [24].

3.3 Approach

SPOT aims to help ITS developers by focusing their attention on design issues that hamper effective tutoring. SPOT automates the discovery of these design issues based on different criteria and quantitative metrics. Figure 3 illustrates the SPOT work flow.

Table 1. Features from performance logs

Granularity	Features
Attempt	Duration (sec), Tutor Name, Tutor Level, Problem Name, Student Input, Activity Status, Expected Answer, Hiatus, Content Area, Outcome, Student Repeated Activity (binary), Student Took Hint (binary), Tablet Name, Attempt Number
Activity or item	Tutor Name, Tutor Level, Tutor Matrix (Category), Tutor Sequence Session (for Activity), Tutor Sequence User (for Activity), Problem Name (for Item), # of attempts, # of items (for Activity), Duration (sec), Avg/min/max duration of attempt, # of correct attempts, % of correct attempts, # of bail-outs (activity quit), # of scaffolds (i.e hints received), # of re-attempts, # of activity repeat, Activity/Item Hiatus

1. The SPOT user first specifies a design criterion, and picks metric(s) to operationalize it. A design criterion is any property of an Intelligent Tutor that we wish to analyze. SPOT currently addresses four design criteria: Reliability, Recognition, Usability and Engagement. Using pre-defined metrics (heuristics; see Table 2), SPOT approximately labels each row of the data as a suspicious instance indicating a design issue, or a non-suspicious instance. For example, the rejection rate of responses in writing activities is a metric of writing recognition, and a suspicious instance is any written response other than the expected answer. Our labeling is approximate because a metric does not guarantee finding a design issue. We also used metrics when we did not have ground truth (crash logs for reliability or field observations for engagement [6]) to label the data. In our example, suspicious instances may also include genuinely incorrect student responses, not just correct responses rejected due to mis-recognition. Only correct responses falsely rejected or accepted are recognition issues; frequent rejected incorrect responses may indicate a usability issue such as an excessive level of difficulty. Such issues are informative in their own right, because they may indicate that students are being given problems beyond their zone of proximal development.

2. SPOT trains a decision tree on the labeled data and selects the top N hot spots using the F_1 score for each subtree. Decision trees have been extensively used in the past for summarizing and generalizing data [2,22,26] due to their comprehensibility. We chose decision trees in the interest of discovering simple, understandable rules in contrast to more opaque learning methods such as neural networks. Since the focus of this work is not to identify the optimal learning method but rather to demonstrate the viability of the general approach, we utilised a popular off-the-shelf Decision Tree implementation. We used scikit-learn's [23] Decision Tree API that uses an optimised version of the CART algorithm [8] for constructing the tree.

Our intuition suggests that suspicious events in the same sub-tree are likely to have the same underlying cause, and a large cluster of suspicious instances might suggest an underlying design issue. SPOT uses different combinations of features such as activity name, student input, expected answer, attempt duration etc, with different features for attempt, item and activity-level data (see Table 1). We selected these features to enhance the comprehensibility of hot spot characterizations. Consequently, we chose categorical features like Tutor Name and Content Area since they facilitate discovery of hot spots in activities.

For a subtree to be counted as a hot spot it must balance both purity (concentrating suspicious instances) and number of suspicious instances. Thus, leaves of a decision tree (which don't have many instances despite being pure) and the root (which is significantly impure despite including all the logged events) aren't hot spots. Figure 4 illustrates a particular hot spot in a portion of the decision tree. For striking this balance, SPOT uses the F_1 score and gets the score ('heat') of every subtree for discovering hot spots:

$$F_1 = \frac{2PC}{P + C}$$

where:

$$P(precision) = \frac{\text{number of suspicious instances in hot spot}}{\text{total number of instances in hot spot}}$$

$$C(coverage) = \frac{\text{number of suspicious instances in hot spot}}{\text{total number of suspicious instances in data}}$$

The higher this F_1 score of a subtree, the hotter it is. SPOT chooses and prioritizes the top N hottest subtrees as the hot spots.

3. SPOT automatically characterizes each of the hot spots by conjoining every test on the path from the root node of the tree to the hot spot. This 'characterisation' is the possible 'underlying cause' mentioned in Point 2. We use Decision Trees **strictly** for finding and characterizing hot spots, not for classifying unseen data.

4. SPOT picks a random sample of suspicious instances of a hot spot, and presents their characterisation along with their corresponding screen capture videos (indexed in Google Drive) scrolled to 10 s before the start of the instances. These screen capture videos may expose the design issue or its underlying cause, and inspire ideas on how to address it by modifying Robo-Tutor.

Figure 5 illustrates a screen shot from the screen capture videos for the hot spot characterized as: "expected answer = 7". After examining the sample of screen capture videos from this hot spot, we realized that the writing recognizer often mis-recognizes 7 in the absence of its middle stick as 1, leading children to confuse the two.

Table 2. SPOT metrics and labeling criteria

Criterion	SPOT metrics	Positive labeling criteria
Reliability	Crashes (quick session changes)	Session ID changes in successive attempts on the same tablet within a very short period (indicative of a crash), mark the first session change attempt as positive
Recognition	Rejection rate	Rejection (Outcome = INCORRECT) for spoken and written input
Usability	Hiatus: time between end of previous attempt/item/activity, and start of the current one	Above mean hiatus
Engagement	# of bailouts in activities	Activities with at least one bailout
	% correct attempts in activities	Activities with % correct <0.5 and # of incorrect attempts ≥2
	Average duration per attempt	Above mean average duration
	Number of reattempts per item	Number of reattempts ≥2

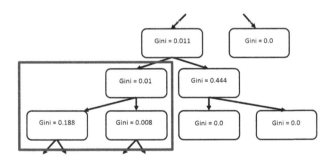

Fig. 4. Hot spot illustration. Nodes in decision tree with high F_1 scores are taken as hot spots. CART uses the Gini Impurity for getting the optimal split at every node in the tree.

4   Discoveries

We used SPOT to identify hot spots in the criteria previously described. Based on SPOT's findings (characterizations and extracted screen video clips), we proposed several design changes to RoboTutor. Table 3 presents some of the findings and their corresponding design implications.

Table 3. SPOT findings and implications

Criterion	SPOT findings	Design implications
Reliability	Crashes especially occur in Counting and Story Reading activities	Implement crash logging, Examine Counting and Story Reading activities for bugs
Recognition	Children confuse 1 with 7 (often forget the middle dash in 7) and 3 with 5	Bias item sequences for number copying and dictation to give more practice on such frequently confused digits
Usability	Children spend unusually long time per story in Story Reading activities	Add a timeout for story reading
Engagement	Children tend to bail out of an activity when given the same problems to solve repeatedly	Do not give the same items twice in a row

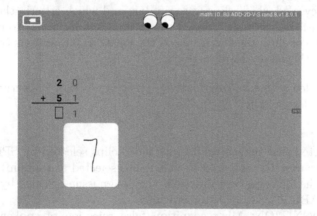

Fig. 5. Recognizer mis-recognizing 7 as 1 because of absence of middle stick. This instance and its characterisation and corresponding screen video was provided by SPOT

5   Evaluation

How should SPOT be evaluated? Since its purpose is to guide the improvement of an intelligent tutor (in particular RoboTutor), an evaluation with unlimited time and resources might start with the same version of RoboTutor, give some software developers some period of time to improve it using SPOT, kid-test the resulting software, and compare it to software produced without SPOT.

Developers and children vary, so a valid comparison would require multiple versions developed by statistically comparable teams of developers and tested on statistically comparable samples of children. Even then the comparison would be of limited use because the idiosyncrasies of teams and the contexts in which they operate make software development projects irreproducible by their very nature.

5.1   Experiment Design

Instead, we conducted an experiment to test SPOT's impact on the process of identifying design issues. To evaluate the impact of Steps 2–4 in our approach (Sect. 3.3), we conducted a panel-of-judges experiment to measure the informativeness of 8 groups of 2–3 video clips picked by SPOT compared to 4 random groups of clips. To represent SPOT's target users, we recruited four judges (a faculty member, a PhD student, an undergraduate, and a staff programmer) who were familiar with RoboTutor and knowledgeable about intelligent tutors and HCI, but did not work on SPOT.

To make the comparison fair, we had to blind the judges to how the video clips were selected. We therefore showed them only the 12 groups of video clips, without the hot spot descriptions output by SPOT, because including them would have revealed which clips were random and which criteria the other clips illustrated. Instead, the survey listed the criteria and metric(s) to choose from. For each group of clips, the survey asked the following 3 questions (worded more completely in the survey introduction):

1. Are the video segments informative – that is, do they collectively reveal a significant design issue?
   - Yes
   - No
   (we predicted that judges would rate video clips selected by SPOT as more informative overall, compared to video clips selected at random)
2. Which of the following best describes the design issue revealed by these clips?
   - RELIABILITY: RoboTutor crashing
   - RECOGNITION: Mis-recognition/false rejection of spoken or written input
   - USABILITY: Long hesitations, time on menus/debugger/irrelevant screens
   - ENGAGEMENT: Repeated bailouts, quitting RoboTutor after multiple bailouts
   - OTHER: None of the above
   (we predicted that judges would rate video clips as more informative regarding the criteria that led SPOT to select them, compared to the other criteria. That is, SPOT is sensitive to the specified criteria)
3. Briefly describe the specific issue in more detail.
   (we used this question to elicit judges' perceptions of the design issues).

5.2 Experiment Results

The four judges each rated the same 12 groups of video clips. In deciding which groups were informative overall, they showed substantial agreement with each other (Fleiss' Kappa = 0.71). Unbeknownst to the judges, the 12 groups comprised 8 chosen by SPOT, which 97% of their responses rated as informative, and 4 chosen randomly, which only 3% of their responses rated as informative.

To quantify judges' agreement as to which of the 12 sets of video clips revealed which issues, we computed Fleiss' Kappa for the 5 issues RELIABILITY, RECOGNITION, USABILITY, ENGAGEMENT, and OTHER. Their agreement was almost perfect (Kappa = 0.86). This higher value may seem counter-intuitive but reflects the lower probability of chance agreement in choosing among 5 issues than between informative and uninformative. 90% (45/48) of the judges ratings matched SPOT's categorization as RELIABILITY, RECOGNITION, USABILITY, or ENGAGEMENT or categorization of the randomly chosen groups as OTHER. The exceptions may have been caused by insufficiently clear instructions, ambiguity due to overlap between criteria (usability and engagement), or judges' incomplete knowledge of RoboTutor, e.g. which of its activities listen to spoken input and may therefore mis-recognize it.

Finally, the open-ended question was useful in exposing some responses to the first question as anomalous, i.e., contradicted by judges' descriptions of the issues. Correcting their Yes/No responses to match their descriptions yielded almost perfect agreement among judges as to which groups of videos were informative (Kappa = 0.90).

The practical significance of these results is the time saved. For the data set used, SPOT took only 4–5 min to identify hot spots for a given criterion, generate charactersations of them, and provide links to specific instances of them in the form of video links scrolled to 10 s before the precise moment where they occurred. This capability saves an enormous amount of time compared to finding issues by browsing randomly through screen videos. Issues may be important to fix yet infrequent to occur. The rarer they are, the more video one would have to watch in order to discover them. One would have to watch on average 5 h of random video to encounter an instance of an issue that occurs once every 10 h. In contrast, SPOT can identify such issues in minutes, characterize them in terms of features of the hot spot, and immediately locate and present instances of them in video clips.

The purpose of this evaluation was to test SPOT's ability to discover phenomena interesting to humans representative of its intended target users, namely ITS developers. Better heuristics for identifying such phenomena might well be developed by hand, but at considerable cost. Although SPOT's methods are intended to be general, however, we acknowledge that so far they have been applied only to data logged by various versions of RoboTutor at 28 XPRIZE sites and two beta sites.

6 Conclusion and Future Work

In this paper we introduced SPOT, a tool to simplify data driven iterative design of an Intelligent Tutoring System by focusing on criteria such as reliability, recognition, usability and engagement. SPOT discovers, prioritizes and characterises design issues (hot spots) in pre-defined design criteria using decision trees and heuristic metrics. SPOT automatically characterizes hot spots using the features associated with the item, and presents their corresponding screen-capture videos to help diagnose the problem. We described some hot spots that SPOT discovered and their design implications. A panel-of-judges experiment demonstrated that the videos chosen by SPOT are significantly more informative and relevant to the specified criteria than videos chosen randomly.

As future work, we plan to experiment with other scoring functions. We also plan to develop SPOT into a web application and integrate it into LearnSphere [20], one of the world's largest learning analytics platforms. We believe SPOT would benefit many tutor designers, help them iteratively improve their tutors, and in turn save a lot of time.

References

1. Aleven, V., McLaren, B.M., Sewall, J., Koedinger, K.R.: The cognitive tutor authoring tools (CTAT): preliminary evaluation of efficiency gains. In: Ikeda, M., Ashley, K.D., Chan, T.-W. (eds.) ITS 2006. LNCS, vol. 4053, pp. 61–70. Springer, Heidelberg (2006). https://doi.org/10.1007/11774303_7
2. Apté, C., Weiss, S.: Data mining with decision trees and decision rules. Future Gener. Comput. Syst. **13**(2–3), 197–210 (1997)
3. Baker, R.S., Corbett, A.T., Koedinger, K.R., Wagner, A.Z.: Off-task behavior in the cognitive tutor classroom: when students game the system. In: Proceedings of the SIGCHI Conference on Human Factors in Computing Systems, pp. 383–390. ACM (2004)
4. Baker, R.S.J., Corbett, A.T., Koedinger, K.R., Roll, I.: Generalizing detection of gaming the system across a tutoring curriculum. In: Ikeda, M., Ashley, K.D., Chan, T.-W. (eds.) ITS 2006. LNCS, vol. 4053, pp. 402–411. Springer, Heidelberg (2006). https://doi.org/10.1007/11774303_40
5. d Baker, R.S., et al.: Towards sensor-free affect detection in cognitive tutor algebra. Int. Educ. Data Min. Soc. (2012)
6. Baker, R.S.: Modeling and understanding students' off-task behavior in intelligent tutoring systems. In: Proceedings of the SIGCHI Conference on Human Factors in Computing Systems, pp. 1059–1068. ACM (2007)
7. Baker, R.S., de Carvalho, A., Raspat, J., Aleven, V., Corbett, A.T., Koedinger, K.R.: Educational software features that encourage and discourage "gaming the system". In: Proceedings of the 14th International Conference on Artificial Intelligence in Education, pp. 475–482 (2009)
8. Breiman, L., Friedman, J., Olshen, R., Stone, C.: Classification and regression trees (1984)

9. Chrysafiadi, K., Virvou, M.: Usability factors for an intelligent tutoring system on computer programming. In: Damiani, E., Jeong, J., Howlett, R.J., Jain, L.C. (eds.) New Directions in Intelligent Interactive Multimedia Systems and Services-2, vol. 226, pp. 339–347. Springer, Heidelberg (2009). https://doi.org/10.1007/978-3-642-02937-0_31

10. Chughtai, R., Zhang, S., Craig, S.D.: Usability evaluation of intelligent tutoring system: its from a usability perspective. In: Proceedings of the Human Factors and Ergonomics Society Annual Meeting, vol. 59, pp. 367–371, SAGE Publications, Los Angeles (2015)

11. Cocea, M., Hershkovitz, A., Baker, R.S.: The impact of off-task and gaming behaviors on learning: immediate or aggregate? In: Proceeding of the 2009 Conference on Artificial Intelligence in Education: Building Learning Systems that Care: From Knowledge Representation to Affective Modelling. IOS Press (2009)

12. Cocea, M., Weibelzahl, S.: Can log files analysis estimate learners' level of motivation? In: LWA. University of Hildesheim, Institute of Computer Science (2006)

13. Cocea, M., Weibelzahl, S.: Cross-system validation of engagement prediction from log files. In: Duval, E., Klamma, R., Wolpers, M. (eds.) EC-TEL 2007. LNCS, vol. 4753, pp. 14–25. Springer, Heidelberg (2007). https://doi.org/10.1007/978-3-540-75195-3_2

14. Goel, A.L.: Software reliability models: assumptions, limitations, and applicability. IEEE Trans. Softw. Eng. **12**, 1411–1423 (1985)

15. Granić, A., Glavinić, V.: An approach to usability evaluation of an intelligent tutoring system. In: Mastorakis, N., Kluev, V. (eds.) Advances in Multimedia, Video and Signal Processing Systems (2002)

16. Henry, G., Lin, J., Park, C.Y.: SPOT: refining robotutor. HCI senior capstone project presentation, Carnegie Mellon University (2017)

17. Joseph, E.: Engagement tracing: using response times to model student disengagement. Artif. Intell. Educ.: Support. Learn. Through Intell. Soc. Inf. Technol. **125**, 88 (2005)

18. Karunanithi, N., Whitley, D., Malaiya, Y.K.: Prediction of software reliability using connectionist models. IEEE Trans. Softw. Eng. **18**(7), 563–574 (1992)

19. Koedinger, K.R., Booth, J.L., Klahr, D.: Instructional complexity and the science needed to constrain it. Science **342**, 935–937 (2014)

20. LearnSphere: LearnSphere (2017). http://learnsphere.org/

21. Mostow, J., Beck, J., Cen, H., Cuneo, A., Gouvea, E., Heiner, C.: An educational data mining tool to browse tutor-student interactions: time will tell. In: Proceedings of the Workshop on Educational Data Mining, National Conference on Artificial Intelligence, pp. 15–22. AAAI Press (2005)

22. Murthy, S.K.: Automatic construction of decision trees from data: a multidisciplinary survey. Data Min. Knowl. Discov. **2**(4), 345–389 (1998)

23. Pedregosa, F., et al.: Scikit-learn: machine learning in python. J. Mach. Learn. Res. **12**(Oct), 2825–2830 (2011)

24. Play, G.: AZ Screen Recorder (2018). https://play.google.com/store/apps/details?id=com.hecorat.screenrecorder.free

25. RoboTutor: RoboTutor (2015). http://robotutor.org

26. Romero, C., Ventura, S.: Educational data mining: a review of the state of the art. IEEE Trans. Syst. Man Cybern. Part C (Appl. Rev.) **40**(6), 601–618 (2010)

27. XPRIZE: Global Learning XPRIZE (2015). http://learning.xprize.org

Reducing Mind-Wandering During Vicarious Learning from an Intelligent Tutoring System

Caitlin Mills[1]([✉]), Nigel Bosch[2], Kristina Krasich[3],
and Sidney K. D'Mello[4]

[1] University of New Hampshire, Durham, NH, USA
caitlin.mills@unh.edu
[2] University of Illinois at Urbana Champaign, Champaign, IL, USA
pnb@illinois.edu
[3] University of Notre Dame, Notre Dame, IN, USA
kristina.krasich.l@nd.edu
[4] University of Colorado Boulder, Boulder, CO, USA
sidney.dmello@colorado.edu

Abstract. Mind-wandering is a ubiquitous phenomenon that is negatively related to learning. The purpose of the current study is to examine mind-wandering during vicarious learning, where participants observed another student engage in a learning session with an intelligent tutoring system (ITS). Participants ($N = 118$) watched a prerecorded learning session with GuruTutor, a dialogue-based ITS for biology. The response accuracy of the student interacting with the tutor (i.e., the firsthand student) was manipulated across three conditions: Correct (100% accurate responses), Incorrect (0% accurate), and Mixed (50% accurate). Results indicated that Firsthand Student Expertise influenced the frequency of mind-wandering in the participants who engaged vicariously (secondhand students), such that viewing a moderately-skilled firsthand learner (Mixed correctness) reduced the rate of mind-wandering ($M = 25.4\%$) compared to the Correct ($M = 33.9\%$) and Incorrect conditions ($M = 35.6\%$). Firsthand Student Expertise did not impact learning, and we also found no evidence of an indirect effect of Firsthand Student Expertise on learning through mind-wandering (Firsthand Student Expertise $\rightarrow$ Mind-wandering $\rightarrow$ Learning). Our findings provide evidence that mind-wandering is a frequent experience during online vicarious learning and offer initial suggestions for the design of vicarious learning experiences that aim to maintain learners' attentional focus.

Keywords: Mind-wandering · Vicarious learning ·
Intelligent tutoring systems · Attention · Task-unrelated thought

1 Introduction

It is rather fascinating that simply viewing another student engage in an interactive tutoring session can yield (in the observer) approximately two-thirds of the learning gains obtained by the student who actually engaged in the session ($d = 1.20$) [1]. This observation-based learning method, called vicarious learning (defined as learning

© Springer Nature Switzerland AG 2019
S. Isotani et al. (Eds.): AIED 2019, LNAI 11625, pp. 296–307, 2019.
https://doi.org/10.1007/978-3-030-23204-7_25

through observation without overt behaviors [2, 3]), produces robust learning gains in a number of observational contexts including computer-based instruction and peer-to-peer interactions [2–9]. The benefits of vicarious learning highlight its potential as an educational paradigm, particularly given the scalability and cost-effectiveness of learning vicariously through video compared to, for example, engaging with an intelligent tutoring system (ITS). Despite its potential, there are many questions about which features of a vicarious learning session make it effective. This research gap is especially wide for the assessment of moment-to-moment process variables such as attention—a gap we address in the current paper.

1.1   Theoretical Background and Motivation for Current Study

Vicarious learning activities provide a more interactive alternative to traditional online learning (e.g., MOOCs). Here, we are interested in a particular form of vicarious learning involving observing one-on-one tutoring sessions (e.g., [2, 3, 5]). In this context, it is necessary for the vicarious learner (i.e., the secondhand student) to *actively* process a dialogue between a tutor and another student (i.e., the firsthand student). This activity requires the student to engage in a number of complex processes, such as the integration of multiple perspectives, as well as the evaluation of the credibility and accuracy of each perspective [6, 8, 10]. This form of active processing contrasts with more passive learning activities, like monologues (e.g., video lectures), which remain a popular method of information delivery in online learning contexts (e.g., MOOCs and online courses).

Dialogue-based vicarious learning activities are an effective educational tool that promote active learning, particularly in comparison to similar monologue-based activities [1, 3, 4, 6, 8, 10]. The effectiveness of such vicarious tasks can be explained through the ICAP framework [11] (Interactive > Constructive > Active > Passive), which suggests that while interactive tasks are the most effective for learning, followed by construct, active, and passive tasks. Olney et al. [12] recently extended ICAP to highlight the role of attention. According to their ICAP-A framework, students' attentional processes (i.e. mind-wandering, or off-task thought) would follow the same general ICAP pattern, such that students would be least likely to mind wander during an interactive learning activity, followed by constructive, active, and passive activities. In line with their framework, mind-wandering tends to occur most often during monologues (i.e. video lectures, ($\sim 43\%$ of the time [12, 13]) and least often during interactions with a dialogic intelligent tutoring system ($\sim 23\%$ of the time [14–16]), although these results are correlational.

Notably absent from the literature are studies exploring the frequency and influence of mind-wandering during vicarious learning tasks. Watching a video of a learning session is similar to viewing a video of a lecture; yet, mind-wandering may be less frequent during vicarious learning due to the active processing required by perspective-taking when viewing a dialogue. The current study addresses this gap by asking participants (i.e. secondhand learners) to watch a short video of prerecorded interactive intelligent tutoring sessions to assess the frequency of mind-wandering during vicarious learning.

Two decades ago, Cox et al. originally posed the question, "What are good models for the vicarious learner - experts or novices?" [6] (p. 432). Three hypothetical explanations were laid out for the "best" type of firsthand student: (1) experts can model perfect behavior, which may be preferable because the learning session is "uncluttered" or without error; (2) moderately-skilled learners can make the learning session more student-centered since the secondhand learners may better identify with such learners; (3) unskilled firsthand learners may be effective because the secondhand learner would learn what to avoid, and would be motivated to do so after witnessing any negative feedback.

Some prior work may be in support of the effectiveness of viewing unskilled an firsthand student. For example, viewing erroneous examples can help promote more critical evaluation and deeper learning [17, 18]. However, a study by Chi et al. [5] provided tentative evidence in support of the expert firsthand student. In their study, students learned more from "good" firsthand students (five students were retroactively assigned to be "good" students based on their pretest scores) when secondhand students collaboratively observed a one-on-one human tutoring session. However, the authors acknowledged their small sample size ($N = 20$ secondhand students) and solely focused on learning outcomes. Thus, we also examined whether the expertise of the firsthand student would have an effect on the mind-wandering rates and subsequent learning of the secondhand learner.

1.2   Current Study

We take the first steps toward understanding mind-wandering in the context of vicarious learning from an interactive ITS. We address three research questions: First, what is the overall rate of mind-wandering during vicarious learning from an ITS? This is an important consideration given the overwhelming discrepancy between monologue-based learning activities – which have the highest rates of mind-wandering [19] – and vicarious learning from dialogues, both of which can be disseminated via short videos online.

Second, does the expertise of the firsthand student influence mind-wandering and learning? We operationally defined expertise as correctness of the firsthand student's responses, which we manipulated across three conditions: 100% correct condition, to correspond with Cox et al.'s expert level; 0% correct (Incorrect condition), to correspond with the unskilled student; and 50% correct (Mixed condition), corresponding to the moderately-skilled student. Exploring the impact of the firsthand student's expertise can help inform strategies on how to design effective vicarious dialogues.

Third, we investigated if any main effects of Firsthand Student Expertise on learning are mediated by mind-wandering (Firsthand Student Expertise → Mind-wandering → Learning).

2 Methods

2.1 GuruTutor Overview

Participants viewed a video of a firsthand student interacting with GuruTutor, an ITS modeled after   expert human tutors [20]. GuruTutor is designed to teach biology topics through collaborative conversations in natural language. Throughout the conversation, an animated tutor agent references (using gestures) a multimedia   workspace that displays content relevant to the   conversation (see Fig. 1). GuruTutor analyzes learners' typed responses via natural language processing techniques and the tutor's responses are tailored to each learner's conversational turns. For a more detailed description of GuruTutor, see [20–22]. Participants viewed the firsthand student interacting with the two sections of GuruTutor that involve collaborative dialogue: (1) Common Ground Building Instruction and (2) Scaffolded Dialogue. The Common Ground Building Instruction section—sometimes called collaborative lecture [23]—is where basic information and terminology are covered. This section is critical because many biology topics involve specialized terminology (e.g., thermoregulation, metabolism) that need to be introduced before scaffolding can occur. In the Scaffolded Dialogue section, the tutor prompts the learner to answer questions about key concepts using a Prompt → Feedback → Verification Question → Feedback → Elaboration cycle. Importantly, the tutor elaborates the correct answer after every response.

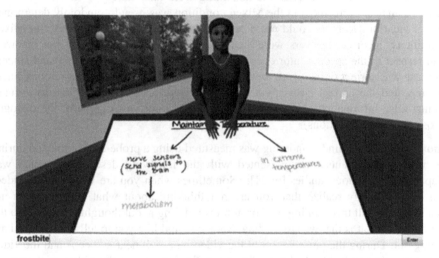

Fig. 1. Screenshot of learning session with GuruTutor.

2.2 Participants and Design

Participants ($N = 118$) were recruited from Amazon's Mechanical Turk, a platform for crowdsourcing and online data collection [24–26]. Participants had to be at least 18 years of age ($M = 35.3$ years, $SD = 20.1$) and their location was limited to the United States. Each participant received $2.75 for completing the study.

Participants were randomly assigned to watch a video of a Guru tutoring session recorded in one of three conditions that varied in terms of the frequency of correct responses provided by the firsthand student (here a simulated student) during the Common Ground Building Instruction and 2) Scaffolded Dialogue phases: 100% correct (Correct Condition), 100% incorrect (Incorrect Condition), and 50% correct (Mixed Condition). Participants in the Mixed Condition were randomly assigned to watch one of two videos that were counterbalanced with respect to which specific questions were answered correctly versus incorrectly. For example, whenever an answer was correct in version A, it was incorrect in version B, and vice versa (see Table 1 for examples). There were no differences in mind-wandering rates ($p = .759$), pretest ($p = .935$), and posttest scores ($p = .338$) as a function of counterbalance.

2.3   Materials and Procedure

Videos of GuruTutor Session. All videos were prerecorded with a screen capture program (Camtasia) while a researcher interacted with GuruTutor using a predetermined script for firsthand student responses. The topic pertained to how animals maintain body temperature. Answer length and video length were consistent across conditions with an average video length of approximately 16-min. with all videos being within 45 s in length from the others. Each video had the same number ($n = 142$) of dialogue turns, with the firsthand student's responses comprising 18% (answering 21 questions) of the dialogue turns; the remaining were tutor turns.

Order of answer correctness in the Mixed condition was pseudo-randomly determined so that vicarious learners could not detect a pattern. In both the Incorrect and Mixed Conditions, incorrect answers were thematically-related to the content but incorrect with respect to the specific tutor question. Regardless of whether the firsthand student response was correct or incorrect, the tutor provided feedback about answer correctness and repeated the correct answer via elaborated feedback. This was done as a guard against false information being retained (see Table 1 for an example of the dialogue across the three conditions).

Thought Probes. Mind-wandering was measured using a probe-caught method during the video. Participants were presented with the following description, which was adapted from previous studies [6, 21]: "Sometimes when you are watching the video, you may suddenly realize that you are not thinking about what it is that you are watching. We call this "zoning out" or mind wandering about thoughts unrelated to the content of what it is that we are reading. So, we would like you to tell us when you are zoning out. During the presentation of the video, you will hear a "beep" and the video will stop. We would like to know if you are thinking about the video or if you are thinking about something else (e.g., what you will be eating for dinner, your plans for the week). When you hear the tone and you are zoning out, please indicate "Yes" by pressing the "Y" key on your keyboard. If you hear the tone and you are not zoning out, please indicate "No" by pressing the "N" key on your keyboard."

The instructions also emphasized that participants should be as honest as possible when reporting mind-wandering and that their responses would have no influence on their progress and compensation. There were nine probes per video with probe timings approximately evenly interspersed and set to align with the same events across conditions (e.g., after the tutor completed a specific turn).

Table 1. Example dialogue across the three conditions.

	Correct	Incorrect	Mixed A	Mixed B
Example from collaborative lecture section				
Tutor	Our bodies inevitably get too hot			
Tutor	They release a watery substance onto its surface which serves to cool skin down during evaporation processes			
Tutor	Hint. Recollected that when you are hot it occurs			
	Do you foresee what this substance is called?			
*Firsthand Student**	*Sweat*	*Blood*	*Sweat*	*Blood*
*Tutor**	*OK, Good*	*That's not it*	*OK, Good*	*That's not it*
Tutor	It's sweat	It's sweat	It's sweat	It's sweat
Example from scaffolding section				
Tutor	The brain changes the body's metabolism in order to change the body's temperature. Here is a related question			
Tutor	What is metabolism?			
*Firsthand Student**	*Rate of chemical reactions*	*Energy used to pump blood*	*Energy used to pump blood*	*Rate of chemical reactions*
*Tutor**	*Very good*	*Nope*	*Nope*	*Very good*
Tutor	Metabolism is the rate of chemical reactions in the body. It can be slowing down or speeding up			

Notes. Italics * = manipulated dialogue. Mixed 1 and 2 represent the two different counterbalanced videos in the Mixed condition.

Learning Measures. We used 16 four-foil multiple-choice questions to assess learning. The questions were derived from previously administered standardized test items or from researcher-created items (see [27]). The questions targeted specific concepts mentioned during the session, such as: *Which of the following is true about blood temperature?: a. it is cooled as it is pumped near the brain: b. it is heated as it is pumped near the extremities: c. it is heated as it is pumped near the core (correct answer): d. blood temperature generally stays about the same.* Two parallel versions of the test were created (8 items each) by randomly dividing the questions, which were counterbalanced as pre- and posttest.

Procedure. After providing electronic consent, participants completed a pretest to gauge prior knowledge. They then received instructions for the thought probes and were informed they would watch a prerecorded video of a student interacting with a computer tutor called GuruTutor. They were instructed that their task was to watch the video in order to understand the concepts being taught and that they would be subsequently assessed on their learning. At this point, the video was presented along with the thought probes. Finally, participants completed the posttest and were debriefed.

3 Results and Discussion

Table 2 presents descriptive statistics for key variables. An analysis of variance (ANOVA) revealed no differences across conditions with respect to prior knowledge, $F(2,115) = 2.29$, $p = .106$, thereby confirming successful random assignment.

Table 2. Means and standard deviation (in parentheses) for key variables across the conditions.

	Correct	Incorrect	Mixed
	M (SD)	M (SD)	M (SD)
n participants	40	40	38
Mind-wandering proportion	.339 (.325)	.356 (.340)	.254 (.288)
Pretest scores	.278 (.207)	.353 (.189)	.272 (.163)
Posttest scores	.605 (.274)	.574 (.223)	.554 (.274)

3.1 How Often Did Participants' Mind-Wander?

We first explored the frequency of mind-wandering during the vicarious learning session. Participants reported mind-wandering 31.7% of the time ($SD = 31.9$; or 2.9 mind-wandering episodes on average during the session). This finding parallels the rates found in other active online learning activities, such as reading [12, 19, 28].

3.2 Did Firsthand Student Expertise Influence Mind-Wandering in Secondhand Learners (Participants in Current Study)?

Mind-wandering rates were analyzed using a Poisson regression which is suitable for count data (i.e. the count of the number of probes with positive mind-wandering responses). We first assessed the main effect of Firsthand Student Expertise by including it as the only independent variable. A significant omnibus test indicated that model fit improved after including Firsthand Student Expertise in comparison to the intercept-only model, $\chi^2(2) = 6.69$, $p = .035$. Comparisons of parameter estimates revealed that participants in the Mixed condition reported significantly less mind-wandering compared to both the Incorrect ($B = .335$, SE $= .139$, Wald $\chi^2(1) = 5.81$, $p = .016$) and Correct conditions ($B = .287$, SE $= .140$, Wald $\chi^2(1) = 4.18$, $p = .041$). Rates of mind-wandering across the Correct and Incorrect conditions were on par with one another, $p = .704$, yielding the following pattern of results (Mixed < [Correct = Incorrect]).

We tested whether the main effect of Firsthand Student Expertise was robust after adding prior knowledge as a covariate. The omnibus test was significant, $\chi^2(3) = 8.95$, $p = .030$. The tests of model effects indicated that pretest was not a significant predictor of mind-wandering, $B = -.445$, SE $= .299$, Wald $\chi^2(1) = 2.22$, $p = .137$. The effect of Firsthand Student Expertise was still significant after including the covariate, Wald $\chi^2(2) = 7.32$, $p = .026$, with the same pattern of effects: Participants reported less mind-wandering in the Mixed condition compared to the Correct ($p = .040$) and Incorrect ($p = .009$) conditions, which were on par with one another ($p = .524$).

3.3   Did Firsthand Student Expertise Influence Learning?

We first assessed participants learned from the vicarious learning session using a paired samples t-test. There was a significant increase from pre- to posttest, $t(117) = 9.99$, $p < .001$, $d = 1.22$ after pooling across conditions, suggesting that vicarious learning was effective in our context. We then tested whether Firsthand Student Expertise predicted post-test scores after controlling for pre-test in an ANCOVA, but found no main effect of Firsthand Student Expertise, $F(2,114) = .434$, $p = .649$.

3.4   Did Firsthand Student Expertise Influence Learning Through Mind-Wandering?

Although there was no evidence for a main effect, it is possible that Firsthand Student Expertise may influence learning indirectly through mind-wandering (Firsthand Student Expertise → mind-wandering → learning) [29] – particularly given that mind-wandering was negatively related to posttest scores, $rho = -.173$, $p = .061$. We tested indirect effects using the 'mediation' package in R [30]. We specified two models: (1) a mediator model, which was a Poisson model regressing mind-wandering on Firsthand Student Expertise, controlling for pretest scores; and (2) an outcome variable model, which was a linear model regressing posttest scores on mind-wandering and Firsthand Student Expertise including the same covariate. We obtained causal estimates for the indirect effect over 10,000 quasi-Bayesian Monte Carlo simulations; however, there was no evidence of mediation, $p = .190$, 95% CI $= -.005, .014$.

4   General Discussion and Conclusion

Until now, mind-wandering had not been explored in the context of vicarious learning from an ITS—an important context given the effectiveness of vicariously observing dialogues [3, 4] combined with the cost-effectiveness of delivering vicarious learning sessions online. The current study addressed this gap while also examining whether Firsthand Student Expertise influenced the rate of mind-wandering, learning, including both direct and indirect effects.

4.1   Main Findings

Participants reported mind-wandering approximately 32% of the time, underscoring its frequency during vicarious learning activities [19]. These rates are considerably lower than those typically observed by students viewing a monologue – e.g., recorded classroom lectures (rates around 40% [13, 31]). At the same time, these rates are slightly higher in comparison to rates produced by interacting with an ITS (23% reported in [14]), perhaps because ITSs afford a more interactive experience. These general patterns are in line with predictions made by ICAP-A [12] in that participants may be more likely to mind wander in passive contexts compared to active (e.g., vicariously listening to a dialogue or interactive (e.g., engaging with an ITS) contexts.

We also examined how the expertise of the firsthand student influenced mind-wandering rates. Secondhand learners reported mind-wandering less often when the firsthand student's answers included a mix of both incorrect and correct answers (i.e. Cox et al.'s [6] version of a moderately-skilled student). This pattern is consistent with Cox et al.'s prediction that secondhand learners may identify with a moderately-skilled student and therefore attend more closely to both perspectives in the dialogue. Another plausible explanation is that uncertainty about the firsthand student's answers held participants' attention in the Mixed condition, whereas correctness was predictable in the other two conditions. Future work, however, will be needed to determine which of these accounts explains *why* participants were on task more often in the Mixed condition.

All three conditions performed equally well on the posttest and Firsthand Student Expertise did not indirectly influence learning through mind-wandering. This may indicate that participants in condition adopted a different strategy for processing the dialogue – by paying attention more overall (Mixed condition), or perhaps only to certain parts of the dialogue (Incorrect/Correct conditions). For example, once participants understood the firsthand student's level expertise, they may have guessed which pieces of information required more focused. Recent evidence from a sustained attention task suggests that participants indeed develop strategies to alter off-task behaviors based on motivation to perform well on the task [32], but future studies should assess the specific strategies employed in vicarious dialog-based learning contexts.

4.2   Limitations and Future Directions

It is important to note the limitations of this study. First, this study was conducted online, so we had no control over the participants' environment. However, this may also be reflective of vicarious learning in ecologically valid scenarios during online learning. Further, although the use of Mechanical Turk has been validated as a reliable source of data [24], replication with actual students is warranted. Second, our sample size was limited to 118 participants. It is therefore possible that we did not have adequate power to detect an indirect effect of mind-wandering (see Sect. 3.4). Third, in contrast to prior work on vicarious learning [7], we used experimenter-generated learning sessions instead of authentic learning sessions to implement the key manipulation with high internal validity. Future work should, therefore, attempt to use authentic learning sessions by first having an actual student interacting with the ITS, then assigning a second participant to watch their video. This method could provide a broader range of student expertise rather than two extremes used here (100% and 0% accuracy). Fourth, we only explored one topic (maintaining body temperature) in a single ITS; therefore, follow up studies are needed to determine if results generalize more broadly.

Finally, some people may object to the intentional use of incorrect responses. We acknowledge this limitation, but we feel that they are less of a concern in the current study for the following reasons: (1) all incorrect responses were corrected immediately after the firsthand student's response; (2) all three conditions performed equally well on the posttest; (3) all protocols were approved by the appropriate ethics board; (4) secondhand learners were consenting participants instead of actual students.

Our findings can help inform the design of vicarious learning systems that aim to promote engagement and learning. For example, GuruTutor could be strategically modified so that the firsthand student introduces and resolves specific misconceptions [33] or asks deep-reasoning questions [7]—both of which have been shown to be effective for learning. Additional characteristics of the firsthand student can be manipulated, including factors like affective tone, length of responses, or amount of turn-taking in the dialogue. It is also possible to build detectors of mind-wandering (e.g., using eye-gaze [15, 34, 35]) during vicarious learning so that real-time interventions can be deployed to steer participants back on task. Such systems could dynamically adjust the correctness of firsthand student answers depending on mind-wandering, while also ensuring that correct answers are repeated after a mind-wandering episode.

4.3   Conclusion

This study provides a foundation for examining the role of attention in vicarious learning contexts. Although online vicarious learning sessions are a time- and cost-effective learning method [2], mind-wandering still occurs with some regularity (approximately 30% of the time) during vicarious learning. The current study sheds light on how the expertise of the firsthand student can influence mind-wandering. However, more work is needed to explore ways to design and optimize online vicarious learning tasks to promote attention and learning.

Acknowledgments. This research was supported by the National Science Foundation (NSF) DRL 1235958 and IIS 1523091. Any opinions, findings and conclusions, or recommendations expressed in this paper are those of the authors and do not necessarily reflect the views of NSF.

References

1. Craig, S.D., Driscoll, D.M., Gholson, B.: Constructing knowledge from dialog in an intelligent tutoring system: Interactive learning, vicarious learning, and pedagogical agents. J. Educ. Multimedia Hypermedia **13**, 163 (2004)
2. Gholson, B., Craig, S.D.: Promoting constructive activities that support vicarious learning during computer-based instruction. Educ. Psychol. Rev. **18**, 119–139 (2006)
3. Driscoll, D.M., Craig, S.D., Gholson, B., et al.: Vicarious learning: effects of overhearing dialog and monologue-like discourse in a virtual tutoring session. J. Educ. Comput. Res. **29**, 431–450 (2003)
4. Chi, M.T.H., Kang, S., Yaghmourian, D.L.: Why students learn more from dialogue- than monologue-videos: analyses of peer interactions. J. Learn. Sci. **26**, 10–50 (2017)
5. Chi, M.T.H., Roy, M., Hausmann, R.G.M.: Observing tutorial dialogues collaboratively: insights about human tutoring effectiveness from vicarious learning. Cogn. Sci. **32**, 301–341 (2008)
6. Cox, R., McKendree, J., Tobin, R., et al.: Vicarious learning from dialogue and discourse. Instr. Sci. **27**, 431–458 (1999)

7. Craig, S.D., Sullins, J., Witherspoon, A., Gholson, B.: The deep-level-reasoning-question effect: the role of dialogue and deep-level-reasoning questions during vicarious learning. Cogn. Instr. **24**, 565–591 (2006)
8. Tree, J.E.F.: Listening in on monologues and dialogues. Discourse Processes **27**, 35–53 (1999)
9. Twyford, J., Craig, S.D.: Modeling goal setting within a multimedia environment on complex physics content. J. Educ. Comput. Res. **55**, 374–394 (2017)
10. Tree, J.E.F., Mayer, S.A.: Overhearing single and multiple perspectives. Discourse Processes **45**, 160–179 (2008)
11. Chi, M., Wylie, R.: The ICAP framework: linking cognitive engagement to active learning outcomes. Educ. Psychol. **49**, 219–243 (2014)
12. Olney, A.M., Risko, E.F., D'Mello, S.K., Graesser, A.C.: Attention in educational contexts: the role of the learning task in guiding attention. In: Fawcett, J.M., Risko, E.F., Kingstone, A., et al. (eds.) The Handbook of Attention, pp. 623–641. MIT Press, Cambridge (2015)
13. Risko, E.F., Anderson, N., Sarwal, A., et al.: Everyday attention: variation in mind wandering and memory in a lecture. Appl. Cogn. Psychol. **26**, 234–242 (2012)
14. Hutt, S., Mills, C., Bosch, N., et al.: "Out of the fr-eye-ing pan": towards gaze-based models of attention during learning with technology in the classroom. In: Proceedings of the 25th Conference on User Modeling, Adaptation and Personalization, pp. 94–103. ACM, New York (2017)
15. Hutt, S., Mills, C., White, S., et al.: The eyes have it: Gaze-based detection of mind wandering during learning with an intelligent tutoring system. In: Proceedings of the 9th International Conference on Educational Data Mining, International Educational Data Mining Society, EDM, pp. 86–93 (2016)
16. Mills, C., D'Mello, S., Bosch, N., Olney, Andrew M.: Mind wandering during learning with an intelligent tutoring system. In: Conati, C., Heffernan, N., Mitrovic, A., Verdejo, M.F. (eds.) AIED 2015. LNCS (LNAI), vol. 9112, pp. 267–276. Springer, Cham (2015). https://doi.org/10.1007/978-3-319-19773-9_27
17. Adams, D.M., McLaren, B.M., Durkin, K., et al.: Using erroneous examples to improve mathematics learning with a web-based tutoring system. Comput. Hum. Behav. **36**, 401–411 (2014)
18. Tsovaltzi, D., Melis, E., McLaren, B.M., Meyer, A.-K., Dietrich, M., Goguadze, G.: Learning from erroneous examples: when and how do students benefit from them? In: Wolpers, M., Kirschner, Paul A., Scheffel, M., Lindstaedt, S., Dimitrova, V. (eds.) EC-TEL 2010. LNCS, vol. 6383, pp. 357–373. Springer, Heidelberg (2010). https://doi.org/10.1007/978-3-642-16020-2_24
19. D'Mello, S.K.: What do we think about when we learn? In: Millis, K., Magliano, J., Long, D.L., Weimer, K. (eds.) Understanding Deep Learning, Educational Technologies and Deep Learning, and Assessing Deep Learning, pp. 52–67. Routledge/Taylor and Francis (2018)
20. Olney, A., Person, N.K., Graesser, A.C.: Guru: designing a conversational expert intelligent tutoring system. In: Boonthum-Denecke, C., McCarthy, P., Lamkin, T. (eds.) Cross-Disciplinary Advances in Applied Natural Language Processing: Issues and Approaches, pp. 156–171. IGI Global, Hershey (2012)
21. Olney, Andrew M., et al.: Guru: a computer tutor that models expert human tutors. In: Cerri, Stefano A., Clancey, William J., Papadourakis, G., Panourgia, K. (eds.) ITS 2012. LNCS, vol. 7315, pp. 256–261. Springer, Heidelberg (2012). https://doi.org/10.1007/978-3-642-30950-2_32
22. Person, N.K., Olney, A., D'Mello, S.K., Lehman, B.: Interactive concept maps and learning outcomes in guru. In: Florida Artificial Intelligence Research Society (FLAIRS) Conference, pp. 456-461. AAAI Press (2012)

23. D'Mello, S., Hays, P., Williams, C., Cade, W., Brown, J., Olney, A.: Collaborative lecturing by human and computer tutors. In: Aleven, V., Kay, J., Mostow, J. (eds.) ITS 2010. LNCS, vol. 6095, pp. 178–187. Springer, Heidelberg (2010). https://doi.org/10.1007/978-3-642-13437-1_18

24. Mason, W., Suri, S.: Conducting behavioral research on Amazon's Mechanical Turk. Behav. Res. Methods **44**, 1–23 (2012)

25. Rand, D.G.: The promise of Mechanical Turk: how online labor markets can help theorists run behavioral experiments. J. Theor. Biol. **299**, 172–179 (2012)

26. Sprouse, J.: A validation of Amazon Mechanical Turk for the collection of acceptability judgments in linguistic theory. Behav. Res. Methods **43**, 155–167 (2011)

27. Mills, C., Fridman, I., Soussou, W., et al.: Put your thinking cap on: detecting cognitive load using EEG during learning. In: Proceedings of the Seventh International Learning Analytics & Knowledge Conference, pp. 80–89. ACM (2017)

28. Mills, C., Graesser, A., Risko, E.F., D'Mello, S.K.: Cognitive coupling during reading. J. Exp. Psychol. Gen. **146**, 872–883 (2017)

29. Zhao, X., Lynch Jr., J.G., Chen, Q.: Reconsidering Baron and Kenny: Myths and truths about mediation analysis. J. Consum. Res. **37**, 197–206 (2010)

30. Tingley, D., Yamamoto, T., Hirose, K., et al.: Mediation: R package for causal mediation analysis UCLA Statistics/American Statistical Association, pp. 1–40 (2014)

31. Risko, E.F., Buchanan, D., Medimorec, S., Kingstone, A.: Everyday attention: mind wandering and computer use during lectures. Comput. Educ. **68**, 275–283 (2013)

32. Seli, P., Carriere, J.S., Wammes, J.D., et al.: On the clock: evidence for the rapid and strategic modulation of mind wandering. Psychol. Sci. **29**, 1247–1256 (2018)

33. Muller, D.A., Bewes, J., Sharma, M.D.: Reimann P Saying the wrong thing: improving learning with multimedia by including misconceptions. J. Comput. Assist. Learn. **24**, 144–155 (2008)

34. Bixler, R., D'Mello, S.: Automatic gaze-based user-independent detection of mind wandering during computerized reading. User Model. User-Adap. Inter. **26**, 33–68 (2016)

35. Mills, C., Bixler, R., Wang, X., D'Mello, S.K. Automatic gaze-based detection of mind wandering during film viewing. In: Proceedings of the International Conference on Educational Data Mining. International Educational Data Mining Society, pp. 30–37 (2016)

Annotated Examples and Parameterized Exercises: Analyzing Students' Behavior Patterns

Mehrdad Mirzaei[1]([✉]), Shaghayegh Sahebi[1], and Peter Brusilovsky[2]

[1] Department of Computer Science, University at Albany - SUNY, Albany, USA
{mmirzaei,ssahebi}@albany.edu
[2] School of Computing and Information, University of Pittsburgh, Pittsburgh, USA
peterb@pitt.edu

Abstract. Recent studies of student problem-solving behavior have shown stable behavior patterns within student groups. In this work, we study patterns of student behavior in a richer self-organized practice context where student worked with a combination of problems to solve and worked examples to study. We model student behavior in the form of vectors of micro-patterns and examine student behavior stability in various ways via these vectors. To discover and examine global behavior patterns associated with groups of students, we cluster students according to their behavior patterns and evaluate these clusters in accordance with student performance.

Keywords: Student sequence analysis · Frequent pattern mining

1 Introduction

With the improvement of online learning systems, students are provided with more opportunities for learning. In modern learning systems, students are usually free to choose to access multiple learning material types. While some systems provide restrictions on order of accessing learning content, in many systems, there are no predefined activity sequences and students are free to choose to work with any learning materials in any order. This choice provides students with more freedom to learn according to their own pace and background knowledge, and to repeat their past activities as they seem fit. For example, students can skip some learning materials and work on the more advanced ones if they believe that they have already mastered the prerequisite concepts. Similarly, they can go back and repeat some learning materials.

Despite these advantages, this freedom could lead to some inefficient and non-productive behavior. For example, past research on students' problem-solving behavior has found that students tend to practice the same set of concepts, well after mastering them, instead of moving to new concepts and more difficult problems [7,10,19]. While past research on behavior patterns has mainly focused on problem-solving behavior, student behavior can get more complex as other types

S. Isotani et al. (Eds.): AIED 2019, LNAI 11625, pp. 308–319, 2019.
https://doi.org/10.1007/978-3-030-23204-7_26

of learning materials are introduced. For example, consider a learning system that includes both reading materials and self-assessment problems. Here, a student can spend a significant amount of time persisting in reading on an advanced concept and failing in related problems, without having the prerequisites.

Ideally, a learning system should be able to detect inefficient behavior and guide the students towards efficient ones. To do this, the main challenge is to understand the relationship between students' behavior and performance. This challenge, translates into two main questions: (1) could we discover stable student behavior patterns which could be recognized in-time to react? are them persistent, or they happen at random; and (2) could we recognize efficient and inefficient student behavioral patterns by associating it with their learning performance?

Past research in the area of student problem-solving behavior indicated that stable patterns of student behavior do exist, however, theses patterns might not be directly related to their performance [7]. Instead, different patterns characterize students' individual ways to learn and approach a problem. To find stable patterns of student behavior, Guerra et al. [7] built student "problem-solving genomes" from micro-patterns ("genes") and grouped the students based on their "genomes" into clusters. While students belonging to the same cluster tend to show the same behavior patterns, these clusters included both high and low performing students. However, there was an indication that within each cluster, the "genomes" could help to discover efficient and inefficient behaviors.

In this paper, we attempt to apply the sequence mining-approach suggested in [7] to a more complex case, where students are working with two types of learning material, and are repeating their attempts within the same topics. The two learning material types we focus on are: parameterized problems and annotated examples. Our research questions within this more complex context are still the same: (1) do individual students exhibit stable behavioral patterns in their work with learning content, or their approach to learn vary by factors, such as time in the semester or learning material difficulty? (2) to what extent student behavioral patterns are associated with their learning performance?

2   Related Work

Online educational systems collect increasing volumes of information from students' interactions with various kinds of learning content. The process of data collection has been followed by a rapid increase in research, which focused on using this data to better understand and improve the learning process. Among the explored topics was the early prediction of student success or failure [2], which could be helpful to identify and support students-at-risk [20]. While early work focused mostly on cumulative factors such as frequency of watching videos or using discussion forums [18], recent work attempted to build more complex models of student behavior and identify various kinds of behavior patterns to help students make progress and improve education outcomes. Analyzing students' sequences and trajectories have been of increased interest recently.

Since student behavior is commonly considered as a sequence of students' actions or interactions with the system, various kind of sequence-oriented Markov models were explored for behavior analysis. For example, Hansen et al. [8] analyzed log data of an online education system, and modeled student behavior as interpretable Markov chains. The model compares action sequences across different lengths, focusing on the flow of actions. A two-layer Hidden Markov Model is used in [5] to automatically detect student behavior patterns from logged data of a MOOC platform. They have shown that the extracted patterns are meaningful and have a correlation with students' learning outcome. In such works, all of the students' activities are observed to generate latent states.

Another popular group of approaches used matrix factorization to transform a student learning traces to a smaller number of "soft clusters". For example, Gelman et al. [6] segmented student use of content and assessment, into weeks. Then, they used non-negative matrix factorization (NMF) to distill five basic behaviors of students ("Deep", "Consistent", "Bursty", "Performance", and "Response") and built vectors specifying how much each student shows each behavior. Similar approaches were used successfully with different kind of data in [13,15] that can be incorporated with ranking techniques from social networks [3].

Some work in the literature focused on student trajectories [11,21]. For example, Boubekki et al. [1] compared the navigation behavior of students in reading textbooks and discovered student clusters that were indicators of student performance. Sawyer et al. [17] proposed a time-series representation of student problem-solving trajectories in a learning game. They used Euclidean distance and trajectory slope to measure students' distance with "expert paths", which was correlated with students' learning gain.

The least explored, but potentially very powerful group of approaches focused on using sequence mining to identify behavior patterns. In [14], authors have extracted frequent action sequences in a collaborative learning environment to distinguish high achieving student from low achieving students. The analysis in this work is on interaction with resources on a tabletop. Students' actions on the tabletop are logged and coded into events to create sequences. Then, frequent sub-sequence sequential patterns are extracted using n-grams. The patterns are clustered to build higher-level patterns and students are compared based on them. In [7] patterns of student work with parameterized exercises are modeled and analyzed. In this work, micro-patterns are extracted using a sequential pattern mining algorithm and used to build student behavior profiles ("genomes"). Then, students with similar genomes are clustered into behavior groups.

3   System and Dataset

In our experiments, we use the student interaction data with learning materials in the "Introduction to object-oriented programming" course using *Progressor+* interface [9]. The system includes parameterized exercises and worked-out code examples as two different types of learning material. The learning material in the practice system were grouped into topics. For each topic, multiple problems and examples are available. Although the order of topics was shown to the students, they could choose any topic, problem, or example to practice at any time in any preferred order. The parameterized exercises are small problems focused on program behavior prediction. Each exercise is a template with a parameter, which is generated randomly every time a student chooses to work on them. Consequently, students can use the same exercise template for practice multiple times with different parameters. Worked-out annotated code examples are small complete programs annotated with short explanations for each line of code. Students can click on the lines of code in any order to read the explanation. The dataset includes three semesters of student activities in Java classes in a large public US university. After data cleaning, the dataset contains 83 students. There are 103 parameterized problems and 42 annotated examples in the dataset. The number of correct attempts to solve problems is 13796, the number of incorrect attempts is 6233, and the number of clicks on examples is 12713. In addition to the student behavior log, the dataset includes pre-test and post-test scores for each student. The pre- and post-tests included the same set of program behavior prediction questions administered at the beginning and at the end of each semester correspondingly. The minimum and maximum score in pre-test are 0 and 14 and in post-test are 5 and 24 respectively. To measure students' improvement over the course of the semester, learning gain is calculated for each student as the normalized difference between post-test and pre-test.

4   Modeling Student Behavior

To extract micro-patterns from student logs, we code them into sequences and analyze them using a frequent pattern mining algorithm. We build macro-pattern vectors or *genome* as a representation of each individual student's behavior.

4.1   Coding Student Behavior

To discover behavioral patterns of students, we first label student attempts. Inspired by [7], we focus on two aspects when labeling student problem solving attempts: whether the student succeeds (or fails) in solving the problem, and whether the student spent shorter or longer time to answer a problem, compared to a median answering time. The median answering time is calculated separately for each problem, considering all attempts on it[1]. If a student solves a problem correctly in less time than the median, the attempt is labeled as 's' (short success). Likewise, if a student's successful attempt takes longer than the median it will be labeled as 'S' (long success). Similarly, if the student solves a problem incorrectly in a short time (vs. long time), her attempt will be labeled as 'f' (vs. 'F'). In total, the dataset included 760 short successes, 6030 long successes, 2242 short failures, and 3991 long failures.

In addition to students' problem-solving, we code their example-reading behavior. Unlike the problem-solving attempts, working on annotated examples is not associated with correctness. Thus, we only measure the time spent by each student

---

[1] The median split can be calculated within each students also. Since we are interested in capturing content access differences between students, and since time-spent variance among problems is larger than among students, we chose to split the data according to problem-answering medians.

on each annotated example. To do this, we sum up all sequential student clicks on example lines of one annotated example, as the time spent on that example. We first calculate the median time spent on each annotated example by all students. For each example-reading activity of a student, if the time spent on the annotated example is less than its median, it is labeled 'e', otherwise as 'E'. The dataset included 6348 short example attempts and 6365 long attempts.

The students continue to work in the system during the whole semester. To chunk the large sequence of student actions into smaller comparable sequences, we define a "session" as a consecutive set of student activities within one topic. In other words, a session is a sequence of attempts on parameterized problems and annotated examples inside the same topic. An attempt on an example or problem from another topic starts another session. To indicate session borders within each student's sequence, we insert '_' between two consecutive sessions. For instance, the sequence '_ffSsee_' means that the student has a long success after two short failures, then a short success and finally is quickly examined two annotated examples within the same session.

4.2   Sequential Pattern Mining

To discover the most frequent micro patterns of student behavior, we use CM-SPAM [4] sequential pattern mining algorithm. We set the minimum support to 1% (i.e., we are interested in patterns that could be found in at least 1% of sequences) and require no gap between encoded attempts. Besides that, we only consider the patterns with more than one sequential attempts. In total, 111 frequent patterns are discovered using this approach. The top 30 frequent patterns are illustrated in Table 1.

Interestingly, we can see that the top frequent patterns are either problem-solving micro-patterns or example-reading micro-patterns. In other words, there are no mixed activity patterns (such as 'eF') among the top frequent ones. From this, we conclude that switching from one type of activity to another was considerably more rare than continuing with the same kind of activity.

Table 1. Top 30 extracted patterns ordered by support

	Pattern	Support		Pattern	Support		Pattern	Support
1	ss	1516	11	Fs_	680	21	_FF	486
2	Ss	1456	12	Ff	680	22	_Sss	449
3	ss_	1378	13	sss_	668	23	FS_	449
4	Fs	1153	14	Sss	663	24	_Ss_	443
5	_Ss	974	15	_FS	630	25	_ee	431
6	_Fs	901	16	ee	593	26	fs_	393
7	FS	828	17	ee_	552	27	ssss	373
8	fs	788	18	FF	546	28	ff	367
9	sss	735	19	_Fs_	539	29	Fss	361
10	Ss_	692	20	_Ff	515	30	_FS_	351

4.3 Building Pattern Vectors

The top frequency patterns found in Sect. 4.2 represent a variety of patterns used by all students. Each student could use each micro-pattern with different frequency or not at all. To model the behavior of an individual student, we build a behavioral pattern vector for each student. We use the top 60 of frequent patterns to build this vector. This pattern vector includes normalized frequencies of observing each of the top frequent patterns in the behavior log of the modeled student. To build it, we first count the number of times each frequent pattern occurred in the student's sequence. These *absolute* frequencies, however, could vary depending on the total length of student behavior sequence, i.e., how much the modeled student interacted with the system. To capture the relative importance of each micro-pattern regardless of total sequence length, we normalized the count vectors (i.e., the frequency of patterns are summing to one for each vector). These vectors represent the behavior of individual students and are used to discover macro-patterns by clustering student vectors.

5 Behavior Stability Analysis

Before establishing a relationship between students' micro-patterns and their performance, we should make sure that the patterns are representative of students' behavioral traits, and not other environmental factors. To do this, we analyze the *stability* of student patterns in three different setups: randomized, longitudinal, and complexity-based. In each of these setups, we split student sequences into two equal sets. Then, we independently build a pair of two pattern vectors for each student: one for each set. If our model of student behavior is stable, the vectors in each pair should be more similar to each other than to vectors form other pairs. Thus, in each of the setups, we test whether the students' behavior vector built from the first set is significantly more similar to their own behavior vector in the second set than to the behavior vectors of the rest of the students. To measure the similarity, we use Jensen-Shannon divergence [12].

Table 2. Comparing average of students' pattern vector distances with themselves vs. other students according to various splits

	Self distance		Distance to others			
	Mean	SE	Mean	SE	t Stat	P-value
Random split	0.2082	0.0207	0.4639	0.0105	−16.0279	<0.0001
First half/second half	0.2995	0.0211	0.5207	0.0113	−12.3501	<0.0001
Random split (Easy)	0.3644	0.0258	0.5769	0.0110	−9.9099	<0.0001
Random split (Medium)	0.3266	0.0246	0.5465	0.0092	−11.1404	<0.0001
Random split (Hard)	0.4219	0.0266	0.5703	0.0106	−6.4266	<0.0001

5.1 Randomized Analysis

In the randomized analysis, our goal is to examine whether a student's pattern vector is stable across all sessions, if we split them randomly. It is to test if we can distinguish a student from other students according to their pattern vectors. To do this, we randomly split student sequences into two halves and build a pattern vector for each half. If a student's pattern vector from the first half is significantly more similar to her own pattern vector in the second half – compared to being similar to other students – then, we conclude that the student's patterns are stable and do not change randomly. To test the significance, we run paired sample t-test. The results are shown in Table 2. As we can see, the distance between the two pattern halves for the same student (0.2082 on average) is significantly smaller than the distance to other students (0.4639).

5.2 Longitudinal Analysis

Here, we are interested to see if the student patterns change as the semester advances. To study this, we split each student's activity sequence according to a mid-semester point: we build pattern vectors for the first half and the second half of the semester. Similar to randomized analysis, we compare the distance between halves within each student's vector and between student vectors. As shown in Table 2, we see that the distance between first half and second half of one student is significantly smaller than the distance to other students. Individual student behavior pattern changes slightly over the semester, yet this change is by far not sufficient to cross the difference from other students.

5.3 Complexity Analysis

Another factor that can affect students' behavior is activity complexity. Each learning material is labeled with "easy", "medium" or "hard" in our dataset. Accordingly, we build separate pattern vectors for each group of learning activities for each student. E.g., in each topic and session, we separate the "easy" problems and examples as one "easy" session. We assess pattern vector stabilities by comparing the difference within a student (comparing according to complexity) and between students. Table 2 represents the distances and statistical tests that show student pattern vectors are stable across learning material complexities.

6 Behavior Cluster Analysis

Having stable student pattern vectors, we aim to distinguish efficient patterns. To do this, we study if student behavioral patterns are associated with student performance. Namely, we would like to understand if students with similar behavior have a similar performance. First, we cluster the students based on their behavior patterns to have students with similar patterns together. Afterward, we analyze the patterns to recognize useful patterns in each cluster.

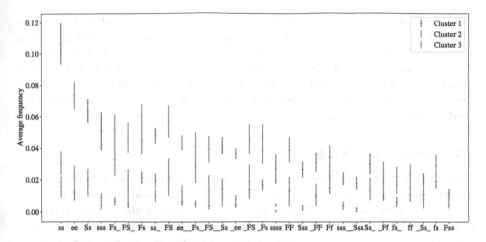

Fig. 1. Top 30 patterns and their frequencies in 3 clusters. Patterns are ordered by the maximum difference of frequencies between three clusters.

6.1 Pattern Analysis

We apply Spectral clustering [16] on student pattern vectors. Our cluster and interpretation analysis showed that 3 clusters will provide the best results with 23, 28, and 33 students in each cluster. To understand the student differences in each cluster, we compare their average pattern frequencies in the top 30 micro-patterns. Figure 1 shows these top 30 frequent patterns and their average frequencies in each cluster. The error bars show 95% confidence interval for each micro-pattern. The micro-patterns are ordered based on the maximum frequency

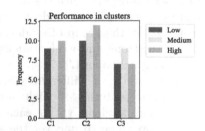

Fig. 2. High, medium, and low-performance student frequencies in three clusters generated based on student patterns

difference in three clusters. As we can see in the figure, in cluster 1, patterns like 'ss', 'Ss' and 'sss' are significantly more frequent than the other two clusters. We can say, students in this cluster tend to repeat practicing an exercise within a topic even if they succeed in it. Significantly frequent patterns in cluster 2 are '_FS_', 'FS' and 'FS_' that demonstrate longer failures and longer successes afterwards. We can conclude that the students in this cluster tend to spend more time on solving a problem, and then succeed afterwards. They do not attempt the problems randomly and do not answer them by chance. Students in cluster 3 read more examples since patterns such as 'ee' and 'ee_' are frequent in this cluster. In [7] students were grouped into "confirmers" and "non-confirmers" according to their patterns. "confirmers" were the students who preferred to confirm their success by repeating it. "non-confirmers" were the ones who ended their session right after having a short success. Here, we see a "confirmers" type of pattern in cluster 1. However, in cluster 2 students are mostly "thinkers" rather than "non-confirmers". They fail and then succeed, but with thinking and spending time on the activity. Cluster 3 students are mostly "readers". They tend to spend more time on reading the annotated examples. Therefore, we can recognize 3 types of behaviors with the extracted patterns. We should mention that these labels are provided to distinguish the discovered clusters, rather than exactly describing student behaviors in them.

6.2 Performance Analysis

Here we examine the clusters to detect whether we can associate the macro-behavioral patterns represented by each cluster with students' learning performance, measured by normalized learning gain. Figure 2 shows the number of students with low, medium and high performance (learning gain) in each cluster. As we can see (also by our statistical tests) the clusters do not show a significant difference in the number of high, medium, or low performance students. The similar conclusion holds for pre-test and post-test performance of students. We can conclude that the macro-patterns represented by the clusters are neither related to students' past performance nor to their course-level performance. In other words, the patterns do not separate weak students from strong ones. Instead, they represent students' different approaches to work with learning content. Within the group of students using the same approach, however, we can find both strong and week students. The results are similar to the observation in the original paper [7].

Next, we study the differences in behavioral micro-patterns of high and low-performance students within each cluster. By this, we hope to uncover the efficient and inefficient micro-patterns that happens within students with the same studying traits. To achieve this, we examine the average frequencies of micro-patterns for low and high performance students in each cluster and select the ones with a significant difference. The results are shown in Figs. 3, 4 and 5. As presented in Fig. 3, in cluster 1 ("confirmers"), patterns such as

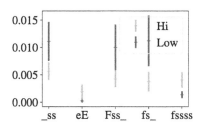

Fig. 3. Patterns with significant difference of frequency for low performance and high performance (learning gain) students in Cluster 1

'fssss' and 'eE' are found to be significantly more in high performance students. On the other hand, patterns 'Fss_', 'Fs_', and '_ss' appear more in low performance students. According to this, we can conclude that, the "confirmer" group students do repeat their success. But their approach to this repetition determines their performance in the course: (1) repeat after an initial success ('_ss') is associated with weaker students; (2) more repetition after an initial failure ('fssss') is associated with stronger students, as short repetitions and quitting after failure ('Fss_' and 'Fs_') is associated with weaker students; and (3) repeat reading examples is associated with stronger students. We can see that in cluster

2 ("thinkers", Fig. 4), high performance students have patterns such as '_FF', 'FF', and 'Sss', while low performance ones have higher rate of patterns with short failure ('f') in them. This shows that high-performance thinkers think each time they try a problem, until it is sufficiently understood. In contrast, weaker students frequently try to guess and fail in solving problems. Interestingly, low-performance thinkers also have a high frequency of 'Fff' pattern. It can be concluded that they start with serious intentions, but then start to guess the answers.

For the "reader" students (cluster 3, Fig. 5), we see longer attempts (e.g., 'EE', '_FS_', and 'FS') for high-performance students, compared to shorter attempts (e.g., 'ffs' and 'Fs') for low-performance ones. We can see that (1) high-performance students work with examples more carefully; (2) they do not rush after failure, but think and most always get it right; and (3) in contrast, low-performance

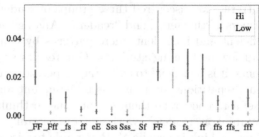

Fig. 4. Patterns with significant difference of frequency for low performance and high performance (learning gain) students in Cluster 2

students do not spend enough time on their attempts, whether it is a success or failure. In general, having patterns that include long attempts among high performance students and short attempts in low-performance ones demonstrate the impact of spending time on the performance.

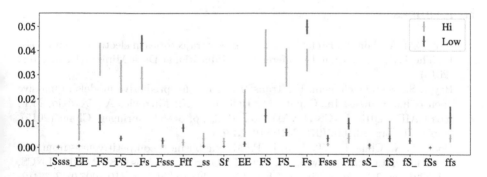

Fig. 5. Patterns with significant difference of frequency for low performance and high performance (learning gain) students in Cluster 3

7   Conclusions

In this paper we analyzed students' behavior patterns in working with parameterized exercises and annotated examples. Using frequent pattern mining, we discovered frequent *micro-patterns* of student behavior and used them to construct *macro-pattern* behavior vectors for students. Using data driven approaches, we analyzed the stability of these macro-patterns and showed that these are results of students' behavioral traits. Clustering students according to these macro-patterns, we discovered three groups of students, which we nicknamed as "confirmers", "thinkers", and "readers". Among these groups, we identified students' efficient and inefficient micro-patterns by comparing frequent patterns of high and low-performing students. Our results suggested that for "confirmer" students, it is beneficial to encourage repetitions after they fail in solving a problem. But, repetitions after success is redundant and inefficient. For "thinkers", it is useful to encourage them to continue to think deeper each problem, even after failure. For "readers", working more carefully with examples and spending more time to think is beneficial. Being able to discover a few behavioral clusters that represent different ways of learning is a promising step towards personalization: if learning behavior diversity among students is not that large, we can nudge different student groups towards the optimal behavior in different ways. In future, these results can be extended to be used as encouragements or recommendations to help students of each group to take on more efficient behaviors.

Acknowledgement. This work is partially supported by the National Science Foundation, under grant IIS-1755910.

References

1. Boubekki, A., Jain, S., Brefeld, U.: Mining user trajectories in electronic text books. In: The 11th International Conference on Educational Data Mining (EDM 2018) (2018)
2. Boyer, S., Veeramachaneni, K.: Transfer learning for predictive models in massive open online courses. In: Conati, C., Heffernan, N., Mitrovic, A., Verdejo, M.F. (eds.) AIED 2015. LNCS (LNAI), vol. 9112, pp. 54–63. Springer, Cham (2015). https://doi.org/10.1007/978-3-319-19773-9_6
3. Doan, T.-N., Chua, F.C.T., Lim, E.-P.: Mining business competitiveness from user visitation data. In: Agarwal, N., Xu, K., Osgood, N. (eds.) SBP 2015. LNCS, vol. 9021, pp. 283–289. Springer, Cham (2015). https://doi.org/10.1007/978-3-319-16268-3_31
4. Fournier-Viger, P., Gomariz, A., Campos, M., Thomas, R.: Fast vertical mining of sequential patterns using co-occurrence information. In: Tseng, V.S., Ho, T.B., Zhou, Z.-H., Chen, A.L.P., Kao, H.-Y. (eds.) PAKDD 2014. LNCS (LNAI), vol. 8443, pp. 40–52. Springer, Cham (2014). https://doi.org/10.1007/978-3-319-06608-0_4
5. Geigle, C., Zhai, C.: Modeling student behavior with two-layer hidden markov models. J. Educ. Data Min. **9**(1), 1–24 (2017)

6. Gelman, B., Revelle, M., Domeniconi, C., Veeramachaneni, K., Johri, A.: Acting the same differently: a cross-course comparison of user behavior in MOOCs. In: Barnes, T., Chi, M., Feng, M. (eds.) The 9th International Conference on Educational Data Mining (EDM 2016), pp. 376–381 (2016)
7. Guerra, J., Sahebi, S., Brusilovsky, P., Lin, Y.: The problem solving genome: analyzing sequential patterns of student work with parameterized exercises. In: 7th International Conference on Educational Data Mining, pp. 153–160 (2014)
8. Hansen, C., Hansen, C., Hjuler, N., Alstrup, S., Lioma, C.: Sequence modelling for analysing student interaction with educational systems. In: The 10th International Conference on Educational Data Mining, pp. 232–237 (2017)
9. Hsiao, I.-H., Brusilovsky, P.: Motivational social visualizations for personalized E-learning. In: Ravenscroft, A., Lindstaedt, S., Kloos, C.D., Hernández-Leo, D. (eds.) EC-TEL 2012. LNCS, vol. 7563, pp. 153–165. Springer, Heidelberg (2012). https://doi.org/10.1007/978-3-642-33263-0_13
10. Hsiao, I.-H., Sosnovsky, S., Brusilovsky, P.: Adaptive navigation support for parameterized questions in object-oriented programming. In: Cress, U., Dimitrova, V., Specht, M. (eds.) EC-TEL 2009. LNCS, vol. 5794, pp. 88–98. Springer, Heidelberg (2009). https://doi.org/10.1007/978-3-642-04636-0_10
11. Kinnebrew, J.S., Segedy, J.R., Biswas, G.: Analyzing the temporal evolution of students' behaviors in open-ended learning environments. Metacognition Learn. 9(2), 187–215 (2014)
12. Lin, J.: Divergence measures based on the shannon entropy. IEEE Trans. Inf. Theory 37(1), 145–151 (1991)
13. Lorenzen, S., Hjuler, N., Alstrup, S.: Tracking behavioral patterns among students in an online educational system. In: The 11th International Conference on Educational Data Mining (EDM 2018), pp. 280–285 (2018)
14. Martinez, R., Yacef, K., Kay, J.: Analysing frequent sequential patterns of collaborative learning activity around an interactive tabletop. In: Educational Data Mining 2011, pp. 111–120, June 2010
15. Mouri, K., Shimada, A., Yin, C., Kaneko, K.: Discovering hidden browsing patterns using non-negative matrix factorization. In: The 11th International Conference on Educational Data Mining (EDM 2018), pp. 568–571 (2018)
16. Ng, A.Y., Jordan, M.I., Weiss, Y.: On spectral clustering: analysis and an algorithm. In: Advances in Neural Information Processing Systems, pp. 849–856 (2002)
17. Sawyer, R., Rowe, J., Azevedo, R., Lester, J.: Filtered time series analyses of student problem-solving behaviors in game-based learning. In: The 11th International Conference on Educational Data Mining (EDM 2018) (2018)
18. Sharma, K., Jermann, P., Dillenbourg, P.: Identifying styles and paths toward success in MOOCs. In: Santos, O., et al. (eds.) The 8th International Conference on Educational Data Mining (EDM 2015) (2015)
19. Sosnovsky, S., Brusilovsky, P., Lee, D.H., Zadorozhny, V., Zhou, X.: Re-assessing the value of adaptive navigation support in e-learning context. In: Nejdl, W., Kay, J., Pu, P., Herder, E. (eds.) AH 2008. LNCS, vol. 5149, pp. 193–203. Springer, Heidelberg (2008). https://doi.org/10.1007/978-3-540-70987-9_22
20. Whitehill, J., Williams, J., Lopez, G., Coleman, C., Reich, J.: Beyond prediction: first steps toward automatic intervention in MOOC student stopout. In: The 8th International Conference on Educational Data Mining (EDM 2015) (2015)
21. Yin, C., Okubo, F., Shimada, A., Oi, M., Hirokawa, S., Ogata, H.: Identifying and analyzing the learning behaviors of students using e-books. In: 23rd International Conference on Computers in Education ICCE 2015. Asia-Pacific Society for Computers in Education (2015)

Investigating the Effect of Adding Nudges to Increase Engagement in Active Video Watching

Antonija Mitrovic[1]([⊠]) [iD], Matthew Gordon[1], Alicja Piotrkowicz[2] [iD],
and Vania Dimitrova[2] [iD]

[1] University of Canterbury, Christchurch, New Zealand
tanja.mitrovic@canterbury.ac.nz
[2] University of Leeds, Leeds, UK

Abstract. In order for videos to be a powerful medium for learning, it is crucial that learners engage in constructive learning. Historic interactions of previous learners can provide a rich resource to enhance interaction and promote engagement fostering constructive learning. This paper proposes such a novel approach of adding nudges to AVW-Space, a platform for video-based learning. We present the enhancements implemented in AVW-Space in the form of interactive visualizations and personalized prompts. A study focusing on presentation skills was conducted in a large first-year engineering course, in which AVW-Space provided an online resource for the students to use as they wish. The students were randomly divided into the control and experimental groups, which had access to the original and enhanced version of AVW-Space respectively. Our findings show that nudging is effective in fostering constructive learning: there was a significant difference in the percentage of constructive students in the two groups. The experimental group students wrote more comments, found AVW-Space easier to use, reported less frustration when commenting, and had higher confidence in their performance on commenting.

Keywords: Video-Based Learning · Intelligent support · Personalized nudges · Experimental study · Soft skill learning · Engagement

1 Introduction

Videos have become the predominant delivery method in both formal and informal online learning. However, research shows that watching videos can be a passive activity and result in limited learning [4, 5, 15, 22, 30]. A number of problems have been identified with Video-Based Learning (VBL), including limited interactivity with videos, the lack of human interaction, personalization, assessment and feedback [4]. New research strands related to VBL have appeared. There is established work on developing guidelines for producing effective videos (e.g. [10, 21]). Significant work has been done on increasing engagement with videos, by adding annotation tools, quizzes, examples and interactive exercises [8, 9, 14–16, 28, 30]. Data-driven approaches using interaction traces from VBL have been proposed to improve techniques for video navigation, such as visualizations of collective navigation traces,

© Springer Nature Switzerland AG 2019
S. Isotani et al. (Eds.): AIED 2019, LNAI 11625, pp. 320–332, 2019.
https://doi.org/10.1007/978-3-030-23204-7_27

dynamic timeline scrubbing and enhanced in-video searches [3, 4, 13, 17]. There are also approaches using students' ratings, annotations and forum contributions to support social navigation and collaborative learning [2, 4, 25, 29]. While efforts on augmenting the interaction to enhance VBL exist, there is no research that explicitly considers personalization, i.e. tailoring the interaction to the engagement profile of the learner.

Our approach is to encourage engagement during VBL via interactive note taking, tapping into learners' familiarity with commenting on videos in social networking sites. In previous work [7, 18, 19], we developed AVW-Space, a VBL platform, which supports reflection during interactive note taking, and also supports social learning through rating of comments. We categorized participants in previous studies into inactive, passive and constructive learners, based on the ICAP framework [5]. Our previous studies [7, 18, 19] show that only constructive learners, who wrote comments on videos and rated comments written by their peers, improved their conceptual knowledge.

To promote engagement with videos that leads to better learning, while at the same time preserving the learners' freedom to interact with videos in a way they prefer, we proposed the use of nudges [7, 19]. Nudges were introduced in decision support [27] as a form of interventions which influence people's behavior to make choices that lead to better lives (paternalism), but in a non-compulsory manner (libertarian). Behavior change is complex and so are the corresponding interventions. Choice architecture, which defines the ways to select and present choices that can lead to better behavior, is the core when designing nudges [20, 27]. To the best of our knowledge, our research is the first attempt to evaluate a choice architecture for personalized nudges in VBL.

In this paper, we describe how we implemented two types of nudges we previously proposed [19]: signposting through interactive visualizations and personalized prompts. The closest to our approach is the work by Shin et al. [26] who proposed in-video prompts as a way to assess students' comprehension and information about their learning experience. Such prompts are given to all students, at the pre-defined times during videos [26]. Our personalized prompts differ because they are *adaptive*: prompts given to a particular student depend on the student's interactions with the current video.

This paper also presents the study we conducted in a large first-year engineering course, in which students were required to give presentations but received no formal training on presentation skills. AVW-Space provided an online resource for the students to use as they wish. The students were randomly divided into the control and experimental groups, which had access to the original and enhanced version of AVW-Space respectively. The goal of our study was to investigate the effect of the nudges on students' engagement, learning and subjective impressions of the platform.

2 Previous Work on AVW-Space

AVW-Space is a controlled video-watching environment designed for self-study, which supports interactive notetaking and rating of comments written by peers, tapping into students' familiarity with social networking sites. We developed AVW-Space with transferable skills in mind, but the platform is general purpose and can be used for other types of skills. The environment supports engagement by providing micro-scaffolds to facilitate the commenting on videos and the reviewing of comments made by others. In order to create a learning space, the teacher needs to select a set of publicly available YouTube videos for students to watch, and specify micro-scaffolds.

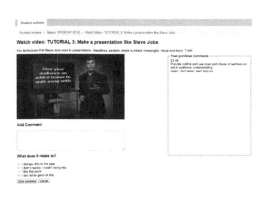

Fig. 1. The commenting interface

Learning starts with students watching and commenting on the videos individually. In our studies [7, 18, 19] we used eight short videos, four of which were tutorials on how to give presentations, while the remaining four videos were example presentations. The student can enter a comment at any time during the video, and needs to select an aspect, which indicates the intention of the comment (Fig. 1). For the tutorials, aspects aimed at stimulating reflection included: "*I didn't realize I wasn't doing it*", "*I am rather good at this*", and "*I did/saw this in the past*". There was one additional aspect, "*I like this point*", to encourage the learner to externalize relevant learning points. For the example videos, the aspects corresponded to presentation skills covered in the tutorials, which included "*Delivery*", "*Speech*", "*Structure*", and "*Visual aids*".

In the second phase, the teacher selects the comments that will be open to the whole class, so that students can review and rate each other's anonymized comments. As this second phase is not relevant for the research presented in this paper, we refer the interested reader to [7, 19] for further details.

We operationalized the ICAP framework [5] in the context of AVW-Space as follows [7, 19]. **Passive Learners** are those who watched videos, but had minimal other interactions with them. On the other hand, **Constructive Learners** show higher levels of engagement by commenting on videos. Comments contain remarks on important events in videos, and contain statements showing reflection and self-explanation. As AVW-Space does not currently support collaboration between students, we do not consider the Interactive mode of ICAP. In addition to passive and constructive learners, we have also added another mode to characterize students who do not engage in learning at all, i.e. do not watch videos; we refer to them as **Inactive Learners** (IL).

The findings from our previous studies showed that only constructive students improved their knowledge when interacting with AVW-Space [7, 18, 19]. We presented a set of requirements for fostering constructive behavior in [19]. In this paper, we focus on adding intelligent nudging to the platform.

3 Adding Nudging to AVW-Space

As a first step towards more intelligent support for active video watching, we implemented nudging in the form of signposting through interactive visualisations and personalized prompts to AVW-Space.

3.1 Interactive Visualizations

Interactive visualisations (shown below the video in Fig. 2) are used to support social learning. The top visualization is the **comment timeline**; it provides signposts in terms of comments written by previous learners. Each comment is represented as a coloured dot, representing the time when the comment was made. The colour of the dot depends on the aspect used, with the legend shown on the side. We selected the best comments from previous studies to use in the comment timeline [24]. The comment timeline also allows the learner to inspect comments written by previous learners. When the mouse is positioned over a particular dot, the student can see the comment (as in Fig. 2). Dots are slightly transparent, so that comments made in temporal proximity to each other can be differentiated. Clicking on a dot begins playing the video from that point.

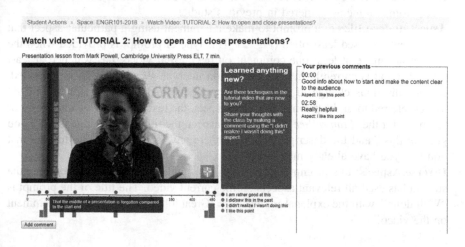

Fig. 2. Interactive visualizations and a prompt. The interactive visualizations are modified to show only comments written with the "I didn't realize I wasn't doing this" aspect.

The bottom visualization is the **comment histogram,** representing the number of comments written for various segments of the video. This visualization allows the student to quickly identify important parts of a video, where other students have made many comments. The two visualizations meet two identified needs: (1) providing social reference points so that students can observe others' comments, and (2) indicating important parts of a video and what kind of content can be expected in those parts, differentiated by aspect colours.

3.2   Personalized Prompts

Personalized prompts are designed to encourage students to write comments. An example of a personalized prompt is shown in Fig. 2, to the right of the video. AVW-Space maintains a profile for each student, and uses it to select prompts adaptively.

Prompts are provided when the student is in a *high-attention interval*, which is a part of a video during which previous students wrote many comments. To identify high attention intervals for the videos, we used interaction traces collected previously, and identified parts of videos with high user interest and relevant comments [7]. We designed four types of prompts:

1. **No comment reminder** is a prompt encouraging the student to make a comment. This prompt is offered when the student has watched at least 30% of the video without making any comments, and is currently in a high-attention interval.
2. **No comment reference point** prompt reminds the student to make a comment, and offers an example as stimulus. The prompt is only shown if the *No comment reminder* prompt has not resulted in a comment. Such prompts are provided when the student has watched at least 70% of the video without comments, the student is in a high-attention interval, and this type of prompt has not been issued on the current video. The comments used as stimuli have been manually selected for each video from comments gathered in previous studies.
3. **Aspect under-utilized**: a prompt to make a comment using a particular aspect that the student has used least often (Fig. 2). This type of prompt is provided when the student has made at least one comment on the current video, has watched at least 30% of it and is currently in a high-attention interval. When the prompt is issued, the visualizations change to only show comments made using the under-utilized aspect referred to in the prompt. For each aspect, the text of the nudge changes. For example, for the *'I am rather good at this'* aspect, the title of the nudge is "Are you good at this", and the description is "Are there any techniques in the tutorial that you feel you have already mastered?"
4. **Diverse Aspects**: this prompt provides positive reinforcement, displayed when the student has used all relevant aspects on the current video. The title of the prompt is "Well done!" with the explanatory message "Great job using all aspects to comment on the video!"

4   Experimental Design

The study was conducted in a first-year course mandatory for all engineering students at the University of Canterbury. The students worked on a group project and gave a presentation, during which each student presented for one minute. Due to an already full curriculum, students received no formal training on giving presentations. Instead, they were invited to use AVW-Space for online training. The students who watched at least one video in AVW-Space received 1% of the final course grade. Ethical approval was obtained from the University of Canterbury.

The goal of our study was to evaluate the impact of intelligent nudging. The students were randomly allocated to the control or experimental group. The control group interacted with the original version of AVW-Space (Fig. 1), while the experimental group interacted with the enhanced version (Fig. 2). We defined three **research questions:**

RQ1. Are nudges effective in fostering constructive behavior? We expect to see a higher proportion of students from the experimental group engaging in constructive behavior in comparison to the control group (Hypothesis H1).

RQ2. What features of AVW-Space influence learning? Can we infer causal relationships between the use of AVW-Space's features and learning? Our previous studies showed that only constructive students improved their knowledge after interacting with AVW-Space. We anticipate that intelligent nudging will have a positive effect on the number of comments written, which will in turn have a positive effect on learning (H2).

RQ3. Do students in control/experimental group have different opinions about the usefulness of AVW-Space and cognitive load? We expect that the students in the experimental group would find the environment more useful and report smaller cognitive load (H3).

Materials. The videos used in the study were the same ones as those described in Sect. 3. We designed two surveys, similar to those used in the previous studies [7, 18, 19]. Survey 1 collected participant's profile (demographic information, background experiences, Motivated Strategies for Learning Questionnaire (MSLQ) [23]). The survey also contained three questions on the participants' knowledge of presentations (we refer to those questions as *conceptual knowledge questions*). The student was asked to list as many concepts related to (1) Structure, (2) Delivery and Speech, and (3) Visual Aids. For each of those three questions, students had one minute to write responses. Survey 2 included the same conceptual knowledge questions; NASA-TLX instrument [11] to check participants' perception of cognitive load; Technology Acceptance Model (TAM) [6] to check participants' perceived usefulness of AVW-Space. Additionally, Survey 2 contained questions on usability related to commenting on videos and rating of comments. The experimental group also received questions related to interactive visualizations and personalized nudges.

Procedure. The students were invited to participate in the study on 3 May 2018. After completing Survey 1, the participants were instructed to log on to AVW-Space, watch the four tutorial videos first and then to proceed to critique the example videos. The rating of comments was enabled on May 16. Invitations to complete Survey 2 were emailed on May 24, and the survey was closed on 3 June 2018.

5   Results

Table 1 presents the number of participants from the two groups who completed various parts of the study. Out of 1039 students enrolled in the course, 449 completed Survey 1. Of those, 347 have used AVW-Space, while the remaining 102 participants were inactive learners. We received 263 responses for Survey 2, but that number included some inactive students. After removing those responses, we had 119 students from the control and 102 responses from the experimental group who completed both surveys and interacted with AVW-Space.

Table 1. Number of participants who completed various parts of the study.

Group	Survey 1	Inactive	AVW-Space	Survey 2 (all)	Survey 2 (excl. IL)
Control	234	54	180	138	119
Experim.	215	48	167	125	102

Table 2 presents the demographic data for the 347 students who interacted with AVW-Space. As typical for engineering courses, there were more males than females. The majority of participants (79.83%) were native English speakers. Most participants (95.39%) were aged between 18 and 23. The questions related to training on giving presentations, experience in giving presentations, using YouTube and using YouTube for learning were based on the Likert scale from 1 (Low) to 5 (High). There were no significant differences between the two groups on these features, as well as on MSLQ scales, with the exclusion of Task Value ($U = 15,066.5$, $p = .043$).

Table 2. Demographic data for the participants who completed Survey 1. Apart from the first three rows, the remaining rows present the mean and standard deviation in parentheses.

	Control (180)	Experimental (167)
Gender	124 males, 55 females, 1 other	118 males, 49 females
Aged 18–23	175	156
Native English speakers	135	142
Training	1.64 (.76)	1.66 (.82)
Experience	2.17 (.81)	2.19 (.79)
YouTube	4.22 (1.08)	4.22 (1.03)
YouTube for learning	3.36 (1.14)	3.28 (1.12)
Task value	5.47 (.85)	5.22 (.79)

5.1   Do Nudges Foster Constructive Behavior?

We divided the students (post-hoc) into Constructive and Passive, using the median number of comments written by the class (median = 1). We expected to see a higher number of constructive students in the experimental group. The numbers of constructive and passive students in the two groups are given in Table 3 (for all students who completed Survey 1). A chi-square test of homogeneity between group and behavior type (i.e. Constructive or Passive) revealed a significant difference (Chi-square = 4.463, p = .035), with the effect size of (Phi) of .142. Therefore, hypothesis H1 is confirmed.

Table 3. Numbers of constructive and passive students in the two groups

	Passive (187)	Constructive (160)
Control (180)	107 (59.44%)	73 (40.56%)
Experimental (167)	80 (47.90%)	87 (52.10%)

It is also interesting to compare constructive and passive students in the experimental group (Table 4). Both subsets received nudges – why did only some participants respond to nudges? The only significant difference between passive/constructive students in the experimental group on the variables from Survey 1 is on Training (U = 2,906.5, p = .042). During interaction with AVW-Space, in addition to a significant difference on the number of comments written, these two subgroups differed significantly on the number of videos watched (t = 4.61, p < .001) and prompts received (t = 2.33, p = .022). Please note that some students watched the same video multiple times, so the average number of videos watched can be higher than 8.

Table 4. Differences between passive and constructive students from the experimental group.

	Training	Videos	Prompts
Passive (80)	1.53 (.71)	5.60 (3.50)	8.21 (5.329)
Constructive (87)	1.78 (.88)	8.41 (4.37)	12.44(8.21)

5.2 What Features of AVW-Space Improve Students' Knowledge?

There were no significant differences between the two groups on either the number of sessions with AVW-Space (control: mean = 2.58, sd = 2.05; experimental: mean = 2.53, sd = 2.56), or the number of videos watched (control: mean = 7.03, sd = 4.34; experimental: mean = 7.03, sd = 4.22). The only significant difference (U = 17,608, p = .004) was on the number of comments written (control: mean = 4.31, sd = 7.76; experimental: mean = 6.34, sd = 9.60). We used the Mann-Whitney test as the number of comments is not normally distributed. Figure 3 shows the number of comments per video for the two groups. The distributions of comments are

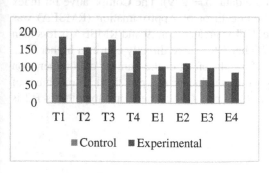

Fig. 3. Number of comments per video

significantly different for the two groups (U = 51.5, p = .038).

In Surveys 1 and 2, the participants had one minute to list all concepts they knew related to the structure, visual aids, and delivery and speech. The students' replies were marked automatically, using the ontology of presentation skills we developed in previous work [1, 7, 24]. Table 5 presents the resulting conceptual knowledge scores from Surveys 1 and 2 (CK1 and CK2 respectively) for constructive/passive students in the two groups. A two-way ANCOVA found no significant interaction between group and category (i.e. Constructive vs Passive), but there was a significant main effect of Category, $F(1, 216) = 3.872$, $p = .05$, partial $\eta^2 = .018$. As in previous studies, constructive students improved their knowledge of presentation skills significantly.

Table 5. Conceptual knowledge scores for the two groups.

Group		CK1	CK2
Control (119)	Constructive (59)	13.56 (5.65)	15.76 (5.66)
	Passive (60)	12.25 (4.16)	12.88 (5.95)
	All (119)	12.28 (5.51)	14.31 (5.96)
Experimental (102)	Constructive (65)	14.00 (5.66)	14.98 (6.36)
	Passive (37)	13.35 (5.29)	13.89 (6.00)
	All	13.12 (5.50)	14.59 (6.22)
All (221)	Constructive (124)	13.79 (5.63)	15.35 (6.03)
	Passive (102)	12.67 (4.63)	13.27 (5.95)

We used IBM SPSS Amos to infer the causal relationships between CK1, CK2 and variables showing how students used AVW-Space, such as the number videos watched, the number of comments made, the number of prompts received (for the experimental group) and the number of ratings made. All these variables are observed and measured without errors. We were unable to find any well-fitting path models for the control group, except the simplest one, which shows the correlation between CK1 and CK2 (.60, p < .001).

Figure 4 illustrates the best fitting model for the experimental group. The chi-square test (2.55) for this model (df = 2) shows that the model's predictions are not statistically significantly different from the data (p = .279). The Comparative Fit Index (CFI) was .988, and the Root Mean Square Error of Approximation (RMSEA) was .052. Therefore the model is acceptable: CFI is greater than .9 and RMSEA is less than .06 [12]. The model indicates that the higher CK1 score directly causes a higher CK2 score (coefficient = .44, p < .001). Therefore, the effect of the number of comments on CK2 is adjusted for and above and beyond this influence (.2, p = .024). The number of nudges affects the number of comments (.41, p < .001). Therefore, hypothesis H2 is confirmed.

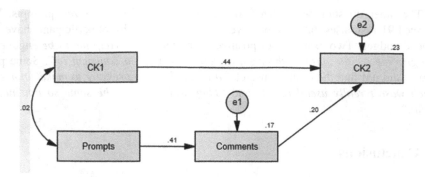

Fig. 4. Path diagram for the experimental group.

5.3 Do Students in the Two Conditions Differ in Their Opinions About the Usefulness of AVW-Space and Cognitive Load?

The participants replied to the TAM questionnaire [6] using the Likert scale from 1 (highest) to 7 (lowest). We analyzed the replies using the two factor ANOVA (group and category), and found no significant interaction of the two factors. For question 8 (*My interaction with* AVW-Space *would be clear and understandable*), there was a main effect of category, $F(1,211) = 7.19$, $p = .008$, partial $\eta^2 = .033$. The average score of the 92 passive students was 3.67 (1.73), while the average score of the 123 constructive students was better, 3.05 (1.43). There was also a significant main effect of group for question 9 (*I would find* AVW-Space *easy to use*), $F = 4.86$, $p = .029$, partial $\eta^2 = .023$. The average score of the 115 control group students was 3.30 (sd = 1.68), while the average score of the 100 students from the experimental group was better, 2.78 (sd = 1.20).

Table 6 reports the scores on the TLX-NASA questions related to writing comments. Constructive students reported significantly lower frustration ($F(1,220) = 8.62$, $p = .004$, partial $\eta^2 = .038$), and significantly higher performance on commenting ($F(1,220) = 7.99$, $p = .005$, partial $\eta^2 = .035$). These analyses provide evidence supporting hypothesis H3.

Table 6. TLX-NASA scores on commenting. Effort and Demand: Likert scale 1 (very easy) to 20 (very hard); Frustration and Performance: 1 (not at all) to 20 (very much)

	Demand	Effort	Frustration	Performance
Passive (98)	8.96 (4.64)	8.10 (4.44)	8.76 (6.02)	10.61 (5.27)
Constructive (124)	8.13 (4.33)	7.38 (4.42)	6.54 (5.21)	12.40 (4.13)

The experimental group received two additional questions in Survey 2, the first of which asked for feedback on the usefulness of interactive visualizations. We received 100 responses, 85 of whom were positive, such as "*See which parts of the video other people find useful*" and "*To compare yourself with the rest of the class.*" One student wrote "*I didn't understand them till id finished most of the videos.*"

The other question was related to the usefulness of personalized prompts. We received 91 responses, 62 were positive, and 21 negative. Eight participants have not noticed nudges. Two examples of positive opinions were: *"Help me to be engaged"*, *"To give me a little push in the right direction of what to comment on"*. Some participant did not find the prompts useful: *"It created subtle pressure to make comments which wasn't really useful at all"* and *"They were always the same so not hugely useful."*

6 Conclusions

We proposed the use of nudges (signposting through interactive visualizations and personalized prompts) to encourage constructive behavior during VBL. We found that nudging was effective in fostering constructive behavior and resulted in the students in the experimental condition making more comments, found AVW-Space easier to use, reported less frustration when commenting, and had higher confidence in their performance on commenting. No differences between passive/constructive students in the experimental group suggests that nudging seems to work all types of students.

The work presented here is part of a larger research stream on adding intelligent features to augment interaction with videos for informal learning. In our future work we plan to implement more types of nudges, following the formal framework defined in [7, 19]. This will take into account not just the engagement with videos but individual profiles (e.g. MSLQ scores or previous experience). Future work also includes extending the support for the rating phase.

Our research opens a new avenue in developing intelligent learning environments which adapt established interventions for behavior change in the form of nudges. This can be applied in a range of domains to foster informal learning where one can learn from their experience and that of others.

Acknowledgements. This research was supported by two Southern Hub Ako Aotearoa grants, a teaching development grant from the University of Canterbury, and the EU-FP7-ICT-257184 ImREAL grant. We thank Lydia Lay, Amali Weerasinghe and Jay Holland for their contributions to AVW-Space, as well as Peter Gostomski and Alfred Herritsch for helping with the study.

References

1. Abolkasim, E., Lau, L., Mitrovic, A., Dimitrova, V.: Ontology-based domain diversity profiling of user comments. In: Penstein Rosé, C., et al. (eds.) AIED 2018. LNAI, vol. 10948, pp. 3–8. Springer, Cham (2018). https://doi.org/10.1007/978-3-319-93846-2_1
2. Agarwala, M., Hsiao, I.H., Chae, H.S., Natriello, G.: Vialogues: videos and dialogues based social learning environment. In: Proceedings of the 12th IEEE International Conference on Advanced Learning Technologies, pp. 629–633. IEEE (2012)
3. Carlier, A., Charvillat, V., Ooi, W.T., Grigoras, R., Morin, G.: Crowdsourced automatic zoom and scroll for video retargeting. In: Multimedia 2010, pp. 201–210 (2010)
4. Chatti, M.A., et al.: Video annotation and analytics in CourseMapper. Smart Learn. Environ. 3(1), 10 (2016)

5. Chi, M.T., Wylie, R.: The ICAP framework: linking cognitive engagement to active learning outcomes. Educ. Psychol. **49**(4), 219–243 (2014)
6. Davis, F.D.: Perceived usefulness, perceived ease of use, and user acceptance of information technology. MIS Q., 319–340 (1989)
7. Dimitrova, V., Mitrovic, A., Piotrkowicz, A., Lau, L., Weerasinghe, A.: Using learning analytics to devise interactive personalised nudges for active video watching. In: Bielikova, M., Herder, E., Cena, F., Desmarais, M. (eds.) Proceedings 25th ACM UMAP Conference, Bratislava, Slovakia (2017)
8. Dodson, S., Roll, I., Fong, M., Yoon, D., Harandi, N.M., Fels, S.: Active viewing: a study of video highlighting in the classroom. In: Proceedings of the 2018 Conference on Human Information Interaction and Retrieval, pp. 237–240. ACM (2018)
9. Giannakos, M., Sampson, D., Kidziński, Ł.: Introduction to smart learning analytics: foundations and developments in video-based learning. Smart Learn. Environ. **3**(1), 1–9 (2016)
10. Guo, P.J., Kim, J., Rubin, R.: How video production affects student engagement: an empirical study of MOOC videos. In: Proceedings 1st ACM Conference Learning at Scale, pp. 41–50 (2014)
11. Hart, S.G.: NASA-task load index (NASA-TLX); 20 years later. In: Proceedings the Human Factors and Ergonomics Society Annual Meeting, vol. 50, no. 9, pp. 904–908. Sage Publications (2006)
12. Hu, L.T., Bentler, P.M.: Cutoff criteria for fit indexes in covariance structure analysis: conventional criteria versus new alternatives. Struct. Equ. Model. Multi. J. **6**(1), 1–55 (1999)
13. Kim, J., Guo, P.J., Cai, C.J., Li, S.W.D., Gajos, K.Z., Miller, R.C.: Data-driven interaction techniques for improving navigation of educational videos. In: Proceedings of the 27th ACM Symposium on User Interface Software and Technology, pp. 563–572. ACM (2014)
14. Kleftodimos, A., Evangelidis, G.: Using open source technologies and open internet resources for building an interactive video based learning environment that supports learning analytics. Smart Learn. Environ. **3**(1), 1–23 (2016)
15. Koedinger, K.R., Kim, J., Jia, Z., McLaughlin, E., Bier, N.: Learning is not a spectator sport: doing is better than watching for learning from a MOOC. In: Proceedings 2nd ACM Conference Learning @ Scale, pp. 111–120 (2015)
16. Kovacs, G.: Effects of in-video quizzes on MOOC lecture viewing. In: Proceedings 3rd Learning @ Scale, pp. 31–40 (2016)
17. Liu, C., Kim, J., Wang, H.C.: ConceptScape: collaborative concept mapping for video learning. In: Proceedings of the 2018 CHI Conference on Human Factors in Computing Systems. ACM (2018). https://doi.org/10.1145/3173574.3173961
18. Mitrovic, A., Dimitrova, V., Weerasinghe, A., Lau, L.: Reflexive experiential learning using active video watching for soft skills training. In: Chen, W. et al. (eds.) Proceedings 24th International Conference Computers in Education, Mumbai, India, APSCE, pp. 192–201 (2016)
19. Mitrovic, A., Dimitrova, V., Lau, L., Weerasinghe, A., Mathews, M.: Supporting constructive video-based learning: requirements elicitation from exploratory studies. In: André, E., Baker, R., Hu, X., Rodrigo, M.M.T., du Boulay, B. (eds.) AIED 2017. LNAI, vol. 10331, pp. 224–237. Springer, Cham (2017). https://doi.org/10.1007/978-3-319-61425-0_19
20. Münscher, R., Vetter, M., Scheuerle, T.: A review and taxonomy of choice architecture techniques. J. Behav. Decis. Making **29**, 511–524 (2015)
21. Ou, C., Goel, A.K., Joyner, D.A., Haynes, D.F.: Designing videos with pedagogical strategies: online students' perceptions of their effectiveness. In: Proceedings of the 3rd ACM Conference on Learning@ Scale, pp. 141–144. ACM (2016)

22. Pardo, A., Mirriahi, N., Dawson, S., Zhao, Y., Zhao, A., Gašević, D.: Identifying learning strategies associated with active use of video annotation software. In: Proceedings 5th International Conference Learning Analytics and Knowledge, pp. 255–259. ACM (2015)

23. Pintrich, P.R., De Groot, E.V.: Motivational and self-regulated learning components of classroom academic performance. J. Educ. Psychol. **82**(1), 33 (1990)

24. Piotrkowicz, A., Dimitrova, V., Mitrovic, A., Lau, L.: Self-regulation, knowledge, experience: which characteristics are useful to predict user engagement? In: HAAPIE Workshop, Adjunct Proceedings of the 26th ACM UMAP Conference, Singapore, 8–11 July 2018, pp. 63–68 (2018)

25. Risko, E., Foulsham, T., Dawson, S., Kingstone, A.: The collaborative lecture annotation system (CLAS): a new tool for distributed learning. IEEE Trans. Learn. Technol. **6**(1), 4–13 (2013)

26. Shin, H., Ko, E.Y., Williams, J.J., Kim, J.: Understanding the effect of in-video prompting on learners and instructors. In: Proceedings of the CHI Conference on Human Factors in Computing Systems, p. 319. ACM (2018)

27. Thaler, R.H., Sunstein, C.R.: Nudge: Improving Decisions about Health, Wealth, and Happiness. Yale University Press, New Haven (2008)

28. Wachtler, J., Hubmann, M., Zöhrer, H., Ebner, M.: An analysis of the use and effect of questions in interactive learning-videos. Smart Learn. Environ. **3**(1), 13 (2016)

29. Wang, X., Wen, M., Rosé, C.P.: Towards triggering higher-order thinking behaviors in MOOCs. In: Proceedings 6th International Conference Learning Analytics & Knowledge, pp. 398–407. ACM (2016)

30. Yousef, A.M.F., Chatti, M.A., Schroeder, U.: The state of video-based learning: a review and future perspectives. Int. J. Adv. Life Sci. **6**(3/4), 122–135 (2014)

Behavioural Cloning of Teachers
for Automatic Homework Selection

Russell Moore[✉], Andrew Caines, Andrew Rice, and Paula Buttery

ALTA Institute, Department of Computer Science and Technology,
University of Cambridge, 15 J. J. Thomson Avenue, Cambridge, UK
{rjm49,apc38,acr31,pjb48}@cam.ac.uk
http://alta.cambridgeenglish.org

Abstract. We describe a machine-learning system for supporting teachers through the selection of homework assignments. Our system uses behavioural cloning of teacher activity to generate personalised homework assignments for students. Classroom use is then supported through additional mechanisms to combine these predictions into group assignments. We train and evaluate our system against 50,065 homework assignments collected over two years by the *Isaac Physics* platform. We use baseline policies incorporating expert curriculum knowledge for evaluation and find that our technique improves on the strongest baseline policy by 18.5% in Year 1 and by 13.3% in Year 2.

Keywords: Homework selection · Behavioural cloning · Deep learning

1 Introduction

Tutoring by human teachers is known to produce large learning gains for students, whether one-to-one or in groups [5]. For larger groups, a teacher's attention must inevitably be spread more thinly across their students. We seek to develop learning technologies which can support teachers in this scenario.

We focus on setting homework assignments. The setting of good-quality homework is of recognised pedagogical importance [14,30], but how work should be chosen is less clear: human teachers have been shown to adopt varied strategies of teaching even for very simple lessons [16]. In light of this we developed a system, HWGen that learns from expert actions using behavioural cloning.

HWGen generates homework assignments at both the individual and group level which we compare to assignments derived from a selection of naive and curriculum-aware baseline policies.

We trained our system with two years of student data from *Isaac Physics*, a major online teaching platform and find that our model is able to suggest assignments that are closely aligned with the choices made by human teachers. At the individual level, HWGen improves on the strongest baseline policy by 17.8% in Year 1 and by 12.9% in Year 2. Making homework selections per group, HWGen improves on the strongest baseline policy by 18.5% in Year 1 and by 13.3% in

© Springer Nature Switzerland AG 2019
S. Isotani et al. (Eds.): AIED 2019, LNAI 11625, pp. 333–344, 2019.
https://doi.org/10.1007/978-3-030-23204-7_28

Year 2. These results reflect the way that human teachers make group-level homework decisions based on class progress as a whole, but also that setting personalised assignments stays fairly close to teacher decisions while also taking into account individual factors.

In this paper we make the following contributions: (1) we give a neural-network architecture for suggesting personalised homework assignments; (2) we propose mechanisms to combine personalised predictions in order to generate group assignments; (3) we use data on student and teacher behaviour from the *Isaac Physics* platform to show that our technique significantly outperforms a variety of baseline policies.

2   Related Work

Personalisation in educational technology is an exciting prospect since learners are known to progress at different rates and in different styles [4,8,28] and personalised tutoring has been shown to have beneficial effects on student learning [27]. Lindsey and colleagues found that a personalised review system for course content yielded a 16.5% boost in retention rates over standard practice (massed study) and a 10% improvement over a one-size-fits-all strategy for spaced study [18]. Advancing personalised learning is recognised as one of the National Academy of Engineering's *Grand Challenges* for the 21st Century.[1] Such systems traditionally involve considerable effort and pedagogical knowledge to design, author, and structure content [10,20,33]. One example is the use of concept maps [22] to represent the structure of skills and knowledge; these can help guide learning [2] and are used in tutoring systems [12] but they require expert knowledge and careful design.

There has been work on intelligent tutoring systems to imitate teachers and widen educational access since the 1960's [34], but successful autonomous selection of the right homework task at the right time remains elusive. Rather than developing an analytical solution, we hope instead to learn from the usage data logged by education platforms.

Teaching data combined with student data provide an appropriate setting for *imitation learning*. In imitation learning, a system learns appropriate responses to its environment from a human actor. The archetypal scenario is the self-driving car (in which the vehicle controller learns from a human driver) [25], but it has broader applications, and it has recently been used to provide personalised navigation for web-based learners [23].

In particular *behavioural cloning* is a simple but widely used form of imitation learning [6], useful in situations where non-interactive data is available for training, where actions are judged immediately (*i.e.* no long term reward information is available), and where imperfect actions are unlikely to lead to cascading errors. In a classroom setting there is no risk of physical mishap, and teachers are able to exercise expert judgement with respect to any suggestions a system might make: this makes behavioural cloning a viable candidate as an approach.

---

[1] http://www.engineeringchallenges.org/challenges/learning.aspx.

3   Experiments

This work is based on user data from *Isaac Physics*,[2] a UK government funded project aimed at pre-university students. Launched in 2015, *Isaac* serves physics and mathematics exercises to over 120,000 registered users globally. As well as for private study, the platform is designed for use by school classes with homework assignments set by their teachers. It is the homework-setting actions of these teachers that we aim to imitate.

There are two related sub-tasks in this work: (1) the task of setting homework for an individual, and (2) of setting homework for a group. The homework selection should be available 'on demand', so that a teacher setting work for a class could use the software as a teaching aid, or so that a student can obtain personalised suggestions for independent study.

3.1   Data

This work uses a dataset from *Isaac* activity logs collected from March 2015 to March 2018, from non-affiliated UK secondary schools. These general-purpose logs record when users visit a page, answer a question (correctly or otherwise), view a hint, and so on. The logs also track when teachers assign homework to their class (a *SET_ASSIGNMENT* actions) and it is this behaviour we aim to clone.

Questions on Isaac are organised into *pages*: a page being the smallest teaching unit which may be assigned as homework. Students are organised into *groups*, which can be thought of as virtual classes. A teacher's *SET_ASSIGNMENT* action points a group of students to one or more question pages.

The *Isaac* project publishes a number of textbooks that accompany the platform[3]. Of these, the earliest book *Mastering Essential Pre-University Physics* ('the textbook'), currently accounts for around 80% of the homework set on the platform, with material across 73 pages forming the curriculum from which HWGen selects work. There are hundreds more question pages in *Isaac* but they are not all ordered for difficulty. By using only the textbook questions we have a controlled experiment in which the questions have an implicit canonical ordering – the order of page numbers from 1 to 73 – which we can use for baseline selection policies. If HWGen proves to be successful at this task, it can then be extended to the wider unordered curriculum.

Data was filtered to students in the 16–18 year age range (the target range for the textbook). We wanted to imitate teachers who are engaged and who know their students well, so we removed teachers who rarely set work (< 5 assignments) and excluded large groups (≥ 30 students). We ranked teachers by activity, choosing the most prolific first. We measure each student's time on the platform from the date of their first assignment, and we split the students into a Year 1 group and Year 2 group at 365 days of use.

---

[2] https://isaacphysics.org.

[3] https://www.isaacbooks.org.

After filtering, 6672 instances of group-level assignments (50,065 assignments to individual students) were available for training and validation. For testing, 970 group-level assignments (6028 individual assignments) were chosen from a separate held-out set of teachers. A box-and-whisker plot of the training data is given in Fig. 1, showing a noisy upward trend between students' time on the platform in days and the position of homework assignments in the textbook in pages. From this we can see that although there is evidence of the curriculum being followed, we should not expect a linear predictor to perform well due to noise, especially in the tail of the data where homework level becomes less consistent.

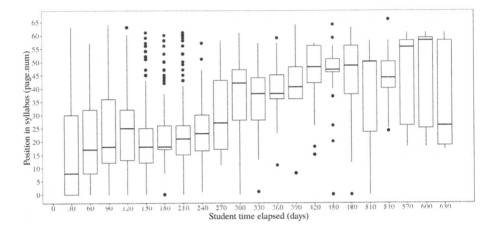

Fig. 1. The training dataset summarised in two dimensions: student duration on platform (days) is on the x-axis, position in textbook of homework selected (page number) is on the y-axis. The boxes show the range between first and third quartiles, the horizontal stripe indicates the median, the whiskers represent $1.5 * IQR$ where IQR is the inter-quartile range (third minus first quartile); other points beyond the whiskers are outliers. There are 12,489 features in the full dataset.

3.2   Student Features

The student is represented as a combination of four vectors: x for practice, u for success, a for previously assigned work, and a small set of real-valued features s. There are 2605 elements in x and a, one for each page on *Isaac*, and 7274 in u (for each page sub-part) and these take binary values depending on whether a page is attempted/assigned, or a sub-part is completed. Note that although we only set homework from the textbook, we track students' encounters with all pages on the platform: many of the pages outside the book are more difficult (or more involved) and so a student who has attempted such questions may be ready for more advanced material.

The s vector contains the following statistics about the student: *age* (in years, rounded to 1 decimal place), *days on platform, number of distinct questions attempted, number of attempts at all questions, number of questions passed.*

Other features were measured but not included in the model as they provided no benefit to performance, including: questions passed per day, question pass-rates, and the number of attempts the student had taken at each question (effectively this was clipped to binary in the x vector). Finer grained breakdowns of attempts, and 'recent history' versions of these features also made no improvement. Several of the profile features were tried in combination (e.g. question attempts per day as a proxy for motivation) but these were less successful than allowing the neural network to learn its own internal representations.

3.3 Implementation

The homework-selection task was formulated as a multiclass classification problem, using a feed-forward neural network (Fig. 2). Hidden layers are all rectified linear units (ReLU), except for a concatenation layer to merge encodings. These are followed by two more ReLU layers each suffixed by a dropout layer ($rate = 0.2$). Layer structure, dropout, learning rate and batch size, were found by randomised parameter search.

The network output is passed through a softmax activation function (1) to convert internal scores into a probability distribution. The function transforms the jth element of the raw neural network output, z, into a probability conditioned on the input (s, x, u, a), $Pr(y = j \mid (s, x, u, a))$. The output of the network consists of 73 units, one for each possible homework choice.

$$softmax(z)_j = \frac{\exp(z_j)}{\sum_{k=1}^{K} \exp(z_k)} \tag{1}$$

The software used in this work was implemented in `Python 3.6` using `Keras` [11] with a `TensorFlow` [1] back-end, and `scikit-learn` [24].

The network was trained with a categorical cross-entropy loss function using the Adam optimiser [17] with best results at the default $\alpha = 0.001$, $batch = 32$. Real-valued inputs were centred and scaled to have unit variance and zero mean. Binary inputs were not transformed. From the 50,065 samples in the training data, 10,013 (20%) were used for validation and to trigger early-stopping.

3.4 Setting Individual Work

The homework-setting decision is treated as an action-selection task, modelled as multiclass classification. The neural network outputs a softmax vector, the jth element of which is the probability of taking the jth action as determined from training. We choose the action given the highest probability by the network.

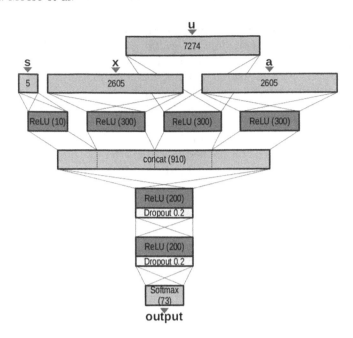

Fig. 2. Outline of the neural architecture. Input layers are at the top. Cross-hatched areas denote densely connected layers.

3.5 Modifications to Set Group Work

For setting homework at the group level, two alternative methods were applied to select a page for the group's next assignment:

- *Softmax averaging*: the softmax vectors for all students in the group are summed, and normalised by the group size. This produces a new softmax vector, and the top-scoring element is selected.
- *Voting*: the top-scoring candidates are chosen from each student's softmax vector. These are then counted, and the candidate that receives the most 'votes' is selected.

Both approaches can be implemented with simple control logic.

3.6 Baseline Policies

The performance of HWGen (in both individual and group modes) was judged against three baseline policies: a random policy and two *oracle* policies. The random policy randomly chooses an assignment from those that the student has not yet attempted.

The two oracle policies are 'curriculum aware'—they have access to ordering knowledge about the material that is not available to HWGen. The ordering was developed by the authors of the textbook as the logical progression of content.

The *linear* oracle policy selects the material in book order at isochronal intervals; the *step* oracle policy selects the next item from the book that has not yet been attempted by the student.

3.7 Evaluation

Our investigation into the Isaac data (Fig. 1) showed that teachers only broadly adhere to the curriculum order in the textbook (as captured by the oracle policies). We therefore hypothesise that HWGen should outperform these policies by capturing the context of each assignment choice.

We are interested to know how closely a homework selection policy will match a human teacher in a given situation. Using the textbook page numbers to give an ordering, the policy is a map $\pi : (s, x, u, a) \rightarrow \hat{y}$ from the student's vectors onto the index of the page of the assignment to be set. The true index, y, giving the choice actually made by the human teacher in this situation, is not known to the policies and is only discovered during testing.

Each $\hat{y}$ is evaluated against the real selection y, by taking $(\hat{y} - y)$. This gives an integer score in the range $[-72, 72]$, where zero is the target. Having values in a range allows partial credit to be assigned. Note that a negative or positive score indicates a policy that lags or leads the human teacher, respectively.

The above steps are repeated for every $((s, x, u, a), y)$ pairing in the test data. The policy is summarised by a standard metric, root mean squared error, which provides interpretable performance measures in terms of number of pages deviated (see Fig. 3).

$$RMSE = \sqrt{\frac{1}{N} \sum_{i=1}^{N} (\hat{y}_i - y_i)^2}$$

Fig. 3. Root mean squared error (RMSE) of a predicted set of target values, $\{\hat{y}\}_1^N$, from their true values, $\{y\}_1^N$.

4 Results

Results from the experiment are presented in Table 1. In Year 1, HWGen with group voting has the highest accuracy in terms of the proportion of matches with teacher selections. The other HWGen policies perform similarly on this measure, as does oracle-step. However, oracle-step is on average slightly ahead of the teacher's point in the textbook (a positive lead/lag value) whereas the HWGen policies all lag slightly behind (negative values). However, all HWGen policies have lower RMSE than oracle-step, indicative of a tighter fit around the gold-standard teacher selections.

Table 1. Accuracy (prop. exact matches), mean lead (+) or lag (-) compared to teacher selections, and root mean squared error for HWGen and baseline homework-setting policies (*significantly different to best baseline, $p < 0.001$; [†]significantly different to HWGen for individuals, $p < 0.001$).

Policy	Year 1 (n = 6111)			Year 2 (n = 572)		
	Acc.	Lead/Lag	RMSE	Acc.	Lead/Lag	RMSE
Human teacher (target)	1	0	0	1	0	0
Random	.015	+17.2	31.4	.009	**+.091**	25.4
Oracle linear	.095	−6.24	20.2	0	+42.3	48.6
Oracle step	.137	**+1.18**	17.4	**.094**	+2.49	12.5
HWGen	.132	−2.37	14.3*	.017	−4.98	10.9*
HWGen group vote	**.141**	−2.81	14.4[†]	.019	−5.06	**10.8**[†]
HWGen group softmax ave.	.133	−2.82	**14.2**[†]	.014	−5.51	11.2

The same is true for Year 2, though in this case the policy with the highest proportion of exact matches is oracle-step, likely because as the choice of outstanding items in the curriculum narrows, a policy which draws from those remaining items in sequence will tend towards the ground truth. Note that HWGen is allowed to select from all textbook pages and so does not have this advantage of a narrowing pool. Being able to re-select items as revision work is a desirable mechanism however. Again, we see that oracle-step stays slightly ahead of the teacher while HWGen lags behind. The random baseline has the best lead/lag score (closest to zero), but this masks the high variance in its choices, shown by an RMSE much higher than HWGen.

In both Year 1 and Year 2, the lowest RMSE (our primary evaluation metric) is one of the HWGen group policies – softmax averaging in the first year, voting in the second year. This is expected, as we train on group level decisions made by the teachers. What is encouraging though is that HWGen for individuals outperforms all baseline policies too, indicating that it will serve individual users of intelligent tutoring systems as well as class groups.

In Fig. 4 we show a density plot of the different spreads of HWGen and the baseline policies in terms of number of pages difference from the teacher's choices ($x = 0$). We see that in Year 1 HWGen closely mimics teacher behaviour, with the tallest peak around zero on the x-axis, while in Year 2 it still clusters around ground-truth but much of its decisions lag behind zero. In contrast, oracle step, while also grouped close to zero, tends to lead the teacher by a few pages.

We find HWGen and oracle step homework differences ($true.homework − predicted.homework$) are statistically separable by paired sample t-tests both for Year 1 ($t = 21.4, df = 6110, p < 2.2e{-}16$) and Year 2 ($t = 14.8, df = 571, p < 2.2e{-}16$). Meanwhile the same is true of HWGen versus HWGen group vote in ($t = 5.79, df = 6110, p = 7.6e{-}09$) and HWGen group softmax average in Year 1 ($t = 5.78, df = 6110, p = 7.94e{-}09$), and of HWGen versus HWGen group softmax average in Year 2 ($t = 4.74, df = 571, p = 2.61e{-}06$). However, HWGen and HWGen group vote were not found to be significantly different in Year 2, therefore we infer the order of performance to be HWGen group vote < HWGen < HWGen softmax averaging in Year 1, and HWGen softmax ave. < HWGen & HWGen group vote in Year 2.

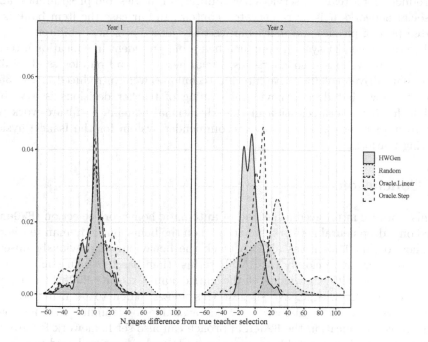

Fig. 4. Density plot of differences for each homework selection by HWGen and baselines: difference from the true selection on the x-axis (n.pages), density on the y-axis (sums to 1 for each curve).

5   Discussion

In its various modes, HWGen was able to more closely approximate the homework-setting actions of teachers in the test dataset, when compared to the other policies tested here. This is despite the fact that both oracle policies have access to expert knowledge through the book ordering of material. In particular, the oracle-step policy makes use of both this ordering data and of user history to inform its selections. In contrast, for the HWGen policy, the neural network knows very little in advance. Instead it learns the ordering and pace of delivery from observations in training. Nonetheless the policy outperforms both oracles. This gives the approach considerable flexibility to be fitted to systems where a canonical ordering is not specified, since it can learn a suitable ordering itself.

In all, these results suggest that even when material is ordered linearly, an automated homework selection system like HWGen could be used to set homework in a more human-like way, both for individual students and for class groups. Thus our work may be viewed as a type of *recommender system* – concerned with selecting items for an individual based on their history in relation to others' histories [7, 15, 29] in a manner which will be familiar to users of many online services [9, 13, 19]. For online retailers the item bank in need of filtering is a set of products, for streaming services it is a library of movies and programmes, and for social networks it is user-generated content. In our case the item bank is a curated pool of physics and mathematics tasks aimed at school-children.

The recommender system approach has some precedent in educational technology. Early systems used heuristics, social networks and ontologies [3, 26, 32] before data-driven collaborative filtering techniques were introduced [21, 31, 35]. Here we show that deep behavioural cloning of teacher decisions is a viable method for homework selection in the educational domain. In future work we plan to implement HWGen as a live recommender system for the Isaac Physics tutoring platform.

6   Conclusion

In this work we introduced a method of automatic homework selection (HWGen), based on a deep neural network and trained on the behaviour of human teachers. We showed that HWGen was able to track the behaviour of previously unseen human teachers more closely (in RMSE terms) than baseline heuristic policies, including those with knowledge of the curriculum – despite HWGen having no access to such knowledge. We suggest this allows HWGen to be fitted to pre-existing systems, where historical data is available for training. Furthermore, with simple modification the HWGen approach can also set homework for groups of students, making it suitable for private study and classroom-based use.

It remains to be seen whether HWGen, either in individual or group mode, leads to improved learning for students. We will seek to address this question in future work with reward-based models.

Acknowledgements. This paper reports on research supported by Cambridge Assessment, University of Cambridge. We thank members of the Isaac Physics team, our colleagues in the ALTA Institute, and the three anonymous reviewers for their valuable feedback.

References

1. Abadi, M., et al.: TensorFlow: Large-scale machine learning on heterogeneous systems (2015). www.tensorflow.org
2. Aboalela, R., Khan, J.: Model of learning assessment to measure student learning: inferring of concept state of cognitive skill level in concept space. In: 2016 3rd International Conference on Soft Computing & Machine Intelligence (ISCMI), pp. 189–195. IEEE (2016)
3. Anderson, M., et al.: RACOFI: a rule-applying collaborative filtering system. In Proceedings of IEEE/WIC COLA 2003 (2003)

4. Ba-Omar, H., Petrounias, I., Anwar, F.: A framework for using web usage mining to personalise e-learning. In: Proceedings of the 7th IEEE International Conference on Advanced Learning Technologies (2007)
5. Bloom, B.: The 2 sigma problem: the search for methods of group instruction as effective as one-to-one tutoring. Educ. Res. **13**, 4–16 (1984)
6. Bratko, I., Urbančič, T., Sammut, C.: Behavioural cloning: phenomena, results and problems. IFAC Proc. Volumes **28**(21), 143–149 (1995)
7. Breese, J., Heckerman, D., Kadie, C.: Empirical analysis of predictive algorithms for collaborative filtering. In: Proceedings of the 14th Conference on Uncertainty in Artificial Intelligence (1998)
8. Brinton, C., Rill, R., Ha, S., Chiang, M., Smith, R., Ju, W.: Individualization for education at scale: MIIC design and preliminary evaluation. IEEE Trans. Learn. Technol. **8**, 136–148 (2015)
9. Burke, R.: Hybrid recommender systems: survey and experiments. User Model. User-Adap. Inter. **12**, 331–370 (2002)
10. Chi, M., VanLehn, K.: Porting an intelligent tutoring system across domains. Front. Artif. Intell. Appl. **158**, 551 (2007)
11. Chollet, F., et al.: Keras (2015). https://keras.io
12. Craig, S.D., et al.: Learning with ALEKS: the impact of students' attendance in a mathematics after-school program. In: Biswas, G., Bull, S., Kay, J., Mitrovic, A. (eds.) AIED 2011. LNCS (LNAI), vol. 6738, pp. 435–437. Springer, Heidelberg (2011). https://doi.org/10.1007/978-3-642-21869-9_61
13. Deshpande, M., Karypis, G.: Item-based top-n recommendation algorithms. ACM Trans. Inf. Syst. **22**, 143–177 (2004)
14. Dettmers, S., Trautwein, U., Lüdtke, O., Kunter, M., Baumert, J.: Homework works if homework quality is high: Using multilevel modeling to predict the development of achievement in mathematics. J. Educ. Psychol. **102**(2), 467 (2010)
15. Dron, J., Mitchell, R., Siviter, P., Boyne, C.: CoFIND – an experiment in n-dimensional collaborative filtering. J. Netw. Comput. Appl. **23**, 131–142 (2000)
16. Khan, F., Mutlu, B., Zhu, X.: How do humans teach: on curriculum learning and teaching dimension. In: Advances in Neural Information Processing Systems (NeurIPS), pp. 1449–1457 (2011)
17. Kingma, D.P., Ba, J.: Adam: a method for stochastic optimization. arXiv preprint arXiv:1412.6980 (2014)
18. Lindsey, R.V., Shroyer, J.D., Pashler, H., Mozer, M.C.: Improving students' long-term knowledge retention through personalized review. Psychol. Sci. **25**(3), 639–647 (2014)
19. Munro, A., Höök, K., Benyon, D.: Personal and Social Navigation of Information Space. Springer, London (1999). https://doi.org/10.1007/978-1-4471-0837-5
20. Murray, T.: Authoring intelligent tutoring systems: an analysis of the state of the art. Int. J. Artif. Intell. Educ. (IJAIED) **10**, 98–129 (1999)
21. Nadolski, R., et al.: Simulating lightweight personalised recommender systems in learning networks: a case for pedagogy-oriented and rating based hybrid recommendation strategies. J. Artif. Soc. Soc. Simul. **12** (2009)
22. Novak, J.D., Cañas, A.J.: The theory underlying concept maps and how to construct and use them. Technical Report, Institute for Human and Machine Cognition (2008)
23. Pardos, Z.A., Tang, S., Davis, D., Le, C.V.: Enabling real-time adaptivity in MOOCs with a personalized next-step recommendation framework. In: Proceedings of the Fourth (2017) ACM Conference on Learning@ Scale, pp. 23–32. ACM (2017)

24. Pedregosa, F., et al.: Scikit-learn: machine learning in Python. J. Mach. Learn. Res. **12**, 2825–2830 (2011)
25. Pomerleau, D.A.: Alvinn: an autonomous land vehicle in a neural network. In: Advances in Neural Information Processing Systems, pp. 305–313 (1989)
26. Recker, M., Walker, A., Lawless, K.: What do you recommend? implementation and analyses of collaborative filtering of web resources for education. Instr. Sci. **31**, 229–316 (2003)
27. Rosen, Y., et al.: The effects of adaptive learning in a massive open online course on learners' skill development. In: Proceedings of Learning @ Scale (2018)
28. Sampson, D., Karagiannidis, C.: Personalised learning: educational, technological and standardisation perspective. Interact. Educ. Multimedia **4**, 24–39 (2002)
29. Sarwar, B., Karypis, G., Konstan, J., Riedl, J.: Item-based collaborative filtering recommendation algorithms. In: WWW 2001 (2001)
30. Sharp, C.: Should schools set homework. Nat. Found. Educ. Res. **27**, 1–4 (2002)
31. Tang, T.Y., McCalla, G.: Smart recommendation for an evolving e-learning system. Int. J. E-learn. **4**, 105–129 (2005)
32. Tarus, J., Niu, Z., Mustafa, G.: Knowledge-based recommendation: a review of ontology-based recommender systems for e-learning. Artif. Intell. Rev. **50**, 21–48 (2018)
33. VanLehn, K.: The behavior of tutoring systems. Int. J. Artif. Intell. Educ. (IJAIED) **16**, 227–265 (2006)
34. VanLehn, K., et al.: The Andes physics tutoring system: five years of evaluations. In: Proceedings of the 12th International Conference on Artificial Intelligence in Education, pp. 678–685 (2005)
35. Wang, P.-Y., Yang, H.-C.: Using collaborative filtering to support college students' use of online forum for english learning. Comput. Educ. **59**, 628–637 (2012)

Integrating Students' Behavioral Signals and Academic Profiles in Early Warning System

SungJin Nam[(✉)] and Perry Samson

University of Michigan, Ann Arbor, USA
{sjnam,samson}@umich.edu

Abstract. In this paper, we investigated how students' behavioral signals and incoming profiles can be integrated to describe and predict student success in a higher education's STEM course. The results include three major findings. First, we found behavioral signals like the number of correct responses to in-class questions, the number of confusing slides, and the number of viewed slides and videos are stable predictors of student success across different periods of a semester. Second, from the mixed-effect modeling results, we could identify significant gender gaps between mid-level incoming GPA student groups. We also showed some possible course advising scenarios based on the interaction between student behaviors and incoming profile factors. Third, using both behavioral signals and incoming profiles, our weekly forecast model on student success achieved a 72% prediction accuracy. We believe these findings can set the stage for subsequent early warning system studies that use different types of student data. Further investigations on the causal relationships for suggested results and developing other novel predictive features for student success would be beneficial for designing a better early warning system.

Keywords: Early warning system · Behavioral logs ·
Incoming profile · Grade prediction · Mixed effect model

1 Introduction

Several Early Warning Systems (EWSs) have been developed that harness the predictive power of Learning Management System (LMS) and Student Information System (SIS) data to identify at-risk students and allow for more timely pedagogical interventions [1,4,15,16]. For example, data on student online activity in a web-based LMS may provide an early indicator of students' academic performance [28]. Other studies also found that there is a strong relationship between LMS usage patterns and student exam scores [7,10].

Recent studies suggest that EWSs can benefit from using relevant signals from LMS data, such as access to course resources [25], the usage patterns of

© Springer Nature Switzerland AG 2019
S. Isotani et al. (Eds.): AIED 2019, LNAI 11625, pp. 345–357, 2019.
https://doi.org/10.1007/978-3-030-23204-7_29

a digital coaching application [6], and students' incoming prior academic performance [6, 25]. Similarly, studies like [8, 11] showed applications for identifying at-risk students for dropping out from higher education institutes, using student activity data collected from a software lecture environment [11] or incorporating the incoming data with semester-wise enrollment information [8]. Although many studies claimed that students' behavioral signals and incoming academic profile are important predictors of academic outcome, not many of them addressed how these different types of signals are related to each other.

In this paper, we explored how students' behavioral signals and incoming profiles can be integrated to describe and predict student success in a higher education STEM course. First, we identified the list of significant behavioral variables collected from a lecture software to describe students' academic success from a course. Second, we compared the different likelihood of succeed in a course between student groups. We also investigated how behavioral patterns are different in each student group. Third, we compared the predictive performance of statistical models with weekly accumulated training data, including each model's *day-one* performance and how the performance changes by adding students' weekly behavioral data. The findings of this paper can set the ground for integrating different types of behavioral signals and incoming academic profiles for building an EWS. Detailed investigation of these signals can also provide data-driven evidence for developing customized course-taking strategies.

2   Related Works

2.1   Predicting Student Performance

In many studies, behavioral signals are significant predictors of the student's academic outcome [2, 22]. In a higher education context, studies used behavioral data to predict the student's performance in a course. Studies like [27, 28, 30] suggested how different behavioral signals observed in LMS, such as frequency or length of interactions, can be significant predictors of student success in higher education courses. Studies from Waddington et al. [25, 26] specifically focused on using different types of course resources accesses to predict student success in entry-level STEM courses. They showed that course resources related to exam preparations and lecture materials are more significantly related to the course grade than other resource types, such as course information or assignments. Real-time stream data from courses can illustrate more details of student behaviors during learning. Studies like [5] showed behavioral signals, such as the number of interactions with video, exercise, or assignment, can be useful predictors of student engagement in MOOC courses. Other studies on MOOC suggested the use of implicit signals, such as click patterns or use of language in the forum group [23, 29], and interaction with video lectures or timely completion of assignments [20, 21] to predict different cognitive states of students.

Behavioral signals contain rich information on how students interact with the learning materials. However, other factors, such as students' incoming academic profile [17] and temporal conditions [18], can add more contextual information about students' learning. Many studies included non-behavioral factors, such as demographic information, in their models to control the effect of behavioral variables with student learning [6,25]. The non-behavioral factors also can be used to capture more complex patterns in student behaviors. Studies showed that interaction with peers [9] or intervention from the course [13] need to be designed differently by students' cultural background or the learning style that they are accustomed to [12]. To provide customized advice for each student's learning, identifying interactions between non-behavioral factors and behavioral variables would be important. In our study, we will investigate how different types of information, such as students' behavioral interactions with the lecture software and their incoming profile factors, can be used to describe and predict student success in a higher education course.

2.2 Early Warning Systems

EWS is a computerized system that focuses on identifying at-risk students early and providing data-driven evidence for developing strategies that can maximize students' academic success [1,15]. Many existing studies on early warning systems are designed to predict the student's retention in a course. Studies like [11,24] evaluated the quality of signals from earlier weeks and model's performance by making a prediction on the later weeks. More broadly, [8] modeled students' dropout from an institution in a semester level. The authors used various types of data. They used semester-level variables like GPA, credit hour, or enrolled year, along with incoming academic profiles like demographic factors and entrance exam information. Earlier identification of at-risk students can help institutions to improve students' course retention [11] and degree completion rates [8]. However, it can also benefit high achieving students to keep their success in a course. In our study, we will suggest an EWS that can be used in higher education institutions where the retention rate can be a lesser problem, and focus on the earlier prediction of students' *success* in a course.

2.3 Overview of the Current Study

Based on previous studies, we formulated the following research questions:

RQ1: Can we identify significant behavioral predictors of student success in a course across different weekly ranges? Answering the first research question will examine the significance and stability of behavioral predictors collected from a lecture software for describing and predicting student success in a course. Selected behavioral variables will also be used to answer later research questions.

RQ2: What is the benefit of including incoming profile factors in models to describe student success? The second research question will compare the model developed from RQ1 with a mixed-effect model, which can address variances between student groups for describing student success in a course.

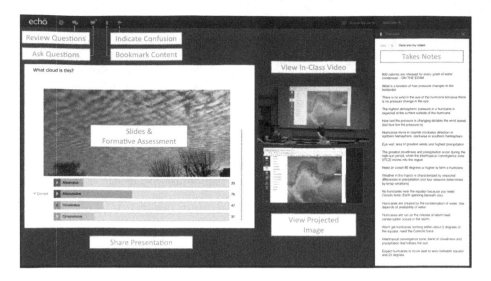

Fig. 1. A screenshot of Active Learning Platform. Students could attend or review the lecture by seeing the lecture video and slides simultaneously. Also, they could take notes or ask questions related to the lecture.

RQ3: Can we identify distinctive behavioral patterns between student groups in terms of success in a course? The third research question will aim to expand the model from RQ2 by adding random slope structures. The results will identify if there are significantly different behavioral patterns exist between student groups.

RQ4: Can we predict student success at a weekly level by using the findings from RQ1-3? The last research question will investigate the predictive power of models suggested in RQ1-3, especially in a weekly-level early warning scenario.

3   Methods

3.1   Data Source

Behavioral Signals. For this study, we collected students' behavioral signals from lecture software. Data were collected across seven semesters, from the year 2015 to 2018. Students were enrolled in an entry-level STEM course, *CLIM 999*. *CLIM 999* was a survey course, and part of 'science distribution' requirement at the university. The course covered the physics of extreme weather events and potential relationships with a changing climate.

   CLIM 999 was a blended learning course [30] using *Echo 360*'s Active Learning Platform (ALP) software[1]. By using ALP, students could access the physical classroom session through a web browser. It included video recordings and

---

[1] One of the authors, Dr. Samson, is a consultant to Echo360 Inc. and uses the Echo360 Active Learning Platform in his class.

Table 1. Distribution of students' incoming GPA and gender profiles. None represents the students without incoming GPA information (e.g., the first semester Freshmen or transfer students).

Incoming GPA	Female	Male	All
None	50 (3.9%)	83 (6.5%)	133 (10.4%)
(0.0,2.5]	21 (1.6%)	59 (4.6%)	80 (6.2%)
(2.5,3.0]	75 (5.8%)	157 (12.3%)	232 (18.1%)
(3.0,3.5]	180 (14.1%)	279 (21.7%)	459 (35.8%)
(3.5,4.0]	153 (11.9%)	226 (17.6%)	379 (29.5%)
All	479 (37.3%)	804 (62.7%)	1283 (100.0%)

slides of the lecture. It also provided a virtual learning environment, where students could answer in-class activity questions, take notes, ask questions to other students and instructors, and mark where they felt confused from the lecture (Fig. 1). ALP recorded students' interactions with the system. In this study, we used frequencies of different behavioral signals as predictors (Table 2). Following the previous studies [25, 26], we normalized recorded frequencies from ALP by using a percentile rank method for each semester. We expected this would normalize the outlier data-points, and make behavioral signals collected from multiple semesters easily comparable.

Academic Profiles. For the study, we also collected students' incoming profile factors to represent their different academic profiles. To do this, we used a combination of incoming GPA and gender information, retrieved from the university's data-warehouse. To help with better convergence of mixed-effect modeling results, we filtered out data from 63 students since they did not have gender information. As results, we used data from 1283 students for the analysis. These students' profiles were represented by the combination of 5-level incoming GPA labels and 2-level gender labels (Table 1).

Student Success. We labeled the student's performance as *successful* if she achieved 80% or better on the average of three exams from the course. As results, we labeled 53.4% (685) of students as *successful*[2].

The threshold for *successful* label was decided based on the nature of our dataset and previous studies. Our data were collected from students who enrolled in the R1 university of the U.S.[3]. Students in these institutions may aim higher than just passing the course. For example, in our dataset, only 4.53% of students

---

[2] The distribution of grades were: 178 students with A (90 or higher, 13.87%), 507 with B (80–90, 39.52%), 373 with C (70–80, 29.07%), 167 with D (60–70, 13.02%), and 58 with F (60 or lower, 4.52%).

[3] We follow the definition of R1 university in here: https://en.wikipedia.org/wiki/List_of_research_universities_in_the_United_States.

Table 2. List of behavioral variables (**Beh.**) and student profile factors (**Aca.**). Numbers in parentheses represent the number of times each behavioral variable was selected from the step-wise process.

Type	Name	Description
Beh.	`ActivitiesParticipation(9)`	Activities they submitted an answer to
	`ActivitiesCorrect(10)`	Students get score for correct answer on quiz
	`Attendance(0)`	Students entered the SW interface during classtime
	`NotesCount(0)`	Notes word count
	`NotesInteraction(0)`	Number of times user interacted/edited notes
	`QnA(3)`	Each time they created a question or response
	`SlideDeckView(10)`	Viewed over 5% of a presentation
	`SlideConfused(10)`	Marked confusing in slide
	`VideoView(10)`	Viewed over 5% of a video
	`VideoConfused(0)`	Marked a scene confusing in video
Aca.	`GPA`	Students' average GPA from previous semesters
	`Gender`	Binary label (female or male) for students' gender.

achieved the 60 or less average score (F). Previous works also specified that academic advisers in these institutions consider getting B (i.e., 80% of total grade) or better is an important goal for their students, to pursue entering graduate schools or getting a job easily in the STEM field [17,25,26]. We realize that the results of our study may not be generalized for preventing students from failing the course or degree [8,11]. However, we believe our findings can provide unique insights on designing an EWS for high achieving students to keep succeeding in their institution, and sets the stage for subsequent work that includes the analysis on other types of higher education courses.

3.2   Building Models

Selecting Behavioral Predictors. To answer the first research question (RQ1), we used a cross-validation method to identify more meaningful behavioral predictors. In each cross-validation fold, a generalized linear regression model (GLM) model was initially fitted using all 10 behavioral predictors from the data excluding the held-outs. Then the backward and forward step-wise process was applied to the model, using Akaike Information Criterion (AIC) with R's [19] `step` function. After the cross-validation process is finished, we counted the number of times that each variable was selected in individual folds (Table 2).

Descriptive and Predictive Models. Using selected behavioral predictors as fixed-effect predictors, we firstly built descriptive models using the whole dataset without cross-validation. The GLM descriptive model expected to show how the entire dataset can be described solely using behavioral signals (RQ1). We also built generalized linear mixed-effect regression models (GLMER) to address variances across incoming profile factors, incoming GPA and gender (RQ2), and different behavioral patterns between these student groups (RQ3). We used `glmer` function with `nAGQ=0` argument from `lme4` [3] to fit GLMER models. Like the `GLM` model, GLMER descriptive models were also fitted with the whole dataset.

We also developed prediction models that can provide a weekly forecast of student success in the course (RQ4). The structure of predictive models followed GLM and GLMER models that used for the previous research questions (RQ1-3). The performance of predictive models was evaluated with held-out sets from 10-fold cross-validation. Average scores of evaluation metrics, such as prediction accuracy, area under the curve scores of receiver operation characteristics (ROCAUC) and precision-recall curve (PRAUC), were used to compare the prediction performance of each model. The evaluation included how early the model can effectively predict student success in the course. For example, testing models' performance with `Week 0` subset was considered as a *day-one* scenario, which allowed models to predict student success without any behavioral signals. Testing with `Week 1-5` subset used behavioral signals accumulated until the fifth week of the semester. Similarly, `Week 1-15` subset included all students' behavioral data recorded across the entire semester.

4   Results

4.1   Selecting Behavioral Variables

In Table 2, we noted how many times each fixed-effect variable were selected from the variable selection process. It shows that behavioral variables like `ActivitiesCorrect`, `SlideDeckView`, `SlideConfused`, and `VideoView` were significant predictors across all cross-validation folds. We call these variables as `Unanimous` set. Another variable set is `MoreOnce` set, which contains additional variables of `ActivitiesParticipation` and `QnA` that were selected more than once from the variable selection process.

From the ANOVA analysis using residual deviance scores with the GLM descriptive model setting, we found that the model trained with `MoreOnce` variable set (1570.8) show similar performance from the model using all 10 behavioral variables (1568.8, $p = 0.738$). This model also showed significantly better performance than the model using `Unanimous` variable set (1577.5, $p < 0.05$). Thus, we used `MoreOnce` variable set for further analyses, keeping the model structure simpler but minimizing the performance sacrifice.

To answer the first research question (RQ1), we built descriptive GLM and GLMER models, using `MoreOnce` variable set, and see how coefficients change across different week ranges. Figure 2 shows that coefficients for variables in `MoreOnce` set were most significant in different week and model conditions, except `QnA` and `ActivitiesParticipation` variables.

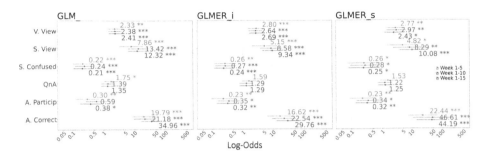

Fig. 2. Coefficients for behavioral variables were relatively consistent across different models and weekly accumulation ranges. Error bars indicate the 95% confidence interval of coefficient estimates. (* : $p < 0.05$, ** : $p < 0.01$, *** : $p < 0.001$)

4.2   Integrating Student Profile Information

To answer RQ2 and RQ3, we built GLMER models that include additional random intercepts from incoming profile factors (`GLMER_i`) and random slopes from behavioral variables (`GLMER_s`). From each GLMER model, we expected to see if the model captures different starting points (RQ2) and some behavioral pattern differences (RQ3) between different student groups.

Figure 3 shows variances between different student groups (random intercepts) and behavioral patterns (random slopes) from the `GLMER_s` model[4]. First, in the most left panel for random intercepts, we could find significant gender differences among mid-level incoming GPA groups. While there were positive relationships between students' success in incoming GPA levels, the estimated random intercept for female students in the `(3.0,3.5]` incoming GPA group was almost as similar as the intercept of lower incoming GPA male students (`(2.5,3.0]`). This pattern was also observed between female students in the `(2.5,3.0]` incoming GPA group and male students in the `(0.0,2.5]` group. These gender gaps were not observed in the highest (`(3.5,4.0]`) or the lowest (`(0.0,2.5]`) incoming GPA groups, or the first semester students without incoming GPA information (`none`).

Second, other panels in Fig. 3 provide some potential customized advising scenarios. For example, we could see some significant random slope coefficients. For students with the highest incoming GPA group (`(3.5,4.0]`), providing correct

---

[4] The results for random intercepts were similar between `GLMER_i` and `GLMER_s` models.

answers to in-class activity questions (`ActivitiesCorrect`) and more reviewing of the slides (`SlideDeckView`) were less important for predicting their success in the course. Additionally, in the highest incoming GPA group of female students, marking confusion (`SlideDeckConfused`) was more significantly related to their success. These results show that different student groups may need different study strategies to succeed in a course.

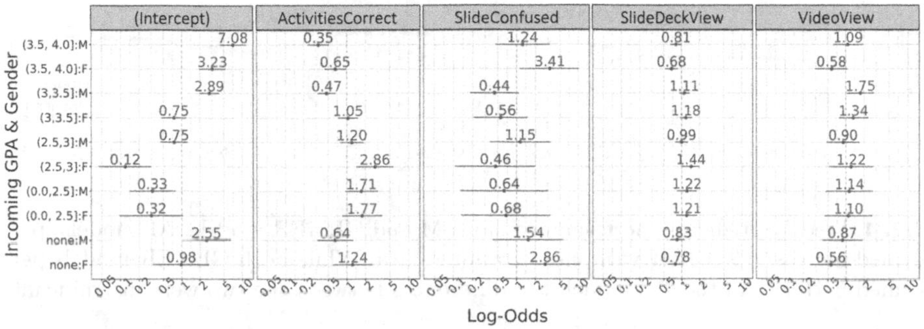

Fig. 3. Random effect results from the `GLMER_s` model, using Week 1–15 accumulated data. Significant gender gaps in random intercepts were observed in (3.0,3.5] and (2.5,3.0] incoming GPA groups. Also, some random slope coefficients for `ActivitiesCorrect`, `SlideDeckView`, and `SlideDeckConfused` showed significantly different patterns by incoming profiles.

4.3   Predictive Modeling: Weekly Performance

Lastly, we examined the prediction performance of GLM and GLMER models (RQ4)[5]. Figure 4 shows that all models reach better performances when training data is accumulated through more weeks. All models quickly achieved meaningful ROCAUC scores of grater than 0.5 (maximum scores were 0.724 (`GLM`), 0.791 (`GLMER_i`), and 0.789 (`GLMER_s`)). PRAUC results were similar to ROCAUC results. Across the semester weeks, both GLMER models using random effect structures achieved the maximum accuracy near 72%. Both models performed consistently better than the `GLM` model, which used behavioral predictors only (66.3% maximum accuracy; 8.7% relative worse than GLMER models).

To measure the *day-one* performance in `Week` 0, models predicted the majority label (`GLM` model) or relied on the incoming profile information only (GLMER models). For the `GLM` model, the average accuracy was 53.33%, which is equivalent to the likelihood of students achieving an average of 80 or better in exams. However, GLMER models could predict students' success in the course with 70% accuracy using incoming profiles alone (Fig. 4). The `GLM` model showed maximum accuracy in week 11 and 12. For both GLMER models, the maximum accuracy scores were achieved in week 15. However, adding behavioral predictors to GLMER models did not provide meaningful gains to their predictive performances.

---

[5] Detailed prediction results can be found at http://bit.ly/nam-EWSpreds.

We also examined a prediction performance for models with a single important behavioral variable (`ActivitiesCorrect`) from Sect. 4.2. A GLM model only achieved 59.1% accuracy and 0.67 ROCAUC scores. Both GLMER models showed around 70% accuracy and 0.77 ROCAUC scores, which are not so different from the *day-one* performance. All these best performing scores were observed in week 12 or 13.

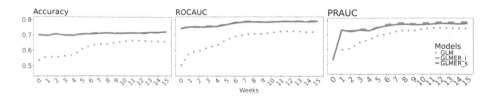

Fig. 4. Weekly prediction performance of GLM and GLMER models. All models' performed increasingly better with more training data. While both GLMER models performed better than the GLM model, the differences between the two were insignificant.

5   Conclusion and Discussion

In this paper, we identified significant behavioral predictors of student success from a lecture software. We also found gender gaps in the mid-level incoming GPA groups, and differences in behavioral patterns between student groups. By combining behavioral signals and incoming profile factors, our early warning prediction model achieved up to a 72% accuracy for predicting student success in an entry-level STEM course. We believe these findings can provide more contexts to student behaviors in higher education settings. Practically, it can give data-driven evidence for developing more personalized course-taking strategies, which would be helpful for academic advisers, instructors, and students.

Based on our findings, we have a few discussion points for future studies. First, like other studies in entry-level STEM courses [14], we could observe the gender gap between students. As specified in Sect. 4.2, it was surprising that the gender gap existed even among students that share similar GPAs from the previous semesters. However, in this study, we were not able to provide further explanations of what are the relevant causes of this gap. Including additional demographic or enrollment information to the analysis, or comparing student behaviors between multiple STEM courses may provide deeper contexts for our current findings. Second, further investigation on why different behavioral patterns were observed between student groups would be interesting. For the results in Sect. 4.2, we only suspect that higher incoming GPA students might have explored more on incorrect answers with in-class questions, or spent lesser time on lecture slides because they tend to have their own study notes. Using qualitative methodologies, such as interviewing students or observing them during class, may provide more detailed insights on these behavioral differences between students. Lastly, developing more effective predictive features would improve the models' prediction performances. As we saw in Sect. 4.3, adding behavioral signals to GLMER models provided only marginal increases to the prediction performances from Week 0. Adding multiple behavioral variables to the model was more effective than adding a single important behavioral variable. However, since the improvement was not very significant, we may need to explore other exogenous features that can provide additional information to the model that are not addressed with current incoming profile factors. Developing non-frequency based variables, such as a semantic representation of student notes or predicted student engagement states during the lecture, can be also helpful options for developing more sophisticated and accurate predictive models.

References

1. Aguilar, S., Lonn, S., Teasley, S.D.: Perceptions and use of an early warning system during a higher education transition program. In: Proceedings of the Fourth International Conference on Learning Analytics and Knowledge, pp. 113–117. ACM (2014)
2. Baker, R.S., Corbett, A.T., Koedinger, K.R., Wagner, A.Z.: Off-task behavior in the cognitive tutor classroom: when students game the system. In: Proceedings of the SIGCHI Conference on Human Factors in Computing Systems, pp. 383–390. ACM (2004)
3. Bates, D., Máchler, M., Bolker, B., Walker, S.: Fitting linear mixed-effects models using lme4. J. Stat. Softwa. **67**(1), 1–48 (2015). https://doi.org/10.18637/jss.v067.i01
4. Beck, H.P., Davidson, W.D.: Establishing an early warning system: predicting low grades in college students from survey of academic orientations scores. Res. High. Educ. **42**(6), 709–723 (2001)
5. Bote-Lorenzo, M.L., Gómez-Sánchez, E.: Predicting the decrease of engagement indicators in a MOOC. In: Proceedings of the Seventh International Learning Analytics and Knowledge Conference, pp. 143–147. ACM (2017)
6. Brown, M.G., DeMonbrun, R.M., Teasley, S.D.: Don't call it a comeback: academic recovery and the timing of educational technology adoption. In: Proceedings of the Seventh International Learning Analytics and Knowledge Conference, pp. 489–493. ACM (2017)
7. Campbell, J.P., DeBlois, P.B., Oblinger, D.G.: Academic analytics: a new tool for a new era. Educase Rev. **42**(4), 40 (2007)
8. Chen, Y., Johri, A., Rangwala, H.: Running out of stem: a comparative study across stem majors of college students at-risk of dropping out early. In: Proceedings of the Eighth International Learning Analytics and Knowledge Conference, pp. 270–279. ACM (2018)
9. Davis, D., Jivet, I., Kizilcec, R.F., Chen, G., Hauff, C., Houben, G.J.: Follow the successful crowd: raising mooc completion rates through social comparison at scale. In: Proceedings of the Seventh International Learning Analytics and Knowledge Conference, pp. 454–463. ACM (2017)
10. Goldstein, P.J., Katz, R.N.: Academic analytics: the uses of management information and technology in higher education, vol. 8. Educause (2005)
11. Hlosta, M., Zdrahal, Z., Zendulka, J.: Ouroboros: early identification of at-risk students without models based on legacy data. In: Proceedings of the Seventh International Learning Analytics and Knowledge Conference, pp. 6–15. ACM (2017)

12. Hoffmann, B., Ritchie, D.: Using multimedia to overcome the problems with problem based learning. Instr. Sci. **25**(2), 97–115 (1997)
13. Kizilcec, R.F., Cohen, G.L.: Eight-minute self-regulation intervention raises educational attainment at scale in individualist but not collectivist cultures. Proc. Natl. Acad. Sci. U.S.A. **114**(17), 4348–4353 (2017)
14. Koester, B.P., Grom, G., McKay, T.A.: Patterns of gendered performance difference in introductory stem courses. arXiv preprint arXiv:1608.07565 (2016)
15. Krumm, A.E., Waddington, R.J., Teasley, S.D., Lonn, S.: A learning management system-based early warning system for academic advising in undergraduate engineering. In: Larusson, J.A., White, B. (eds.) Learn. Anal., pp. 103–119. Springer, New York (2014). https://doi.org/10.1007/978-1-4614-3305-7_6
16. Macfadyen, L.P., Dawson, S.: Mining LMS data to develop an "early warning system" for educators: a proof of concept. Comput. Educ. **54**(2), 588–599 (2010)
17. Nam, S., Lonn, S., Brown, T., Davis, C.S., Koch, D.: Customized course advising: Investigating engineering student success with incoming profiles and patterns of concurrent course enrollment. In: Proceedings of the Fourth International Conference on Learning Analytics and Knowledge, pp. 16–25. ACM (2014)
18. Park, J., Yu, R., Rodriguez, F., Baker, R., Smyth, P., Warschauer, M.: Understanding student procrastination via mixture models. In: Proceedings of the International Conference on Educational Data Mining (2018)
19. R Core Team: R: A Language and Environment for Statistical Computing. R Foundation for Statistical Computing, Vienna, Austria (2018). https://www.R-project.org/
20. Ramesh, A., Goldwasser, D., Huang, B., Daume III, H., Getoor, L.: Learning latent engagement patterns of students in online courses. In: Twenty-Eighth AAAI Conference on Artificial Intelligence (2014)
21. Sinha, T., Li, N., Jermann, P., Dillenbourg, P.: Capturing "attrition intensifying" structural traits from didactic interaction sequences of MOOC learners. In: Proceedings of the EMNLP 2014 Workshop on Modeling Large Scale Social Interaction in Massively Open Online Courses, pp. 42–49 (2014)
22. Skinner, E.A., Kindermann, T.A., Furrer, C.J.: A motivational perspective on engagement and disaffection: conceptualization and assessment of children's behavioral and emotional participation in academic activities in the classroom. Educ. Psychol. Meas. **69**(3), 493–525 (2009)
23. Tomkins, S., Ramesh, A., Getoor, L.: Predicting post-test performance from online student behavior: a high school mooc case study. In: Proceedings of the International Conference on Educational Data Mining (2016)
24. Veeramachaneni, K., O'Reilly, U.M., Taylor, C.: Towards feature engineering at scale for data from massive open online courses. arXiv preprint arXiv:1407.5238 (2014)
25. Waddington, R.J., Nam, S., Lonn, S., Teasley, S.D.: Improving early warning systems with categorized course resource usage. J. Learn. Anal. **3**(3), 263–290 (2016)
26. Waddington, R.J., Nam, S.: Practice exams make perfect: incorporating course resource use into an early warning system. In: Proceedings of the Fourth International Learning Analytics And Knowledge Conference, pp. 188–192. ACM (2014)
27. Wang, A.Y., Newlin, M.H.: Characteristics of students who enroll and succeed in psychology web-based classes. J. Educ. Psychol. **92**(1), 137 (2000)
28. Wang, A.Y., Newlin, M.H.: Predictors of performance in the virtual classroom: identifying and helping at-risk cyber-students. J. (Technol. Horiz. Educ.) **29**(10), 21 (2002)

29. Yang, D., Kraut, R., Rose, C.P.: Exploring the effect of student confusion in massive open online courses. J. Educ. Data Min. **8**(1), 52–83 (2016)
30. Zacharis, N.Z.: A multivariate approach to predicting student outcomes in web-enabled blended learning courses. Internet High. Educ. **27**, 44–53 (2015)

Predicting Multi-document Comprehension: Cohesion Network Analysis

Bogdan Nicula[1], Cecile A. Perret[3], Mihai Dascalu[1,2(✉)], and Danielle S. McNamara[3]

[1] University Politehnica of Bucharest,
313 Splaiul Independentei, 060042 Bucharest, Romania
bogdan.nicula@cti.pub.ro, mihai.dascalu@cs.pub.ro
[2] Cognos Business Consulting S.R.L.,
32 Bd. Regina Maria, Bucharest, Romania
[3] Department of Psychology, Arizona State University,
PO Box 871104, Tempe, AZ 85287, USA
{cperret,dsmcnama}@asu.edu

Abstract. Theories of discourse comprehension assume that understanding is a process of making connections between new information (e.g., in a text) and prior knowledge, and that the quality of comprehension is a function of the coherence of the mental representation. When readers are exposed to multiple sources of information, they must make connections both within and between the texts. One challenge is how to represent this coherence and in turn how to predict readers' levels of comprehension. In this study, we represent coherence using Cohesion Network Analysis (CNA) in which we model a global cohesion graph that semantically links reference texts to different student verbal productions. Our aim is to create an automated model of comprehension prediction based on features extracted from the CNA graph. We examine the cohesion links between the four texts read by 146 students and their (a) self-explanations generated on target sentences and (b) responses to open-ended questions. We analyze the degree to which features derived from the cohesive links from the extended CNA graph are predictive of students' comprehension scores (on a [0 to 12] scale) using either (a) students' self-explanations, (b) responses to comprehension questions, or (c) both. We compared the use of Linear Regression, Extra Trees Regressor, Support Vector Regression, and Multi-Layer Perceptron. Our best model used Linear Regression, obtaining a 1.29 mean absolute error when predicting comprehension scores using both sources of verbal responses (i.e., self-explanations and question answers).

Keywords: Multi-document comprehension and integration ·
Natural Language Processing · Cohesion Network Analysis ·
Comprehension modeling · Machine learning

S. Isotani et al. (Eds.): AIED 2019, LNAI 11625, pp. 358–369, 2019.
https://doi.org/10.1007/978-3-030-23204-7_30

1 Introduction

Comprehension is challenging. The process involves understanding the words and sentences within the text (or discourse), connecting the ideas within the text, and linking the ideas to prior knowledge, in order to generate a coherent mental representation of the content. Comprehension processes are further challenged when faced with multiple sources of information. Multiple document comprehension adds on the need to make connections both within and between texts to generate a coherent mental representation of the disparate sources of information. We are faced with these challenges on a regular basis, when reading separate documents, papers, news, blogs, emails, and so on.

One question is how to simulate the coherence of a reader's mental representation and in turn, the extent to which that coherence predicts comprehension. In this study, we examine that extent to which the semantic connections (i.e., coherence) between a text and a reader's constructed responses while reading and after reading multiple documents predict comprehension. Similar modeling and linguistic techniques have been applied in the context of single text comprehension [1, 2]. Techniques evaluating reading comprehension for multiple document scenarios were previously researched by Hastings, Hughes, Magliano, Goldman and Lawless [3]; however, there is a dearth of research attempting to model how individuals integrate information across texts to form a coherent representation of information from separate sources. Cohesion Network Analysis (CNA) [2] is a technique that combines Social Network Analysis (SNA) [4] and Natural Language Processing (NLP) [5] techniques to identify semantic similarities between various sources of discourse and the levels of semantic cohesion within and between networks. This paper applies CNA to multiple document discourse to predict comprehension as well as to better understand the underlying cognitive processes of integrating information from multiple texts. Students' self-explanations and their responses to open-ended questions after reading multiple documents are analyzed in order to evaluate semantic connections between the documents and the students' productions.

1.1 Comprehension of Multiple Documents

Reading comprehension is a difficult and complex task that requires connecting ideas in a text in order to produce a coherent mental representation of the information [6]. Such a task not only requires understanding the semantic relations between words and sentences, but also necessitates connecting ideas from various sentences throughout a text in order to produce a coherent understanding [7]. Thus, successful comprehension of single texts requires an ability to comprehend textbase content (explicit information derived from a single sentence) as well as develop intra-textual inferences that connect adjacent or distal textbase content from that same text.

This is a dynamic process between the reader and the text requiring the integration of information from the immediate sentence with previous sections of the text as well as the reader's own prior knowledge [6]. This continuous construction of a mental representation of textual materials can be enhanced by a reader's ability to integrate information across texts, thus developing a coherent knowledge base about a specific topic [8]. This can in turn aid in developing mental representations of future texts on related topics. However, comprehension becomes increasingly challenging when readers are expected to combine information from disparate sources.

Each text generally adheres to a consistent style, however, texts from different sources are highly variable in these characteristics and are not typically presented as a set [9]. These features can vary across genres and individual texts potentially creating an additional obstacle for integration. Individual texts contain discourse markers of cohesion that signal relations between ideas, whereas these features are not available between texts thus complicating the integration task for readers [10, 11]. Without these connectors to help guide inferencing, the integration of concepts relies on the reader's prior knowledge. This diversity and lack of clear connections may impose additional challenges for comprehension and integration of multiple texts.

1.2 Assessing and Evaluating Comprehension

Writing tasks during online and offline comprehension have been employed as a means of aiding students in making textual inferences. Both online and offline tasks enhance a reader's ability to process information and potentially integrate ideas across texts.

Offline comprehension tasks, such as essays, recall tasks, and comprehension questions, are often used to assess comprehension. However, they can also be used to support comprehension through the reactivation of relevant concepts. In particular, the recall-cues present in the questions combined with generating responses to convey understanding prompts readers to reactivate concepts, in turn aiding comprehension [12].

Online tasks, such as self-explanations and think-alouds, prompt readers to actively process text information. Self-explanation, the process of explaining information to oneself while employing reading comprehension strategies, is a valuable reading strategy that encourages deeper comprehension throughout the reading process, thus facilitating the construction of a more coherent mental model [13, 14].

Self-explanations also provide insights into a reader's cognitive processing of the text. When students generate responses to sequential text sections as they do in self-explanation tasks, their aggregated responses reveal semantic overlap across sections as well as connectives and other signaling devices that indicate specific connections of causal events. The cohesive devices expressed within reader's self-explanations provide insight into their coherence building processes because they can inform on the reader's depth of comprehension. For example, surface level processing is associated with the overlap of specific words across sentences or the amount of semantic information that can be traced back to previous portions of the text. Deeper comprehension processes also contain semantic overlap, but also have greater lexical diversity of the content relating to the text, suggesting the use of external information such as prior knowledge [1].

This study includes both students' self-explanations during reading and their responses to open-ended questions after reading multiple documents. Our objective here is to examine the semantic connections between the documents and (a) students' self-explanations, and (b) students' responses to questions. These semantic connections are assumed to represent the coherence of students' mental representations of the content. Students' constructed responses provide a glimpse into their processing of text, and thus a potential means of predicting students' comprehension. Here, we represent comprehension via students' score on the comprehension questions (i.e., expert ratings), and we assess the coherence of students' comprehension via semantic links between the documents and students' responses during and after reading. We do so by combining computational linguistics and SNA using CNA.

1.3 Cohesion Network Analysis

Cohesion Network Analysis (CNA) [2] was first introduced to assess participation in Computer Supported Collaborative Learning, but its underlying representation is suitable for any type of discourse. CNA relies on cohesion that is estimated using multiple semantic similarity metrics [15], combines advanced NLP techniques, and integrates SNA measurements applied on the resulting cohesion graph [16, 17]. The cohesion graph can be perceived as a proxy for the underlying semantic content of discourse within a document. It is represented as a multi-layered graph that considers both macro-level and micro-level constituents present at different levels (i.e., sentences, paragraphs, or the entire text). A document is decomposed into its paragraphs and, subsequently, into the underlying sentences and words. Cohesive links are defined between different layers of the hierarchy in order to measure the strength of the inclusion, represented as the relevance of a sentence with regards to the entire document or the impact of a word within each sentence. Cohesive links are also introduced between adjacent sentences and paragraphs in order to model the information flow throughout the discourse; these links are also indicative of cohesion gaps that are often caused by changes in topics. In addition, cohesive links are introduced between highly related discourse constituents in order to better reflect both high local or global text cohesion.

2 Method

We propose a method that extends CNA [2] for performing multi-document evaluations in order to predict students' comprehension of information presented in multiple texts. CNA considers text content and discourse structure in terms of cohesive links that are defined between multiple levels (i.e., sentences, paragraphs and the entire text). CNA can be used to quantify both local and global cohesion while relying on multiple semantic similarity models.

2.1 Dataset

Undergraduate students ($n = 146$) from a southwestern university in the United States participated in the study. Students first completed a demographics survey followed by a reading task composed of four texts about green living (i.e., lifestyle centered on balancing the usage, as well as preserving Earth's natural resources). As they read, each student wrote 30 self-explanations on specific target sentences distributed throughout the four texts. Target sentences were presented every two to four sentences and were selected on the basis that self-explanations could support inference generation of the content. After reading all of the texts, students answered 12 open-ended comprehension questions covering information from one or multiple texts followed by a reading skill test and a prior knowledge test. The questions are categorized under three types (textbase, intra-textual, and inter-textual) with four questions per category so as to cover the different comprehension and inferencing tasks in which readers engage. Each of the 12 questions were assigned a score of 0 to 1.0 and then summed to provide an assessment of the overall performance on a [0 to 12] scale. The final dataset consists of four independent texts (labeled A, B, C and D), 30 self-explanations, and 12 question responses per student (labeled from 1 to 12).

Table 1. Question identifiers (Questions 1 to 12) as a function of question type

Question type	Number of questions	Question identifiers
Text-base	4	Q4, Q7, Q8, Q10
Intra-textual	4	Q1, Q2, Q5, Q11
Inter-textual	43	Q3, Q6, Q9, Q12

2.2 Multi-document Cohesion Network Analysis

Figure 1 introduces an extension for CNA that considers multiple texts and student responses. Our aim is to build an overarching undirected cohesion graph for each student that semantically links the initial texts as a whole, or specific paragraphs or sequences from them, to individual representations of students' self-explanations or their question responses. This CNA network graph addresses coherence by building a global cohesion map in which we semantically link reference texts to different student constructed responses. Thus, the extended CNA network graph contains as nodes individual cohesion graphs generated for each target text level, as well as for each student response. The cohesive links within the extended graph are established based on the instructional setup and denote semantic relatedness between nodes of interest. For example, textbase and intra-textual questions are related to a specific text, whereas inter-textual questions are related to all four texts. Self-explanations are linked to sequences from the corresponding text (e.g., all prior text, adjacent text). The semantic distances were computed using the *ReaderBench* framework [18], which allowed us to experiment with several semantic models (i.e., LSA, LDA, and word2vec) and semantic distances in WordNet [19].

We extracted features describing the semantic relatedness between the reference texts and students' self-explanations or question responses to provide comparisons on what information most accurately predicts students' comprehension. Our feature extraction approach has slight differences in the way we process the self-explanations and the question answers based on the generated cohesive links, namely the granularity of the reference texts, as well as the consideration of one versus all texts.

In addition, we group together the cohesive links between a question answer/self-explanation and the corresponding paragraphs, and compute aggregate statistical metrics such as mean, median, max, and standard deviation when analyzing the links in the extended CNA graph. In the case of inter-textual questions, we also compute an average of the semantic distances between the question answer and all of the existing texts. For a given student, we obtained 42 sets of features (30 self-explanations and 12 question responses). These features were then grouped into question-related and self-explanation-related features, together with their corresponding aggregated statistical metrics.

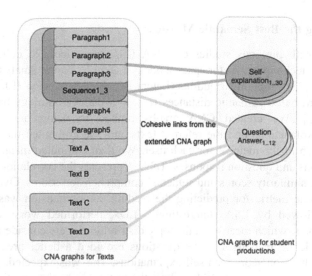

Fig. 1. The CNA multi-document graph.

2.3 Classification Methods

In order to predict the comprehension scores, we used regressor models which are statistical models aimed at making predictions based on a set of features. The models chosen for this experiment are the ones which are known to fare well on a dataset with a small number of examples. We used standard implementations, present in the Python library Scikit-learn [20], for the following models: Linear Regression, Extra Trees Regressor, Support Vector Regression (SVR), and Multi-Layer Perceptron (MLP). The four models were chosen in order to have a varied set of prediction tools, ranging from the least-sophisticated (Linear Regression) to the most complex (Extra Trees Regressor, or SVR). Existing neural network models are unsuited for a regression task with so few data points; thus, from that family of models we opted to solely examine the accuracy of an MLP model.

3 Results

The ReaderBench framework offers several semantic distances, which are related to one another. For each of those, around 300 possible features could be extracted from the CNA graph, meaning that the set of possible features could easily be of the order of thousands. This is why a multiple-step approach was required in order to keep only the most useful features. First, we determined the most suitable semantic distance for our dataset. Afterwards, we filtered the features to retain only the most relevant ones. Third, once we settled on a metric and a restricted group of features; we trained multiple models to observe their predictive accuracy with regards to student comprehension scores.

3.1 Selecting the Best Semantic Measures

In alignment with previous studies on CNA [21], we calculated cohesion using a variety of NLP techniques: vector space models (Latent Semantic Analysis (LSA) [22] and word2vec [23]), topic distributions (Latent Dirichlet Allocation (LDA) [24]), and non-latent word-based semantic distances (i.e., Wu-Palmer ontology-based semantic similarity) [25]. We created CNA graphs limited to using only the question answers/self-explanations and the referred texts, and we computed the semantic distances with each of the metrics, for each user. We then computed mean scores for all self-explanations and question responses. Table 1 presents the correlations between the mean semantic similarity scores and students' comprehension scores. Overall, the most relevant semantic metric for predicting the reading comprehension was provided by word2vec, followed by LSA. Interestingly, LDA performed worst with negative relatedness scores, which means that the topic distributions were considerably different. Moreover, students' responses to the questions provided a better predictor for estimating comprehension score than self-explanations. This was expected, given that the comprehension score was directly based on the responses to the questions. Nonetheless, this indicates that the semantic connections estimated using CNA correlated highly with scores.

Table 2. Pearson correlations between comprehension scores and SE/QA average semantic similarities.

	Score and SE average	Score and QA average
WU-Palmer	.014	.529
LSA	.034	.591
LDA	−.033	.433
word2vec	**.019**	**.675**

3.2 Features Filtering

By employing all the strategies presented in Sect. 2.3, we computed a total of 362 features based on word2vec semantic distances, 272 covering self-explanations and 90 covering question answers. To reduce multicollinearity, a **baseline filtering** step removed indices with inter-correlations above .9, leaving 126 features (34 for question answers and 92 for the self-explanations). A **second filtering step** consisted in eliminating all the features that had a correlation lower than 0.4 with the comprehension score. The resulting set consisted of 20 features (13 for question answers and 7 for self-explanations). After the second filtering step, the features relating to question answers were almost twice as many as those relating to self-explanation, despite being drawn from a much smaller pool of features. One reason for this is the fact that the questions cover an entire text or a group of texts, while the self-explanations are always centered on a small set of paragraphs.

As displayed in Table 2, six of these features were aggregated features over all exercises in a specific task (question answering or self-explanation), and 13 features were related to the student's performance on a particular task. The notation SE_X_Py considers the cohesive link between the first y paragraphs from text X to the self-explanation (denoted as SE), where as $SE\_X\_Py\_z$ reflects the cohesive link between paragraphs y to z from text X and the SE. The most highly correlated feature score is the mean of the averaged distances between each question and all texts. The best particular task feature is the median over the distances between the answer to question 10 which required intra-textual integration ("Explain how and why these claims might be misleading") and all the paragraphs from the referred text.

Table 3. Correlation between the best features and the comprehension scores.

Aggregated features	r
Links between Qs and all texts (M)	.672
Links between Qs and primary text targeted (SD)	.557
Links between SEs and the median of their links to target paragraphs (M)	.527
Links between Qs and the max of their links to target paragraphs (Med)	.515
Links between SEs and target sentence (SD)	.470
Links between SEs for current and prior targeted sentences (Med)	.418
Particular task features	r
Links between Q11 and target paragraphs (Med)	.560
Links between Q6 and all texts (M)	.531
Links between Q2 and target paragraphs (Maximum)	.521
Links between Q4 and target paragraphs (Med)	.504
Links between Q6 and target paragraphs (M)	.462
Links between SE_A_P1_3 and target paragraphs (M)	.451
Links between SE_B_P3_4 and target paragraphs (Med)	.448
Links between Q3 and all texts (M)	.432
Link between Q2 and target text	.430
Link between Q7 and target text	.425
Links between Q8 and target paragraphs (Maximum)	.412
Links between Q10 and target paragraphs (M)	.410
Links between SE_B_P4_6 and target paragraphs (Med)	.410
Links between SE_A_P4_7 and target paragraphs (Maximum)	.403

Note: Q = question;   SE = self-explanation;   M = mean;   Med = median;   SD = standard deviation.

When analyzing the most important feature for a question in relation to the question type (textbase, intra-textual, inter-textual), we observed that the feature type depends on the question type. The best predicting features for 2 out of 3 inter-textual questions (Q3, Q6) evaluated the semantic similarity between the answer and all the texts. This is in line with how those questions were constructed (as queries for information appearing throughout the 4 texts). In the case of textbase and intra-textual questions which considered information found in a single text, the main features are the aggregating ones (mean, median, or max) applied on the semantic similarity between the answers and all the paragraphs of the text. A second observation is that some question answers are much better predictors for the overall comprehension task than others. The main features for Q11, Q6, Q2, Q4 have a correlation coefficient with the final score above .5, while the main features for Q1, Q5, Q9, and Q12 have a correlation coefficient of around .35 or slightly below. This result is likely due to the complexity of the task, as the four latter questions required inter-textual or intra-textual inferences, which are more complex than textbase questions.

3.3 Predicting Reading Comprehension

We used 5-fold cross-validation as our dataset only has 146 examples. For each model, we trained and tested 5 independent models and report the average and minimum values for mean absolute error (i.e., the measure of difference between the predicted and observed comprehension scores). We examined the models based on the baseline filter (filtering based on multicollinearity) and models using features correlated above .4 with the comprehension score. Table 3 indicates that models using fewer and more highly correlated features were more predictive. This is notably circular given that our ultimate objective is to provide predictions without having the score. Nonetheless, this provides some evidence that the CNA provides good estimates of comprehension scores (Table 4).

Table 4. Prediction performance for the chosen models.

	Classifier	Filtered		Filtered over 0.4	
		MAE average	MAE min	MAE average	MAE min
SE	Linear regression	3.230	2.907	**1.612**	**1.317**
	Extra trees	**1.679**	**1.525**	1.664	1.361
	SVR	1.828	1.497	1.701	1.359
	MLP	1.813	1.401	1.771	1.426
QA	Linear regression	1.551	1.302	**1.434**	**1.096**
	Extra trees	**1.466**	**1.142**	1.508	1.228
	SVR	1.569	1.333	1.435	1.163
	MLP	1.668	1.357	1.600	1.280
Both	Linear regression	5.335	4.372	**1.298**	**0.886**
	Extra trees	**1.480**	**1.221**	1.446	1.133
	SVR	1.721	1.425	1.415	1.097
	MLP	1.853	1.425	1.632	1.259

In addition, results indicate that question-related features are overall more predictive than the self-explanation ones, which was expected, given that the comprehension scores were based on the question answering task. The large discrepancy between question answers and self-explanations that was identified in the first analysis (see Table 2) was considerably lower for comprehension predictions (i.e., a 0.2 MAE difference between the best models for self-explanations and questions answers). This is normal taking into account that self-explanations relate only to one text and they provide a reduced contextualization in contrast to a more detailed question answer. Overall, the best results are obtained using a Linear Regression model on the most highly filtered set of features from both question answers and self-explanations. This shows that even though the question response features are more predictive, self-explanations provide extra information that improves model performance.

Regarding the regressor models, we observed that Extra Trees obtained the best results when trained using a large set of features. However, when switching to the small feature set, the linear regression model narrowly outmatched Extra Trees in all three cases (question answers, self-explanations, and both), despite the fact that its poor performance without filtering based on correlations above .4.

4 Conclusions and Future Work

In this paper, we represent coherence using Cohesion Network Analysis (CNA) in which we model a global cohesion graph that semantically links reference texts to different student constructed responses in order to predict comprehension. We modeled performance using a dataset containing four documents for which students provided self-explanations and answers to open-ended comprehension questions addressing both individual documents as well as aggregated information from multiple sources. Several features were extracted and then filtered by eliminating those that were highly correlated among themselves, or those with weak correlations with the comprehension scores. Four regressor models were trained based on these features, side-by-side comparisons were made in order to highlight which models displayed the lowest MAEs for scores between 1 and 12. The best model without filtering based on correlations with the score was the Extra Trees model, providing between 1.1 and 1.7 MAE. The best model using the added correlation-based filter was Linear Regression, providing between 0.9 and 1.6 MAE. Both outcomes are encouraging - demonstrating that the features extracted from an extended CNA cohesion graph are capable of estimating student's comprehension scores within acceptable margins of error.

Our results showed that answers to some questions may be more suitable predictors than others and question complexity decreased performance. For example, three questions for which the answers were not good predictors of comprehension required inter-textual or intra-textual inferences. Self-explanations also offered valuable insights regarding the students' comprehension. When training a model with self-explanation-related features, the model without filtering provided a close proximation to comprehension scores (i.e., 1.5 MAE). This means that even without having students answer comprehension questions, we can estimate comprehension with relatively good accuracy.

As future developments, this experiment needs to be replicated on various datasets with different text sets and populations. Ultimately, our objective is to twofold: (a) simulate comprehension of multiple documents on line, thus providing the means for feedback, and (b) model the coherence of students' comprehension of multiple documents. The current study is our initial foray toward reaching these objectives.

Acknowledgments. This research was supported by the ReadME project "Interactive and Innovative application for evaluating the readability of texts in Romanian Language and for improving users' writing styles", contract no. 114/15.09.2017, MySMIS 2014 code 119286, the FP7 2008-212578 LTfLL project, the Institute of Education Sciences (R305A180144 and R305A180261), and the Office of Naval Research (N00014-17-1-2300).

References

1. Allen, L.K., Jacovina, M., McNamara, D.S.: Cohesive features of deep text comprehension processes. In: 38th Annual Meeting of the Cognitive Science Society, pp. 2681–2686. Cognitive Science Society, Philadelphia, PA (2016)
2. Dascalu, M., McNamara, D.S., Trausan-Matu, S., Allen, L.K.: Cohesion network analysis of CSCL participation. Behav. Res. Methods **50**(2), 604–619 (2018)
3. Hastings, P., Hughes, S., Magliano, J.P., Goldman, S.R., Lawless, K.: Assessing the use of multiple sources in student essays. Behav. Res. Methods **44**(3), 622–633 (2012)
4. Wasserman, S., Faust, K.: Social Network Analysis: Methods and Applications. Cambridge University Press, Cambridge (1994)
5. Jurafsky, D., Martin, J.H.: An Introduction to Natural Language Processing. Computational Linguistics, and Speech Recognition. Pearson Prentice Hall, London (2009)
6. McNamara, D.S., Magliano, J.: Toward a comprehensive model of comprehension. Psychol. Learn. Motiv. **51**, 297–384 (2009)
7. Kintsch, W.: The role of knowledge in discourse comprehension: a construction-integration model. Psychol. Rev. **95**(2), 163 (1988)
8. Wiley, J., Goldman, S.R., Graesser, A.C., Sanchez, C.A., Ash, I.K., Hemmerich, J.A.: Source evaluation, comprehension, and learning in Internet science in-quiry tasks. Am. Educ. Res. J. **46**(4), 1060–1106 (2009)
9. Magliano, J., McCrudden, M.T., Rouet, J.F., Sabatini, J.: The modern reader: should changes to how we read affect research and theory? In: Schober, M.F., Rapp, D.N., Britt, M. A. (eds.) Handbook of Discourse Processes, 2nd Ed. Taylor & Francis, New York (2018)
10. Goldman, S.R., Braasch, J.L., Wiley, J., Graesser, A.C., Brodowinska, K.: Comprehending and learning from Internet sources: processing patterns of better and poorer learners. Read. Res. Q. **47**(4), 356–381 (2012)
11. Goldman, S.R., Rakestraw, J.A.: Structural aspects of constructing meaning from text. In: Kamil, M.L., Mosenthal, P.B., Pearson, P.D., Barr, R. (eds.) Handbook of Reading Research, vol. 3, pp. 311–335. Erlbaum, Mahwah (2000)
12. Ozuru, Y., Briner, S., Kurby, C.A., McNamara, D.S.: Comparing text comprehension measured by multiple-choice and open-ended questions. Can. J. Exp. Psychol. **67**, 215–227 (2013)
13. Chi, M.T., De Leeuw, N., Chiu, M.-H., LaVancher, C.: Eliciting self-explanations improves understanding. Cogn. Sci. **18**(3), 439–477 (1994)

14. McNamara, D.S.: Self-explanation and reading strategy training (SERT) improves low-knowledge students' science course performance. Discourse Processes **54**(7), 479–492 (2017)

15. Dascalu, M., Trausan-Matu, S., McNamara, D.S., Dessus, P.: ReaderBench – automated evaluation of collaboration based on cohesion and dialogism. Int. J. Comput.-Support. Collaborative Learn. **10**(4), 395–423 (2015)

16. Trausan-Matu, S., Dascalu, M., Dessus, P.: Textual complexity and discourse structure in computer-supported collaborative learning. In: Cerri, S.A., Clancey, W.J., Papadourakis, G., Panourgia, K. (eds.) ITS 2012. LNCS, vol. 7315, pp. 352–357. Springer, Heidelberg (2012). https://doi.org/10.1007/978-3-642-30950-2_46

17. Dascalu, M., Trausan-Matu, S., Dessus, P.: Cohesion-based analysis of CSCL conversations: holistic and individual perspectives. In: 10th International Conference on Computer-Supported Collaborative Learning (CSCL 2013), vol. 1, pp. 145–152. ISLS, Madison, USA (2013)

18. Dascalu, M., Dessus, P., Bianco, M., Trausan-Matu, S., Nardy, A.: Mining texts, learner productions and strategies with ReaderBench. In: Peña-Ayala, A. (ed.) Educational Data Mining. SCI, vol. 524, pp. 345–377. Springer, Cham (2014). https://doi.org/10.1007/978-3-319-02738-8_13

19. Budanitsky, A., Hirst, G.: Evaluating WordNet-based measures of lexical semantic relatedness. Comput. Linguist. **32**(1), 13–47 (2006)

20. Pedregosa, F., et al.: Scikit-learn: machine learning in Python. **12**, 2825–2830 (2011)

21. Dascalu, M.: Analyzing Discourse and Text Complexity for Learning and Collaborating. Studies in Computational Intelligence, vol. 534. Springer, Switzerland (2014). https://doi.org/10.1007/978-3-319-03419-5

22. Landauer, T.K., Foltz, P.W., Laham, D.: An introduction to latent semantic analysis. Discourse Processes **25**(2/3), 259–284 (1998)

23. Mikolov, T., Chen, K., Corrado, G., Dean, J.: Efficient estimation of word representation in vector space. In: Workshop at ICLR, Scottsdale, AZ (2013)

24. Blei, D.M., Ng, A.Y., Jordan, M.I.: Latent dirichlet allocation. J. Mach. Learn. Res. **3**(4–5), 993–1022 (2003)

25. Wu, Z., Palmer, M.: Verb semantics and lexical selection. In: 32nd Annual Meeting of the Association for Computational Linguistics ACL 1994, pp. 133–138. ACL, New Mexico (1994)

Student Network Analysis: A Novel Way to Predict Delayed Graduation in Higher Education

Nasheen Nur[1], Noseong Park[2(✉)], Mohsen Dorodchi[1], Wenwen Dou[1], Mohammad Javad Mahzoon[1], Xi Niu[1], and Mary Lou Maher[1]

[1] University of North Carolina at Charlotte, Charlotte, NC, USA
{nnur,mdorodch,wdou1,mmahzoon,xniu2,m.maher}@uncc.edu
[2] George Mason University, Fairfax, VA, USA
npark9@gmu.edu

Abstract. We present a prediction model to detect delayed graduation cases based on student network analysis. In the U.S. only 60% of undergraduate students finish their bachelors' degrees in 6 years [1]. We present many features based on student networks and activity records. To our knowledge, our feature design, which includes conventional academic performance features, student network features, and fix-point features, is one of the most comprehensive ones. We achieved the F-1 score of 0.85 and AUCROC of 0.86.

Keywords: Network analysis · Student data · Risk prediction

1 Introduction

One of major strategic challenges that the U.S. higher education faces is timely completion of degree for college students [2]. Recent data from the National Center for Education Statistics shows that the majority (60%) of full-time undergraduate students take 6 years to earn a bachelor's degree [1]. As a result, higher education is under increasing pressure to demonstrate institutional effectiveness across a range of complicated factors [3]. According to [4], for instance, the U.S. government emphasizes the need of producing successful Science, Technology, Engineering, and Mathematics (STEM) graduates in a timely manner.

We propose a novel network analytic approach to predict at-risk students who fail to complete their degrees on time. Our approach is distinct from others due to the following two features: (1) We predict at-risk students early after 5-th semester; (2) In addition to classical academic features such as GPA and earned credits, we use various data from students' (extracurricular) activities to calculate student network features. We also define another type of features based on the same data, called fix-point features in our paper (see Fig. 1(a)). Throughout the analysis process, we have some interesting observations. At-risk students tend to have many weak connections, rather than a selected small

© Springer Nature Switzerland AG 2019
S. Isotani et al. (Eds.): AIED 2019, LNAI 11625, pp. 370–382, 2019.
https://doi.org/10.1007/978-3-030-23204-7_31

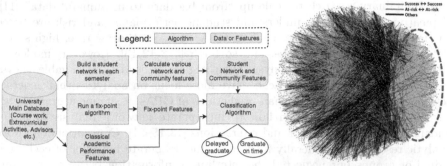

(a) The overall work-flow of our method   (b) A student network example

Fig. 1. (a) The overall work-flow of our presented framework. (b) Edge-colored student network after filtering out weak (i.e., edge weight <90th percentile) connections. Note that many at-risk students inside the dotted red circle do not have strong connections with successful students. (Color figure online)

number of strong connections so their network features (such as degree centrality, ego-network density and so forth) are distinctly different from that of successful students. In the future, we plan to intervene by strategically selecting at-risk students to consolidate their connections with successful students and to enhance their network features. We answer the following research questions in this paper:

1. Does including students' network features help predict at-risk students?
2. Does including students' network features help predict at-risk students in an earlier stage?
3. To what extent are the student network connections of successful students different from those of at-risks students over time?
4. Who are active participants of student communities & who are peripheral participants? How differently do successful and at-risk students behave in communities?

2   Related Work

During the past five years there has been an increase in research for improving learning and educational environments by leveraging analytics and the vast amount of data collected about the interaction of students with learning management systems (LMS). Course Signals [5–8] is an example of a learning analytic tool that not only classifies and identifies students at-risk but also provides interventions to improve student learning based on the analyzed data. This system processes student data from the Blackboard LMS to provide an early warning for students at-risk. Latest trends in learning analytics and knowledge (LAK) also shows a move towards sense-making from broad and general predictive models [9,10]. The LAK community is expanding and including broader interdisciplinary research to scale up "from big data to meaningful data" [11]. Typical models to better understand student performance and risk are based on classical academic features such as GPA, course withdraw rate, high school GPA, standardized test scores, and so forth [5–8,12,13]. However, models of retention and risk/success analysis often neglect to lead to actionable knowledge [14] whereas some approaches have more focus on analytics that generate actionable knowledge rather than predictions of GPAs, assignment grades, etc.

There are also reports on how to incorporate network analysis to better understand student behavior and interactions [13,15–21]. Conventional network analytic research has generally focused on the deliberate behavior of each individual or groups but neglected the interlinked information between or among individuals or groups [22]. These network models are based on students' LMS or social media logs, e.g., who responds to whom in LMS tools or who likes whom in social media. The purpose of the network analyses includes identifying student groups and social networking behaviors that lead to risk or success [15,18]. For example, Romero [13] investigated interaction patterns among students in their LMS tools and created an unsupervised clustering method to detect course failures. In [18], they analyzed participation patterns in online discussions in order to reveal student clusters with leaders and peripherals. Authors of [21] presented that students' social involvement accumulated through academic activities is positively related to their academic performance. In [23], it was shown that students' co-enrollment networks follow power-law degree distributions and they predict course grades with simple network features whereas we predict a longer-term success with more comprehensive network features. The related works have clearly laid out the functionality of social network analysis and provided guidance for our study.

Our approach is distinct in two aspects: (i) We construct student networks using student records other than interactions recorded in the LMS or social media logs, and (ii) Our success measure is on-time graduation whereas some existing models focus only on course success and GPA.

3   Our Dataset

We collected data from a 13-year period and limited our analysis to undergraduate students who spent 8 or more semesters in our school and have selected computer science as their major at some point in their academic career. We chose on-time graduation as the measure of success, and built predictive models to identify students being at-risk of not graduating in six years. After excluding on-going students who enrolled in the past six years, the total number of students in our analysis is 2,552 where approximately 30% are at risk. We did not use self-reported data such as social media data and LMS logs. The benefit of not using LMS data is that not all Professors use the LMS in the same way so the analysis for all students in a major will not have consistent data. We collected student background information such as demographics and tests taken before admission; academic information such as major, courses taken, transferred courses, and advisers; extracurricular activities and participation in student organizations.

4   Proposed Prediction Method

We design and extract three types of features: academic features, student network features, and fix-point features. The first type is already widely used in the field of learning analytics, but to our knowledge, there are few works using the second and third types. Our main contributions lie in the feature engineering based on student networks and academic activities.

4.1   Academic Features

Academic features such as grade-point average (GPA) are very effective to identify at-risk students and had been used very widely in many works [5–8,12,13]. We adopted features related to personal information (age, citizenship, gender, primary ethnicity, etc.), high-school record (school rank, percentile, etc.), and academic progress (GPA, success rate in earning credits, the number of course withdraws, etc.).

4.2   Student Network Features

How to Build Student Network. We build a weighted student network in each semester. Edge weight value between two students represents the *cumulative intensity* of the connection by the time point we draw the network. Because it is cumulative, their intensity will increase as time goes by. We calculate the edge weight, denoted as $w(x, y, t)$ hereinafter, between two students x and y at a certain semester t as follows:

$$w(x, y, t) = \exp\left(\sum_i rescale(normalize(w_i(x, y, t)))\right),$$

$$w_1(x, y, t) = \sum_{t' \leq t} \frac{C(x, t') \cap C(y, t')}{C(x, t') \cup C(y, t')}, \quad w_2(x, y, t) = \sum_{t' \leq t} same\_activity(x, y, t'),$$

$$w_3(x, y, t) = \sum_{t' \leq t} same\_advisor(x, y, t'), \quad w_4(x, y, t) = same\_dept(x, y), \tag{1}$$

$$w_5(x, y, t) = same\_major(x, y), \quad w_6(x, y, t) = same\_age\_high\_school(x, y),$$

where $C(x, t)$ is a set of courses taken by student x at semester t, $w_i(x, y, t)$ in the left-hand side means a cumulative value by t between two students x and y, and $same(x, y)$ or $same(x, y, t)$ in the right-hand side is an indicator function that returns 1 if two students x and y have the same (i) activity, (ii) advisor, (iii) department, (iv) major, or (v) high school with the same age and otherwise 0. Note that we do not consider time for high school record. Others do depend on time. For instance, $w_1(x, y, t)$, inspired from the Jaccard index[1], is to calculate the sum of the common course ratio between x and y until t. After normalization, w_i ranges in $[0, 1]$ and after re-scaling, the mean of w_i at t becomes 0.5.

---

[1] The Jaccard index is a popular node similarity metric in networks based on the number of common neighbors divided by the sum of all neighbors.

Some weights have larger scales than others and may dominate the final weight value without the normalization and re-scaling. We prevent it by standardizing edge weights. In our definitions, thus, *each w_i becomes equally important*[2]. Activity describes most student-focused extracurricular clubs, sports, and programs at the college. The exponential function makes strong (i.e., large final weight) connections stronger. Therefore, a student network at t may consist of many edges that have relatively small weights and a small number of edges that have large weights. We draw a student network for each semester t and there are more than 40 networks created from 13 years of student records. In Fig. 1(b), we draw a student network using three edge colors: orange (among successful) and blue (among at-risks) between the same classes and dark green between different classes. Note that many at-risk students do not maintain strong connections with other successful students.

Basic Network Features. In network analysis, *centrality* comprises methods to measure the relative importance of nodes (i.e., students in our case). There exist many different centrality concepts such as degree centrality, closeness centrality [24], clustering coefficient [25, 26], betweenness centrality [27, 28], PageRank [29], and so forth. Among many, we select centrality measures that have enough discriminatory information to identify at-risk students after some statistical analyses, such as t-test and histograms. For almost all centrality concepts, there exist both unweighted and weighted versions. All centrality metrics used in this paper are weighted, unless otherwise stated.

Community-Based Features. Community detection is a long-standing research topic in network analysis. Sometimes it is used as a subroutine to solve other problems similar to our case [30–33]. We use overlapping community detection methods because one student can join multiple communities. We choose SLPA [34] as our base community detection method, considering its accuracy and popularity. After finding many overlapping communities in each N_t, we calculate the following features:

1. Let $Com(x) = \{Com_1, Com_2, \cdots\}$ be a set of communities that student x belongs to. Finally, we do MIN/MAX/AVG aggregations over the communities in $Com(x)$ for each type of network features(x).
2. In each N_t, a giant component means the biggest community. In many cases, the giant component is one of the most influential student groups and we check if a student is its member. After that, we calculate the ratio of such cases over time. The ratio of 1 for a student means that the student is a stable member of the giant components for his or her entire academic period.

---

[2] This is a very important fact about our network dentition. We do not focus on only courses but also many other aspects of academic life.

Ego-Network-Based Features. Ego-network(also called node-centric network) means an induced sub-network by one node and its neighbors [35]. From each node's (student's) ego-network, we extract its ego-network density and clustering coefficient [25, 26] as network features. The density is formally defined as follows:

$$Density(x,t) = \frac{2 \sum_{(a,b) \in Ego(x,t)} w(a,b,t)}{|Nei(x,t)| \cdot (|Nei(x,t)| - 1)}, \tag{2}$$

where $Ego(.,t)$ returns an edge set of one's ego-network in N_t and $Nei(.,t)$ means a set of one's neighbors in N_t.

4.3   Fix-Point Features

Given a function $f(\cdot)$, a fix-point x means $x = f(x)$. fix-point calculation is used in various domains. One representative example is the stationary distribution of Markov chain, i.e., $\pi = \pi \mathbf{P}$, where $\mathbf{P}$ is transition matrix. In [29, 36–41], authors defined a mutually recursive complex system of variables and their fix-point values are used to understand vertices. Defining a complex variable system requires domain dependent knowledge. We first introduce domain knowledge we gain from our educational experience and available data:

1. We think that courses/activities/undergraduate advisors *simultaneously* taken by many students share common characteristics. Suppose that course A and course B have many overlapping students in a semester. Those two courses may have some common characteristics.
2. A student's characteristic can be described by courses/activities/advisors that she/he had taken [42–44]. In the network features, we analyzed the interactions among students. Our fix-point features describe students from their course/activity/advisor records without considering other students.

Based on those intuitions, we define several variables that are mutually recursive as follows:

$$val(c_i, t) = \sum_{c_j} \frac{\#stu(c_i, c_j, t)}{\sum_{c_k} \#stu(c_k, c_j, t)} val(c_j, t), \quad val(s_i, t) = \sum_{c_j} take(c_j, s_i, t) \frac{1}{\#stu(c_j, t)} val(c_j, t)$$

$$val(a_i, t) = \sum_{a_j} \frac{\#stu(a_i, a_j, t)}{\sum_{a_k} \#stu(a_k, a_j, t)} val(a_j, t), \quad + \sum_{a_j} take(a_j, s_i, t) \frac{1}{\#stu(a_j, t)} val(a_j, t)$$

$$val(v_i, t) = \sum_{v_j} \frac{\#stu(v_i, v_j, t)}{\sum_{v_k} \#stu(v_k, v_j, t)} val(v_j, t), \quad + \sum_{ad_j} take(v_j, s_i, t) \frac{1}{\#stu(v_j, t)} val(v_j, t),$$

where c_i, a_i, v_i, and s_i represent course, activity, adviser, and student, respectively. $\#stu(x, y, t)$ returns the number of students who took two courses, activities, or advisers x and y together at semester t and $take(x, s, t) \in \{0, 1\}$ is an indicator variable to denote if course, activity, or adviser x is taken by student s at semester t. Thus, $val(x, t)$ means an influence value each entity x has at semester t. As we construct a network N_t in each semester, these variables are defined for each semester too. We ignore department, major, degree information because they are too broad to be used in the variable definition.

If two courses, activities, or advisers have largely overlapping students, their values will be very similar because of the coefficient based on normalization. A student value is an aggregation of all the course/activity/advisor values so it will be solely decided by the courses/activities/advisers taken by the student (See Algorithm 1). It is an iterative method to update values. We check the convergence only for the student variables because they are what we are interested in. The converged student values are used as additional features. The proof of its convergence is removed in this paper due to space reasons.

4.4   Experiments

The time point of the data is Spring 2004 and the time point is Fall 2016. All students who graduated on or before Spring 2013 are in the train set and others are in the test set. The ratio of train:test is 77:23. We perform the grid search with 10-fold cross validation to find the best model. Many classifiers (including SVM, Random Forest, Decision Tree, AdaBoost, RBM, Bagging, Multi-Layer Perceptron, etc.) are tested. In the training set, two classes are slightly imbalanced, i.e., 63% successful and 37% at-risk, so we apply under/oversampling techniques [45] to make them balanced. In general, Random Forest works very well and all of the reported values were produced by it.

Input: Student network $N_t = (V, E)$, Course and Activity Records
Output: $val(s_i, t)$ for each student

1 Initialize $val(x, t) = \frac{1}{n}$ where n is the total number of courses, activities, advisers, or students depending on the type of x.
2 **while** *until the convergence of* $val(s_i, t)$ **do**
3 | Update $val(c_i, t)$; Update $val(a_i, t)$; Update $val(v_i, t)$; Update $val(s_i, t)$
4 **return** $val(s_i, t)$

Algorithm 1: Fix-point calculation algorithm for our complex variable system

Table 1. Prediction results

	F-1 Overall	AUCROC	Recall of At-risk	Recall of Successful	F-1 of Successful
	All students				
Academic Features	0.78	0.76	0.56	0.67	0.78
Network Features (w_1 only)	0.76	0.72	0.62	0.77	0.83
Network Features (w_2 only)	0.73	0.66	0.51	0.70	0.81
Network Features (w_3 only)	0.74	0.70	0.61	0.75	0.81
Network Features (w)	0.81	0.8	0.64	0.85	0.87
Academic + Network Features	0.84	0.86	0.69	0.87	0.89
Academic + Network + fix-point Features	**0.85**	**0.86**	**0.70**	**0.90**	**0.89**
	Early phase students				
Academic Features	0.8	0.75	0.5	0.89	0.8
Network Features (w_1 only)	0.8	0.71	0.54	0.85	0.86
Network Features (w_2 only)	0.75	0.66	0.51	0.84	0.82
Network Features (w_3 only)	0.76	0.68	0.55	0.87	0.83
Network Features (w)	0.8	0.77	0.55	0.79	0.79
Academic + Network Features	0.84	0.85	0.56	0.86	0.9
Academic + Network + fix-point Features	**0.85**	**0.85**	**0.58**	**0.88**	**0.9**

Prediction Results. We calculated our network features in various types of student networks whose edge weights are calculated with w_i only or with the combined weight w marked with "$(w_i$ only)" or "(w)" as shown in Table 1, respectively. Note that some w_i is omitted in the table due to their performance inferior to others. Network features based on the combined weight w shows the best performance among them. Using only the academic features, we could recall 56% and 50% of at-risk students among all and early phase students. After adding network features, we could achieve the recall of 69% and 56% for the at-risk student class and after using all available features, they are improved to 70% and 58%. For other measures such as F-1 and AUCROC, our predictive model including all academic, network, and fix-point features outperform others in non-trivial margins. These results strongly teach us answers on our research questions **1** and **2**. That is, network features improve the overall prediction and in particular, during earlier periods.

Network Analysis. The degree centrality of a student in N_t is the sum of the edge weights to neighbors. At the beginning, we expected that successful students have more friends, thereby higher degree values. However, our observations disprove the hypothesis. In Table 2, we show degrees in various perspectives. We calculate average values for the top 50% and bottom 50% students in terms of $avg\_degree(\cdot)$ in each prediction class. Their average values are quite different, i.e., 2522.4 for successful v.s. **5258.1** for at-risk students (with p-value < 0.01).

Table 2. Average centrality, average community-based features and of two student classes. For space reasons, we list selected values. P-values are smaller than 0.01 only except the cases marked in boldface.

		Successful (entire period)	At-risk (entire period)	Successful (at 5th sem.)	At-risk (at 5th sem.)
Degree	All	2522.4	**5258.1**	**3106.5**	2856.6
	Top 50%	4487.3	**10143.0**	**5575.5**	5392.7
	Bottom 50%	**557.4**	373.1	**634.8**	311.7
Page Rank	All	0.00075	**0.00202**	0.00072	**0.00193**
	Top 50%	0.00118	**0.00365**	0.0011	**0.00346**
	Bottom 50%	0.00031	**0.00038**	0.00032	**0.00038**
Eigen.	All	0.0068	**0.02212**	0.0058	**0.02323**
Betw.	All	0.00079	**0.00456**	0.0006	**0.00391**
Close.	All	**0.4357**	0.4281	**0.4372**	0.4214
Min Degree	All	**553.3**	498.8	**568.8**	423.1
Min Eigen.	All	**0.5348**	0.5145	**0.5419**	0.4719
Giant.	All	0.0674	**0.087**	0.0665	**0.0929**
	Top 50%	0.1268	**0.1606**	0.122	**0.1713**
	Bottom 50%	0.0074	**0.0131**	0.0077	**0.0137**
Fix Point	All	0.000601	**0.000722**	0.001303	**0.001793**
	Top 50%	0.00115	**0.00178**	0.002344	**0.00541**
	Bottom 50%	**2.144450e-81**	1.561194e-81	**2.511162e-131**	1.818025e-131

Some at-risk students are exposed to more interactions than the majority of successful students. This can be interpreted in multiple ways, e.g., some at-risk students have too many student activities. At the same time, the bottom 50% at-risk students have much lower degrees than the bottom 50% successful students, i.e. **557.4** v.s. 373.1 (with p-value <0.01). This means that there also exist many at-risk students who do not interact with other students as much.

In the top 50% and bottom 50% cases of early phase students, successful students have higher average values than at-risk students but their significance level is low (p-value >0.01). However, successful students' average degree at 5th semester is larger than that of the entire period, i.e., 3106.5 v.s. 2522.4. This is possible when successful students (i) quickly stabilize their connections during their early academic periods and (ii) do not make many new connections in their late academic periods. However, at risk students need to take some courses many times in order to pass the course, and this shows up in our network as having many more connections because they take the same course more times than the successful student. At 5th semester, the average degree of the top 50% at-risk students is 5392.7 but it is improved to 10143 when considering their entire academic periods. This means that they interacted with many new students even in their late academic periods for student activities, courses, and advisers. Interestingly the bottom 50% of the at-risk students' average degree does not change significantly over time, i.e., 311.7 at 5th semester v.s. 373.1 during entire academic periods. They are consistently isolated from others. All these findings support research question **3**, that successful and at-risk student show different behavior over time. For other network features, we also hypothesized before calculating them that successful students have better values. However, our results show counter-evidences in some features. Because some at-risk students (e.g., the top 50% at-risk students in the previous degree centrality analysis) keep making new connections (rather than staying in a community), they are bridges over communities and as a result, their PageRank, betweenness, and eigenvector centrality values will be higher than other successful students. In the bottom 50% case, we could not observe significant differences.

We also tested many other centrality metrics such as closeness centrality [24], leverage centrality [46], clustering coefficient [25], and so on. For some of them, we did not observe significant differences between the two student classes. In Table 2, the higher average values of community features for successful students implies they play more important roles in communities than at-risk students. This is also well matched with the at-risk students' high betweenness centrality results which means that they are bridges over communities rather than core members. Moreover, they are more likely to be members of the giant components than successful students since they interact with many communities and thus end up with peripheral positions in the communities. Interestingly, for the bottom 50% at-risk cases, their average degree centrality is lower than that of the bottom 50% successful students but their average percentage of the membership to giant components is higher. This implies that those at-risk students may visit many communities but do not make many connections in the communities.

The ego-network density and clustering coefficient are complementary to each other. Ego-network density can be high if some edge weights are large. For clustering coefficient, however, some large edge weights cannot solely lead to the clustering coefficient of 1. Only when ego-network is a complete network, its clustering coefficient becomes 1. Because of this property, we expected larger clustering coefficients from successful than at-risk students. Successful students' ego-networks tend to be more complete than that of at-risk students. Ego-network density will be small if one does not maintain long-term connections in various activities. Thus, we hypothesized that successful students may have dense ego-networks. In all cases, successful students have better ego-network density than at-risk students, i.e., **3.634** vs. 3.073 and **3.707** vs. 2.706. This observation is well aligned with all the previous analyses because the ego-network density results also imply that successful students and their neighbors maintain strong connections. These evidences provide support for our research question **3** and **4**.

Our result also shows that at-risk students have higher fix-point values in general. We think that it is because of the same reason that at-risk students make many short-term connections with others. In early phase, the pattern is unique (supporting research question **3**). In the degree centrality, for instance, successful students' degree values are higher than that of at-risk students in the early period. However, for fix-point values at-risk students already show higher values. This means that we can catch students exposed to too many early connections. In all cases, p-values are smaller than 0.01.

5 Key Observations, Future Work and Conclusions

We start by building a network that represents the connections between students and a mutually recursive variable system using data collected by the university to solve a problem in higher education. The network is constructed using the data stored in the University student management system and does not rely on access to social media data or consistent use of LMS data. Our prediction of success or risk achieves F-1 = 0.85 and AUCROC = 0.86. Our student network analysis teaches us two very important insights, that is (i) At-risk students establish disorderly connections while successful students keep strengthening their existing connections, (ii) Successful students have high GPA neighbors and their ties are strong. The density of successful students' ego-networks are stable regardless of time period. Our degree centrality results say that some at-risk students keep making new connections and their ego-networks do not become dense or complete. Our community-based features also support that successful students are core community members where at-risks students reside in periphery.

We think that this network model of students can identify effective intervention points at an early stage for at-risk students. It might be their natural characteristics to make connections in such a way. But by helping them maintain long-term and stable connections, we believe that at-risk students can improve their success probabilities. Moreover, considering aspects of gender, race, ethnicity, generational social class, student body demographics, geographic location of institution, and socio-economic status of students, are also large factors when determining how long it takes a student to graduate. We are in process of collecting these data and also the data from our LMS logs. We think that it is essential to consider those factors when setting up intervention plans and creating compact student networks.

Acknowledgements. This work is supported by the National Science Foundation under Grant No. 1820862. Noseong Park and Mohsen Dorodchi are the co-corresponding authors.

References

1. National Center for Education Statistics: Table 326, 10 (2016)
2. Tinto, V.: Research and practice of student retention: what next? J. Coll. Stud. Retent. Res. Theory Pract. **8**, 1–19 (2006)
3. Suskie, L.: How can we address ongoing accreditation challenges? Assess. Updat. **28**, 3–14 (2016)
4. Acton, R.K.: Characteristics of STEM success: a survival analysis model of factors influencing time to graduation among undergraduate stem majors (2015)
5. Arnold, K.E.: Signals: applying academic analytics. Educ. Q. **33**, n1 (2010)
6. Arnold, K.E., Pistilli, M.D.: Course signals at purdue: using learning analytics to increase student success. In: Proceedings of the 2nd International Conference on Learning Analytics and Knowledge, pp. 267–270. ACM (2012)
7. Campbell, J.P., Oblinger, D.G., et al.: Academic analytics. EDUCAUSE Rev. **42**, 40–57 (2007)
8. Campbell, J.P.: Utilizing student data within the course management system to determine undergraduate student academic success: an exploratory study. ProQuest (2007)
9. Dawson, S., Gašević, D., Mirriahi, N.: Challenging assumptions in learning analytics. J. Learn. Anal. **2**, 1–3 (2015)
10. Dorodchi, M., Bendict, A., Desai, D., Mahzoon, M.J.: Reflections are good!: analysis of combination of grades and students' reflections using learning analytics. In: Proceedings of the 49th ACM Technical Symposium on Computer Science Education, pp. 1077–1077. ACM (2018)
11. Merceron, A., Blikstein, P., Siemens, G.: Learning analytics: from big data to meaningful data. J. Learn. Anal. **2**, 4–8 (2015)
12. Jayaprakash, S.M., Moody, E.W., Lauría, E.J.M., Regan, J.R., Baron, J.D.: Early alert of academically at-risk students: an open source analytics initiative. J. Learn. Anal. **1**, 6–47 (2014)
13. Romero, C., López, M.I., Luna, J.-M., Ventura, S.: Predicting students' final performance from participation in on-line discussion forums. Compu. Educ. **68**, 458–472 (2013)
14. Gašević, D., Dawson, S., Siemens, G.: Let's not forget: learning analytics are about learning. TechTrends **59**, 64–71 (2015)
15. De Laat, M., Lally, V., Lipponen, L., Simons, R.-J.: Investigating patterns of interaction in networked learning and computer-supported collaborative learning: a role for social network analysis. Int. J. Comput.-Support. Collab. Learn. **2**, 87–103 (2007)

16. Blackmore, C.: Social Learning Systems and Communities of Practice. Springer, London (2010). https://doi.org/10.1007/978-1-84996-133-2
17. Shum, S.B., Ferguson, R.: Social learning analytics. J. Educ. Technol. Soc. **15**, 3–26 (2012)
18. Takaffoli, M., Zaïane, O.R., et al.: Social network analysis and mining to support the assessment of on-line student participation. ACM SIGKDD Explor. Newsl. **13**, 20–29 (2012)
19. Mohamad, S.K., Tasir, Z.: Educational data mining: a review. Procedia - Soc. Behav. Sci. **97**, 320–324 (2013). The 9th International Conference on Cognitive Science
20. Adraoui, M., Retbi, A., Idrissi, M.K., Bennani, S.: Social learning analytics to describe the learners' interaction in online discussion forum in moodle. In: The 16th International Conference on Information Technology Based Higher Education and Training (2017)
21. Gašević, D., Zouaq, A., Janzen, R.: Choose your classmates, your GPA is at stake! The association of cross-class social ties and academic performance. Am. Behav. Sci. **57**, 1460–1479 (2013)
22. Carolan, B.V.: Social Network Analysis and Education: Theory, Methods & Applications. Sage Publications, Thousand Oaks (2013)
23. Gardner, J., Brooks, C.: Coenrollment networks and their relationship to grades in undergraduate education. In Proceedings of the 8th International Conference on Learning Analytics and Knowledge, pp. 295–304 (2018)
24. Freeman, L.C.: Centrality in social networks conceptual clarification. Soc. Netw. **1**, 215–239 (1978)
25. Barrat, A., Barthelemy, M., Pastor-Satorras, R., Vespignani, A.: The architecture of complex weighted networks. Proc. Natl. Acad. Sci. U.S.A. **101**, 3747–3752 (2004)
26. Wasserman, S., Faust, K.: Social Network Analysis: Methods and Applications. Cambridge University Press, Cambridge (1994)
27. Anthonisse, J.M.: The rush in a directed graph. Stichting Mathematisch Centrum. Mathematische Besliskunde, pp. 1–10 (1971)
28. Freeman, L.C.: A set of measures of centrality based on betweenness. Sociometry **40**, 35–41 (1977)
29. Lawrence, P., Sergey, B., Rajeev, M., Terry, W.: The PageRank citation ranking: bringing order to the web. Technical report, Stanford InfoLab (1999)
30. Li, Y., Martinez, O., Chen, X., Li, Y., Hopcroft, J.E.: In a world that counts: clustering and detecting fake social engagement at scale. In Proceedings of the 25th International Conference on World Wide Web, pp. 111–120. International World Wide Web Conferences Steering Committee (2016)
31. Mavroforakis, C., Valera, I., Gomez-Rodriguez, M.: Modeling the dynamics of learning activity on the web. In: Proceedings of the 26th International Conference on World Wide Web, pp. 1421–1430. International World Wide Web Conferences Steering Committee (2017)
32. Hoang, M.X., Dang, X.-H., Wu, X., Yan, Z., Singh, A.K.: GPOP: scalable group-level popularity prediction for online content in social networks. In: Proceedings of the 26th International Conference on World Wide Web, pp. 725–733. International World Wide Web Conferences Steering Committee (2017)
33. Chaturvedi, S., Castelli, V., Florian, R., Nallapati, R.M., Raghavan, H.: Joint question clustering and relevance prediction for open domain non-factoid question answering. In: Proceedings of the 23rd International Conference on World Wide Web, pp. 503–514. ACM (2014)

34. Xie, J., Szymanski, B.K., Liu, X.: SLPA: uncovering overlapping communities in social networks via a speaker-listener interaction dynamic process. In: Data Mining Workshops, pp. 344–349. IEEE (2011)

35. Burt, R.S.: Models of network structure. Ann. Rev. Sociol. **6**, 79–141 (1980)

36. Gibson, D., Kleinberg, J., Raghavan, P.: Inferring web communities from link topology. In: Proceedings of the Ninth ACM Conference on Hypertext and Hypermedia: Links, Objects, Time and Space–Structure in Hypermedia Systems: Links, Objects, Time and Space–Structure in Hypermedia Systems, HYPERTEXT 1998, pp. 225–234. ACM (1998)

37. Kleinberg, J.M.: Authoritative sources in a hyperlinked environment. In: Proceedings of the Ninth Annual ACM-SIAM Symposium on Discrete Algorithms, SODA 1998, pp. 668–677 (1998)

38. Miller, J.C., Rae, G., Schaefer, F., Ward, L.A., LoFaro, T., Farahat, A.: Modifications of kleinberg's hits algorithm using matrix exponentiation and web log records. In: Proceedings of the 24th Annual International ACM SIGIR Conference on Research and Development in Information Retrieval, SIGIR 2001, pp. 444–445. ACM (2001)

39. Li, L., Shang, Y., Zhang, W.: Improvement of hits-based algorithms on web documents. In: Proceedings of the 11th International Conference on World Wide Web, WWW 2002, pp. 527–535. ACM (2002)

40. Kang, C., Park, N., Prakash, B.A., Serra, E., Subrahmanian, V.S.: Ensemble models for data-driven prediction of malware infections. In: Proceedings of the Ninth ACM International Conference on Web Search and Data Mining, WSDM 2016, pp. 583–592. ACM (2016)

41. Kumar, S., Hooi, B., Makhija, D., Kumar, M., Faloutsos, C., Subrahmanian, V.S.: Rev2: fraudulent user prediction in rating platforms. In: Proceedings of the Eleventh ACM International Conference on Web Search and Data Mining, WSDM 2018, pp. 333–341. ACM (2018)

42. Diaz, D.P., Cartnal, R.B.: Students' learning styles in two classes: online distance learning and equivalent on-campus. Coll. Teach. **47**, 130–135 (1999)

43. Picciano, A.G.: Beyond student perceptions: issues of interaction, presence, and performance in an online course. J. Asynchronous Learn. Netw. **6**, 21–40 (2002)

44. Astin, A.W.: Student involvement: a developmental theory for higher education. J. Coll. Stud. Pers. **25**, 297–308 (1984)

45. Lemaître, G., Nogueira, F., Aridas, C.K.: Imbalanced-learn: a python toolbox to tackle the curse of imbalanced datasets in machine learning. J. Mach. Learn. Res. **18**, 1–5 (2017)

46. Joyce, K.E., Laurienti, P.J., Burdette, J.H., Hayasaka, S.: A new measure of centrality for brain networks. PLoS One **5**, e12200 (2010)

Automatic Generation of Problems and Explanations for an Intelligent Algebra Tutor

Eleanor O'Rourke[1][(✉)], Eric Butler[2], Armando Díaz Tolentino[2], and Zoran Popović[2]

[1] Northwestern University, Evanston, IL, USA
eorourke@northwestern.edu
[2] University of Washington, Seattle, WA, USA
{edbutler,ajdt,zoran}@cs.washington.edu

Abstract. Intelligent tutors that emulate one-on-one tutoring with a human have been shown to effectively support student learning, but these systems are often challenging to build. Most methods for implementing tutors focus on generating intelligent explanations, rather than generating practice problems and problem progressions. In this work, we explore the possibility of using a single model of a learning domain to support the generation of both practice problems and intelligent explanations. In the domain of algebra, we show how problem generation can be supported by modeling if-then production rules in the logic programming language *answer set programming*. We also show how this model can be authored such that explanations can be generated directly from the rules, facilitating both worked examples and real-time feedback during independent problem-solving. We evaluate this approach through a proof-of-concept implementation and two formative user studies, showing that our generated content is of appropriate quality. We believe this approach to modeling learning domains has many exciting advantages.

Keywords: ITS · Problem generation · Answer set programming

1 Introduction

Over the past fifty years, researchers have developed robust artificial intelligence systems that can emulate one-on-one tutoring with a human [3,6,8,29]. These *intelligent tutoring systems* provide adaptive problem progressions and personalized feedback in a variety of domains, and have been shown to produce strong learning gains in classroom studies [10,30]. However, these systems are often challenging and time-consuming to build. Researchers have explored a variety of approaches for modeling learning domains, resulting in the development of cognitive tutors [3,6], constraint-based tutors [16,18], example-tracing tutors [9], and ASSISTments [8]. However, these approaches all focus on optimizing the modeling and authoring of intelligent feedback, rather than of problems and problem

© Springer Nature Switzerland AG 2019
S. Isotani et al. (Eds.): AIED 2019, LNAI 11625, pp. 383–395, 2019.
https://doi.org/10.1007/978-3-030-23204-7_32

progressions, an area which has been highlighted as interesting for future development [11]. While a variety of problem-generation approaches have been developed and studied [14, 19, 26], most depend on models of the learning domain that are very different than those used to generate intelligent explanations, making it difficult to integrate them into existing tutoring systems.

In this work, we explore the possibility of using a single underlying model to generate both practice problems and intelligent explanations. We build on prior work in problem generation [2, 5, 19, 28] by using the logic programming language *answer set programming* (ASP) to model if-then production rules similar to those used by cognitive tutors in the domain of algebra. We show how this model can be used to generate algebra problems and all valid solutions to those problems. We also present a new method for automatically generating step-by-step explanations directly from the ASP model. We show how our explanation content can be used to create worked examples, feedback during tutored problem-solving, and a progression that gradually fades between the two.

We evaluate our approach through a proof-of-concept implementation and two formative user studies. First, we developed an application called the *Algebra Notepad* that embeds the problems, solutions, and explanations generated from our model, demonstrating how our content can be used to implement an intelligent tutor. Next, we evaluated the application through two user studies. In a study with 57 Mechanical Turk workers, we found that participants solve problems more accurately and efficiently after practicing with our tutor, demonstrating that the generated solutions and explanations are understandable. In a study with seven eighth-grade students, we found that the tutor helps learners in our target population solve problems successfully. This approach for generating problems and explanations from a single domain model has many advantages, and could support robust content generation for tutoring systems in the future.

2   Background

2.1   Modeling Learning Domains

A variety of approaches for designing and implementing intelligent tutoring systems (ITS) have been explored. Cognitive tutors provide the most sophisticated form of intelligent feedback. They represent knowledge using production rules that define if-then relationships which capture all knowledge needed to solve problems in a target learning domain, allowing the computer to solve problems step-by-step along with the student [3, 6]. Cognitive tutors detect errors when the student's action does not match any production rule in the model, and most include explicitly programmed "buggy" production rules that match common mistakes and misconceptions so that these can be explained [3]. Cognitive tutors are complex, and typically include as many as 500 production rules [4].

Constraint-based tutors are another type of system designed to help students identify and learn from mistakes [18]. These tutors model learning domains as sets of pedagogically important constraints [14, 16, 17]. Rather than tracing student actions, constraint-based tutors analyze the student's current state to identify violations of model constraints [17]. These tutors typically require less authoring effort than cognitive tutors, but can only provide feedback about constraint violations rather than also providing goal-oriented feedback [16].

Finally, *peudo tutors* exhibit many of the behaviors of ITS without requiring complex modeling. Example-tracing tutors are created by demonstrating correct solutions and common mistakes for specific types of problems. These demonstrations used to create a behavior graph that can trace learner behavior and provide feedback [9]. A downside of this approach is that content must be demonstrated or authored for each problem type; in contrast, a cognitive tutor's production rules can generalize across many different problem types [9].

2.2 Explanation Generation

Most tutors use hand-authored templates to generate explanations. Cognitive tutors produce next-step hints and feedback by associating an explanation template with each correct and buggy production rule in the model [3,4,29]. To fill the templates with appropriate problem-specific content, each problem must also be labeled to indicate which phrases should be inserted into the template [4]. The Cognitive Tutor Authoring Tools (CTAT) were developed to support efficient authoring of both model-tracing cognitive tutors [12] and example-tracing tutors [9]. Example-tracing tutor authors can annotate hints and feedback messages for specific problems in the system [9]. In constraint-based tutors, hand-authored feedback messages are attached directly to the constraints [17], and can either be given after each student action or at the end of the problem [16]. In contrast to cognitive tutors, these explanations are problem-independent.

2.3 Problem Generation

Researchers have explored problem generation for a variety of domains including word problems [19], natural deduction [1], procedural problems [2], and embedded systems [23]. Most approaches are template-based; given a general template for a type of problem, they generate more problems that fit the template. These templates can be generated automatically [1], semi-automatically [2], or manually [23]. While many of these approaches use exhaustive search or logical reasoning to generate problems, others use logic programming languages to model domains and generate problems. For example, Andersen et al. use the code coverage toolkit Pex, built on the Z3 SMT solver, to generate problems for procedural mathematics (e.g., long division) [2]. Others use ASP directly for domains such as word problems [19] or educational math puzzles [5,28].

Despite this extensive research on problem generation, most intelligent tutoring systems still rely on hand-authored problems. In a recent paper discussing areas for ITS improvement, Koedinger et al. highlight automated problem generation as an interesting area for future development [11]. In the domain of constraint-based tutors, Martin and Mitrovic developed an algorithm that can generate problems from a set of target constraints [13,14]. However, since the domain models used to generate problems are different than those used to generate explanations, integrating problem generation into tutors is non-trivial.

3   Implementation Approach

In this work, we explore using a single underlying model to generate both problems and explanations for an intelligent tutor in the domain of algebra. Our core approach is to model valid algebraic operations using answer set programming (ASP), which facilitates generating both new problems and all valid solutions to those problems. During the modeling process, we structure the ASP program such that explanations for each solution step can be generated directly from the code itself. We note that if a tutoring system has access to all valid solutions to a target problem, it can trace a learner's steps and compare them to those in the solutions to detect errors. Furthermore, given step-by-step explanations of each solution, a tutor can provide worked examples and also use explanations of specific steps to provide feedback in response to learner mistakes. In this section, we first describe our approach for modeling algebra in ASP, then discuss how we generate problems, solutions, and explanations from this model. Finally, we describe how we can use the ASP program to automatically detect and explain a class of misconceptions related to applying algebraic operators incorrectly.

3.1   Modeling Algebra in ASP

ASP programs define facts and rules that are represented in first-order logic. Answer set solvers search the space of truth assignments for each logical statement in an ASP program to produce satisfying solutions called *answer sets*, which define a set of self-consistent statements that identify a valid state of the world. ASP programs typically include three types of rules: *choice rules* that allow the solver to guess facts that might be true, *deductive rules* that allow the solver to deduce new facts from established or guessed facts, and finally *integrity constraints* that forbid certain solutions.

To solve an algebraic equation, a learner must isolate a variable on one side by applying a sequence of operators, such as combining terms or dividing both sides by a constant. In ASP, we model operators using deductive rules and integrity constraints. Then, we use event calculus [25], a logical formulation that can represent the state of the world at multiple time steps, to model the problem-solving process. For each operator, we use deductive rules to define a set of precondition predicates that must hold for that operator to applicable at a given time step. Then, we use additional deductive rules to describe how the equation will change on the next time step if that operator is applied. For example, consider the operator for adding two like terms on the same side of the equation. This operator is only applicable when a set of preconditions hold: two terms must be on the same side of the equation, they must be added, and they must be monomial terms of the same degree. If these hold, the operator can be applied by adding the coefficients of the two monomial terms to produce a single term.

Most algebra problems have many valid solutions. In general, textbooks recommend first simplifying by canceling and combining terms, and then rearranging terms to isolate a variable on one side of the equation. We therefore group operators into five classes ordered by precedence – *cancel, combine, rearrange, move*, and *expand* – and define integrity constraints that force the program to explore the classes of operators in this priority order. Finally, integrity constraints are used to ensure that the final step is a valid solution.

3.2   Problem and Solution Generation

To generate new problems and their solutions from the ASP program, we define choice rules that set the initial problem configuration, the operators used in the solution, and predicates describing which operator is chosen at which time. Answer set solves also require that you define a finite search space, so we constrain both the size of the equations and the number of steps in the solution. In practice, novices focus on relatively simple problems, so we constrained generation to equations with a maximum of six terms per side and a maximum solution length of four steps. An answer set calculated on our ASP program produces both a problem and a sequence of valid operators that solves the problem.

This allows us to generate problem-solution pairs, but we want to generate all valid solutions to each problem. This requires some subtlety because generating all solutions is in a higher complexity class than generating a single solution. ASP is capable of solving this class of problem, and previous work has explored the specific challenge of generating math puzzles with all solutions [27], but implementing this type of model is technically challenging. Since we do not need to enforce any constraints over all problem solutions, we can take a simpler two-step approach. First, we generate a set of problem-solution pairs, and then we use a second ASP program to generate all shortest solutions for each problem.

3.3   Explanation Generation

Our approach for generating step-by-step explanations for each problem is to name the rules and predicates in the ASP program in such a way that explanations can be generated directly from the program itself. This allows us to produce explanations without having to write any problem- or solution-specific content, but requires structuring the ASP program differently than we would if we were not generating explanations from the code. Our explanations have two parts: we describe each operator that is applied to the equation, and we also provide strategy explanations that describe the priority of operator classes.

For operators, we first describe *why* an operator can be applied to the equation at this step, and then describe *how* to modify the equation to apply the operator. We generate explanation text from the declarative rules defined for each operator the ASP program. The precondition predicates in the operator rule define precisely why the operator can be applied, but we typically do not want to explain all predicates to the learner. For example, a predicate that states that a term cannot be added to itself is necessary for the solver, but not for the

---

Algorithm 1. ASP rules used to define the *add two terms* operator. The preconditions are separated into two tiers, one used to generate a high-level description of the precondition and a second used to provide a more detailed description. The second tier is also used to generate explanations for rules that *almost apply*.

```
applicable(T, weCanSimplifyByAddingTheseTwoTermsTogether(L, R))
    :- weAreAddingTwoTermsWithVariablesThatHaveTheSameDegree(T, L, R).

weAreAddingTwoTermsWithVariablesThatHaveTheSameDegree(T, L, R)
    :- _areDistinctNodes(L, R), isMonomial(T, L), isMonomial(T, R),
       areOnTheSameSideOfTheEquation(L, R),
       areBeingAdded(T, L, R), haveEqualDegrees(T, L, R),
       _isNotZero(T, L), _isNotZero(T, R).
```

---

learner. To handle these cases, we add an underscore to the beginning of predicate names that should not be explained. More importantly, we may want to describe multiple predicates through one high-level explanation. To handle these cases, we define our rules in ASP using a tiered approach that defines multiple levels of explanation detail for each operator. The first level produces a general explanation of why the rule applies, while the second level describes each of the predicates that must hold for the rule to apply. Algorithm 1 shows the ASP code that we used to model the *add like terms* operator in our system, which includes multiple rule definitions that provide different levels of explanation.

We also wanted to explain the problem-solving strategy of first simplifying the equation and then rearranging terms, which is represented through the priority of the five operator classes. We designed a dialog that presents these classes through a sequence of question-answer pairs, which start by asking whether an operator in each class can be applied (e.g. "can we combine?"). This question is answer either *no* ("no we cannot combine") or *yes* ("yes we can combine these terms"). In cases where an operator class can be applied multiple times, we note this in the response ("we can combine multiple terms"). This text is generated from five templates, one for the question and one for each possible response. The templates are populated with the current operator class and the ids of the terms to which the operator is applied. Figure 1 shows the sequence of explanations generated for applying the *combine like terms* operator to an example equation.

Our approach for automatically generating explanations requires authoring the ASP program with these explanations in mind, by abstracting predicates into multiple levels and naming the rules and predicates such that they will produce clear and understandable descriptions. While this requires significant up-front authoring effort, once the program is written explanations can be automatically generated for any problem and solution generated by the model.

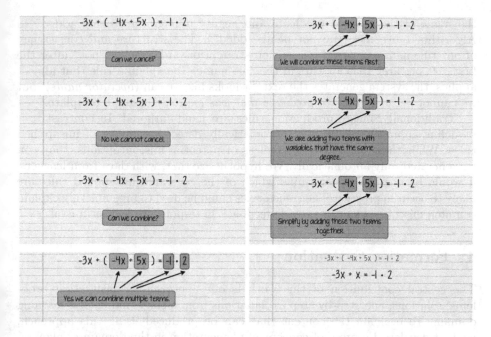

Fig. 1. An explanation sequence generated for the *add like terms* operator that describes problem-solving strategy and how and why the operator can be applied.

3.4 Misapplied Rules

One potential benefit of this modeling approach is that it provides an opportunity to automatically generate rules that describe learner misconceptions. Many intelligent tutors detect and respond to common misconceptions, typically using hand-authored "buggy" production rules [3,17]. One class of misconceptions is misapplied rules. For example, given the equation $2x * 5x = 100$, a student could mistakenly think the add-like-terms operator applies to $2x$ and $5x$, since they are monomials of the same degree on the same side of the equation. However, they are being multiplied, not added, so the operation is not valid. Modeling algebraic operators in ASP allows us to automatically detect and reason about a subset of such misapplied rules. Each operator has several predicates which make up the precondition. If nearly all the predicates hold (e.g., there are two terms, both monomials of same degree) but one such predicate is missing (e.g., terms are not being added), then such a rule *almost applies*.

Given the set of predicates that define when an operator is applicable, we can produce an exhaustive list of all rules that almost apply for a given equation by searching for those that apply when a single predicate is omitted from the rule body. As with correctly applied operators, we can automatically generate explanations for operators that almost apply from the rules themselves. We take the name of the omitted predicate and negate it (e.g., "are being added" in Algorithm 1 becomes "are not being added"). The structure and consistency of our rule names makes this negation straightforward, placing "not" after the first "is" or "are" that appears in the rule name. To explain a rule that almost applies, the system generates the text "it looks like we can ⟨*operator name*⟩, but we cannot because ⟨*negated predicate that was omitted*⟩" using a template.

Traditionally, buggy production rules are hand-authored. In contrast, our modeling approach allows us to automatically detect a wide set of misapplied rules. While these only cover a subclass of potential misconceptions, they can be generated fully automatically. We hypothesize that, with data from learners, it would be possible to determine which of the generated misapplied rules are likely to occur in practice. Then, such rules could be used to preemptively explain common misconceptions or provide feedback in response to learner mistakes.

4   Formative Evaluation

The central contribution of this work is our approach for generating problems, solutions, and step-by-step explanations from a single model of a learning domain. To evaluate the content that can be produced using this approach, we first developed a proof-of-concept implementation in the domain of algebra. We modeled algebraic problem solving in ASP as described above, and then developed an interactive algebra tutor on top of the content generated by this model. We show that our tutor can provide step-by-step worked examples, can give real-time feedback during independent problem solving, and can support a problem progression that gradually fades between the two types of scaffolding. To evaluate the proof-of-concept tutor and its content, we conducted formative user studies with Mechanical Turk workers and eighth-grade students.

4.1   Proof-of-Concept Implementation

Many design decisions go into the development of any tutoring system [29]. Our goal in this work is not to study any particular instructional approach, but rather to show the variety of scaffolds that can be implemented with our generated content. We developed an interactive tutor we call the *Algebra Notepad* that uses a gesture-based interface to emulate solving equations on paper (see Fig. 2). The application uses problems, solutions, and explanations that were generated by our ASP model offline. We implemented a scaffolded problem progression that gradually fades between step-by-step demonstrations of correct solutions [7,15,20] and independent problem solving with mistake feedback [3,6,29], a pedagogical approach known as faded worked examples [21,22,24]. Our fading policy has five levels. In *Level 1*, learners walk through example solutions step-by-step, viewing our generated strategy and operator explanations (see Fig. 1). In *Level 5*, learners solve problems independently while the system compares their steps to the correct solutions. The system displays sparkly stars in response to correctly applied operators, and gives tiered feedback messages when steps are incorrect. The remaining three levels blend these extremes, for example showing explanations but requiring the learner to perform the operations on their own.

4.2 Mechanical Turk Study

First, we conducted study on Mechanical Turk with the goal of recruiting a relatively large number of users to try the *Algebra Notepad* application. We used this study to evaluate whether our scaffolds helped participants solve problems, and whether the generated explanations are understandable. We generated a static progression of nine problems for this study, six of which required a minimum of four steps to reach a solution and three of which required three steps. We used the fading policy described in the previous section; the progression started with one worked example at *Level 1*, followed by two problems each at *Level 2, 3, 4* and *5*. Participants took a three-problem test before and after using the *Algebra Notepad*, and completed a short survey about the experience at the very end.

We collected data from 57 Mechanical Turk workers, who provided informed consent and were paid for their time. First, we measured whether their problem-solving performance improved after practicing with the *Algebra Notepad*. A repeated measures ANOVA showed that participants performed better on the post-test ($F(1, 56) = 8.02$, $p < 0.01$), with a mean score of 2.4 out of three correct on the pre and 2.6 on the post. We also counted the number of steps used in correct solutions, and found that participants used fewer steps on the post-test ($F(1, 50) = 80.06$, $p < 0.0001$), with a mean of 5.1 steps per problem on the pre and 4.3 steps on the post. These findings suggest that practicing with the algebra tutor improved the correctness and efficiency of participant's solutions.

We also analyzed log data from the *Algebra Notepad* to measure how participants performed during independent problem solving (fading *Level 5*). We found that participants applied the correct operator on the first try in 84.5%

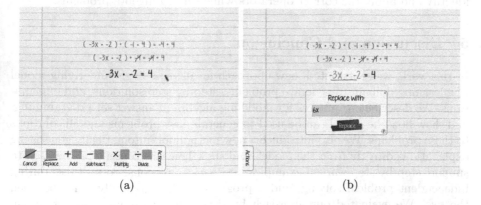

(a)   (b)

Fig. 2. Screenshots of the *Algebra Notepad* application. The interface displays each step on a separate line, and learners manipulate equations using gestures, as shown in the actions bar in (a). (b) shows a *replace* gesture.

of cases. Using tiered feedback, they applied the correct operator on the second or third try in 10.3% of cases, and only needed the system to perform the operation for them in 5.1% of cases. This shows that the generated feedback helped participants reach correct solutions most of the time. On the final survey, participants agreed that the explanations *"were clear and understandable"* and *"helped me solve problems"*, rating these statements an average of 4.8 and 4.7 on a six-point Likert scale respectively. When responding to a question asking what they thought of the explanations, one participant said *"I thought they were great. It has been years since I've done algebra and the explanations on the notepad refreshed by memory and improved my ability to solve problems correctly."*

4.3   Student Study

Since adults are not the target population of our *Algebra Notepad*, we conducted a second informal user study with seven eighth-grade students at a local Boys & Girls club to confirm that learners can successfully interact with the content generated with our model. In this study, students used the *Algebra Notepad* application to complete a static progression of 20 problems. The first nine problems in the progression were identical to those used in the Mechanical Turk study, and the same fading policy was used. However, we added an additional 11 problems at *Level 5*, where students worked on problems independently and were given feedback in response to mistakes as needed. The seven students all completed the progression of 20 problems. We analyzed their log data, and saw that for the 13 problems in *Level 5* that required independent problem-solving, students applied the correct operator on the first try in 81.4% of cases. The applied the correct operator on the second or third try, with the help of the tiered feedback, in 16.3% of cases. They only needed the system to apply the correct operator in 2.3% of cases. This suggests that most students were able to use the explanations and corrective feedback generated through our model to identify and apply the correct operators while solving algebra problems.

5   Discussion and Conclusion

In this work, we contribute a new approach for using a single underlying model of a learning domain to generate problems, step-by-step solutions, and explanations. We describe our process for modeling algebra in answer set programming, and show how the model can be used to generate new problems and all solutions to those problems. We also introduce a new method for automatically generating explanations directly from the model, and show how this content can be used to support step-by-step worked examples, feedback in response to mistakes during independent problem solving, and a progression that gradually fades between the two. We evaluated our approach by developing a proof-of-concept implementation of an intelligent tutor that uses content generated by our model, and we show that both adult users and eighth-grade students can interact with our explanations to successfully solve algebra problems.

We believe this modeling approach has exciting potential for supporting robust automated content generation for intelligent tutoring systems in the future. While this work focuses on the domain of algebra, ASP can be used to model any domain that can be represented through if-then relationships, where learners determine when rules or conditions apply and take actions in response. Logic programming languages have already been used to model diverse learning domains such as math procedures [2], word problems [19], and game puzzles [5,28]. While this work takes an important first step towards understanding how to construct an intelligent tutor around an ASP model, it has a number of limitations. We hope future research can continue this line of work, in particular expanding on our formative evaluation to determine whether content generated using this approach can effectively support student learning in real-world contexts.

References

1. Ahmed, U.Z., Gulwani, S., Karkare, A.: Automatically generating problems and solutions for natural deduction. In: Proceedings of the Twenty-Third International Joint Conference on Artificial Intelligence, IJCAI 2013, pp. 1968–1975. AAAI Press (2013). http://dl.acm.org/citation.cfm?id=2540128.2540411
2. Andersen, E., Gulwani, S., Popović, Z.: A trace-based framework for analyzing and synthesizing educational progressions. In: Proceedings of the SIGCHI Conference on Human Factors in Computing Systems, CHI 2013, pp. 773–782. ACM, New York (2013). https://doi.org/10.1145/2470654.2470764
3. Anderson, J.R., Corbett, A.T., Koedinger, K.R., Pelletier, R.: Cognitive tutors: lessons learned. J. Learn. Sci. 4(2), 167–207 (1995)
4. Anderson, J.R., Pelletier, R.: A development system for model-tracing tutors. In: Proceedings of the International Conference of the Learning Sciences, pp. 1–8 (1991)
5. Butler, E., Andersen, E., Smith, A.M., Gulwani, S., Popovic, Z.: Automatic game progression design through analysis of solution features (2015)
6. Corbett, A., Koedinger, K.R., Anderson, J.R.: Intelligent tutoring systems. In: Helander, M., Landauer, T.K., Prahu, P. (eds.) Handbook of Human-Computer Interaction, 2nd edn, pp. 849–874. Elsevier Science, Amsterdam (1997)
7. van Gog, T., Paas, F., van Merriënboer, J.J.: Process-oriented worked examples: improving transfer performance through enhanced understanding. Instr. Sci. 32(1–2), 83–98 (2004). https://doi.org/10.1023/B:TRUC.0000021810.70784.b0
8. Heffernan, N.T., Heffernan, C.L.: The assistments ecosystem: building a platform that brings scientists and teachers together for minimally invasive research on human learning and teaching. Int. J. Artif. Intell. Educ. 24(4), 470–497 (2014). https://doi.org/10.1007/s40593-014-0024-x
9. Koedinger, K.R., Aleven, V., Heffernan, N., McLaren, B., Hockenberry, M.: Opening the door to non-programmers: authoring intelligent tutor behavior by demonstration. In: Lester, J.C., Vicari, R.M., Paraguaçu, F. (eds.) ITS 2004. LNCS, vol. 3220, pp. 162–174. Springer, Heidelberg (2004). https://doi.org/10.1007/978-3-540-30139-4_16
10. Koedinger, K.R., Anderson, J.R., Hadley, W.H., Mark, M.A.: Intelligent tutoring goes to school in the big city. Int. J. Artif. Intell. Educ. 8, 30–43 (1997)

11. Koedinger, K.R., Brunskill, E., de Baker, R.S.J., McLaughlin, E.A., Stamper, J.C.: New potentials for data-driven intelligent tutoring system development and optimization. AI Mag. **34**(3), 27–41 (2013)

12. Koedinger, K.R., Heffernan, N.: Toward a rapid development environment for cognitive tutors. In: Proceedigns of the International Conference on Artificial Intelligence in Education, pp. 455–457. IOS Press (2003)

13. Martin, B., Mitrovic, A.: Automatic problem generation in constraint-based tutors. In: Cerri, S.A., Gouardères, G., Paraguaçu, F. (eds.) ITS 2002. LNCS, vol. 2363, pp. 388–398. Springer, Heidelberg (2002). https://doi.org/10.1007/3-540-47987-2_42

14. Martin, B.I.: Intelligent tutoring systems: the practical implementation of constraint-based modelling. Ph.D. thesis, University of Canterbury (2001)

15. McLaren, B.M., Lim, S.J., Koedinger, K.R.: When and how often should worked examples be given to students? New results and a summary of the current state of research. In: Love, B.C., McRae, K., Sloutsky, V.M. (eds.) Cognitive Science Society, pp. 2176–2181. Cognitive Science Society, Austin (2008)

16. Mitrovic, A.: Fifteen years of constraint-based tutors: what we have achieved and where we are going. User Model. User-Adapt. Interact. **22**(1–2), 39–72 (2012). https://doi.org/10.1007/s11257-011-9105-9

17. Mitrovic, A., Koedinger, K.R., Martin, B.: A comparative analysis of cognitive tutoring and constraint-based modeling. In: Brusilovsky, P., Corbett, A., de Rosis, F. (eds.) UM 2003. LNCS (LNAI), vol. 2702, pp. 313–322. Springer, Heidelberg (2003). https://doi.org/10.1007/3-540-44963-9_42

18. Ohlsson, S.: Constraint-based student modeling. In: Greer, J.E., McCalla, G.I. (eds.) Student Modelling: The Key to Individualized Knowledge-Based Instruction, vol. 125, pp. 167–189. Springer, Heidelberg (1994). https://doi.org/10.1007/978-3-662-03037-0_7

19. Polozov, O., O'Rourke, E., Smith, A., Zettlemoyer, L., Gulwani, S., Popović, Z.: Personalized mathematical word problem generation. In: Proceedings of the 24th International Joint Conference on Artificial Intelligence, IJCAI (2015)

20. Renkl, A.: Learning from worked-out examples: a study on individual differences. Cogn. Sci. **21**(1), 1–29 (1997)

21. Renkl, A., Atkinson, R.K., Große, C.S.: How fading worked solution steps works - a cognitive load perspective. Instr. Sci. **32**(1–2), 59–82 (2004). https://doi.org/10.1023/B:TRUC.0000021815.74806.f6

22. Renkl, A., Atkinson, R.K., Maier, U.H., Staley, R.: From example study to problem solving: smooth transitions help learning. J. Exp. Educ. **70**(4), 293–315 (2002). http://www.jstor.org/stable/20152687

23. Sadigh, D., Seshia, S.A., Gupta, M.: Automating exercise generation: a step towards meeting the MOOC challenge for embedded systems. In: Proceedings of the Workshop on Embedded and Cyber-Physical Systems Education, WESE 2012, pp. 2:1–2:8. ACM, New York (2013). https://doi.org/10.1145/2530544.2530546

24. Salden, R.J.C.M., Aleven, V.A.W.M.M., Renkl, A., Schwonke, R.: Worked examples and tutored problem solving: redundant or synergistic forms of support? Top. Cogn. Sci. **1**(1), 203–213 (2008). https://doi.org/10.1111/j.1756-8765.2008.01011.x

25. Shanahan, M.: The event calculus explained. In: Wooldridge, M.J., Veloso, M. (eds.) Artificial Intelligence Today. LNCS (LNAI), vol. 1600, pp. 409–430. Springer, Heidelberg (1999). https://doi.org/10.1007/3-540-48317-9_17

26. Singh, R., Gulwani, S., Rajamani, S.: Automatically generating algebra problems. In: Proceedings of the Twenty-Sixth AAAI Conference on Artificial Intelligence (2012)
27. Smith, A., Butler, E., Popović, Z.: Quantifying over play: constraining undesirable solutions in puzzle design. In: Proceedings of the 8th International Conference on the Foundations of Digital Games (2013)
28. Smith, A.M., Andersen, E., Mateas, M., Popović, Z.: A case study of expressively constrainable level design automation tools for a puzzle game. In: FDG 2012: Proceedings of the Seventh International Conference on the Foundations of Digital Games. ACM, New York (2012)
29. VanLehn, K.: The behavior of tutoring systems. Int. J. Artif. Intell. Educ. **16**, 227–265 (2006)
30. Vanlehn, K., et al.: The Andes physics tutoring system: five years of evaluations. In: In Proceedings of the 12th International Conference on Artificial Intelligence in Education, pp. 678–685. IOS Press (2005)

Generalizability of Methods for Imputing Mathematical Skills Needed to Solve Problems from Texts

Thanaporn Patikorn[(✉)] ⓘ, David Deisadze, Leo Grande, Ziyang Yu, and Neil Heffernan ⓘ

Worcester Polytechnic Institute, Worcester, MA 01609, USA
{tpatikorn,dodeisadze,lgrande,zyu,nth}@wpi.edu

Abstract. Identifying the mathematical skills or knowledge components needed to solve a math problem is a laborious task. In our preliminary work, we had two expert teachers identified knowledge components of a state-wide math test and they only agreed only on 35% of the items. Previous research showed that machine learning could be used to correctly tag math problems with knowledge components at about 90% accuracy over more than 100 different skills with five-fold cross-validation. In this work, we first attempted to replicate that result with a similar dataset and were able to achieve a similar cross-validation classification accuracy. We applied the learned model to our test set, which contains problems in the same set of knowledge component definitions, but are from different sources. To our surprise, the classification accuracy dropped drastically from near-perfect to near-chance. We identified two major issues that cause of the original model to overfit to the training set. After addressing the issues, we were able to significantly improve the test accuracy. However, the classification accuracy is still far from being usable in a real-world application.

Keywords: Natural language processing · Knowledge component · Multiclass classification · Generalizability

1 Introduction

One of the most important skills teachers need to have is the ability to recognize which sets of skills are needed to solve specific problems. While teaching, many teachers solve practice problems as an example. Those problems are also often used as homework practices to help students learn and to allow teachers to be able to measure student knowledge. For students who are unable to reach satisfactory level of knowledge, it is also common for teachers to give a student a few more problems as a chance to show their improvements [9]. Thus, it is important for teachers, educators, content providers, publishers, and researchers to use the same categorization of skills, also known as knowledge components (KCs), as

© Springer Nature Switzerland AG 2019
S. Isotani et al. (Eds.): AIED 2019, LNAI 11625, pp. 396–405, 2019.
https://doi.org/10.1007/978-3-030-23204-7_33

vocabularies for their communication and interaction. Common Core State Standard (CCSS, www.corestandards.org) is one of the most common categorizations of knowledge components skills in English language arts and mathematics from kindergarten to high school in United States. CCSS provides both broad and in-depth specific descriptions of skills, which is often accompanied by an example problem belonging to the skill. Figure 1 shows Common Core definitions of two 8th grade skills in expressions and equations.

CCSS.MATH.CONTENT.8.EE.A.1	CCSS.MATH.CONTENT.8.EE.A.2	CCSS.MATH.CONTENT.8.EE.A.3
CCSS.MATH.CONTENT.8.EE.A.4	CCSS.MATH.CONTENT.8.EE.B.5	CCSS.MATH.CONTENT.8.EE.B.6
CCSS.MATH.CONTENT.8.EE.C.7	CCSS.MATH.CONTENT.8.EE.C.8	

Expressions and Equations Work with radicals and integer exponents.
CCSS.MATH.CONTENT.8.EE.A.1
Know and apply the properties of integer exponents to generate equivalent numerical expressions. For example, $3^2 \times 3^{-5} = 3^{-3} = 1/3^3 = 1/27$.

CCSS.MATH.CONTENT.8.EE.A.2
Use square root and cube root symbols to represent solutions to equations of the form $x^2 = p$ and $x^3 = p$, where p is a positive rational number. Evaluate square roots of small perfect squares and cube roots of small perfect cubes. Know that $\sqrt{2}$ is irrational.

Grade 3
Grade 4
Grade 5
Grade 6
Grade 7
Grade 8
   Introduction
   The Number System
 ▸ **Expressions & Equations**
   Functions
   Geometry
   Statistics & Probability

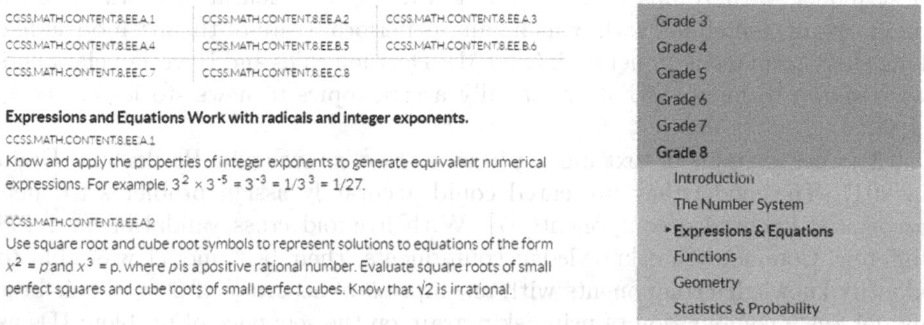

Fig. 1. An example of Common Core definitions from 8th grade skills in expressions and equations.

In recent years, there has been a significant increase of digital devices in the classroom. Such devices enable teachers and students to use learning management systems (LMSs), such as Google Classroom (classroom.google.com) and Schoology (www.schoology.com). These LMS tools are designed to help teachers organize their classrooms and classwork, improve communication between teachers and students, and provide students with help such as instant feedback for their homework. In addition, LMSs allow teachers to easily access course materials provided by other teachers, content creators, or publishers through Learning Tools Interoperability (LTI). As such, skill standards such as Common Core are now more important than ever, as they reduce miscommunication and ensures that teachers can navigate to the right materials, especially through content sharing such as LTI.

Tagging problems with their associated skills or knowledge components are usually done manually, often by experienced teachers and experts in the fields of learning. Identifying knowledge components is also hard, even for experts. In our preliminary study, we had two expert teachers identify knowledge components of a 37-item state-wide math test called TerraNova and they only agreed only on 13 items (35%). In addition, skill tagging is a very tedious and time consuming task [5]. With the growing pools of content created and shared through online systems, there is a need for a method that can identify knowledge components quickly and accurately, using the only information that is consistent across problems from different systems: the content of problems. In the field of machine learning, techniques used to extract information from text are called text mining and natural language processing (NLP). With the interactions of many systems through LTI, generalizability of models is the top priority since the input problems in real applications may be from different sources, authored for different intents, and come in different formats. Generalizability will be the main focus of this work.

In the field of education, there have been multiple usages of text mining and NLP to help automate laborious tasks. Zhao et al. applied several types of neural network-based models to automatically grade English essays [10]. Their best model, with Kappa of 0.78, was a variant of a neural network called a memory-augmented network, which can outperform state-of-the-art models like long short-term memory network (LSTM). Decision rules and Bayesian classifiers were shown to be able to automatically assign topics to news stories correctly [1, 6].

Another example of text mining in education is a work by Pardos and Dadu in 2017. The model they presented could accurately assign problems to their associated knowledge components [7]. With five-fold cross validation and 198 different Common Core knowledge components, their best model was able to identify knowledge components with an impressive accuracy of 90%. Their best model was a combination of using skip-gram on the sequence of problem IDs as they are encountered by students, and a neural network on bag of words of the problem text.

In this work, we replicated the methodology presented in [7] on a similar set of problems from the same source they used. In addition, as our work is driven by the need for the models to be accurate both problems within the same system and problems from other sources, i.e. problems created by different teachers, textbooks and publishers. Thus, the main focus of this work is applying the trained models to different sets of problems, and find the best hyper-parameters and preprocessing that allow the model to generalize effectively.

2   Replicability

Our first step to replicate the results from [7] was to obtain a dataset. We chose to use problems from a web-based LMS called ASSISTments (www.assistments. org). We decided to use only certified Skill Builder problems for K-12 mathematics as our training sets because these problems are officially curated and maintained by experts from the ASSISTments system. In addition, each ASSISTments Skill Builder is problem set of a large number of similar problems specifically created to help students learn a specific Common Core State Standard. Thus, these problems match ground truth labels of KCs. The dataset used in [7] also came from ASSISTments, which led us to expect results similar to theirs. Our final dataset includes 65,120 problems from 336 problem sets belonging to 173 different skill standards. The minimum number of problems belonging to a single skill standard is 14 problems and the maximum is 6,480 problems. All problems are formatted using HTML, which is a standard markup for text display in web pages. An example of a problem formatted using html is shown in Fig. 2.

2.1   Text Representation and Preprocessing

For text representation, we used the bag-of-word technique similarly to [7]. Bag-of-word is a technique that converts text into a vector representation of the size of the vocabulary. Each element of the vector represents how many times each word in appears in the text. After we get the bag-of-word representation of each problem text, we divide each element of the vector by the sum of all elements. This process is also called L1 normalization.

We decided to keep all html tags and entities in our bag of words. However, unlike in [7] where html elements are used directly as input to the models, we transformed html tags and many symbols that are often used in mathematical problems such as less than ($<$), greater than ($>$), and equal ($=$) into special "words" that corresponds to each html tag and entity. For instance, $<$img$>$ π are encoded as htmltag_img and htmlentity_pi. All html attributes (such as links, text colors, and text sizes) and html syntax elements (such as closing tags) are removed. The main idea of this text replacement is to keep relevant special formatting and symbols, such as superscript and subscript, with discarding other information that are not directly related to math knowledge such as text colors and font sizes. In addition, some content sources may use different schemes to format their contents other than html. This transformation would allow models to be used on problems with equivalent formatting.

2.2   Replicated Model

After preprocessing the dataset, we applied various machine learning models. We evaluated each model by calculating its prediction accuracy with five-fold cross-validation. We explored using three different common machine learning models: artificial neural networks, decision trees, and random forests. Artificial neural networks, or ANNs, are models that are inspired by how human nerve cells (neurons) connect and communicate. ANNs have been shown to work very well on text processing, including in [7].

Decision trees are another type of model that have been shown in [1] and [6] to be able to do well in text classification tasks. The main benefit of decision trees are simplicity and interpretability of the models. Since in our dataset, certain skill contains only one sample, we chose to train all decision trees with minimum leaf size of 1.

Random forests are tree-based, ensemble machine learning models [2]. Random forests utilize feature bagging and bootstrapping to achieve both flexibility and prediction powers, while being resistant to overfitting. Random forests have been widely used in many fields from biology [4], text mining [8], and UMAP [3]. We chose the minimum leaf size of 1, similarly to decision trees, with 10 random trees in a random forest.

2.3   Replicability Result

We were able to successfully replicate the results from [7] using the methodology they described. For each method, only the model with the best 5-fold

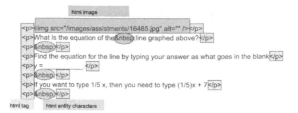

Fig. 2. An example of a problem formatted using html

cross-validation accuracy is included. Our best model, which is shown in Table 1, which uses all the html information of all the problems in the dataset, was able to achieve 92.47% accuracy with five-fold cross-validation in our dataset. Removing html markups reduces the 5-fold cross-validation. While Random Forest is the best model, the 5-fold cross-validation accuracy for the other two models are only 1%–2% lower than that of the Random Forest.

Table 1. Results from using bag-of-word approach with different models using all problems and transformed html markups.

Preprocessing	Model	5-Fold cross-validation accuracy	Illustrative math. problem accuracy
Keep all html markups	Decision tree	92.47%	12.14%
Keep all html markups	Neural network	92.07%	13.67%
Keep all html markups	Random forest	92.74%	6.98%

2.4   Does It Generalize?

Since our goal is to develop a model that can identify knowledge components of problems from different sources and authors, we chose to obtain a second dataset for our generalizability test. We chose problems from Illustrative Mathematics as our test set. Illustrative Mathematics (www.illustrativemathematics.org) is an open and free mathematics curriculum. We chose problems from Illustrative Mathematics because (1) the problems are created with Common Core State Standard in mind, meaning we have the ground truth KCs, (2) it is an open and free educational resource widely used by teachers across the United States, and (3) all problems from Illustrative Mathematics we compiled are in html format similar to the training set. We compiled together 1,581 problems from grade 7 and 8, belonging to 114 different skill standards. The number of problem per skill standard ranged from 4 to 71. We removed all problems belonging to skills outside of our training set. Our final test set contains 392 problems from 23 skills. Afterward, we retrained the model using all training data, and uses on the problems from Illustrative Mathematics. To our surprise, the accuracy dropped drastically as shown in Table 1. While the performance of all models is still better than chance, it is far from usable in a real-world application.

2.5   Causes of Overfitting

We investigated what could have caused such a massive overfitting by looking through our dataset and models. We found two potential causes of the overfitting. The first cause stems from a large number of near-identical problems. In order to create a large number of math problems, it is common for content creators to create a few templates and substitute different numbers and keywords in the problems and answers. For instance, a teacher might create "Train A from New Mexico to Nevada leaves at TIME_A at SPEED_A mph. Train B from Nevada to New Mexico leaves at TIME_B at SPEED_B mph. The two stations are 900 miles apart. What time will the two trains meet each other?" With this template, the teacher can substitute TIME_A, SPEED_A, TIME_B, SPEED_B with different numbers to create a massive number of practice problems. Specifically, Out of 65,120 problems in our dataset, only 2,523 problems are created without using templates. For the other 62,597 problems, there are at least 1,193 different templates, each of which has been used for more than 10 problems.

Using cross-validation without regard of templates could potentially mislead models to remember the specific words in templates rather than to learn the terminologies of skills, causing their cross-validation accuracy to inflate. In our train example, the model may choose to remember the word "New Mexico" and "Nevada", instead of "mph", "miles", and "time". Pardos and Dadu were aware of this issue of their best model and attempted to solve it by doing cross-validation in such a way that all problems from the same problem set were in the same fold, which significantly reduced their accuracy to around 70%. This approach did not resolve all the issue with templates, since problems created using the same template also exist outside of the problem sets.

The second cause of overfitting stems from the html elements and formatting included in the models. While the html elements are shown to improve cross-validation accuracy in [7], we found that it causes the decision trees to over-prioritize the formatting in the decision. In addition, templates also contribute to overfitting here since the formatting is also copied over to each problem from the same template. This causes the models to be unable to identify the knowledge components once the formatting "styles" have been changed.

3   Towards Generalizability

After we have identified potential causes of overfitting, we came up with multiple ways to address the two issues.

3.1   Near-Identical Problems

In order to address the issue with templates, we removed all but one problems that are created using the same templates. Luckily, the problems inside our dataset also contain the information on creation, specifically if it is a (modified) copy of another problem. We used that information to remove all but one problem derived from each template. As a result, the size of our dataset is significantly reduced to 2,474 problems. The number of problem sets and skills remain the same (336 problem sets, 173 different skills). The number of problems per skill standard is also reduced to a minimum of 1 and a maximum of 198 problems per skill.

3.2   HTML Element and Formatting

In order to address the html formatting issue, we introduced two approaches to process the html elements. The first approach is to remove all html tags and entities. The goal of this approach is that some formatting schemes may not be equivalent or convertible to html, rendering our first approach unusable. In fact, for some math topics, problems can be written entirely in plain text (i.e. no formatting). This approach is advantageous because the model will be usable on problems of any formats (or no format), albeit often with some loss of information.

The second approach is keep only important html elements. The goal of this approach is to have certain html elements act as keywords which, when combined with other words, can be indicative of the knowledge components. For instance, the words "read" and "graph" together with an image could indicate that in this problem, the student needs to be able to extract information from a visual representation of a graph. In this work, we only include tables and graphs as important keywords.

Table 2. Results from using bag-of-word approach with different preprocessing and models using only non-template problems.

Preprocessing	Model	5-Fold cross-validation accuracy	Illustrative math. problem accuracy
Keep all html markups	Decision tree	62.53%	10.85%
Keep all html markups	Neural network	68.23%	16.26%
Keep all html markups	Random forest	65.24%	12.91%
Keep only image and table	Decision tree	59.71%	12.67%
Keep only image and table	Neural network	63.62%	22.19%
Keep only image and table	Random forest	60.97%	9.56%
Remove all html markups	Decision tree	58.73%	10.33%
Remove all html markups	Neural network	63.80%	22.47%
Remove all html markups	Random forest	61.22%	4.92%

4   Results

In order to compare with the models we have in the replication section, we followed the same methodology we used there on each of the different preprocessing approaches to the dataset. The results are shown in Table 2. In general, the 5-fold cross-validation accuracy is much lower than that of the replication, which is to be expected since a large number of near-identical problems are removed. The best model for both training set and test set is a neural network model trained on no html markup at all. The best test accuracy is 22.47%, which almost that of the best replication model. This confirms our suspicion that the near-identical problems and the excessive/irrelevant formatting markups cause the model to overfit.

We also investigated what caused our model to fail. Figure 3 shows the confusion matrix of the our best model (no html markup, neural network) on the training set during 5-fold cross-validation prediction. The green dots indicate where the model correctly classified the problems, the red dots indicate where the model is wrong, and the intensity of the colors is proportional to the percent of correct/incorrect classification. A large number of strong green dots indicates

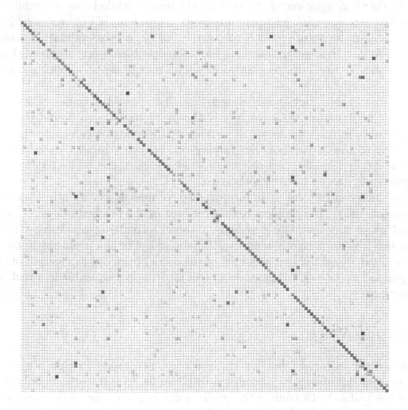

Fig. 3. Confusion Matrix from 5-fold cross validation of models using only non-template problems. Green denotes true positive. Red denotes false positive/negative. The intensity of color is the magnitude of correct/incorrect classification. (Color figure online)

that our models are able recognize the large number different KCs. There are also a large number of red dots, indicating that the model is unable to recognize some KCs, many of which are because there are not enough samples (e.g. only 1 non-template sample from that KC). Interestingly, not many red dots fall on the same columns. This implies that the model is not quite biased toward any specific KCs. The codes we used in this work can be found here: https://drive.google.com/open?id=1yyJgJavBdAyPbKsFSuhI6TRN_cDRcRM7.

5   Conclusion

While cross-validation has been regarded as a gold-standard technique to investigate generalizability of models, in this work we showed that it is also important to investigate generalizability using a separated test dataset that will emulate input data from real application. There are three main contributions of this paper.

The first contribution is the replication of the result from [7]. We were able to successfully replicate their work using the methodology presented. However, after further investigation, we found that the model, while being internally valid and consistent, was unable to generalize well to problems of different sources. We highly recommend future models to not rely solely on cross-validation accuracy for this problem. Instead, researchers should use problem texts obtained through different sources to ensure generalizability. This insight is also applicable to other domains as well.

Our second contribution is the investigation on potential causes of overfitting of models for imputing KCs using only the problem text. We found that near-identical data in the training set as well as html formatting can cause the model to over fit to the "styles" of the training set, reducing its generalizability. We also found that including just the indicators for images and tables do no increase generalizability as we have hypothesized.

The third contribution is an improvement of models for imputing knowledge components using only problem texts. Our best model is a neural network model trained using non-template problems without formatting markups. Our best model was able to correctly identify KCs about 1/4 of the problems in our test set, almost doubled the test accuracy of the model from our replication of [7]. It is important to note that this model is still far from being usable in a real world application.

6   Future Work

There are several areas that we chose not to investigate in this project. For instance, math problems can require multiple skills. In this work, we only choose one of the skills. With conjunctive skill representation, the models can be significantly improved. In this work, we did not individually tune each model. So, it is possible that the models, one finely tuned, may be able to generalize even better than the result presented in this work.

In this work, we did not use any information that Common Core State Standards provided with the skills. For instance, in Fig. 1, we treated 8.EE.A.1 and 8.EE.A.2 are two totally different KCs. However, we believe that a model that can utilize the information on the skill domains and their hierarchies, will perform much better. For instance, a human would be able to recognize that both of those skills are 8th grade Expressions and Equations, and they're also a part of "Expression and Equations Work with radicals and integer exponents".

In addition to improvements to models, another area of future work is to apply the trained models to real applications. One such application that we are currently is a module for ASSISTments that automatically detects the skill standard of problems inside a problem set teachers assigned to students. Then, the system suggests a list of problems belonging to the same set of skill standards as next-day review problems.

Acknowledgements. We thank multiple current NSF grants (IIS-1636782, ACI-1440753, DRL-1252297, DRL-1109483, DRL-1316736, DGE-1535428 & DRL-1031398), the US Dept. of Ed (IES R305A120125 & R305C100024 and GAANN), and the ONR.

References

1. Apté, C., Damerau, F., Weiss, S.M.: Automated learning of decision rules for text categorization. ACM Trans. Inf. Syst. (TOIS) **12**(3), 233–251 (1994)
2. Breiman, L.: Random forests. Mach. Learn. **45**(1), 5–32 (2001)
3. Cheng, H., Rokicki, M., Herder, E.: The influence of city size on dietary choices. In: Adjunct Publication of the 25th Conference on User Modeling, Adaptation and Personalization, pp. 231–236. ACM (2017)
4. Immitzer, M., Atzberger, C., Koukal, T.: Tree species classification with random forest using very high spatial resolution 8-band worldview-2 satellite data. Remote Sens. **4**(9), 2661–2693 (2012)
5. Karlovčec, M., Córdova-Sánchez, M., Pardos, Z.A.: Knowledge component suggestion for untagged content in an intelligent tutoring system. In: Cerri, S.A., Clancey, W.J., Papadourakis, G., Panourgia, K. (eds.) ITS 2012. LNCS, vol. 7315, pp. 195–200. Springer, Heidelberg (2012). https://doi.org/10.1007/978-3-642-30950-2_25
6. Lewis, D.D., Ringuette, M.: A comparison of two learning algorithms for text categorization. In: Third Annual Symposium on Document Analysis and Information Retrieval, vol. 33, pp. 81–93 (1994)
7. Pardos, Z.A., Dadu, A.: Imputing KCS with representations of problem content and context. In: Proceedings of the 25th Conference on User Modeling, Adaptation and Personalization, pp. 148–155. ACM (2017)
8. Svetnik, V., Liaw, A., Tong, C., Culberson, J.C., Sheridan, R.P., Feuston, B.P.: Random forest: a classification and regression tool for compound classification and QSAR modeling. J. Chem. Inf. Comput. Sci. **43**(6), 1947–1958 (2003)
9. Wang, Y., Heffernan, N.T.: The effect of automatic reassessment and relearning on assessing student long-term knowledge in mathematics. In: Trausan-Matu, S., Boyer, K.E., Crosby, M., Panourgia, K. (eds.) ITS 2014. LNCS, vol. 8474, pp. 490–495. Springer, Cham (2014). https://doi.org/10.1007/978-3-319-07221-0_61
10. Zhao, S., Zhang, Y., Xiong, X., Botelho, A., Heffernan, N.: A memory-augmented neural model for automated grading. In: Proceedings of the Fourth (2017) ACM Conference on Learning@ Scale, pp. 189–192. ACM (2017)

Using Machine Learning to Overcome the Expert Blind Spot for Perceptual Fluency Trainings

Martina A. Rau[✉], Ayon Sen, and Xiaojin Zhu

University of Wisconsin – Madison,
1025 W Johnson Street, Madison, WI 53706, USA
marau@wisc.edu

Abstract. Most STEM domains use multiple visual representations to illustrate complex concepts. While much research has focused on helping students make sense of visuals, students also have to become perceptually fluent at translating among visuals fast and effortlessly. Because perceptual fluency is acquired via implicit, nonverbal processes, perceptual fluency trainings provide simple classification tasks that vary visual features across numerous examples. Prior research shows that learning from such trainings is strongly affected by the sequence of the examples. Further, prior research shows that perceptual fluency trainings are most effective for high-performing students but may confuse low-performing students. We propose that a lack of benefits for low-performing students may result from a perceptual expert blind spot of instructors who typically develop perceptual fluency trainings: expert instructors may be unable to anticipate the needs of students who do not see meaningful information in the visuals. In prior work, we used a machine-learning approach to develop a sequence of example visuals of chemical molecules for low-performing students. This study tested the effectiveness of this sequence in comparison to an expert-generated sequence in a randomized experiment as part of an undergraduate chemistry course. We determined students' performance based on log data from an educational technology they used in the course. Results show that the machine-learned sequence was more effective for low-performing students. The expert sequence was more effective for high-performing students. Our results can inform the development of perceptual-fluency trainings for adaptive educational technologies.

Keywords: Multiple visuals · Perceptual fluency · Sequencing · Machine learning

1 Introduction

Visual representations are often used to illustrate concepts in science, technology, engineering, and math (STEM) instruction [1–3]. For example, chemistry instruction on bonding typically uses visuals such as Lewis structures and space-filling models of molecules (see Fig. 1) [4]. Multiple visual representations can enhance learning because they provide complementary information about the to-be-learned concepts [5–7]

© Springer Nature Switzerland AG 2019
S. Isotani et al. (Eds.): AIED 2019, LNAI 11625, pp. 406–418, 2019.
https://doi.org/10.1007/978-3-030-23204-7_34

(e.g., the Lewis structure shows how many electrons form bonds, the space-filling model shows the geometry of the molecule). However, multiple visual representations can impede learning if students are unable to make such connections among them [7, 8].

(a)   **(b)**

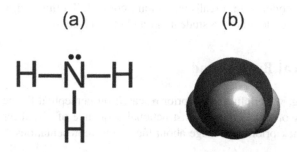

Fig. 1. (a) Lewis structure and (b) space-filling model of ammonia.

Most prior research on connection making among visuals has focused on helping students make sense of the connections by prompting them to explain differences and similarities between visuals [5, 9, 10]. For example, a student has to explain similarities in how the visuals in Fig. 1 show atoms: the H's in the Lewis structure in Fig. 1a correspond to the white spheres in the space-filling model in Fig. 1b because both show hydrogen atoms. They have to explain differences, for example that the dots in the Lewis structure show electrons that are not shown in the space-filling model.

By contrast, little research has focused on the role of *perceptual fluency* in students' learning: that is, the ability to quickly and effortlessly integrate information from multiple visuals [7, 11]. For example, students need to immediately see that the visuals in Fig. 1 show the same molecule and translate between them as fluently as bilinguals translate between languages [7, 12, 13]. Perceptual fluency frees cognitive resources for future learning and effortful conceptual reasoning [13, 14]. Because perceptual fluency is acquired via implicit, inductive, nonverbal processes [15, 16], instructional trainings enhance perceptual fluency by exposing students to a sequence of many simple problems that ask them to quickly judge what a visual shows [7, 11]. Perceptual-fluency training sequences make use of the contrasting cases principle, so that consecutive examples vary visual features so they draw students' attention to relevant features [7, 11]. However, while such trainings have proven effective for high-performing students, they are often ineffective for low-performing students [17, 18].

It is possible that the ineffectiveness of perceptual-fluency trainings for low-performing students is due to a perceptual expert blind spot on the part of the designers of instructional sequences. Sequences that present contrasting cases may be appropriate for high-performing students who already have a preliminary understanding of which visual features are meaningful. However, such sequences may be ineffective for low-performing students who have little prior knowledge about these features. For an instructional designer who is an expert in processing the visuals, it may be difficult to empathize with students who do not "see" meaning in the visuals [3, 7, 19].

In prior work [38], we used a machine-learning algorithm to develop a sequence of visuals for low-performing students. The machine-learned sequence was more effective than an expert sequence for participants from Amazon's Mechanical Turk (MTurk) service. However, it is unclear whether the machine-learned sequence is more effective than an expert sequence in a realistic learning context. To this end, we conducted an experiment with undergraduate students in a chemistry course.

2 Theoretical Background

In the following, we briefly review prior research on perceptual fluency as well as our own prior work on developing an instructional sequence of visual representations for students who lack prior knowledge about the visual representations.

2.1 Inductive Learning of Perceptual Fluency

In contrast to a large body of research on verbally mediated explanation-based sense making of visuals (see [5, 7] for overviews), research on the role of perceptual fluency in education is still relatively novel. This line of research builds on the expert-novice literature, which shows that experts see meaningful connections among visuals like the ones in Fig. 1 quickly and almost automatically [15, 16, 20, 21]. Experts are able to "at a glance" see meaning in visuals because translating and combining information from them takes little or no cognitive effort [11, 22]. This highlevel of efficiency at translating among visuals results from perceptual chunking: visual features of the representations serve to retrieve schemas that describe conceptual information from long-term memory [23, 24]. This high efficiency frees cognitive resources for higher-order thinking [12, 23] and is considered an important learning goal in many STEM domains [3].

Cognitive learning theories suggest that students acquire perceptual fluency via *inductive processes* involved in pattern learning [21, 25]. These processes involve both bottom-up mechanisms (e.g., a visual feature cues the retrieval of a conceptual schema) and top-down processes (e.g., conceptual schemas direct a student's visual attention to relevant visual features). Such inductive processes are considered to be non-verbal [23, 25] because verbal reasoning is not necessary [11, 13] and may even interfere with the acquisition of perceptual fluency [19, 26]. Consequently, students do not require direct instruction to become perceptually fluent, but rather acquire perceptual fluency through experience-based instructional sequences that expose them to many visuals [11, 21].

2.2 Perceptual-Fluency Trainings

In line with cognitive learning theories, perceptual-fluency trainings typically expose students to many examples, for instance in classification problems that ask students to quickly translate between visuals while providing simple feedback on whether the classification is correct or incorrect [27–30]. Because perceptual learning is strongly influenced by the order in which visuals are presented [7], perceptual-fluency trainings purposefully sequence visuals so that consecutive problems vary irrelevant visual features but repeatedly expose students to relevant features [11]. Through experience with many examples, students inductively learn to attend to relevant visual features [11].

The effectiveness of perceptual-fluency trainings has been demonstrated in many STEM domains, including math [31, 32] and chemistry [17, 33]. For example, Fig. 2 shows a perceptual-fluency problem from a training we developed for chemistry students. Each problem asks students to judge whether two visual representations show the same molecule or not. They receive many such problems in a row, sequenced according to the principles just described.

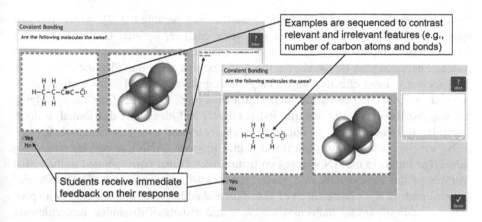

Fig. 2. Example perceptual-fluency problems in the expert-generated training sequence.

However, positive effects of such perceptual-fluency trainings have been confined to students who have substantial prior knowledge about the visual representations and the concepts they show [17, 18]. Indeed, much of the pioneering work on perceptual-fluency trainings in STEM was conducted with students after they had received considerable conceptual instruction and problem-solving practice with the visual representations [32, 34]. Further, recent research shows that students benefit from perceptual-fluency trainings only after they have acquired conceptual knowledge about connections among the visual representations [17, 18].

It is possible that students need to acquire conceptual knowledge about the visual representations before they can become perceptually fluent with them, as proposed by prior research [18]. An alternative explanation is that the lack of effectiveness of perceptual-fluency trainings for students with low prior knowledge may result from an expert blind spot on the part of instructional designers. The expert blind spot is a known phenomenon in the literature on conceptual learning because it can interfere with instructors' anticipation of student difficulties, which may hamper their ability to develop effective instruction [35]. However, we are not aware of any research on perceptual fluency in education that addressed this phenomenon. Specifically, while it is well documented that experts are unaware of why or how they perceive information in a certain way [19, 23, 36], knowledge of this lack of awareness has not informed the instructional design of perceptual-fluency trainings. For example, it is possible that instructors create sequences of visuals that inadvertently assume that students pay attention to specific visual features. In our prior work, which we review next, we used a machine-learning approach that is not subject to expert blind spot biases to create sequences for perceptual trainings for novice students.

2.3 Machine-Learned Sequences for Perceptual-Fluency Trainings

In prior work, we drew on Zhu's machine-teaching paradigm [37], which inverses typical machine-learning approaches to reverse-engineer optimal training sequences (for a detailed report see [38]). Given a cognitive model of a student learning to translate between pairs of visuals, the machine-teaching algorithm identifies a sequence of pairs that is most effective for training the cognitive model. To this end, the algorithm draws possible perceptual-fluency problems (e.g., Figure 2) from an underlying training distribution (not necessarily independently and identically distributed) to form a training sequence. Then, the cognitive model is trained with this sequence. Specifically, we used a feed-forward artificial neural network (ANN) as our learning algorithm. The inputs to the ANN were two feature vectors that corresponded to the visual features of the two visuals in a given perceptual-fluency problem. The ANN had mapped each of the two feature vectors to an embedding that corresponds to a space where visuals of similar molecules are close and visuals of dissimilar molecules are distant. The output was a probability that the two visuals showed the same molecule.

For training, we used back propagation with a history window and multiple back propagation passes, so as to emulate the fact that humans remember past consecutive problems and that humans update their internal models by reviewing the current problem along with the latest problem several times. Then, the effectiveness of the sequence is evaluated based on how well the cognitive model performs on a perceptual-fluency test, which is composed of a sample of perceptual-fluency problems drawn from a separate distribution of problems (i.e., training and test sequences contain different molecules). We use separate test and training distributions to ensure that we optimize the training sequence for learning of mappings among the visual features of the representations, rather than for memorization of translations for specific molecules.

In our prior work [38], we used data from novice undergraduate students in a chemistry course to develop the cognitive model. We then used a modified hill climb search algorithm to find an appropriate training sequence for thatmodel. Next, we compared this sequence to an expert-generated sequence in an experiment with participants from MTurk. Results showed that the machine-learned sequence yielded significantly higher gains in perceptual fluency than the expert-generated sequence.

3   Research Questions

While our prior work showed promising findings for MTurk participants, it remains an open question whether these benefits generalize to low-performing chemistry students. The MTurk participants in our prior study matched our target population because they had low prior knowledge about chemistry and little or no experience with the visual representations. However, their main motivation for participating in the study was to earn money, rather than to learn chemistry. Further, the MTurk study was not conducted in an educational setting. Hence, we address the following open questions:

Research question 1: Does the machine-learned sequence yield higher gains in perceptual fluency than an expert-generated sequence for chemistry students when embedded in instructional materials used in an undergraduate course?

Research question 2: Does the effectiveness of the machine-learned sequence depend on students' prior knowledge?

4 Methods

We address these questions in an experiment that compared the machine-learned sequence to an expert-generated sequence with chemistry undergraduate students with varying levels of prior knowledge enrolled in a chemistry course.

4.1 Participants

We conducted the experiment in a 300-level introductory chemistry course for undergraduates. While the course is open to freshmen and has a prerequisite of students having completed at least one 100-level chemistry course, many students enroll as seniors and have not taken chemistry since their freshman year. Hence, students have highly variable prior knowledge levels. Students received the perceptual-fluency training as a homework assignment with an intelligent tutoring system (ITS) (see Sect. 4.2). Forty students completed the assignment. Two students were excluded because they were statistical outliers on a pretest or posttest (see Sect. 4.4), yielding $N = 38$ students.

4.2 Chem Tutor: An ITS for Undergraduate Chemistry

The chemistry course used the Chem Tutor system for homework. Chem Tutor is an ITS that provides complex problems with individualized step-by-step guidance [4, 39]. Chem Tutor provides interactive instruction that introduces students to how visuals show chemistry concepts (see Fig. 3). In the assignment we used for this experiment, students received four instructional activities prior to the perceptual-fluency training.

The perceptual-fluency training of the assignment was structured as follows. First, students watched a 3-min video explaining that they would receive a large number of single-step problems in a row. The video explained that these problems served to train their perceptual fluency in quickly translating among visuals. Students were instructed not to overthink their answer but to intuitively decide if the two visuals showed the same molecule or not. Further, they were instructed that they would first receive a sequence of problems without feedback (i.e., pretest), then a sequence of problems with feedback (i.e., training), and finally problems without feedback (i.e., posttest).

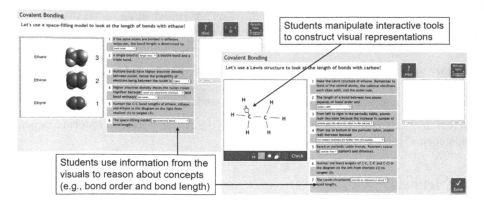

Fig. 3. Example sense-making activity in Chem Tutor.

4.3   Experimental Design

Students were randomly assigned to an expert-generated sequence or a machine-learned sequence that we used in the MTurk study [38]. Each sequence had sixty problems. To control for potential response biases that could affect learning, the number of problems that showed visuals of the same molecules was the same for both sequences.

The *expert-generated sequence* was created by a researcher with a decade of experience with perceptual fluency trainings, using the principles that have been established by prior research on perceptual-fluency trainings we reviewed above. Specifically, problems were sequenced so they would draw attention to relevant visual features. To this end, consecutive problems often repeated one visual while changing the second visual. For example, if one problem presented visuals that showed different molecules (e.g., in the left of Fig. 2, the Lewis structure has more carbon atoms and the wrong bond order), the next problem might present visuals that showed the same molecules (e.g., the right of Fig. 2). To create such sequences, we randomly set the length of the subsequence that retained one visual to be 1–4 problems (i.e., either the first, second, third, or fourth problem would present visuals showing the same molecule). Then, we systematically varied visual features that play a role in chemistry learning, as determined by our prior research with novice students and chemistry experts [4, 40].

The *machine-learned sequence* was constructed using the machine-teaching approach described above. A qualitative inspection of the sequence reveals several differences to the expert-generated sequence that are worth highlighting. First, it does not repeat visuals across consecutive problems. Second, it contains problems that can be solved purely based on knowing which atoms the letters and colors in Lewis structures and space-filling models stand for; which is not one of the visual features the expert-generated sequence systematically varied. Third, it contains problems that can be solved by simply counting the number of atoms in the visuals; which is also not a visual feature that the expert-generated sequence aimed to draw attention to.

4.4   Measures

To assess students' gains in perceptual fluency, we used the same pretest and posttest as in our prior MTurk study [38]. As mentioned, pretest, training, and posttest problems were drawn from separate distributions to ensure that we assess learning of mappings among the visual features of the representations and not of memorization of translations between visuals of specific molecules. For brevity, the pretest contained only 20 problems; the posttest contained 40 problems. Students received no feedback on the test problems. Because perceptual fluency describes students' efficiency in seeing meaningful information in visuals, we computed efficiency scores for each test. Following prior work on efficiency measures [41], we computed perceptual-fluency scores as:

$$\text{perceptual - fluency score} = \frac{Z(\text{average correct responses}) - Z(\text{average time per problem})}{\sqrt{2}} \quad (1)$$

Further, to test if the effect of sequence depends on students' prior knowledge, we used the logs from the four interactive instruction activities that students completed prior to the perceptual-fluency problems. We computed prior-knowledge scores as the number of steps students answered correctly on the first attempt. Because the instruction activities ask students to answer questions about chemistry concepts based on the visuals, this measure assesses students' knowledge about how the visuals show concepts. We treated prior knowledge as a continuous variable in all analyses.

4.5   Procedure

Students were assigned to the Chem Tutor activities as homework in the second week of the semester, including the interactive instruction, pretest, perceptual-fluency training, and posttest. Students were given seven days to complete the assignment online.

5   Results

In the following analyses, we report p. η^2 effect sizes. Following Cohen [42], we consider p. η^2 of .01 to be a small effect, .06 a medium, and .14 a large effect.

5.1   Prior Checks

First, we checked for learning gains using repeated measures ANOVAs with pretest and posttest as dependent measures. Results showed large significant gains in perceptual fluency from the pretest to the posttest, $F(1,36) = 8.762, p = .005$, p. $\eta^2 = .196$.

Second, a multivariate ANOVA showed no significant differences between conditions on the perceptual-fluency pretest or prior knowledge on Chem Tutor's interactive instruction activities ($Fs < 1$). Further, there were no differences between conditions in terms of how much time students spent on the perceptual-fluency training ($F < 1$).

5.2   Effects of Sequence

To test if the machine-learned sequence yielded higher gains in perceptual fluency than an expert-generated sequence (research question 1), we used an ANCOVA model with condition as independent factor, perceptual-fluency pretest and prior knowledge as covariates, and perceptual-fluency posttest as dependent measure. To test if the effects depend on students' prior knowledge (research question 2), we added an interaction between condition and prior knowledge to the model. In line with prior research on aptitude-treatment interactions [43], we did not dichotomize prior knowledge but modeled the interaction between condition and the continuous prior-knowledge variable.

Results showed a medium-sized significant main effect of condition, F (1,33) = 4.699, p = .037, p. η^2 = .125, such that the machine-learned sequence yielded higher gains in perceptual fluency than the expert-generated sequence. The main effect was qualified by a medium-sized significant interaction of condition with prior knowledge, $F(1, 33) = 4.788$, $p = .036$, p. $\eta^2 = .127$. As shown in Fig. 4, the machine-generated sequence was more effective for students with lower prior knowledge, but the expert-generated sequence was more effective for students with higher prior knowledge.

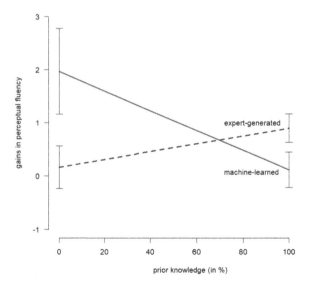

Fig. 4. Effect of machine-learned (red-solid) vs expert-generated (blue-dashed) sequence. The y-axis shows pre-post gains in perceptual-fluency scores based on the efficiency measure in Eq. (1). The x-axis shows prior knowledge. Error bars show standard errors of the mean. (Color figure online)

6 Discussion

Our goal was to investigate if a machine-learning approach enhances the effectiveness of perceptual-fluency trainings for low-performing students in a realistic educational context. We drew on our prior work on a machine-learning approach that was not subject to a potential blind spot bias due to experts' perceptual fluency in seeing meaning in visuals. Instead, it used a bottom-up approach to machine-learn a sequence of visuals based on data from novice chemistry students. Our prior work had established the effectiveness of this sequence for MTurk participants, who were not representative of students in a realistic educational context. The present findings replicate this effect in an undergraduate chemistry course and make several novel contributions.

First, our results show that the machine-learned sequence yield higher gains in perceptual fluency than an expert-generated sequence for students with lower prior knowledge. This finding shows that our machine-learning approach is an effective method for developing perceptual-fluency trainings that are attuned to the needs of students whose needs may not be obvious to instructional designers.

Second, our experiment makes new contributions to the perceptual fluency literature. In contrast to prior research, our findings suggest that perceptual-fluency trainings can be effective for students with low prior knowledge, but that these students require different types of such trainings. Our qualitative comparison of the machine-learned and expert-generated sequences suggests that students with low prior knowledge may benefit from sequences that draw attention to visual features that may seem obvious to experts, such as the mapping between letters and colors. Given that students in our experiment likely had some exposure to the visuals in prior chemistry courses, we think it is unlikely that they did not know that these features are important. Rather, they may not have been efficient at perceiving these features. Further, the machine-learned sequence did not repeat visuals across consecutive problems, whereas the expert-generated sequence did. Such repetitions assume that students recall the visuals from previous problems, which is cognitively demanding. Hence, students with low prior knowledge may benefit more from sequences that reduce cognitive load.

Third, we found that the expert-generated sequence is more effective for students with high prior knowledge. This finding replicates prior research on the effectiveness of expert-generated sequences for advanced students. A new contribution of our findings is that we found that students' performance on prior instructional activities with visuals predicts if they have the prerequisite knowledge to benefit from an expert-generated sequence or if they should receive a sequence that was machine-learned based on data from novice students to prevent expert blind spot biases.

Our findings should be interpreted in the context of the following limitations. First, we focused on a specific set of visuals in chemistry. While we believe that the role of perceptual fluency in chemistry is representative of other STEM domains that rely heavily on visuals, future research should test if our findings generalize to other domains. Second, we did not contrast the characteristics of machine-learned and expert-generated sequences that may account for our results. For example, we did not test if the repetition of visual representations across problems is effective for students with high vs. low prior knowledge. Yet, our findings provide first indications that these characteristics may affect the acquisition of perceptual fluency, which can be systematically tested in future research. Third, because our sample size was relatively small, it is possible that additional smaller effects remained undetected. Finally, we assessed gains of perceptual fluency but not learning of content knowledge. Hence, future research should test whether gains in perceptual fluency for low-performing students translates into an enhanced ability to use the visual representations to learn content knowledge.

In sum, our experiment shows that a bottom-up approach to learn a sequence of visuals for perceptual-fluency trainings can help overcome potential biases resulting from an expert blind spot on the part of instructional designers. Such sequences are particularly effective for students with low prior knowledge. Further, our research provides new directions for future research to systematically investigate which characteristics enhance the acquisition of perceptual fluency for students with low prior knowledge.

Acknowledgements. This research was funded by NSF IIS 1623605 and by NSF ITP 1545481. We thank Yuzi Yu, Blake Mason, John Moore, Rob Nowak, Purav Patel, and Tim Rogers.

References

1. Gilbert, J.K.: Visualization: an emergent field of practice and inquiry in science education. In: Gilbert, J.K., Reiner, M., Nakhleh, M.B. (eds.) Visualization: Theory and Practice in Science Education, vol. 3, pp. 3–24. Springer, Dordrecht (2008). https://doi.org/10.1007/978-1-4020-5267-5_1
2. Ainsworth, S.: The educational value of multiple-representations when learning complex scientific concepts. In: Gilbert, J.K., Reiner, M., Nakama, A. (eds.) Visualization: Theory and Practice in Science Education, pp. 191–208. Springer, Dordrecht (2008). https://doi.org/10.1007/978-1-4020-5267-5_9
3. NRC: Learning to Think Spatially. National Academies Press, Washington, D.C. (2006)
4. Rau, M.A., Michaelis, J.E., Fay, N.: Connection making between multiple graphical representations: a multi-methods approach for domain-specific grounding of an intelligent tutoring system for chemistry. Comput. Educ. **82**, 460–485 (2015)
5. Ainsworth, S.: DeFT: a conceptual framework for considering learning with multiple representations. Learn. Instr. **16**, 183–198 (2006)
6. McElhaney, K.W., Chang, H.Y., Chiu, J.L., Linn, M.C.: Evidence for effective uses of dynamic visualisations in science curriculum materials. Stud. Sci. Educ. **51**, 49–85 (2015)
7. Rau, M.A.: Conditions for the effectiveness of multiple visual representations in enhancing STEM learning. Educ. Psychol. Rev. **29**, 717–761 (2017)
8. Rau, M.A., Aleven, V., Rummel, N.: Successful learning with multiple graphical representations and self-explanation prompts. J. Educ. Psychol. **107**, 30–46 (2015)
9. Seufert, T., Brünken, R.: Cognitive load and the format of instructional aids for coherence formation. Appl. Cogn. Psychol. **20**, 321–331 (2006)
10. Bodemer, D., Faust, U.: External and mental referencing of multiple representations. Comput. Hum. Behav. **22**, 27–42 (2006)
11. Kellman, P.J., Massey, C.M.: Perceptual learning, cognition, and expertise. In: Ross, B.H. (ed.) The Psychology of Learning and Motivation, vol. 558, pp. 117–165. Elsevier Academic Press, New York (2013)

12. Goldstone, R.L., Barsalou, L.W.: Reuniting perception and conception. Cognition **65**, 231–262 (1998)
13. Kellman, P.J., Garrigan, P.B.: Perceptual learning and human expertise. Physics Life Rev. **6**, 53–84 (2009)
14. Goldstone, R.L., Schyns, P.G., Medin, D.L.: Learning to bridge between perception and cognition. Psychol. Learn. Motiv. **36**, 1–14 (1997)
15. Gibson, E.J.: Perceptual learning in development: some basic concepts. Ecol. Psychol. **12**, 295–302 (2000)
16. Goldstone, R.: Perceptual Learning. Academic Press, San Diego (1997)
17. Rau, M.A., Wu, S.P.W.: Support for sense-making processes and inductive processes in connection-making among multiple visual representations. Cogn. Instr. **36**, 361–395 (2018)
18. Rau, M.A.: Sequencing support for sense making and perceptual induction of connections among multiple visual representations. J. Educ. Psychol. **110**, 811–833 (2018)
19. Schooler, J.W., Fiore, S., Brandimonte, M.A.: At a loss from words: verbal overshadowing of perceptual memories. Psychol. Learn. Motiv. Adv. Res. Theor. **37**, 291–340 (1997)
20. Goldstone, R.L., Landy, D.H., Son, J.Y.: The education of perception. Top. Cogn. Sci. **2**, 265–284 (2010)
21. Gibson, E.J.: Principles of Perceptual Learning and Development. Prentice Hall, New York (1969)
22. Chase, W.G., Simon, H.A.: Perception in chess. Cogn. Psychol. **4**, 55–81 (1973)
23. Richman, H.B., Gobet, F., Staszewski, J.J., Simon, H.A.: Perceptual and memory processes in the acquisition of expert performance: the EPAM model. In: Ericsson, K.A. (ed.) The Road to Excellence? The Acquisition of Expert Performance in the Arts and Sciences, Sports and Games, pp. 167–187. Erlbaum Associates, Mahwah (1996)
24. Taber, K.S.: Revisiting the chemistry triplet: drawing upon the nature of chemical knowledge and the psychology of learning to inform chemistry education. Chem. Educ. Res. Pract. **14**, 156–168 (2013)
25. Koedinger, K.R., Corbett, A.T., Perfetti, C.: The knowledge-learning-instruction framework: bridging the science-practice chasm to enhance robust student learning. Cogn. Sci. **36**, 757–798 (2012)
26. Shanks, D.: Implicit learning. In: Lamberts, K., Goldstone, R. (eds.) Handbook of Cognition, pp. 202–220. Sage, London (2005)
27. Massey, C.M., Kellman, P.J., Roth, Z., Burke, T.: Perceptual learning and adaptive learning technology - developing new approaches to mathematics learning in the classroom. In: Stein, N.L., Raudenbush, S.W. (eds.) Developmental Cognitive Science Goes to School, pp. 235–249. Routledge, New York (2011)
28. Rau, M.A., Aleven, V., Rummel, N.: Interleaved practice in multi-dimensional learning tasks: which dimension should we interleave? Learn. Instr. **23**, 98–114 (2013)
29. Rau, M.A., Aleven, V., Rummel, N., Pardos, Z.: How should intelligent tutoring systems sequence multiple graphical representations of fractions? A multi-methods study. Int. J. Artif. Intell. Educ. **24**, 125–161 (2014)
30. Bradley, J.-C., Lancashire, R., Lang, A., Williams, A.: The spectral game: leveraging open data and crowdsourcing for education. J. Cheminform. **1**, 1–10 (2009)
31. Rau, M.A., Aleven, V., Rummel, N.: Supporting students in making sense of connections and in becoming perceptually fluent in making connections among multiple graphical representations. J. Educ. Psychol. **109**, 355–373 (2017)
32. Kellman, P.J., et al.: Perceptual learning and the technology of expertise: studies in fraction learning and algebra. Pragmat. Cogn. **16**, 356–405 (2008)

33. Wise, J.A., Kubose, T., Chang, N., Russell, A., Kellman, P.J.: Perceptual learning modules in mathematics and science instruction. In: Hoffman, P., Lemke, D. (eds.) Teaching and Learning in a Network World, pp. 169–176. IOS Press, Amsterdam (2000)

34. Kellman, P.J., Massey, C.M., Son, J.Y.: Perceptual learning modules in mathematics: enhancing students' pattern recognition, structure extraction, and fluency. Top. Cogn. Sci. **2**, 285–305 (2009)

35. Nathan, M.J., Koedinger, K.R., Alibali, M.W.: Expert blind spot: when content knowledge eclipses pedagogical content knowledge. In: Chen, L., et al. (eds.) Proceedings of the Third International Conference on Cognitive Science, pp. 644–648. USTC Press, Beijing (2001)

36. Chin, J.M., Schooler, J.W.: Why do words hurt? Content, process, and criterion shift accounts of verbal overshadowing. Eur. J. Cogn. Psychol. **20**, 396–413 (2008)

37. Zhu, X.: Machine teaching: an inverse problem to machine learning and an approach toward optimal education. In: AAAI Conference on Artificial Intelligence (2015)

38. Sen, A., et al.: Machine beats human at finding the optimal sequence of visual representations for students learning of perceptual fluency. In: Boyer, K.E., Yudelson, M. (eds.) Proceedings of the 11th International Conference on Educational Data Mining, pp. 137–146. International Educational Data Mining Society, Buffalo (2018)

39. VanLehn, K.: The relative effectiveness of human tutoring, intelligent tutoring systems and other tutoring systems. Educ. Psychol. **46**, 197–221 (2011)

40. Rau, M.A., Mason, B., Nowak, R.: How to model implicit knowledge? Use of metric learning to assess student perceptions of visual representations. In: Barnes, T., Chi, M., Feng, M. (eds.) Proceedings of the 9th International Conference on Educational Data Mining, pp. 199–206. International Educational Data Mining Society, Raleigh (2016)

41. Van Gog, T., Paas, F.: Instructional efficiency: revisiting the original construct in educational research. Educ. Psychol. **43**, 16–26 (2008)

42. Cohen, J.: Statistical Power Analysis for the Behavioral Sciences. Lawrence Erlbaum Associates, Hillsdale (1988)

43. Park, O., Lee, J.: Adaptive instructional systems. In: Jonassen, D.H. (ed.) Handbook of Research for Educational Communications and Technology, pp. 651–658. Erlbaum, Mahwah (2003)

Disentangling Conceptual and Embodied Mechanisms for Learning with Virtual and Physical Representations

Martina A. Rau$^{(\boxtimes)}$ and Tara A. Schmidt

University of Wisconsin – Madison,
1025 W Johnson Street, Madison, WI 53706, USA
marau@wisc.edu

Abstract. Blended educational technologies offer new opportunities for students to interact with physical representations. However, it is not always clear that physical representations yield higher learning gains than virtual ones. Separate lines of prior research yield competing hypotheses about how representation modes affect learning via mechanisms of conceptual salience, embodied schemas, embodied encoding, cognitive load, and physical engagement. To test which representation modes are most effective if they differ in terms of these mechanisms, we conducted a lab experiment on chemistry learning with 119 undergraduate students. We compared four versions of energy diagrams that varied the mode and the actions students used to manipulate the representation. We tested effects on students' learning of three concepts. Representations that induce helpful embodied schemas seem to enhance reproduction. Representations that allow for embodied encoding of haptic cues or makes concepts more salient seem to enhance transfer. Given the high costs of integrating physical representations into blended technologies, these findings may help developers focus on those learning experiences that could most be enhanced by physical interactions.

Keywords: Physical/Virtual modes · Conceptual salience ·
Embodied cognition

1 Introduction

Educational technologies increasingly blend virtual and physical experiences [1–3]. For instance, problem solving in many STEM domains involves virtual and physical representations [4–6]. Virtual representations appear on a screen and are manipulated via mouse or keyboard. For example, chemistry students may construct a virtual energy diagram by clicking to add arrows that show electrons (Fig. 1(left)). By contrast, physical representations are tangible objects that can be manipulated by hand. For example, students may construct a physical energy diagram by hanging arrows on a board (Fig. 1(right)). While much research has compared virtual vs physical representation modes [1, 2], different lines of research focus on different learning mechanisms [1, 7] and hence offer competing hypotheses about which representation mode is more effective. This poses a challenge to developers of blended technologies

© Springer Nature Switzerland AG 2019
S. Isotani et al. (Eds.): AIED 2019, LNAI 11625, pp. 419–431, 2019.
https://doi.org/10.1007/978-3-030-23204-7_35

because they are left with little guidance about which learning experiences can be enhanced by physical interactions.

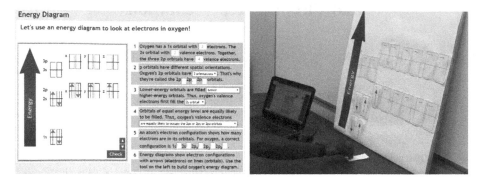

Fig. 1. Energy diagram representations: virtual mode (left); physical mode (right).

To our knowledge, no study has systematically contrasted competing hypotheses about representation modes that emerge from theories on physical engagement, cognitive load, embodied encoding, embodied action schemas, and conceptual salience. We address this gap with an experiment that compared these mechanisms. Our findings advance theory by comparing the relative strength of these mechanisms. Our results yield practical advice for choosing representation modes for blended technologies.

2   Theoretical Background

2.1   Learning with Interactive Visual Representations

Visual representations are powerful tools because they illustrate concepts that are abstract or cannot be directly observed [8–10]. For example, electrons in atoms cannot be observed easily. Scientists often iteratively construct visuals to reflect on difficult and complex phenomena, and then continuously revise them based on their reflections [9]. This iterative representation-reflection process is a key part of STEM practices [11, 12].

Instructional problems with interactive visual representations often mimic iterative representation-reflection processes [2, 5]. Technologies can support such processes by prompting students to construct representations [13], to reflect on how they show concepts [14], and by giving adaptive feedback [15]. While such support is available for virtual and physical representations, it is unclear how to decide whether an instructional activity should include virtual or physical representations.

2.2   Virtual vs Physical Representation Modes

Our review of the literature on learning with virtual and physical representations identified five lines of research that have little overlap and yield competing hypotheses.

Physical Engagement. Proponents of hands-on activities argue that kinesthetic interactions with physical representations are more motivating than virtual ones [16, 17]. Further, physical experiences are concrete, easier to remember, and more connected to real contexts [18]. Hence, physical representations may generally be more effective.

Cognitive Load. In contrast, cognitive load theory recommends eliminating distracting features from the design of visual representations [19, 20]. Because physical representations have richer features that may be distracting, they may increase cognitive load. Further, cognitive load theory recommends designing instructional materials so that students do not have to split their attention between multiple sources of information [19, 20]. In blended educational technologies, students often split their attention between the screen and the physical representation (Fig. 1b). Hence, physical representations have a higher risk of inducing split attention effects. Indeed, studies show that advantages of virtual over physical representations are due to increased cognitive efficiency and attention to target concepts [21–23]. In sum, virtual representations may generally be more effective. However, a limitation of this research is that it has not tested cognitive load effects while systematically varying representation mode.

Embodied Encoding. One line of research on embodied learning proposes that physical experiences provide haptic cues that students can encode through touch, in addition to the visual sense that is engaged in virtual experiences [24, 25]. By experiencing the concepts through additional senses, interactions with physical representations allow for richer, explicit connections between the environment and the concepts [26, 27]. Indeed, embodied experiences that encode haptic cues can reduce cognitive load if students are aware of relations between the cues and the concept [27], which yields higher learning gains than virtual experiences [24, 28]. In sum, physical representations may be more effective if students can explicitly connect embodied experiences to the target concept.

Embodied Schemas. Another line of embodied research focuses on implicit processes that do not require students' awareness [29, 30]. Body actions implicitly affect cognition via metaphors [31, 32] that result from sensory-motor experiences of body movements in the world (e.g., upward movements invoke concepts related to increase [33]). When learning concepts, students form mental simulations that are grounded in embodied schemas [34, 35]. For example, when learning about concepts related to increase, students may mentally simulate upward movements. Indeed, moving the body in ways that are synergistic with mental simulations can enhance learning, even if students are not aware of this relation [36, 37]. Further, virtual representations that are manipulated by synergistic movements enhance learning [3, 34, 38]. In sum, it may not be the representation mode that affects learning. Rather, effects of representation modes may depend on whether they engage students in actions that activate synergistic embodied schemas for the concept. However, this research has not systemically compared modes.

Conceptual Salience. Research on conceptual salience builds on studies that have compared virtual vs physical representations [4, 7, 22]. This research suggests that the effectiveness of a representation does not depend on its mode but on its conceptual salience: the representation that affords an explicit experience of the concept is more effective [4, 7, 39]. For example, research on experimentation skills showed that physical representations make the concept of measurement errors more salient, but virtual representations make concepts of systematic variation more salient [1]. An experiment showed that representations that make the target concept more salient are more effective [1]. However, this research has not tested how effects of conceptual salience compare to effects of embodied schemas. Yet, as we show next, virtual and physical representations often have conflicting advantages for conceptual salience and embodied schemas.

3   Research Questions and Hypotheses

The different theories just reviewed describe mechanisms that may co-occur when students interact with realistic representations. Hence, we investigate: Which representation modes are most effective if they differ in terms of conceptual salience, embodied schemas, embodied encoding, cognitive load, and physical engagement? To this end, we tested hypotheses by the five theories about the effects of two virtual and two physical energy diagrams on learning of three chemistry concepts (see Table 1).

Table 1. Overview of competing hypotheses offered by five theories for the two versions of virtual (V_C/V_E) and physical (P_C/P_E) energy diagrams for each concept.

Theory	Concept A		Concept B		Concept C	
Conceptual salience	$\dfrac{P_C}{V_C} > \dfrac{V_E}{P_E}$	Action effect	$\dfrac{V_C}{P_C} > \dfrac{P_E}{V_E}$	Action effect	NA	Null effect
Embodied schemas	$\dfrac{V_E}{P_E} > \dfrac{P_C}{V_C}$	Action effect	$\dfrac{P_E}{V_E} > \dfrac{V_C}{P_C}$	Action effect	NA	Null effect
Embodied encoding	$\dfrac{P_C}{P_E} > \dfrac{V_C}{V_E}$	Mode effect	NA	Null effect	NA	Null effect
Cognitive load	$\dfrac{V_C}{V_E} > \dfrac{P_C}{P_E}$	Mode effect	$\dfrac{V_C}{V_E} > \dfrac{P_C}{P_E}$	Mode effect	$\dfrac{V_C}{V_E} > \dfrac{P_C}{P_E}$	Mode effect
Physical engagement	$\dfrac{P_C}{P_E} > \dfrac{V_C}{V_E}$	Mode effect	$\dfrac{P_C}{P_E} > \dfrac{V_C}{V_E}$	Mode effect	$\dfrac{P_C}{P_E} > \dfrac{V_C}{V_E}$	Mode effect

3.1   Concept A: Electrons Randomly Fill Equal-Energy Orbitals

An atom's properties are related to its electrons' energy, which is determined by the electrons' positions in subatomic regions called orbitals. Energy diagrams sort orbitals by energy level (bottom to top). Electrons are more likely to fill low-energy orbitals, but they are equally likely to fill equal-energy orbitals. A common misconception is that electrons fill equal-energy orbitals from left to right, rather than randomly.

To construct physical energy diagram P_C, students move cards from the bottom up to put them in orbitals. P_C makes the *concept more salient* because planning the motor action involved in the vertical action requires attention to the height of the orbital when students put a card in an orbital. To construct virtual energy diagram V_E, students click to put electrons in orbitals, moving the mouse horizontally to click in equal-energy orbitals. V_E makes the *concept less salient* because the horizontal action does not require attention to the orbital's height. To test if these effects are due to the action rather than the mode, we created physical energy diagram P_E so that students hold the cards next to the orbitals and move their hands horizontally to put them in orbitals. This horizontal action makes the *concept less salient*. We created virtual energy diagram V_C so that it asks students to click a button at the bottom each time before moving the hand up to put arrows in orbitals. This vertical action makes the *concept more salient*.

V_E induces *beneficial embodied schemas* for this concept because horizontal action induce a metaphor of equality [33]. By contrast, P_C induces a *suboptimal embodied schema* for this concept because vertical action induce a metaphor of increase [33]. By contrast, the vertical action in V_C invokes a *suboptimal embodied schema*, and the horizontal action in P_E invokes a *beneficial embodied schema*.

Both P_C and P_E allow for *embodied encoding* of the height of equal-energy orbitals because they offer haptic cues through features such as the distance from the bottom of the diagram. Hence, they should be more effective than both V_C and V_E.

Both V_C and V_E yield *lower cognitive load* because they contain fewer distracting details than the physical diagrams and do not require split attention between screen and diagram. Hence, they should be more effective than both P_C and P_E.

Both P_C and P_E *engage students physically* and should yield a more motivating experience than both V_C and V_E. Hence, they should be more effective than V_C and V_E.

3.2   Concept B: Up and Down Spins Have Equal Energy

Electrons in the same orbital have opposite spins, shown by up and down arrows. Up and down spin are equally likely because they do not affect an electron's energy level. A common misconception is that an orbital's first electron always has an up spin.

In V_C, students click to add arrows. The first click adds an up arrow, the second click flips it to a down arrow. V_C makes the *concept more salient* because students have to purposefully flip the arrows to show that the spins are equally likely, which requires explicit attention. In P_E, students pick up cards from a stack that is not sorted, so that up and down arrows are random. P_E makes the *concept less salient* because the spin is already random and does not require attention to a related action. To test if these effects are due to the action rather than the mode, we modified the other version of the diagrams to flip the hypotheses: In V_E, the first click creates an arrow with random spin. The second click flips it. This requires no attention to randomness and makes the *concept less salient*. For P_C, the card stack was sorted so that all cards had an up arrow. Now, students have to purposefully flip the cards, which makes the *concept more salient*.

V_E and P_E induce a *beneficial embodied schema* because the random spin means that it takes the same number of actions and hence the same amount of effort to show up or down spin. V_C and P_C induce a *suboptimal embodied schema* because the fixed spin means it takes two clicks and hence more effort to show a down spin than an up spin.

P_C and P_E do not allow for *embodied encoding* as they do not have haptic cues for spin states. Hence, this hypothesis does not predict an effect of mode. V_C and V_E yield *lower cognitive load*, whereas both P_C and P_E yield *more physical engagement*.

3.3   Concept C: Spins Are Rotational Movements

Electron spins are rotational movements of electrons about their own axis that create a small electromagnetic field with a moment that points up or down. A common misconception is that spins are an electron's directional movement towards or away from the nucleus rather than the rotation around their own axes.

The energy diagram does not explicitly show electron rotations. Hence, no representation makes this *concept salient*. The energy diagram does not require rotational movements. Hence, no representation invokes related *embodied schemas*. Also, no representation offers *embodied encoding* of rotational movements.

However, V_C and V_E yield *lower cognitive load*, but P_C and P_E yield more *physical engagement*. Hence, including this concept allows us to estimate the impact of cognitive load and physical engagement mechanisms on students' learning.

4   Methods

4.1   Participants

We recruited 120 undergraduates from a large university in the US Midwest via email, flyers, and posters for monetary compensation. A screening ensured they had not taken chemistry since high school. One student was excluded because a pretest showed considerable prior knowledge of the target concepts, yielding a sample of $N = 119$.

4.2   Experimental Design

Students were randomly assigned to one of four conditions that varied (1) representation mode and (2) actions required to manipulate the diagrams (see Table 2).

Table 2. Experimental conditions with number of participants (n) that vary representation mode and actions: both versions of virtual (V_C/V_E) and physical (P_C/P_E) energy diagrams.

	Conceptually salient action		Embodied action	
Virtual mode	V_C: $n = 30$	Concept A – Vertical	V_E: $n = 30$	Concept A – Horizontal
		Concept B – Random		Concept B – Fixed
		Concept C – No action		Concept C – No action
Physical mode	P_C: $n = 29$	Concept A – Vertical	P_E: $n = 30$	Concept A – Horizontal
		Concept B – Fixed		Concept B – Random
		Concept C – No action		Concept C – No action

4.3   Materials

Intelligent Tutoring System (ITS). Students worked with an ITS for undergraduate chemistry that has proven effective in prior research [40]. The ITS supports iterative representation-reflection practices by asking students to create energy diagrams to illustrate target concepts. Further, it prompts students to reflect on how the diagrams show the concepts by completing fill-in-the-gap sentences. If students make a mistake on a step, the ITS gives adaptive feedback that targets common misconceptions.

Students worked on eight problems. Each covered all three concepts and asked students to build an energy diagram of an atom. For the virtual conditions, V_C or V_E were embedded in the ITS (Fig. 1a). The ITS gave instruction and feedback on all steps. For the physical conditions, P_C or P_E was placed next to the screen (Fig. 1b). The experimenter gave feedback on the diagrams. The ITS gave all other instruction and feedback.

Assessments. We assessed students' learning of each of the three concepts with a pretest that they completed prior to instruction, an immediate posttest given immediately after instruction, and a delayed posttest given 3–6 days after instruction. For each concept, we assessed reproduction (i.e., recall of information given in instruction) and transfer (i.e., the ability to apply the information to problems not covered in the ITS). As the instruction in the ITS was self-paced, we also measured instructional time.

4.4   Procedure

The experiment involved two sessions in a research lab, 3–6 days apart. In session 1, students completed the pretest, the instruction according to their experimental condition, and the immediate posttest. In session 2, students took the delayed posttest.

5   Results

5.1   Prior Checks

First, we checked for learning gains on each concept using repeated measures ANOVAs with pretest, immediate, and delayed posttest as dependent measures. Results showed significant learning gains for all concepts ($ps < .01$) with effect sizes ranging from p. $\eta^2 = .11$ to p. $\eta^2 = .59$. Second, a multivariate ANOVA showed no significant differences between conditions on any of the pretest measures ($ps > .10$). However, mode affected instructional time, such that physical representations took significantly longer, $F(1, 118) = 14.45$, $p < .01$, p. $\eta^2 = .11$. Because instructional time correlated with the learning outcome measures ($r = -.21$ to $-.25$), we included it as covariate in the analyses below.

5.2   Effects of Representation Mode and Movement

We used a repeated measures ANCOVA model to test the hypotheses in Table 1. The model included mode and action as independent factors, pretest scores and instructional time as covariates, and immediate and delayed posttest scores as dependent measures. Figure 2 shows a summary of the results.

For *reproduction of Concept A*, results showed a main effect of action, $F(1, 113) = 4.94$, $p = .03$, p. $\eta^2 = .04$, favoring horizontal over vertical actions. This effect aligns with the embodied schema hypothesis. There was no main effect of mode, $F(1, 113) = 1.65$, $p = .20$, nor an interaction effect ($F < 1$).

For *transfer of Concept A*, there was no significant main effect of action, $F(1, 113) = 1.14$, $p = .29$. A main effect of mode, $F(1, 113) = 6.37$, $p = .01$, p. $\eta^2 = .05$, favored physical over virtual representations. This effect aligns with the embodied encoding and the physical engagement hypotheses. There was no interaction effect ($F < 1$).

For *reproduction of Concept B*, there was a significant main effect of action, $F(1, 113) = 5.30$, $p = .02$, p. $\eta^2 = .05$, favoring a random number of actions over a fixed number of actions. This aligns with the embodied schema hypothesis. There was no main effect of mode, $F(1, 113) = 1.64$, $p = .20$, nor an interaction effect ($F < 1$).

For *transfer of Concept B*, there was a significant effect main of action, $F(1, 113) = 4.40$, $p = .04$, p. $\eta^2 = .04$, such that a fixed number of actions yielded higher gains than a random number of actions. This effect aligns with the conceptual salience hypothesis. There was no effect of mode, $F(1, 113) = 2.60$, $p = .11$, or an interaction effect ($F < 1$).

For *reproduction and transfer of Concept C*, no effects were significant ($Fs < 1$).

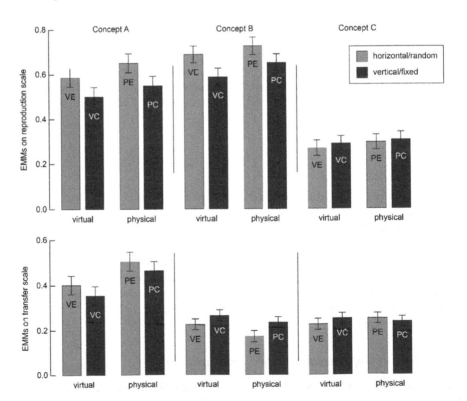

Fig. 2. Estimated marginal means (EMMs) for reproduction and transfer averaged across immediate and delayed posttests, controlling for pretest and instructional time.

6 Discussion and Conclusion

While much prior research has compared virtual vs physical representations, separate lines of research have focused on different mechanisms that yield competing hypotheses for their effectiveness. This leaves developers with little guidance for choosing appropriate representation modes. To address this issue, we investigated which representation modes are most effective if they differ in terms of conceptual salience, embodied schemas, embodied encoding, cognitive load, and physical engagement. Because prior research provides evidence for these mechanisms, it seems likely that they co-occur when students interact with realistic representations. Hence, our goal was not to confirm or refute the theories, but rather to examine which mechanism prevails when students learn abstract concepts. To our knowledge, our study is the first to integrate these theories by systematically comparing effects of representation mode and actions.

Altogether, for *reproduction of knowledge*, our results suggest that the *embodied schema mechanism* outweighs the other mechanisms. The embodied schema hypothesis predicted an advantage of horizontal and random actions for Concepts A and B, and both effects were confirmed for the reproduction scales of these concepts. Hence, representations that are manipulated via body actions that induce beneficial embodied schemas seem to enhance students' ability to recall information covered in instruction.

By contrast, for *transfer of knowledge*, our results suggest that the *embodied encoding mechanism* outweighs the other mechanisms if it applies. The embodied encoding hypothesis predicted an advantage of both physical representations only for Concept A, and this effect was confirmed for the transfer scale of this concept. Hence, physical representations that offer haptic cues for the target concept seem to enhance students' ability to apply their knowledge to novel situations. However, if the representation does not contain haptic cues for the concept, as in the case of Concept B, the *conceptual salience mechanism* appears to outweigh the other mechanisms. This finding suggests that transfer is more affected by conceptual salience than by embodied schemas.

The complexity of embodied schemas, embodied encoding, and conceptual salience mechanisms may explain differences between reproduction and transfer. The embodied schema mechanism describes a simple, implicit process that does not require awareness [36]. Information recall involves simple knowledge structures that have one-on-one question-response mappings [41]. Thus, representations that engage students in simple embodied mechanisms seem to enhance learning of simple knowledge structures.

By contrast, both the conceptual salience and the embodied encoding mechanisms describe complex, explicit learning processes. The conceptual salience mechanism describes how students map visual cues of representations to concepts. Arguably, the embodied encoding mechanism is yet more complex because it describes how students connect haptic *and* visual cues to concepts. Because transfer of knowledge requires many-to-many mappings between question and response, it assesses complex knowledge structures [41]. Thus, representations that engage students in complex mechanisms seem to enhance learning of complex knowledge structures, especially when the representations offer opportunities for embodied encoding of haptic cues.

We found no evidence for the *cognitive load* and *physical engagement* hypotheses. In light of the null effects for Concept C, which were predicted by the other three hypotheses, we can conclude that cognitive load and physical engagement mechanisms either were negligible or cancelled each other out. This also allows us to rule out that cognitive load or physical engagement could have distorted the effects for the other three mechanisms on Concepts A and B. In fact, the only result in line with the physical engagement hypothesis was the advantage of physical representations on transfer of Concept A, but this effect was also predicted by the embodied encoding hypothesis.

In sum, our study suggests that developers may prioritize embodied schema mechanisms if the goal is to enhance reproduction. To enhance transfer, they may choose a physical representation if it offers haptic cues for the concept. Otherwise, they may choose the representation that makes the concept more salient. These considerations should outweigh considerations of cognitive load or physical engagement. Given that the integration of physical representations into educational technologies is costly, these findings may help developers of blended technologies focus on learning experiences where physical representations have the highest impact on learning outcomes.

Our results should be interpreted in light of several limitations. First, we focused on particular concepts, representations, and population. Future research should test if our findings generalize more broadly. Second, while we purposefully selected concepts for which the five theories made conflicting predictions, we did not test all possible conflicts. For example, future research should test cases where conceptual salience and embodied schemas align but conflict with embodied encoding. Third, our intervention was relatively short. Over longer learning periods, it is possible that sequence effects emerge, such that one mechanism prevails at first and another mechanism later. Specifically, we found that embodied schema mechanisms enhance reproduction but embodied encoding and conceptual salience mechanisms enhance transfer. Given that instruction often moves from simple to complex concepts, it is possible that embodied schema mechanisms should be prioritized early and embodied encoding and conceptual salience mechanisms later. Testing such effects may yield new insights into embodied grounding of conceptual knowledge [42] and may provide insights into the concrete-abstract debate [18], which has not accounted for embodied mechanisms.

In conclusion, blended educational technologies offer new opportunities to combine virtual and physical modes, for example, by integrating physical representations into ITSs. However, physical representations are not always more effective than virtual ones. Our study reveals the relative strength and scope of multiple mechanisms that have been examined by thus far separate lines of research even though they likely co-occur when students learn with representations. Further, our results may provide practical advice for developers to choose representation modes for blended technologies.

Acknowledgements. This research was funded by NSF IIS CAREER 1651781. We thank Purav Patel and Tiffany Herder for their help with the study, and Dor Abrahamson, Matthew Dorris, Mary Hegarty, Clark Landis, John Moore, and Mike Stieff for their helpful advice.

References

1. Olympiou, G., Zacharia, Z.C.: Blending physical and virtual manipulatives: an effort to improve students' conceptual understanding through science laboratory experimentation. Sci. Educ. **96**, 21–47 (2012)

2. de Jong, T., Linn, M.C., Zacharia, Z.C.: Physical and virtual laboratories in science and engineering education. Science **340**, 305–308 (2013)

3. Antle, A.N., Corness, G., Droumeva, M.: What the body knows: exploring the benefits of embodied metaphors in hybrid physical digital environments. Interact. Comput. **21**, 66–75 (2009)

4. Chini, J.J., Madsen, A., Gire, E., Rebello, N.S., Puntambekar, S.: Exploration of factors that affect the comparative effectiveness of physical and virtual manipulatives in an undergraduate laboratory. Phys. Rev. Spec. Top.-Phys. Educ. Res. **8**, 010113 (2012)

5. Manches, A., O'Malley, C., Benford, S.: The role of physical representations in solving number problems: a comparison of young children's use of physical and virtual materials. Comput. Educ. **54**, 622–640 (2010)

6. Stull, A.T., Hegarty, M.: Model manipulation and learning: fostering representational competence with virtual and concrete models. J. Educ. Psychol. **108**, 509–527 (2016)

7. Klahr, D., Triona, L.M., Williams, C.: Hands on what? The relative effectiveness of physical versus virtual materials in an engineering design project by middle school children. J. Res. Sci. Teach. **44**, 183–203 (2007)

8. Gilbert, J.K.: Visualization: an emergent field of practice and inquiry in science education. In: Gilbert, J.K., Reiner, M., Nakhleh, M.B. (eds.) Visualization: Theory and Practice in Science Education, vol. 3, pp. 3–24. Springer, Dordrecht (2008). https://doi.org/10.1007/978-1-4020-5267-5_1

9. NRC: Learning to Think Spatially. National Academies Press, Washington, D.C. (2006)

10. Rau, M.A.: Conditions for the effectiveness of multiple visual representations in enhancing STEM learning. Educ. Psychol. Rev. **29**, 717–761 (2017)

11. Fan, J.E.: Drawing to learn: how producing graphical representations enhances scientific thinking. Transl. Issues Psychol. Sci. **1**, 170–181 (2015)

12. Kozma, R., Russell, J.: Students becoming chemists: developing representational competence. In: Gilbert, J. (ed.) Visualization in Science Education, pp. 121–145. Springer, Dordrecht, Netherlands (2005). https://doi.org/10.1007/1-4020-3613-2_8

13. van der Meij, J., de Jong, T.: Supporting students' learning with multiple representations in a dynamic simulation-based learning environment. Learn. Instr. **16**, 199–212 (2006)

14. McElhaney, K.W., Chang, H.Y., Chiu, J.L., Linn, M.C.: Evidence for effective uses of dynamic visualisations in science curriculum materials. Stud. Sci. Educ. **51**, 49–85 (2015)

15. Rau, M.A., Keesler, W., Zhang, Y., Wu, S.: Resolving design tradeoffs of interactive visualization tools for educational technologies. IEEE Trans. Learn. Technol. (in press)

16. Flick, L.B.: The meanings of hands-on science. J. Sci. Teacher Educ. **4**, 1–8 (1993)

17. Deboer, G.: A History of Ideas in Science Education. Teachers College Press, New York (1991)

18. Goldstone, R.L., Son, J.Y.: The transfer of scientific principles using concrete and idealized simulations. J. Learn. Sci. **14**, 69–110 (2005)

19. Sweller, J., van Merrienboër, J.J.G., Paas, F.G.W.C.: Cognitive architecture and instructional design. Educ. Psychol. Rev. **10**, 251–296 (1998)

20. Mayer, R.E.: Cognitive theory of multimedia learning. In: Mayer, R.E. (ed.) The Cambridge Handbook of Multimedia Learning, pp. 31–48. Cambridge University Press, New York (2009)

21. Durmus, S., Karakirik, E.: Virtual manipulatives in mathematics education: a theoretical framework. Turk. Online J. Educ. Technol. **5** (2006)
22. Yuan, Y., Lee, C.Y., Wang, C.H.: A comparison study of polyominoes explorations in a physical and virtual manipulative environment. J. Comput. Assist. Learn. **26**, 307–316 (2010)
23. Barrett, T.J., Stull, A.T., Hsu, T.M., Hegarty, M.: Constrained interactivity for relating multiple representations in science: when virtual is better than real. Comput. Educ. **81**, 69–81 (2015)
24. Magana, A.J., Balachandran, S.: Students' development of representational competence through the sense of touch. J. Sci. Educ. Technol. **26**, 332–346 (2017)
25. Zaman, B., Vanden Abeele, V., Markopoulos, P., Marshall, P.: Editorial: the evolving field of tangible interaction for children: the challenge of empirical validation. Pers. Ubiquit. Comput. **16**, 367–378 (2012)
26. Shaikh, U.A., Magana, A.J., Neri, L., Escobar-Castillejos, D., Noguez, J., Benes, B.: Undergraduate students' conceptual interpretation and perceptions of haptic-enabled learning experiences. Int. J. Educ. Technol. High. Educ. **14**, 1–21 (2017)
27. Skulmowski, A., Pradel, S., Kühnert, T., Brunnett, G., Rey, G.D.: Embodied learning using a tangible user interface: the effects of haptic perception and selective pointing on a spatial learning task. Comput. Educ. **92**, 64–75 (2016)
28. Minaker, G., Schneider, O., Davis, R., MacLean, K.E.: *HandsOn*: enabling embodied, creative STEM e-learning with programming-free force feedback. In: Bello, F., Kajimoto, H., Visell, Y. (eds.) EuroHaptics 2016. LNCS, vol. 9775, pp. 427–437. Springer, Cham (2016). https://doi.org/10.1007/978-3-319-42324-1_42
29. Glenberg, A.M.: Embodiment as a unifying perspective for psychology. Wiley Interdisc. Rev. Cogn. Sci. **1**, 586–596 (2010)
30. Wilson, M.: Six views of embodied cognition. Psychon. Bull. Rev. **9**, 625–636 (2002)
31. Johnson-Glenberg, M.C., Birchfield, D.A., Tolentino, L., Koziupa, T.: Collaborative embodied learning in mixed reality motion-capture environments: two science studies. J. Educ. Psychol. **106**, 86–104 (2014)
32. Black, J.B., Segal, A., Vitale, J., Fadjo, C.L.: Embodied cognition and learning environment design. In: Jonassen, D.H., Land, S.M. (eds.) Theoretical Foundations of Learning Environments, pp. 198–223. Routledge Taylor & Francis Group, New York (2012)
33. Lakoff, G.J., Johnson, M.: Metaphors We Live by. University of Chicago Press, Chicago-London (1980)
34. Abrahamson, D., Lindgren, R.: Embodiment and embodied design. In: Sawyer, R.K. (ed.) The Cambridge Handbook of the Learning Sciences, pp. 358–376. Cambridge University Press, New York (2014)
35. Clark, A.: Whatever next? Predictive brains, situated agents, and the future of cognitive science. Behav. Brain Sci. **36**, 181–204 (2013)
36. Nathan, M.J., Walkington, C.: Grounded and embodied mathematical cognition: promoting mathematical insight and proof using action and language. Cogn. Res. Principles Implications **2**, 1–20 (2017)
37. Hayes, J.C., Kraemer, D.J.: Grounded understanding of abstract concepts: the case of STEM learning. Cogn. Res. Principles Implications **2** (2017)
38. Segal, A., Tversky, B., Black, J.: Conceptually congruent actions can promote thought. J. Appl. Res. Mem. Cogn. **3**, 124–130 (2014)
39. Gire, E., et al.: The effects of physical and virtual manipulatives on students' conceptual learning about pulleys. In: Gomez, K., Lyons, L., Radinsky, J. (eds.) 9th International Conference of the Learning Sciences, vol. 1, pp. 937–943. International Society of the Learning Sciences (2010)

40. Rau, M.A.: A framework for discipline-specific grounding of educational technologies with multiple visual representations. IEEE Trans. Learn. Technol. **10**, 290–305 (2017)
41. Koedinger, K.R., Corbett, A.T., Perfetti, C.: The knowledge-learning-instruction framework: bridging the science-practice chasm to enhance robust student learning. Cogn. Sci. **36**, 757–798 (2012)
42. Nathan, M.J.: An embodied cognition perspective on symbols, grounding, and instructional gesture. In: Symbols and Embodiment: Debates on Meaning and Cognition, pp. 375–396 (2008)

Adaptive Support for Representation Skills in a Chemistry ITS Is More Effective Than Static Support

Martina A. Rau[✉], Miranda Zahn, Edward Misback,
and Judith Burstyn

University of Wisconsin – Madison, 1025 W Johnson Street,
Madison, WI 53706, USA
marau@wisc.edu

Abstract. Multiple visual representations can enhance learning in STEM, provided that students have prerequisite representation skills to make sense of how the visuals show information and to fluently perceive meaning in the visuals. Prior research shows that instructional support for sense-making skills and perceptual fluency enhances STEM learning. This research also shows that students need different types of support, depending on their prior representation skills. Hence, instruction may be most effective if it adaptively assigns students to support for sense-making skills and perceptual fluency. We tested this hypothesis in an experiment with 45 undergraduates in an introductory chemistry course. Students were randomly assigned to a 6-week instructional module of an intelligent tutoring system (ITS) that (1) provided a static sequence of activities that supported sense-making skills and perceptual fluency or (2) adaptively assigned the activities. Results show that the adaptive version yielded significantly higher gains of chemistry knowledge. Our findings expand theories of representation skills and yield recommendations for ITSs with multiple visual representations.

Keywords: Multiple representations · Sense-making skills · Perceptual fluency

1 Introduction

Instruction in most science, technology, engineering, and math (STEM) domains uses multiple visual representations [1–3]. Compared to a single visual, multiple visuals can enhance learning of domain knowledge because they allow students to form more accurate mental models [4–6]. For example, instruction on chemical bonding typically uses the visuals in Fig. 1 that show complementary concepts of atomic structure [7].

But research also shows that multiple visuals can impede learning if students lack representation skills to make sense of how multiple visuals show information and to fluently perceive the information they show [3, 8]. Further, research shows that students need different types of instructional support for representation skills at different times during their learning trajectory [9, 10]. This research suggests that adaptive representation-skills supports may enhance students' learning more so than static supports.

© Springer Nature Switzerland AG 2019
S. Isotani et al. (Eds.): AIED 2019, LNAI 11625, pp. 432–444, 2019.
https://doi.org/10.1007/978-3-030-23204-7_36

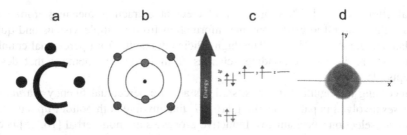

Fig. 1. (a) Lewis structure, (b) shell model, (c) energy diagram, and (d) orbital diagram of carbon.

While it is well known that adaptive support for problem solving enhances learning [11], this hypothesis has not been tested for representation skills. We address this gap in an experiment on chemistry learning. Our results show how different representation skills build on each other and yield guidance for the design of educational technologies.

2 Theoretical Background

Prior research distinguishes two types of representation skills (sense making and perceptual fluency) that are learned via different processes and need different support [6].

2.1 Sense-Making Skills

Students' learning of domain knowledge from multiple visuals depends on their ability to make sense of how the different visuals show complementary concepts [6, 12]. To this end, students map visual features of different representations to the concepts they show [4, 12]. That means students need to distinguish features that show meaningful information (e.g., the number of valence electrons shown in a Lewis structure, Fig. 1a) from incidental features (e.g., the color of electrons in the shell model, Fig. 1b) [13]. Then, students need to compare visuals [4, 14]. That is, they need to understand similarities between visuals (e.g., both Lewis structure and shell model show that carbon has four valence electrons, Fig. 1a–b) and differences (e.g., the shell model in Fig. 1b shows the core electrons, but the Lewis structure in Fig. 1a does not).

According to cognitive theories, students acquire sense-making skills through *learning processes* that are verbally mediated because students explain how visuals show concepts [15, 16]. These processes are explicit because students have to willfully engage in the explanations [17, 18]. Accordingly, *instructional activities* that support sense-making skills engage students in active reasoning about visuals, for example by asking them to self-explain similarities and differences between visuals [19, 20].

2.2 Perceptual Fluency

A largely separate line of research on expertise has focused on a second type of representation skills. Experts "see at a glance" what visuals show without perceived mental effort [21, 22]. They are very efficient at extracting meaning from visuals because they can effortlessly combine information from multiple visuals and quickly translate among them [23, 24]. This high efficiency results from perceptual chunking: visual cues retrieve corresponding schemas from long-term memory that describe concepts [25, 26].

According to cognitive theories, students acquire perceptual fluency via *inductive processes* involved in pattern learning [16, 27] that involve both bottom-up (cuing) and top-down (selection) mechanisms. Inductive processes are non-verbal [16, 25] because verbal reasoning is not necessary [24, 28] and may even interfere with pattern learning [29, 30]. Thus, perceptual induction does not require direct instruction but rather results from experience [24, 27]. Hence, *instructional activities* that support perceptual fluency expose students to many examples, for example in classification tasks [31], interleaved practice [32, 33], or games that require quick translations among visuals [34].

2.3   Combining Support for Sense-Making Skills and Perceptual Fluency

If students need sense-making skills and perceptual fluency, then combining activities that support sense-making skills and perceptual fluency should enhance learning of domain knowledge. Prior studies tested effects of (1) a combination of sense-making and perceptual-fluency activities, (2) only sense-making activities, (3) only perceptual-fluency activities, and (4) a control that received multiple visuals without representation-skills support. Experiments on elementary fractions [35] and undergraduate chemistry [36, 37] show that only the combination is more effective than the control.

Cross-sectional studies [9, 37] show that the effectiveness of specific sequences of sense-making and perceptual-fluency activities depends on students' knowledge about how the visuals show concepts. Specifically, novice students need to become familiar with each visual before they benefit from sense-making and perceptual-fluency activities. Intermediate students benefit from receiving sense-making activities followed by perceptual-fluency activities because sense-making skills enhance perceptual pattern learning by helping students attend to relevant visual features. Finally, advanced students benefit from receiving perceptual-fluency activities followed by sense-making activities because this helps them fluently use information from multiple visuals to make sense of concepts. Thus, the effectiveness of sense-making and perceptual-fluency activities varies depending on students' current skill level, which changes as they learn. Hence, prior research suggests a progression of representation skills [38].

3   Experiment

Based on this prior research, we hypothesize that instruction is most effective if it adaptively assigns sense-making and perceptual-fluency activities depending on a student's current skill level. We tested this hypothesis in an experiment with Chem Tutor, an in ITS for undergraduate chemistry. We first developed an adaptive version of Chem Tutor using data from a prior experiment and then compared it to a static version.

3.1 Chem Tutor: An ITS for Undergraduate Chemistry

Chem Tutor provides complex problems and individualized step-by-step guidance throughout the problem-solving process [11, 39]. It uses a cognitive model of the students' knowledge about how the visuals show chemistry concepts [40]. The model can detect multiple strategies and provides detailed feedback and hints on how to use the visuals [41]. Here, we use Chem Tutor's atomic structure module. It has six units, each with two of the visuals in Fig. 1 (see Table 1 below). Each unit has three problem types.

Regular Activities correspond to chemistry instruction that typically uses one visual at a time [42]. For example, students may be asked to construct an energy diagram of oxygen (see Fig. 4 below). They are first prompted to identify properties of the atom to plan the energy diagram. Next, they use an interactive tool to construct the visual. Students must construct a correct visual before they move on. Finally, they use the visual to make inferences about the atom. Thus, regular activities provide one visual at a time but provide no support for representation skills that involve comparing or translating among multiple visuals. Based on findings that students first have to become familiar with each visual [4], each unit starts with two regular activities, one for each visual.

Sense-Making Activities are designed to help students understand connections among multiple visuals, in line with instructional design principles based on the prior research on sense-making skills reviewed above [19, 20]. Sense-making activities ask students to *actively compare* pairs of visual representations. To this end, students are given two visuals and receive prompts to *self-explain* similarities and differences between the visuals. For example, the activity in Fig. 2 asks students to reflect on similarities between an energy diagram and a Lewis structure for magnesium. Given the energy diagram, students construct the Lewis structure of magnesium. Then, they receive self-explanation prompts to compare the visual representations. This example focuses on similar conceptual aspects shown by both visuals. Other sense-making activities focus on differences between visuals. Alternating activities focus on similarities and differences.

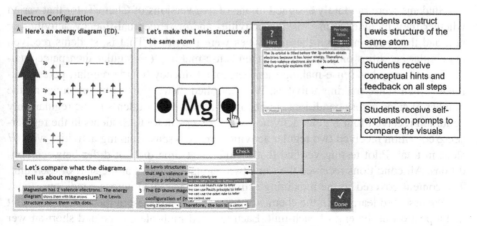

Fig. 2. Example sense-making activity in Chem Tutor.

Perceptual-Fluency Activities embody the principles reviewed in Sect. 2.2 [24, 27]. For example, students may have to select one of four energy diagrams that shows the same atom as a given Lewis structure (Fig. 4). The choices contrast relevant features, such as the number of valence electrons versus the total number of electrons. Each activity has one step, and students solve many of them in a row for a variety of atoms. Students receive immediate correctness feedback. To engage perceptual processing, students are asked to solve the activities fast and intuitively, without fear of mistakes (Fig. 3).

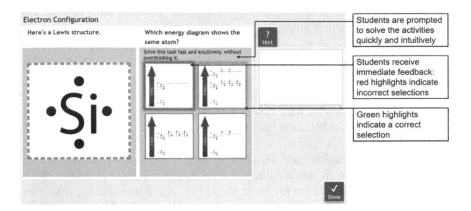

Fig. 3. Example perceptual-fluency activity in Chem Tutor.

3.2   Development of the Adaptive Assignment Algorithm

Data. To develop an algorithm that adaptively assigns these activities, we used data from 129 undergraduate students in an introductory chemistry course at a large university in the US Midwest. They completed one Chem Tutor unit per week, for six weeks.

Students were randomly assigned to one of five versions of Chem Tutor that varied whether they included sense-making activities and/or perceptual-fluency activities or not, and the order in which these activities were provided. That is, students received (1) regular activities only, (2) regular then sense-making, (3) regular then perceptual-fluency, (4) regular, sense-making then perceptual-fluency, or (5) regular, perceptual-fluency then sense-making activities. We controlled instructional time by equating the number of steps across conditions. For example, for unit 1, students in the regular-only condition received four regular activities with 67 steps in total. Students in the regular-sense condition received two regular activities and two sense-making activities with 67 steps in total. Pilot testing verified that instructional time did not differ between conditions. All conditions received the same two regular activities at the start of each unit. The content covered in the remaining activities was identical across conditions.

We assessed learning of the chemistry content with a pretest at the start of each unit and a posttest at the end of each unit. Each test had multiple-choice and short-answer items. Tests and grading scheme were developed by a chemist. Agreement between independent graders on 10% of the responses to the short-answer items was 88.55%.

Further, our goal was to predict benefit from sense-making and perceptual-fluency activities, so we mined the Chem Tutor logs for predictors. We focused on the first two regular activities that all conditions received for each unit. For each step, we computed performance based on whether a student's first attempt at the step was correct or incorrect. In sum, we computed performance measures for 134 steps across six units.

Analyses. To identify problem-solving steps that were predictive of students' benefit from sense-making and perceptual-fluency activities, we conducted linear regression analyses with 10-fold cross-validation. The regression models identified steps for which performance interacted with the experimental factors to predict pre-post gains. Such effects indicate aptitude-treatment interactions (ATIs) [43], such that low-performing students benefit from a different intervention than high-performing students.

We constructed a linear regression model for each unit. In each model, the dependent variable was pre-post gain. Predictors were the experimental factors (i.e., with/without sense-making activities; with/without perceptual-fluency activities, and the order of the activities), performance on each step in the first two regular activities in the unit, and the interaction of performance on each step with the experimental factors. The regression models outputted significance tests and regression coefficients for each predictor.

Adaptive Algorithm. We selected steps for which the regression analysis revealed significant ATIs because this means that the step predicts if a student will benefit from sense-making activities and/or perceptual-fluency activities. For example, a positive regression coefficient for an ATI between performance on a step and the sense-making factor indicates that students who get this step right benefit from sense-making activities. The activity in Fig. 4 shows two example steps for which we found significant positive interactions with the sense-making factor. This suggests that getting these steps

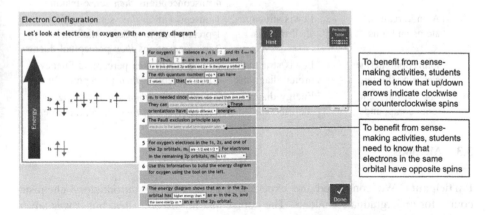

Fig. 4. Regular activity with two steps for which performance interacts positively with sense-making activity. Hence, these steps test prerequisites for sense-making activities.

right indicates prerequisite understanding of how the visual shows the concepts (e.g., how the energy diagram shows electron spins), which predicts benefit from sense-making activities. Hence, *if* a student gets any of these steps right, *then* he/she should receive sense-making activities. By contrast, a negative coefficient suggests that a student who gets this step wrong benefits from sense-making activities, possibly sense-making activities could rectify a misconception (e.g., comparing the energy diagram to another visual may help students interpret the energy diagram). Hence, *if* the student gets the step wrong, *then* he/she should receive sense-making activities. There were no cases that recommended sense-making and perceptual-fluency activities for the same unit.

We then formulated if-then rules so that *if* students exhibited some prerequisite or misconception that indicated sense-making or perceptual-fluency activities would be beneficial, *then* they would be assigned to them. The if-then rules were ordered so that rules corresponding to higher regression coefficients were prioritized. Table 1 summarizes how many steps tested prerequisites or misconceptions that indicated sense-making activities and perceptual-fluency activities. For each unit, a Python algorithm used performance on the first two regular problems to test whether students met conditions for the prerequisites or misconceptions for sense-making or perceptual-fluency activities. If they did, they received either sense-making or perceptual-fluency activities next. If not, they received regular activities on the same content.

Table 1. Topics and visuals in the six Chem Tutor units, and rules (in order of priority).

Unit	Topics	Visuals	Rules: *If* student exhibits at least 1 of...
1	Bohr model, quantum numbers 1 & 2	Shell models, orbital diagrams	4 misconceptions *then* sense-making
2	Quantum numbers 3 & 4, atomic orbitals	Orbital diagrams, energy diagrams	1 prerequisite *then* perceptual-fluency 2 prerequisites *then* sense-making
3	Configurations of atoms and ions	Energy diagrams, Lewis structures	2 misconceptions *then* perceptual-fluency 1 prerequisite *then* sense-making 4 misconceptions *then* sense-making
4	Atomic radii of atoms and ions	Lewis structures, shell models	1 misconception *then* sense-making 1 prerequisite *then* sense-making 2 misconceptions *then* perceptual-fluency
5	Ionization energies, electron affinities	Lewis structures, orbital diagrams	1 prerequisite *then* perceptual-fluency 1 misconception *then* perceptual-fluency
6	Energy, ions, and ionic compounds	Energy diagrams, shell models	2 prerequisites *then* sense-making 1 misconception *then* perceptual-fluency

3.3 Methods

Participants. We conducted the experiment as part of an introductory chemistry course for undergraduate students. The course had no prerequisites, but it was advertised to students in 100- and 300-level courses in chemistry and related programs. Fifty students enrolled, five dropped the course within the first three weeks, yielding $N = 45$ students.

Experimental Design. Students were randomly assigned to the static or the adaptive version of Chem Tutor. For each unit, the static version provided two regular activities, then two sense-making activities, then 16 perceptual-fluency activities. This corresponds to the most effective version in prior studies [9, 44]. The adaptive version provided two regular activities for each unit and then selected sense-making activities and perceptual-fluency activities based on the rules in Table 1. To control instructional time, we equated the number of problem-solving steps across conditions. For example, the static version of unit 1 had 67 steps in total. If—after a student completed the two regular activities—the algorithm indicated they needed more regular activities, the student received two more regular activities, yielding 67 total steps. If the algorithm indicated they needed more sense-making activities, the student received four sense-making activities, yielding 67 total steps. Hence, regardless of what the algorithm assigned, number of steps and content was identical to the static version. What differed was whether students received support for sense-making skills and perceptual fluency for all units (i.e., static version) or only the support that was indicated (i.e., adaptive version).

Measures. We assessed learning of chemistry content with a pretest and posttest as in the prior experiment. Students took a pretest at the start of each unit, an immediate posttest at the end of each unit, and a delayed posttest in the following week before they completed the pretest for the next unit. Agreement among independent graders of short-answer items based on 10% of the responses was 85.91%.

Procedure. At the start of each class, the instructor gave a 3-min overview of the topics. For the next hour, students worked at their own pace on the activities. The instructor and a teaching assistant circulated the class but gave only minimal help and directed students to read the Chem Tutor hints. Then, the instructor led a discussion on the topics. Students who were unable to attend could complete the activities in a lab.

3.4 Results

Prior Checks. First, we checked for learning gains using repeated measures ANOVAs with pretest, immediate, and delayed posttest as dependent measures. Results showed significant learning gains across units, $F(1, 43) = 50.36$, $p < .01$, p. $\eta^2 = .54$. Separate ANOVAs showed significant learning gains for each unit (for units 1, 3–6, $ps < .01$ with effect sizes ranging from p. $\eta^2 = .15$ to p. $\eta^2 = .54$; for unit 2, $p = .05$, p. $\eta^2 = .07$).

Second, a multivariate ANOVA showed no significant differences between conditions on the pretests for units 1–3, 5 ($ps > .10$). However, students in the static condition had marginally higher pretest scores for unit 4, $F(1, 43) = 3.07$, $p = .09$, p. $\eta^2 = .08$, and significantly higher pretest scores for unit 6 $F(1, 43) = 4.37$, $p = .04$, p. $\eta^2 = .09$.

Effects of Adaptive Assignment. To test if the adaptive version of Chem Tutor was overall more effective than the static version, we used a repeated measures ANCOVA model. The model included condition as independent factor, pretest scores across units as covariate, and immediate and delayed posttest scores across units as dependent measures. Results showed a significant effect of condition, $F(1, 42) = 7.52$, $p < .01$, p. $\eta^2 = .15$, such that the adaptive version yielded higher gains than the static version. We also tested if the effects of the adaptive version depended on prior knowledge. Results showed no significant interaction of condition with pretest, $F(1, 41) = 1.74$, $p = .19$.

Next, we tested effects for each unit (Fig. 5). Separate ANCOVAs showed significant advantages of the adaptive version for unit 1, $F(1, 42) = 5.72$, $p = .02$, p. $\eta^2 = .12$, unit 3, $F(1, 42) = 6.27$, $p = .02$, p. $\eta^2 = .13$, unit 4, $F(1, 42) = 7.38$, $p = .01$, p. $\eta^2 = .15$, and unit 6, $F(1, 42) = 7.37$, $p = .01$, p. $\eta^2 = .15$. For units 2 and 5, there was no significant effect of condition ($ps > .10$). For unit 5, we found a significant interaction of condition with pretest, $F(1, 41) = 5.55$, $p = .02$, p. $\eta^2 = .12$, such that the adaptive version was more effective for low-performing students, but the static version was more effective for high-performing students. No other interactions were significant ($ps > .10$).

Finally, we qualitatively explored how the adaptive algorithm assigned sense-making and perceptual-fluency activities. The algorithm assigned regular activities to 65% of the students at unit 1. Among them, 27% of the students received sense-making activities next, starting at unit 2; 73% received perceptual-fluency activities next, starting at unit 2 or 3. The algorithm assigned sense-making activities to 35% at unit 1. Among them, 75% of the students received perceptual-fluency activities next, starting at unit 2 or 3; 25% received more regular or sense-making activities. Finally, 87% of students received more sense-making activities after they completed perceptual-fluency activities for at least one unit, and 13% received more regular or perceptual-fluency activities.

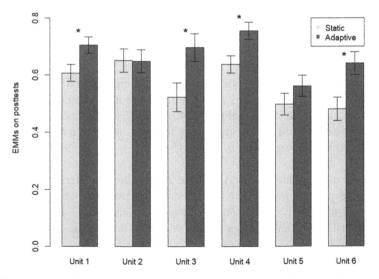

Fig. 5. Estimated marginal means (EMMs) averaged across immediate and delayed posttests, controlling for pretest. Error bars show standard errors of the man * show significant differences.

4 Discussion and Conclusion

Our results show that adaptive support for sense-making skills and perceptual fluency with visual representations is more effective than static supports for these representation skills. This finding expands prior research in several ways. First, prior studies show that students' benefit from different types of representation-skill supports depends on their current skill level, and that adapting support to students' problem-solving skills enhances learning (e.g., by adapting the choice of problems that practice domain knowledge to students' current domain knowledge). Yet, our study is the first to show that adapting support to students' representation skills enhances the effectiveness of problem-solving activities that practice domain knowledge.

Second, our results provide further evidence for a progression of representation skills that had been based on cross-sectional studies. The inspection of ATI effects in the prior study and the adaptive assignments in the current study suggest that students should start with regular activities or sense-making activities before they benefit from perceptual-fluency activities. Further, our findings confirm earlier cross-sectional results that students benefit from sense-making activities again after becoming perceptually fluent.

Third, our results may guide the design of educational technologies with multiple visuals. We used linear regression to identify steps that were predictive of benefit from sense-making and perceptual-fluency activities because they indicates prerequisite knowledge about visuals or misconceptions about visuals. We translated results into if-then rules that assign type of representation-skills support a student needs. This approach can be applied to any set of technology-based activities where students use visual representations to solve domain-relevant problems.

Our findings should be interpreted in light of several limitations. First, we found no advantage of the adaptive over the static version of Chem Tutor for units 2 and 5. For unit 2, this may be due to the lack of learning gains. For unit 5, the adaptive version was more effective only for low-performing students. It is possible that students in the prior experiment that was the basis for the algorithm were low-performers on this unit. For both units, we will use the current data to improve the algorithm. A second set of limitations relates to the sample. Our study was part of a course that involved other activities such as class discussions. While these activities may have affected learning, we do not see why they should have affected differences between conditions. Also, although students matched the target population of Chem Tutor, they were likely highly motivated and may have seen the visuals before. Further, our sample was small due to the small class size of 45 students. Hence, future research should replicate our results in other contexts, with other populations and larger samples. Third, our study did not compare adaptive representation-skills support to a control without representation-skills support. While the effectiveness of static representation-skills support has been established, future research should verify that adaptive representation-skill support is indeed more effective than regular activities alone. Finally, because the use of visuals in chemistry is similar to the use of visuals in other STEM domains, we expect that our findings will generalize broadly. Yet, future research should test if adaptive representation-skill support indeed enhances learning in other domains and with other visuals.

In conclusion, our study is the first to show that adaptive support for representation skills can significantly enhance learning of domain knowledge. Given that multiple visuals are widely used and that lack of representation skills is an obstacle in many STEM domains, such support may significantly enhance STEM learning.

Acknowledgements. This research was funded by NSF DUE-IUSE 1611782. We thank John Moore and Matthew Dorris for their advice and help with this study.

References

1. Gilbert, J.: Visualization: An Emergent Field of Practice and Enquiry in Science Education. In: Gilbert, J., Reiner, M., Nakhleh, M. (eds.) Visualization: Theory and Practice in Science Education, vol. 3, pp. 3–24. Springer, Dordrecht (2008). https://doi.org/10.1007/978-1-4020-5267-5_1
2. Ainsworth, S.: The educational value of multiple-representations when learning complex scientific concepts. In: Gilbert, J.K., Reiner, M., Nakama, A. (eds.) Visualization: Theory and Practice in Science Education, pp. 191–208. Springer, Dordrecht (2008). https://doi.org/10.1007/978-1-4020-5267-5_9
3. NRC: Learning to Think Spatially. National Academies Press, Washington, D.C. (2006)
4. Ainsworth, S.: DeFT: a conceptual framework for considering learning with multiple representations. Learn. Instr. **16**, 183–198 (2006)
5. McElhaney, K., Chang, H., Chiu, J., Linn, M.: Evidence for effective uses of dynamic visualisations in science curriculum materials. Stud. Sci. Educ. **51**, 49–85 (2015)
6. Rau, M.A.: Conditions for the effectiveness of multiple visual representations in enhancing STEM learning. Educ. Psychol. Rev. **29**, 717–761 (2017)
7. Moore, J.W., Stanitski, C.L.: Chemistry: The Molecular Science. Cengage Learning, Stamford (2015)
8. Uttal, D., O'Doherty, K.: Comprehending and learning from visualizations: a developmental perspective. In: Gilbert, J. (ed.) Visualization: Theory and Practice in Science Education, pp. 53–72. Springer, Dordrecht (2008). https://doi.org/10.1007/978-1-4020-5267-5_3
9. Rau, M.A.: Sequencing support for sense making and perceptual induction of connections among multiple visual representations. J. Educ. Psychol. **110**, 811–833 (2018)
10. Rau, M.A., Aleven, V., Rummel, N.: Making connections between multiple graphical representations of fractions: conceptual understanding facilitates perceptual fluency, but not vice versa. Instr. Sci. **45**, 331–357 (2017)
11. VanLehn, K.: The relative effectiveness of human tutoring, intelligent tutoring systems and other tutoring systems. Educ. Psychol. **46**, 197–221 (2011)
12. Schnotz, W.: An integrated model of text and picture comprehension. In: Mayer, R.E. (ed.) The Cambridge Handbook of Multimedia Learning, pp. 72–103. Cambridge University Press, New York (2014)
13. Gentner, D., Markman, A.B.: Structure mapping in analogy and similarity. Am. Psychol. **52**, 45–56 (1997)
14. Ainsworth, S.: The multiple representation principle in multimedia learning. In: Mayer, R.E. (ed.) The Cambridge Handbook of Multimedia Learning, pp. 464–486. Cambridge University Press, New York (2014)
15. Chi, M., Bassok, M., Lewis, M., Reimann, P., Glaser, R.: Self-explanations: how students study and use examples in learning to solve problems. Cogn. Sci. **13**, 145–182 (1989)

16. Koedinger, K.R., Corbett, A.T., Perfetti, C.: The knowledge-learning-instruction framework. Cogn. Sci. **36**, 757–798 (2012)
17. Chi, M.T.H., de Leeuw, N., Chiu, M.H., Lavancher, C.: Eliciting self-explanations improves understanding. Cogn. Sci. **18**, 439–477 (1994)
18. diSessa, A.A., Sherin, B.L.: Meta-representation: an introduction. J. Math. Behav. **19**, 385–398 (2000)
19. Bodemer, D., Faust, U.: External and mental referencing of multiple representations. Comput. Hum. Behav. **22**, 27–42 (2006)
20. Seufert, T., Brünken, R.: Cognitive load and the format of instructional aids for coherence formation. Appl. Cogn. Psychol. **20**, 321–331 (2006)
21. Chi, M.T.H., Feltovitch, P.J., Glaser, R.: Categorization and representation of physics problems by experts and novices. Cogn. Sci. **5**, 121–152 (1981)
22. Gegenfurtner, A., Lehtinen, E., Säljö, R.: Expertise differences in the comprehension of visualizations. Educ. Psychol. Rev. **23**, 523–552 (2011)
23. Chase, W.G., Simon, H.A.: Perception in chess. Cogn. Psychol. **4**, 55–81 (1973)
24. Kellman, P.J., Massey, C.M.: Perceptual learning, cognition, and expertise. In: Ross, B.H. (ed.) The Psychology of Learning and Motivation, vol. 558, pp. 117–165. Elsevier Academic Press, New York (2013)
25. Richman, H., Gobet, F., Staszewski, J., Simon, H.: Perceptual and memory processes in the acquisition of expert performance. In: Ericsson, K. (ed.) The Road to Excellence?, pp. 167–187. Erlbaum Associates, Mahwah (1996)
26. Taber, K.S.: Revisiting the chemistry triplet. Chem. Educ. Res. Pract. **14**, 156–168 (2013)
27. Gibson, E.J.: Principles of Perceptual Learning and Development. Prentice Hall, New York (1969)
28. Kellman, P.J., Garrigan, P.B.: Perceptual learning and human expertise. Phys. Life Rev. **6**, 53–84 (2009)
29. Shanks, D.: Implicit learning. In: Lamberts, K., Goldstone, R. (eds.) Handbook of Cognition, pp. 202–220. Sage, London (2005)
30. Schooler, J.W., Fiore, S., Brandimonte, M.A.: At a loss From words: verbal overshadowing of perceptual memories. Psychol. Learn. Motiv. Adv. Res. Theor. **37**, 291–340 (1997)
31. Massey, C.M., Kellman, P.J., Roth, Z., Burke, T.: Perceptual learning and adaptive learning technology. In: Stein, N.L., Raudenbush, S.W. (eds.) Developmental Cognitive Science Goes to School, pp. 235–249. Routledge, New York (2011)
32. Rau, M.A., Aleven, V., Rummel, N.: Interleaved practice in multi-dimensional learning tasks: which dimension should we interleave? Learn. Instr. **23**, 98–114 (2013)
33. Rau, M.A., Aleven, V., Rummel, N., Pardos, Z.: How should intelligent tutoring systems sequence multiple graphical representations of fractions? A multi-methods study. Int. J. Artif. Intell. Educ. **24**, 125–161 (2014)
34. Bradley, J.-C., Lancashire, R., Lang, A., Williams, A.: The Spectral Game. J. Cheminformatics **1**, 1–10 (2009)
35. Rau, M.A., Aleven, V., Rummel, N.: Supporting students in making sense of connections and in becoming perceptually fluent in making connections among multiple graphical representations. J. Educ. Psychol. **109**, 355–373 (2017)
36. Rau, M.A., Wu, S.P.W.: ITS support for conceptual and perceptual connection making between multiple graphical representations. In: Conati, C., Heffernan, N., Mitrovic, A., Verdejo, M.F. (eds.) AIED 2015. LNCS (LNAI), vol. 9112, pp. 398–407. Springer, Cham (2015). https://doi.org/10.1007/978-3-319-19773-9_40
37. Rau, M.A.: Making connections among multiple visual representations: how do sense-making competencies and perceptual fluency relate to learning of chemistry knowledge? Instr. Sci. (2017)

38. Rau, M.A., Zahn, M.: Sequencing support for sense making and perceptual fluency with visual representations: is there a learning progression? In: Kay, J., Luckin, R. (eds.) Rethinking Learning in the Digital Age. Making the Learning Sciences Count (ICLS) 2018, vol. 1, pp. 264–271. International Society of the Learning Sciences, London (2018)

39. Rau, M.A., Michaelis, J.E., Fay, N.: Connection making between multiple graphical representations: a multi-methods approach for domain-specific grounding of an intelligent tutoring system for chemistry. Comput. Educ. **82**, 460–485 (2015)

40. Blind for review (2017)

41. Corbett, A.T., Koedinger, K., Hadley, W.S.: Cognitive tutors: from the research classroom to all classrooms. In: Goodman, P.S. (ed.) Technology enhanced learning: Opportunities for Change, pp. 235–263. Lawrence Erlbaum Associates Publishers, Mahwah (2001)

42. Rau, M.A.: Do knowledge-component models need to incorporate representational competencies? Int. J. Artif. Intell. Educ. **27**, 298–319 (2017)

43. Park, O., Lee, J.: Adaptive instructional systems. In: Jonassen, D.H. (ed.) Handbook of Research for Educational Communications and Technology, pp. 651–658. Erlbaum, Mahwah (2003)

44. Rau, M.A., Wu, S.P.W.: Support for sense-making processes and inductive processes in connection-making among multiple visual representations. Cogn. Instr. **36**, 361–395 (2018)

Confrustion in Learning from Erroneous Examples: Does Type of Prompted Self-explanation Make a Difference?

J. Elizabeth Richey[1]([⊠]) [iD], Bruce M. McLaren[1],
Miguel Andres-Bray[2], Michael Mogessie[1], Richard Scruggs[2],
Ryan Baker[2] [iD], and Jon Star[3] [iD]

[1] Carnegie Mellon University, Pittsburgh, PA 15213, USA
jrichey@andrew.cmu.edu
[2] University of Pennsylvania, Philadelphia, PA 19104, USA
[3] Harvard University, Cambridge, MA 02138, USA

Abstract. Confrustion, a mix of confusion and frustration sometimes experienced while grappling with instructional materials, is not necessarily detrimental to learning. Prior research has shown that studying erroneous examples can increase students' experiences of confrustion, while at the same time helping them learn and overcome their misconceptions. In the study reported in this paper, we examined students' knowledge and misconceptions about decimal numbers before and after they interacted with an intelligent tutoring system presenting either erroneous examples targeting misconceptions (erroneous example condition) or practice problems targeting the same misconceptions (problem-solving condition). While students in both conditions significantly improved their performance from pretest to posttest, students in the problem-solving condition improved significantly more and experienced significantly less confrustion. When controlling for confrustion levels, there were no differences in performance. This study is interesting in that, unlike prior studies, the higher confrustion that resulted from studying erroneous examples was not associated with better learning outcomes; instead, it was associated with poorer learning. We propose several possible explanations for this different outcome and hypothesize that revisions to the explanation prompts to make them more expert-like may have also made them – and the erroneous examples that they targeted – less understandable and less effective. Whether prompted self-explanation options should be modeled after the shorter, less precise language students tend to use or the longer, more precise language of experts is an open question, and an important one both for understanding the mechanisms of self-explanation and for designing self-explanation options deployed in instructional materials.

Keywords: Erroneous examples · Affect · Confusion · Frustration ·
Affect detection · Learning outcomes

© Springer Nature Switzerland AG 2019
S. Isotani et al. (Eds.): AIED 2019, LNAI 11625, pp. 445–457, 2019.
https://doi.org/10.1007/978-3-030-23204-7_37

1 Introduction

Erroneous examples, or examples that illustrate typical student errors and misconceptions, can support performance, learning, and transfer [1–6]. Researchers have hypothesized that erroneous examples derive their benefits through multiple processes, including helping students to recognize errors in their own work [7] and highlighting the underlying principles necessary for understanding correct solutions [2, 8]. However, no studies to our knowledge have examined the affective consequences of erroneous examples. In particular, it is unclear whether students experience confusion and frustration as they try to identify and understand errors that they might make themselves. To understand how and when erroneous examples are most effective, we examined the effects of erroneous examples on the affective state of confrustion – a combination of confusion and frustration – and its consequences for learning outcomes.

Many previous studies have shown a learning advantage when students are prompted to compare correct and incorrect examples [9] or explain and fix errors in incorrect examples [5, 10], in contrast to more typical worked-example study or problem-solving practice. In particular, studying erroneous examples may highlight for students the common errors that they are likely to make and discourage them from underestimating the difficulty of a problem. Erroneous examples have been shown to be particularly beneficial for supporting long-term learning and transfer [2, 5, 6].

Erroneous examples relate more broadly to research on desirable difficulties [11, 12] and productive failure [13, 14]. Experiencing difficulty on a task can increase engagement and mental effort and improve long-term learning outcomes [12, 13]. Critically, difficulty should directly relate to the concepts or procedures being taught. In other words, simply making a task difficult for the sake of difficulty likely will not improve learning; however, making the task difficult enough to require more effort or some initial failure can ultimately help the learner. Erroneous examples may operate through a similar mechanism by presenting students with solutions that they may have thought were correct, based on their own inaccurate knowledge and misconceptions.

Although emotions likely play a role in learning from difficulty or productive failure, the learner's affective experiences while struggling in such contexts has not, to our knowledge, been explored. The affective states of confusion and frustration are likely to be especially relevant in these contexts. Although confusion and frustration are theoretically distinct constructs, both have strong but mixed connections to learning [15–17]. Confusion is typically viewed as positive if the student believes it can be resolved [18–20], and it has been related to positive motivational experiences such as engagement and flow [21]. Frustration can arise when a student cannot resolve their confusion, and it can lead to disengagement and poor learning outcomes [22].

Despite the differences between confusion and frustration, research on affect detection has suggested there are predictive benefits to combining the two as a measure of confrustion. Affect detectors rely on students' interaction data from a learning system to determine the students' affective states. They can examine affect at a grain-size of about 20-s intervals and can predict immediate performance as well as long-term student outcomes [23, 24]. To create affect detectors, human coders first label data based on the absence or presence of an affective state [25–28]. After acceptable inter-rater reliability is established, machine learning algorithms identify behaviors in the learning system that correspond to the affect judgments made by human coders.

Confusion and frustration are often hard for an observer to distinguish based on the students' interactions with educational technology [22]. Perhaps for this reason, a study comparing the predictive value of confrustion assessed through affect detectors against separate measures of confusion and frustration found that confrustion was a more accurate predictor of learning [22]. Consequently, in this study we combine these constructs and calculate a single measure of confrustion.

To explore the role of confrustion in learning from erroneous examples, we examined students' log interactions with an educational technology platform that presents a series of 32 erroneous examples and 16 practice problems targeting common decimal number misconceptions [29–31]. Previous research with this technology showed that students who corrected and explained erroneous examples performed better than students who solved and explained the same problems, but only on a delayed posttest [1, 5]. This suggests that studying and correcting erroneous examples might hinder – or at least not benefit – immediate performance but does lead to long-term learning.

If erroneous examples support learning by creating difficulty, students who are working through the materials might experience greater confrustion than students completing similar materials without erroneous examples. To examine this, we created affect detectors to assess students' experiences of confrustion [32]. A re-analysis of log files from the experiments reported in [1, 5] indicated that students did experience greater confrustion in the erroneous examples condition. However, confrustion was negatively related to performance. Confrustion thus does not appear to be beneficial on its own, but it may be a necessary consequence of the cognitively demanding learning processes supported by erroneous examples. In other words, students studying erroneous examples might learn more despite experiencing greater confrustion.

The current study aimed to replicate and build on the previous results in several ways. First, the data that was reanalyzed with affect detectors was collected more than six years ago. Use of educational technology has continued to increase in the time since those data were collected, potentially changing the ways students would view and interact with the materials. Erroneous examples have also gained prominence among teachers and instructional designers. As a result, students might be more accustomed to interacting with erroneous examples. For these reasons, we wanted to replicate both the learning and confrustion results with a new group of students.

Second, students in the erroneous example condition previously received several explanation prompts focused on each erroneous example and thus were prompted to do more self-explanation than students in the problem-solving condition [1, 5]. Self-explanation is a robust instructional technique [33, 34], and it is possible some of the benefits experienced by students in the erroneous example condition resulted from extra self-explanation. Additionally, the extra self-explanation may have contributed to students in the erroneous example condition spending nearly twice as much time as students in the problem-solving condition [1, 5]. To reduce the difference in time between conditions and decrease the chance that benefits were being derived from extra self-explanation, were moved one of the additional self-explanation prompts, leaving only one extra self-explanation prompt focused directly on making sense of the erroneous example.

Third, we saw opportunities to revise the self-explanation prompts to improve precision in the mathematical language used, and we worked with a math education expert (the seventh author of this paper) to make these changes. For example, the previous materials referred to the misconception that "longer decimals are larger," while it is more mathematically precise to express the idea as "decimal numbers with more digits to the right of the decimal point are greater in magnitude". While prior research has established that self-explanation can still be beneficial when students select or complete explanations using provided options within a computer-based learning environment [35, 36], instead of generating the explanations themselves, we know of no prior research that has explored the question of whether these provided explanations should be more similar to the less mathematically precise language students typically use or the more mathematically precise language of math experts. In the current study, we investigated whether the same learning benefits would be observed if students selected self-explanation prompts using more mathematically precise language.

Fourth, materials were updated to operate in HTML instead of Flash and to conform with modern look-and-feel instructional technology. For instance, prompted explanation boxes, a common interface feature in this educational technology system, were created with current HTML multiple-choice widgets. Using these revised materials, we sought to replicate previous results by testing the following hypotheses:

H1: Students in both conditions will improve in performance from pretest to posttest and from pretest to delayed test. We do not expect any of the changes made to the materials to disrupt the basic learning benefits of the intervention.

H2: Confrustion will be negatively related to performance, even when controlling for prior knowledge. We do not expect the changes to the materials to change the confrustion students experience, which related negatively to learning in prior studies.

H3: Students in the erroneous example condition will experience greater confrustion than students in the problem-solving condition. We made revisions aimed at simplifying the appearance of the erroneous example materials (modern look-and-feel) and to reduce extra text they had to read (elimination of extra self-explanation prompt text). However, we expect that the greater levels of confrustion come from the erroneous examples themselves and not from other features of the problem interface.

H4: Students in the erroneous example condition will perform better than students in the problem-solving condition on the delayed posttest. We do not expect any of the changes to disrupt the relative benefits of erroneous examples.

While we had no way of empirically testing the effect of revisions on the amount of time students required to complete the materials, we expected them to take less time on the materials overall and to show less of a time difference between conditions.

2   Methods

2.1   Participants

Participants were recruited from a suburban, public middle school (four teachers) and an urban, public elementary school (four teachers) in the metropolitan area of a northeastern U.S. city. Participants' parents provided written consent to collect and analyze students' data. Students completed materials as a part of their regular, in-class instructional activities. A total of 53 fifth-grade students and 134 sixth-grade students participated. Six fifth-grade students and seven sixth-grade students were dropped for failing to complete the materials in the allotted time. The final dataset included 174 students: 47 fifth-graders (30 male, 17 female; mean age 10.4) and 127 sixth graders (69 male, 58 female; mean age 11.2). Students were randomly assigned to conditions at the individual level, with 89 students assigned to the problem-solving condition and 85 students assigned to the erroneous example condition.

All students had previously learned about decimal numbers during their regular math instruction (Common Core standard CCSS.Math.Content.5.NBT.A.3 for fifth grade; CCSS.Math.Content.6.NS.B.3 for sixth grade). To avoid introducing information that could affect students' performance, teachers were asked to refrain from providing decimal-number instruction or practice outside of the intervention during the study.

2.2   Materials

Materials were developed using the Cognitive Tutor Authoring Tool (CTAT) and delivered through Tutorshop, a learning management system for CTAT tutors that supports classroom deployment via web delivery [37]. We followed updated look-and-feel principles to revise the presentation of materials (see Figs. 1 and 2). Materials were aimed at addressing misconceptions while providing feedback and practice to students who had basic knowledge of decimal numbers. Both the erroneous example and problem-solving materials presented the same problems in the same order. Materials were organized into three-item sets of two erroneous example or problem-solving items with self-explanation, followed by one practice item without self-explanation. Practice items were the same across conditions. The materials included a total of 48 problems targeting different decimal number misconceptions. Tasks included number line placement, ordering by magnitude, addition, and completing a decimal number sequence.

Each erroneous example item presented a decimal number word problem with an incorrect solution from a hypothetical student (Fig. 2). Students worked through the problem in three steps and could not advance until they completed each step correctly. First, the student was prompted to explain the error in the example. Second, they corrected the error by solving the problem. Third, they explained the correct solution or relevant principles through two self-explanation prompts with multiple-choice solution options [35, 36]. The tutor provided feedback on all incorrect responses.

Each problem-solving item presented the same decimal number word problem but without an incorrect solution. Students worked through the problem in two steps and

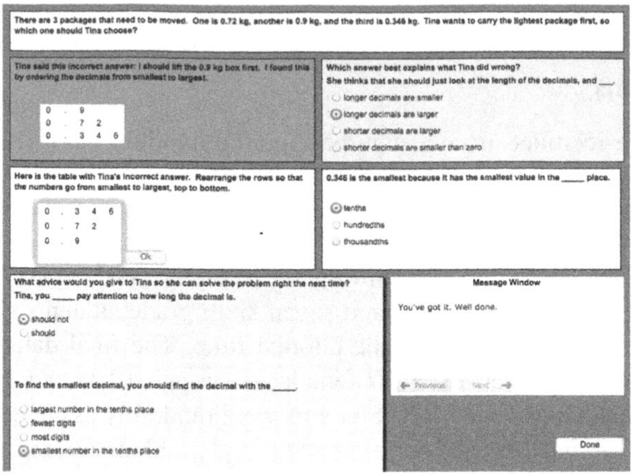

Fig. 1. A sample erroneous example problem from the original materials used in [1, 5]. This problem involves an ordering task and targets the "longer decimals are larger" misconception.

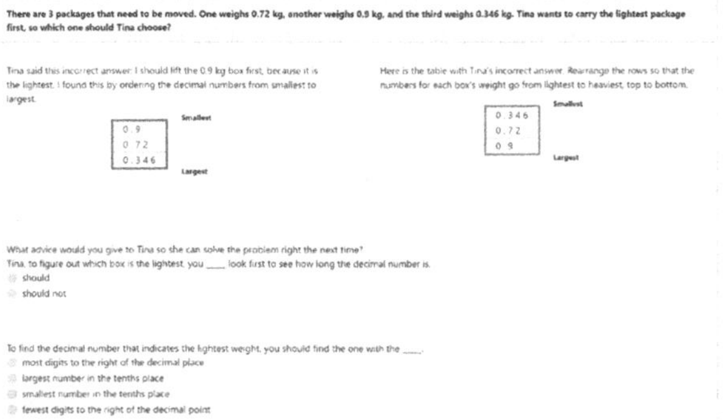

Fig. 2. A sample erroneous example problem from the current study, which has the same decimal content as Fig. 1. Students still received a prompt to explain the error, but the prompt disappeared from the screen after it was answered.

could not advance until they completed each step correctly. They solved the problem and then explained the solution or principle by answering two multiple-choice self-explanation questions. The tutor provided feedback on all incorrect responses.

The third problem in each set was a practice problem targeting the same misconception as the previous two erroneous example or problem-solving items. Practice problems consisted of only one step: solving the problem. Items were identical across conditions and were included to give students additional practice applying what they were learning in the materials. Previous research has shown that including practice problems immediately after example problems can improve learning outcomes [38].

Tests were administered on computers using the same educational technology platform as the intervention. There were 25 items on the pretest, posttest, and delayed posttest, with some items containing multiple parts and points. The tests were worth a total of 61 points, and test scores were computed as the number of points earned out of 61. We deployed three versions of the test with isomorphic problems, and test version order was counterbalanced across students. Items targeted the same misconceptions as the instructional materials and also included some transfer items that did not directly address a specific misconception. Items included tasks such as adding decimal numbers (e.g., 2.41 + 0.6 = ___) and identifying the largest or smallest decimal number from a list (e.g., 5.413, 5.75, 5.6). Transfer items targeted an understanding of decimal number principles (e.g., "Is a longer decimal number larger than a shorter decimal number?") and included new skills not covered in the intervention (e.g., "Select all of the following numbers that are equal to 0.43").

2.3   Procedure

All materials were deployed during students' regular math classes over the course of one week. Members of the research team were present each day to assist in administering the materials and ensure the protocol was followed. Students completed the pretest, instructional materials, and posttest at their own pace and their progress was saved each day. Students who completed the posttest before the end of the week were given math assignments by their teachers that did not target decimal number concepts. One week after the intervention, all students completed a delayed posttest.

2.4   Affect Detection

We developed affect detectors using labels from text replay coding, where segments of log files are pretty-printed and coded by humans. Those codes are input into machine learning algorithms to emulate the coders' judgments, based on prior studies that showed it was feasible to detect confrustion using this approach [27]. The detectors were built on log file data from 598 students across five middle schools collected in previous research with this educational technology platform [1, 5]; data related to the dropped self-explanation step in the previous version of the erroneous example condition were removed from the dataset before developing automated detectors, but remained included during text replay coding. Students' log files were broken down into individual clips for text replay coding, with each problem corresponding to a single clip. Two coders manually labeled 1,600 clips for confrustion based on holistic assessment of confrustion in the current task. For example, for multiple-choice problems, did the student spend a significant amount of time on a first, incorrect attempt and then make a subsequent incorrect attempt? For number line problems, did the student make two substantially distant, incorrect attempts (e.g., 0.3, then 1.1, then 1.8) or multiple incorrect attempts in both directions on the number line (e.g., 0.7, then 0.81, then 0.55)? As with most labels, these text replay labels are imperfect – we do not know if they genuinely capture the affective experience of confusion or frustration in all cases. As these labels are derived only from log files, unlike work that also considers facial expressions or posture [e.g. 25], these labels may in some cases capture only behavior *associated* with confrustion, rather than true confrustion. Agreement was computed after two coders separately labeled the same 129 clips, and results indicated high agreement (κ = .82, p < .001). The remaining 1,471 clips were coded independently by one of the two coders. The set of clips coded was stratified to equally represent both student cohorts, both conditions, and all four problem types.

We built the confrustion detector using the Extreme Gradient Boosting (XGBoost) classifier based on these labeled clips [39]. The classifier uses an ensemble technique that trains an initial, weak decision tree and calculates its prediction errors. It then iteratively trains subsequent decision trees to predict the error of the previous decision tree, with the final prediction representing the sum of the predictions of all the trees in the set. We determined that the detector could effectively infer students' confrustion (κ = .82, AUC = .92) based on 10-fold student-level cross-validation, which involved repeatedly building the model on some students' data and testing it on other students' data. Once effective detection was confirmed, we applied the detector to the new dataset (9,065 clips across 187 students). A total of 30 features were used to predict confrustion, and the importance of each feature was calculated as the proportion it contributed to the final model. The detector reported the probability that a student experienced confrustion on each problem; overall confrustion scores were computed as the average probability of confrustion across all problems.

3   Results

We report results in the order of our hypotheses. To examine whether students' performance improved as a result of the intervention (H1), we conducted a series of paired-samples t tests separately by condition. Results indicated that students in the problem-solving condition improved significantly from pretest to posttest, $t(88)$ – 6.83, p < .001, d = .44, and from pretest to delayed test, $t(88)$ = 8.18, p < .001, d = .48 (Table 1). Likewise, a paired-samples t-test indicated that students in the erroneous example condition improved significantly from pretest to posttest, $t(84)$ = 4.26, p < .001, d = .28, and from pretest to delayed test, $t(84)$ = 5.29, p < .001, d = .37.

Table 1. Proportional confrustion and test scores by condition.

Measure	Problem solving	Erroneous example
Confrustion	M = .22, SD = .12	M = .33, SD = 12
Pretest	M = .53, SD = .22	M = .54, SD = .21
Posttest	M = .62, SD = .21	M = .59, SD = .21
Delayed test	M = .63, SD = .22	M = .62, SD = .23

To test the relation between confrustion and performance (H2), we examined the correlation between the variables. Confrustion was negatively correlated with pretest performance, r = −.64, p < .001, posttest performance, r = −.63, p < .001, and delayed posttest performance, r = −.62, p < .001. To examine the relation between confrustion and performance when controlling for prior knowledge (H2), we tested a multiple regression including pretest and confrustion to predict posttest. The model was significant, R^2 = .70, $F(2,\ 171)$ = 195.66, p < .001. Both confrustion, β = −.186,

$p = .001$, and pretest, $\beta = .69$, $p < .001$, were significant predictors of posttest performance when holding the other factor constant. We applied the same multiple regression model to predict delayed posttest. The model was significant, $R^2 = .69$, $F(2, 171) = 188.22$, $p < .001$. Both confrustion, $\beta = -.15$, $p = .006$, and pretest, $\beta = .72$, $p < .001$, were significant predictors of posttest performance when holding the other factor constant. These results indicate that confrustion predicted test performance even when accounting for students' prior knowledge. In other words, the predictive value of confrustion was not merely a reflection of students' prior knowledge.

To examine the effect of condition on confrustion (H3), we conducted a one-way analysis of variance (ANOVA) that indicated students in the erroneous example condition experienced greater confrustion than students in the problem-solving condition, $F(1, 172) = 41.29$, $p < .001$, $d = 0.38$. To determine whether the relation of confrustion and test performance differed between conditions, we conducted a moderation analyses using PROCESS, an SPSS macro that uses 5000 bootstrap estimates to test mediation and moderation by creating confidence intervals for indirect effects [40]. We tested a PROCESS 1 model using condition as a moderator of the relation between confrustion and posttest performance and, separately, delayed test performance. For the immediate posttest, there was no significant interaction between confrustion and condition, $B = .29$, 95% CI [−.10, .67], and the inclusion of the interaction term did not explain significantly more variance in the model, $\Delta R^2 = .007$, $F(1, 170) = 2.11$, $p = .15$. For the delayed posttest, there was also no significant interaction between confrustion and condition, $B = .27$, 95% CI [-.14, .68], and the inclusion of the interaction term did not explain significantly more variance in the model, $\Delta R^2 = .006$, $F(1, 170) = 2.11$, $p = .19$. These results indicate that the relation between confrustion and performance did not differ between conditions.

To test the effect of condition on performance (H4), we conducted an ANOVA that revealed no differences between conditions on pretest, $F(1, 172) = 0.08$, $p = .77$, $d = 0.04$, posttest $F(1, 172) = 0.82$, $p = .37$, $d = 0.14$, or delayed posttest, $F(1, 172) = 0.17$, $p = .68$, $d = 0.06$ (Table 1). When controlling for pretest, an analysis of co-variance (ANCOVA) indicated that there was a significant effect of condition on posttest, $F(2, 171) = 4.10$, $p = .045$, $\eta_p^2 = .023$, with students in the problem-solving condition performing better. There was no effect of condition on delayed posttest when controlling for pretest, $F(2, 171) = 1.29$, $p = .26$, $\eta_p^2 = .008$. In other words, students in the problem-solving condition improved on the posttest significantly more than students in the erroneous example condition, but there were no differences in improvement on the delayed test. To understand the role of confrustion in this effect, we conducted ANCOVAs testing the effect of condition on test performance controlling for both confrustion and pretest. Results revealed no effect of condition on posttest, $F(3, 170) = 0.02$, $p = .90$, $\eta_p^2 < .001$, or on delayed posttest, $F(3, 170) = 0.35$, $p = .56$, $\eta_p^2 = .002$. This indicates that the variance in confrustion between conditions accounted for the condition effect on posttest improvement.

To understand other potential consequences of the revisions, we examined the amount of time students spent on the materials. An ANOVA indicated a significant difference in total time spent on the instructional materials, $F(1, 172) = 9.95$, $p = .002$, $d = 0.48$, with students in the erroneous example condition ($M = 62.90$, $SD = 23.24$) taking longer to complete the materials than students in the problem-solving condition

($M = 51.59$, $SD = 23.99$). This suggests that the extra self-explanation prompt that was eliminated from the erroneous examples was not responsible for the difference in times across conditions observed in previous studies.

4 Discussion

Unlike prior studies [1, 5], students in the erroneous example condition performed *worse* than students in the problem-solving condition on the immediate posttest when controlling for the pretest, and there were no differences between conditions on the delayed posttest. While the results are reversed in terms of which condition performed better, there is a similar trend in the difference between posttest and delayed posttest. In prior studies, the benefits of erroneous examples emerged only on a delayed posttest, suggesting that students did not experience initial performance benefits but ultimately learned and were able to transfer knowledge better [1, 5]. These previous results were consistent with other research on erroneous examples, which have tended to show the greatest benefit on delayed or transfer tests [2, 6, 10]. In the current study, students in the problem-solving condition showed an immediate performance advantage on the posttest but that advantage did not persist to the delayed posttest, suggesting that benefits from the problem-solving condition primarily affected performance and not the lasting, transferrable learning benefits that are typically most valued as an instructional goal. Thus, while results were inconsistent with prior work in the sense that students in the erroneous examples condition did not perform better on the delayed posttest, it was not a full reversal of effects as would have been seen if students in the problem-solving condition performed better on the delayed posttest.

We predicted that the benefits of erroneous examples would be robust enough to persist despite several changes made to better align the conditions with one another and with the more precise mathematical language used by experts. While this prediction was not upheld, there are several possible explanations for the different outcome. First, students might be more accustomed to using instructional technology than they were when the original materials were tested six years ago. While we would not expect this to change the cognitive benefits of the instructional materials, it might reduce any confusion or frustration students would experience with the interface, such as understanding how to drag numbers to reorder them or select options from a drop-down menu. However, this idea is not supported by the time students spent on the materials. Students in the current study spent on average 50 to 60 min across conditions, while students in the previous studies spent on average 40 to 50 min across conditions [1, 5].

Second, the elimination of the extra self-explanation prompt in the erroneous example condition might have reduced learning in that condition. We think this is unlikely, as students in the erroneous example condition still responded to three self-explanation prompts per question. However, only a direct comparison between versions with and without the additional prompt could provide conclusive evidence.

Third, and we think most likely, the shift to more mathematically precise language may have diminished the benefit of studying and explaining erroneous examples. Students on average spent 10 more minutes on the revised materials compared to the original ones. No other major changes were made to the content of the problem-solving materials, and the other major change to the erroneous examples condition involved *removing* materials. Students' prior knowledge in the current experiment was slightly lower than in prior studies (.53 current, .57 prior), which could cause an increase in the amount of time students needed. Nevertheless, the dramatic increase in time spent on materials supports the idea that students struggled more with reading the new explanation prompts and thus may not have benefitted from them as much. Erroneous example interventions typically instruct students to engage in an evidence-based learning activity to study and understand the erroneous examples, such as comparison [9] or explanation [10]. Without these instructionally robust activities to provide scaffolding, students may not pay as close attention to the erroneous examples or may fail to identify the underlying principles they represent. Put another way, if the mathematically precise language of the new explanation prompts was too difficult for students to understand, then the effect may have been similar to having no explanation prompts at all.

Future research should investigate these possible explanations empirically. We plan to attempt to replicate previous results by randomly assigning students to either the erroneous example or problem-solving conditions using either the original or revised materials, which will also permit a more direct comparison of times and performanceacrossversions. Examination of students' own self-explanations has suggested that the act of engaging in self-explanation is beneficial even when explanations are flawed or mathematically imprecise [41]. Whether provided self-explanation options should be modeled after the imprecise language students tend to use or the more precise language of experts is an open question, and an important one both for understanding the mechanisms of self-explanation and for designing self-explanation options deployed in instructional materials. Since many instructional technologies use the self-explanation method of offering options from which students may choose [35, 36], this is important question to resolve.

References

1. Adams, D., et al.: Using erroneous examples to improve mathematics learning with a web-based tutoring system. Comput. Hum. Behav. **36C**, 401–411 (2014)
2. Booth, J.L., Lange, K.E., Koedinger, K.R., Newton, K.J.: Using example problems to improve student learning in algebra: differentiating between correct and incorrect examples. Learn. Instr. **25**, 24–34 (2013)
3. Borasi, R.: Exploring mathematics through the analysis of errors. Learn. Math. **7**(3), 2–8 (1987)
4. Isotani, S., Adams, D., Mayer, R.E., Durkin, K., Rittle-Johnson, B., McLaren, B.M.: Can erroneous examples help middle-school students learn decimals? In: Kloos, C.D., Gillet, D., Crespo García, R.M., Wild, F., Wolpers, M. (eds.) EC-TEL 2011. LNCS, vol. 6964, pp. 181–195. Springer, Heidelberg (2011). https://doi.org/10.1007/978-3-642-23985-4_15
5. McLaren, B.M., Adams, D.M., Mayer, R.E.: Delayed learning effects with erroneous examples: a study of learning decimals with a web-based tutor. Int. J. Artif. Intell. Educ. **25** (4), 520–542 (2015)

6. Siegler, R.S.: Migrogenetic studies of self-explanation. In: Granott, N., Parziale, J. (eds.) Microdevelopment: Transition Processes in Development and Learning, pp. 31–58. Cambridge University Press, New York City (2002)
7. Siegler, R.S., Chen, Z.: Differentiation and integration: guiding principles for analyzing cognitive change. Dev. Sci. 11(4), 433–448 (2008)
8. Rushton, S.J.: Teaching and learning mathematics through error analysis. Fields Math. Educ. J. 3(1), 4 (2018)
9. Durkin, K., Rittle-Johnson, B.: The effectiveness of using incorrect examples to support learning about decimal magnitude. Learn. Instr. 22(3), 206–214 (2012)
10. Große, C.S., Renkl, A.: Finding and fixing errors in worked examples: can this foster learning outcomes? Learn. Instr. 17(6), 612–634 (2007)
11. Bjork, R.A.: Memory and metamemory considerations in the training of human beings. In: Metcalfe, J., Shimamura, A. (eds.) Metacognition: Knowing about Knowing, pp. 185–205. MIT Press, Cambridge (1994)
12. Soderstrom, N.C., Bjork, R.A.: Learning versus performance: an integrative review. Perspect. Psychol. Sci. 10(2), 176–199 (2015)
13. Kapur, M.: Productive failure in learning math. Cogn. Sci. 38(5), 1008–1022 (2014)
14. Kapur, M., Bielaczyc, K.: Designing for productive failure. J. Learn. Sci. 21(1), 45–83 (2012)
15. D'Mello, S.: A selective meta-analysis on the relative incidence of discrete affective states during learning with technology. J. Educ. Psychol. 105(4), 1082–1099 (2013)
16. Rodrigo, M.M.T., et al.: Affective and behavioral predictors of novice programmer achievement. In: ACM SIGCSE, vol. 41, pp. 156–160. ACM Press, New York (2009)
17. Schneider, B., et al.: Investigating optimal learning moments in U.S. and Finnish science classes. J. Res. Sci. Teach. 53(3), 400–421 (2015)
18. D'Mello, S., Lehman, B., Pekrun, R., Graesser, A.: Confusion can be beneficial for learning. Learn. Instr. 29, 153–170 (2014)
19. Lehman, B., et al.: Inducing and tracking confusion with contradictions during complex learning. Int. J. Artif. Intell. Educ. 22(1-2), 85–105 (2013)
20. VanLehn, K., Siler, S., Murray, C., Yamauchi, T., Baggett, W.B.: Why do only some events cause learning during human tutoring? Cogn. Instr. 21(3), 209–249 (2003)
21. D'Mello, S., Graesser, A.: Dynamics of affective states during complex learning. Learn. Instr. 22(2), 145–157 (2012)
22. Liu, Z., Pataranutaporn, V., Ocumpaugh, J., Baker, R.S.J.D.: Sequences of frustration and confusion, and learning. In: EDM, pp. 114–120 (2013)
23. Kostyuk, V., Almeda, M.V., Baker, R.S.: Correlating affect and behavior in reasoning mind with state test achievement. In: LAK, p. 26 (2018)
24. Pardos, Z.A., Baker, R.S., San Pedro, M., Gowda, S.M., Gowda, S.M.: Affective states and state tests: investigating how affect and engagement during the school year predict end-of-year learning outcomes. J. Learn. Anal. 1(1), 107–128 (2014)
25. Ocumpaugh, J., Baker, R.S., Rodrigo, M.M.T.: Baker rodrigo ocumpaugh monitoring protocol (BROMP) 2.0 Technical and Training Manual. Technical Report. Teachers College, Columbia University, New York, NY. Ateneo Laboratory for the Learning Sciences, Manila, Philippines (2015)
26. Baker, R.S.J.D., Corbett, A.T., Wagner, A.Z.: Human classification of low-fidelity replays of student actions. In: EDM-ITS, pp. 29–36 (2006)
27. Lee, D.M.C., Rodrigo, M.M.T., Baker, R.S.J.D., Sugay, J.O., Coronel, A.: Exploring the relationship between novice programmer confusion and achievement. In: D'Mello, S., Graesser, A., Schuller, B., Martin, J.-C. (eds.) ACII 2011. LNCS, vol. 6974, pp. 175–184. Springer, Heidelberg (2011). https://doi.org/10.1007/978-3-642-24600-5_21

28. Sao Pedro, M.A., de Baker, R.S., Gobert, J.D., Montalvo, O., Nakama, A.: Leveraging machine-learned detectors of systematic inquiry behavior to estimate and predict transfer of inquiry skill. User Model. User-Adap. Inter. **23**(1), 1–39 (2013)
29. Resnick, L.B., Nesher, P., Leonard, F., Magone, M., Omanson, S., Peled, I.: Conceptual bases of arithmetic errors: the case of decimal fractions. J. Res. Math. Educ. **20**, 8–27 (1989)
30. Sackur-Grisvard, C., Léonard, F.: Intermediate cognitive organizations in the process of learning a mathematical concept: the order of positive decimal numbers. Cogn. Instr. **2**, 157–174 (1985)
31. Stacey, K.: Travelling the road to expertise: a longitudinal study of learning. PME **1**, 19–36 (2005)
32. Authors (under review)
33. Chi, M.T.H., de Leeuw, N., Chiu, M., LaVancher, C.: Eliciting self-explanations improves understanding. Cogn. Sci. **18**, 439–477 (1994)
34. Rittle-Johnson, B., Loehr, A.M., Durkin, K.: Promoting self-explanation to improve mathematics learning: a meta-analysis and instructional design principles. ZDM **49**(4), 599–611 (2017)
35. Johnson, C.I., Mayer, R.E.: Adding the self-explanation principle to multimedia learning in a computer-based game-like environment. Comput. Hum. Behav. **26**, 1246–1252 (2010)
36. Mayer, R.E., Johnson, C.I.: Adding instructional features that promote learning in a game-like environment. J. Educ. Comput. Res. **42**, 241–265 (2010)
37. Aleven, V., McLaren, B.M., Sewall, J.: Scaling up programming by demonstration for intelligent tutoring systems development: an open-access website for middle school mathematics learning. IEEE Trans. Learn. Technol. **2**(2), 64–78 (2009)
38. Kalyuga, S., Chandler, P., Tuovinen, J., Sweller, J.: When problem solving is superior to studying worked examples. J. Educ. Psychol. **93**, 579–588 (2001)
39. Chen, T., Guestrin, C.: XGBoost: a scalable tree boosting system. In: ACM-SIGKKD, pp. 785–794 (2016)
40. Hayes, A.F.: Introduction to Mediation, Moderation, and Conditional Process Analysis: A Regression-Based Approach. Guilford Press, New York (2013)
41. Hausmann, R.G., Vanlehn, K.: Explaining self-explaining: a contrast between content and generation. Front. Artif. Intell. Appl. **158**, 417–424 (2007)

Modeling Collaboration in Online Conversations Using Time Series Analysis and Dialogism

Robert-Florian Samoilescu[1], Mihai Dascalu[1,2(✉)],
Maria-Dorinela Sirbu[1], Stefan Trausan-Matu[1,2],
and Scott A. Crossley[3]

[1] University Politehnica of Bucharest,
313 Splaiul Independentei, 060042 Bucharest, Romania
robert.samoilescu@stud.acs.pub.ro, {mihai.dascalu,
stefan.trausan}@cs.pub.ro
[2] Academy of Romanian Scientists,
54 Splaiul Independenţei, 050094 Bucharest, Romania
[3] Department of Applied Linguistics/ESL,
Georgia State University, Atlanta, GA 30303, USA
scrossley@gsu.edu

Abstract. Computer Supported Collaborative Learning (CSCL) environments are frequently employed in various educational scenarios. At the same time, learning analytics tools are frequently used to quantify active learners' participation, collaboration, and evolution over time in CSCL environments. The aim of this paper is to introduce a novel method to cluster utterances from online conversations into zones based on different levels of collaboration. This method depends on time series analyses, grounded in dialogism and focuses on the underlying semantic chains that are encountered in adjacent contributions. Our approach uses Cross-Reference Patterns (CRP) applied on the convergence function between two utterances which captures their semantic relatedness. Two methods for clustering utterances into convergence regions are tested: *clustering by uniformity* and *hierarchical clustering*. We found that *hierarchical clustering* surpasses *clustering by uniformity* by considering only highly related contributions and providing a more straightforward unification mechanism. A validation analysis on the *hierarchical clustering* model was performed on a corpus of 10 chat conversation reporting variance in terms of F1 scores. The model and encountered problems are discussed in detail.

Keywords: Time series analysis ·
Computer Supported Collaborative Learning · Cross-Reference Patterns ·
Clustering

1 Introduction

Technology has facilitated the migration of communication to online environments including simple chat conversation and the exchange of ideas on social platforms. Of particular interest to this study are communicative tools oriented towards educational

© Springer Nature Switzerland AG 2019
S. Isotani et al. (Eds.): AIED 2019, LNAI 11625, pp. 458–468, 2019.
https://doi.org/10.1007/978-3-030-23204-7_38

purposes. These tools have gained increasing usage in the context of Computer Supported Collaborative Learning (CSCL). A specific trait of CSCL communication is the presence of multiple discussion threads, which inter-twine with one with another and may contain direct mapping to the subjects/ideas discussed within the online dialog. Thus, a large number of threads might emerge in a conversation, demonstrating creative stimulations and/or the generation of new ideas [1]. Multiple intertwined viewpoints (i.e., voices) spread across large threads are related to dialogism, which is an important framework for CSCL conversational analysis [2, 3]. Voices are central concepts in dialogism, relate to participants' points of view, and can be operationalized as semantic chains of related concepts [4]. Starting from their distribution and by examining interactions among voices, we can identify collaboration zones in which participants' viewpoints converge [4]. Such collaboration zones play a key role in learning by providing context for elaborating each participant's reasoning [5], as well as for triggering extra cognitive mechanisms (e.g., reduced cognitive load due to social interactions, or knowledge elicitation) [6]. Thus, a crucial element of intelligent tutor systems is to effectively monitor student's interactions and the way students collaborate throughout a conversation or this collaboration allows ideas to converge towards a central topic.

This paper introduces a novel automated method based on time series analyses and semantic similarity to filter online educational dialog for the purpose of identifying the principal moments of collaboration between participants. We test the method in a pilot study using oral language data collected in collaborative educational communication. Our method is aimed at automatically assessing the degree of collaboration between individuals in a conversation, and to potentially provide follow-up feedback to enhance the degree to which participants are involved and collaborate in CSCL environments. Our work is based on time series analysis applied to dialog organized as a series of utterances ordered in time. We use cross-recurrence plots (CRP) [7, 8] to capture the semantic links between all utterance pairs in order to gain an in-depth view of how participants interact and exchange ideas with one another throughout the dialog. Similar approaches have been employed in the past to effectively analyze chaotic signals by identifying different patterns which explain how a dynamic system behaves over time [9].

2   Method

2.1   Corpora

Our experiments are based on the dataset of 10 conversations previously analyzed by Dascalu, Trausan-Matu, McNamara and Dessus [4]. Each conversation was an assignment in which four to five students in a Human-Computer Interaction course in an European Computer Science Department had to debate the advantages and disadvantages of specific CSCL technologies. Afterwards, students propose an integrated alternative that integrates the advantages of each technology, hopefully leading to a convergence of viewpoints. Expert raters then assessed collaboration in the group in terms of ideas exchanged between participants. Moreover, raters were asked to identify continuous segments of collaboration zones within the conversations that exhibited a high degree of collaboration between participants. Each collaboration segment determined by raters was defined as the start and end indexes of utterances within the zone of active collaboration. Based on individual rater scores, we created an overarching histogram of collaboration from which collaboration zones were extracted using a greedy algorithm that grouped contributions around peak/maximum collaboration values [4].

2.2   Method

Our method to automatically assess collaboration and to identify collaboration zones was based on semantically related words spanning throughout the conversations. The three major phases of the method are presented in detail below.

In the *first phase*, we define the concept of convergence between two utterances as a function that computes the similarity between contributions based on their underlying words: $f : Utterances \times Utterances \rightarrow [0, 1]$. As a simple approximation, this function can be computed as lexical convergence [10] which considers the number of common n-gram POS/lexemes (i.e., an index considering pairs of corresponding part of speech tags and stemmed words) identified within pairs of utterances across participants. Although such a function is easy to compute and might offer promising results in certain contexts, it suffers from the fact that an idea must be expressed using the same words, which is seldomly encountered in online conversations. In order to tackle the previous limitation, we start from the concept of "voices" from polyphonic models [3] which can be operationalized as semantic chains [4] or an unification of semantically related lexical chains identified using the method proposed by Galley and McKeown [11]. Lexical chains represent a series of words distributed in multiple utterances, unified either by their common stems or by semantic distances and relations in WordNet (i.e., synonymy, hypernymy, hyponymy and siblings) [11]. A semantic chain is formed by merging multiple lexical chains that share a high semantic similarity between the contained words [4] using similarity functions from different semantic models like vector space models – e.g., Latent Semantic Analysis [12] and word2vec [13] –, as well as topic models – e.g., Latent Dirichlet Allocation [14]. This approach represents a more adequate approximation of how related two utterances are while considering semantically related concepts and their occurrences.

In order to define the semantic convergence between two utterances, we must first reduce the impact of repetitions for words pertaining to the same semantic chain and appearing within a given utterance by considering their logarithmic frequency. Thus, let $l_i : Utterances \rightarrow N$, for $i = 1, 2, \ldots, k$ be the functions that returns the logarithmic number of words that are contained in the i'th semantic chain, given a set of utterances. While considering only two utterances, namely u_l and u_m, the semantic convergence between the two is given by the following formula:

$$f(u_l, u_m) = \frac{\sum_{i=1}^{k} \min(l_i(u_l), l_i(u_m))}{\sum_{i=1}^{k} \max(l_i(u_l), l_i(u_m))}$$

This function also emphasizes the difference between strong and weak convergence while relating to the participants' degree of collaboration – i.e., two utterances will have a strong convergence (equal to one) which can lead to intense collaboration zones, if they have the same number of words in each semantic chain; otherwise, we consider the two utterances to be weakly converging and the resulting collaboration zone to be weaker.

The *second phase* consists of creating cross-recurrence plots based on the previously defined function. We build a matrix in which each row and column corresponds to an utterance, and each cell (i,j) is computed as $f(i,j)$. The main diagonal contains scores of one for all utterances which have content words; the value of one represents the correlation of the utterance with itself. The CRP offers a useful overview of how conversation evolves in time, and also allows us to identify specific regions (squares in the matrix) where there is a high correlation between utterances. Our main idea was to identify collaboration regions as clustered cells near the main diagonal, and then to mark the zones that denote a strong convergence marked with red color in Fig. 3.

In the *third phase* we introduce two methods for clustering utterances into convergence regions, each with its corresponding advantages and drawbacks [15]. The first method is based on clustering by uniformity (see Fig. 1), as we expect a conversation to consist of alternative sequences of zones with strong and weak convergence. Thus, our aim is to find the boundary of such zones. Our algorithm starts with a 2×2 matrix from the first contribution – cell $(0, 0)$ – and it tries to expand this region while the Root Mean Squared Error (RMSE) does not exceed an imposed threshold. When the threshold is exceeded, we start again with a 2×2 matrix from the current position and repeat the above steps. The main diagonal is not taken in consideration when computing the RMSE. Using this method, we identified three main types of regions: zones with weak convergence and a mean near 0, transition zones with a mean near 0.5, and zones with strong convergence with mean near 1. The main problem with this approach is its sensitivity to noise, which makes it very difficult to impose an adequate threshold. In addition, using this method we are faced with a classification problem, namely to classify each cluster as either a relaxed or an intense collaboration zone based on their mean convergence values. Although a constant threshold might represent a straightforward approach, we identified several situations in which this did not generalize well as the conversations from our corpora are quite different, and this approach is problematic even with normalization. Therefore, we applied Expectation Maximization on a Gaussian Mixture Model because this offers the possibility to partition the data in a dynamic way, knowing a priori the number of clusters in which we want to split the utterances. The clustering algorithm produces several zones which are labeled as being either collaborative or non-collaborative. A direct method to measure whether there is a strong connection between the utterance from a zone is to compute the mean value of its cells. Thus, each identified cluster is mapped to a real number for which we applied the EM algorithm into one dimension.

The best method for classification was based on hierarchical clustering (see Fig. 2). With this method, we started with a cluster for each utterance, and we tried to unify them by the mean of their connection area. Therefore, given two consecutive clusters $C_1(entry_1, size_1)$, $C_2(entry_2, size_2)$, the link between them is the region from the CRP matrix with the coordinates $(entry_1, entry_2, size_1, size_2)$ – i.e., it starts at cell $(entry_1, entry_2)$, and has $size_1$ rows and $size_2$ columns. In an iteration, we unify the two clusters if the mean value of convergence is maximal and it exceeds an imposed threshold. The algorithm stops when there are no further unifications which can be performed. The

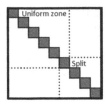

Fig. 1. Visual representation of uniformity clustering.

main advantage of this approach is that no classification is needed after clustering because only highly correlated cells are clustered. Similar to the previous method, a drawback is the threshold hyperparameter which needs to be imposed, for example, via grid search methods that identify the optimal value.

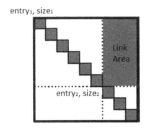

Fig. 2. Visual representation of hierarchical clustering.

3   Results and Discussions

Figure 3 depicts the result of the hierarchical clustering performed on one of our conversations in which rows and columns reflect the participant's pseudonym, the CSCL technology they focused on, and the corresponding utterance ID. Collaboration zones are marked in red with circles as delimitating markers.

From an overarching visual point of view, the clustering algorithms performs quite well by marking squares of interest around the main diagonal. Although student talked about advantages and disadvantages of particular domain-specific topics, utterance pairs still exhibited a high semantic similarity as most contributions contain the same semantic chains centered on the debated technologies. The recurrence of key topics throughout the conversation is essentials for collaboration and, even if students were instructed to represent a certain technology, they frequently made side-by-side comparisons or continued their peers' contributions.

The output of the convergence measure between two utterances, $f(u_l, u_m)$, is a number between 0 and 1, where 1 reflects the strongest correlation. The CRP values are represented on a gray scale and black scales correspond to a similarity value of 1. Note that each utterance from the main diagonal is compared to itself, thus the cells are black. Moreover, white cells denote a lack of semantic overlap and, in some particular cases, can be generated by contributions that do not contain any content words (e.g., a simple "ok"). System generated events were automatically disregarded.

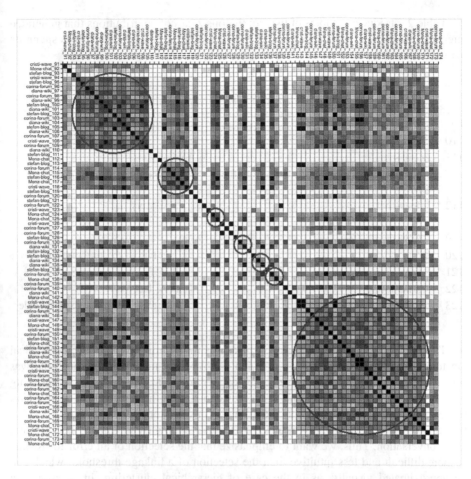

Fig. 3. Cross-recurrence plots after hierarchical clustering (Color figure online).

In order to further exemplify how the convergence function is computed, Table 1 presents two adjacent contributions both containing one semantic chain with the following words marked in grey: "client", "progress", and "project". The values from the cells (103, 104) and (104, 103) are both equal to $\frac{ln(1)+1}{ln(3)+1}$ since the CRP is a symmetric matrix (i.e., a maximum of 3 and a minimum of 1 since utterance 103 contains 3 words from the chain, whereas utterance 104 contains only the word "project").

Table 1. Sample contributions that generate cluster expansion interruptions.

ID	Participant	Utterance
103	corina-forum	well the client must know the progress of a project, no?
104	diana-wiki	and also, during the project, in case some major difficulties

Table 2 introduces samples of two contiguous zones of contributions that generated interruptions after the first and second clusters from Fig. 3, namely utterances spanning between IDs 110–112 and 119–123.

Table 2. Sample contributions that generate cluster expansion interruptions.

ID	Participant	Utterance
110	diana-wiki	ok :)
111	stefan-blog	ok
	Mona-chat (disregarded utterance from the analysis)	*joins the room (system generated event)*
112	Mona-chat	Sorry, guys....
	...	
119	stefan-blog	ok
120	corina-forum	it is a bug in the program :p
121	stefan-blog	kill it :P
122	corina-forum	ok! Perfect
123	cristi-wave	yes... chat is sometimes unreliable

Tables 3 and 4 present the accuracy measures for both uniformity and hierarchical clustering for all 10 conversations in which the hyperparameter for the threshold was selected using grid search. The selection of the threshold for uniformity clustering reports lower results for recall and F1 scores and slightly higher results for precision scores. Although the threshold was fixed for all conversations for evaluation purposes, this value could be dynamically imposed by the user simply by looking at the CRP of a new conversation. Thus, we want to emphasize that the selection of an error for RMSE is more difficult and less intuitive than the selection of a linkage threshold, which can be approximated visually, as in the case of hierarchical clustering. In addition, the results of our method are comparable to previously research – average F1 score of .74 for the Voice Pointwise Mutual Information model and .79 for the Social Knowledge Building model [4]. However, a high variance can be observed in terms of F1 scores which range from .65 to .79, denoting different problems.

The first problem appears because the manually annotated regions from the chat corpus tend to be quite large, which induces a high recall (close to .9 in several cases). This results in a problem for our generative method of collaboration zones based on the cumulative histogram. The regions that we generate are quite large, which interferes with the selection of the hyperparameter. In this specific case, the threshold tends to become smaller in order to generate larger clusters, which in return justifies the large values for recall.

The second problem exists because the baseline annotated collaboration zones generated from the histogram of manual annotations have a hard margin, in accordance to Dascalu, Trausan-Matu, McNamara and Dessus [4]. To be more precise, the original generative algorithm looks at local maxima from the histogram that are greater than the mean value. An intense collaboration zone is created if the distance between two local

Table 3. Accuracy of uniformity clustering for the identification of intense collaboration zones.

Conversation ID	Uniformity clustering (Threshold = .18)		
	Precision	Recall	F1 score
1	.734	.815	.773
2	.844	.551	.667
3	.697	.901	.786
4	.654	.918	.764
5	.608	.774	.681
6	.714	.842	.773
7	.695	.576	.630
8	.630	.812	.710
9	.656	.606	.630
10	.644	.882	.744
Avg.	*.687*	*.767*	*.715*

Table 4. Accuracy of hierarchical clustering for the identification of intense collaboration zones.

Conversation ID	Hierarchical clustering (Threshold = .2)		
	Precision	Recall	F1 score
1	.702	.870	.777
2	.656	.728	.690
3	.716	.866	.784
4	.680	.945	.791
5	.597	.872	.709
6	.752	.829	.789
7	.733	.766	.749
8	.642	.795	.711
9	.638	.667	.652
10	.670	.871	.757
Avg.	*.679*	*.821*	*.741*

maxima is smaller than a slack empirically set at 5% of the length of the conversation. This causes problems because the emerging annotated collaboration zone has both ends local maxima representing an abrupt start and finish, which is not the case for our clustering algorithm.

Third, noise in our CRP influences the clustering results and creates additional segmentations. In Fig. 3 we can observe utterances that do not correlate with others, introducing some gaps. These utterances have no cross-recurrence as they potentially contain automated messages generated by the chat server, or they have no contextual information – e.g., simple replies like "OK", "Yes", etc. – which break down the connection between two clusters. Ignoring those zones, and cumulating clusters around them is quite problematic. At least another hyperparameter must be added, as the maximum size that allows two clusters separated by noise to be merged. However, this is difficult considering how manual annotations are combined, and the overall tendency to have larger annotated zones of collaboration, in contrast to more precise smaller regions.

4 Conclusions and Future Work

In this paper we introduce a novel and flexible method based on time series analysis that clusters utterances from online conversations into collaboration zones. Based on the identified principal moments of collaboration between participants, we are able to evaluate the degree of individuals' collaboration with peers within CSCL conversations. Moreover, our analysis can also highlight potential inconsistencies in communication in terms of zones of contributions that have low semantically relatedness among adjacent utterances. In line with the studies performed by Jermann et al. [16, 17] which considered dual eye-tracking, there is a synergy of perspectives in the sense that collaboration can be identified by evaluating cross-recurrence patterns from diverse learning traces.

The implications of our method are multifold. First, the method facilitates the identification and visual separation of zones in which participants exchange semantically related contributions that create a conversation momentum in which they discuss more central topics. Second, our approach transcends simple semantic similarity measures by considering the occurrence patterns of semantic chains spanning throughout the entire conversation. Third, the clustering algorithm can be also used to observe divergence or out-of-focus/off-topic zones in the conversation that should be closely examined. Although creativity is essential, introducing too many unrelated contributions and points of view may be detrimental to the overall collaboration indicating that perspectives need to be contained by other participants. Fourth, our method enables tutors to easily monitor how students are interacting with their peers while participating in specific CSCL scenarios. This enables tutors to take timely actions and encourage their students to be more actively involved, while also considering the points of view of other members.

We must also take into account the limitations of our approach. First, the clustering algorithm requires proper thresholds that need to be empirically adjusted by the tutors. Currently, there is no automated manner in which these parameters can be set a priori as all analyses are tailored to specific requirements (e.g., dense and narrow convergence regions for targeted conversations versus a wider spread of these regions for longer discussions). Second, our method is most effective while performing a posteriori analyses of conversations because the considered semantic chains emerge when considering concepts from the entire conversation. Although near real-time processing can be performed, shifts in the conversation topics will not be initially perceived as collaboration. Once an automated solution is found to these two problems, our approach can be extended to larger populations like those found in on-line tutoring systems of massive open online classes (MOOCs). Our long-term aim is to create a tool that provides feedback to students on participating and encourages them to get more involved in follow-up conversations, as well as potentially applying different strategies to become better immersed within the conversation, while following the topics introduced by their peers. Such a feedback tool could help learners become more engrained in the learning community and potentially lead to greater learning gains, motivation, and entrenchment.

Considering the identified problems, we aim to test our model on different datasets, with more accurate annotation of collaboration zones in follow up studies. In addition, further improvements can be performed in the clustering method by reducing CRP's noise and considering additional factors when defining the correlation function (for example, speech acts and certain question-answer patterns).

Acknowledgments. This work was supported by a grant of the Romanian Ministry of Research and Innovation, CCCDI - UEFISCDI, project number PN-III-P1-1.2-PCCDI-2017-0689/ "Lib2Life - Revitalizarea bibliotecilor si a patrimoniului cultural prin tehnologii avansate"/ "Revitalizing Libraries and Cultural Heritage through Advanced Technologies", within PNCDI III, as well as the FP7 2008-212578 LTfLL project.

References

1. Trausan-Matu, S.: Computer support for creativity in small groups using chats. Ann. Acad. Rom. Scientists Ser. Sci. Technol. Inf. **3**(2), 81–90 (2010)
2. Koschmann, T.: Toward a dialogic theory of learning: Bakhtin's contribution to understanding learning in settings of collaboration. In: International Conference on Computer Support for Collaborative Learning (CSCL 1999), pp. 308–313. ISLS, Palo Alto (1999)
3. Trausan-Matu, S., Stahl, G., Sarmiento, J.: Polyphonic support for collaborative learning. In: Dimitriadis, Y.A., Zigurs, I., Gómez-Sánchez, E. (eds.) CRIWG 2006. LNCS, vol. 4154, pp. 132–139. Springer, Heidelberg (2006). https://doi.org/10.1007/11853862_11
4. Dascalu, M., Trausan-Matu, S., McNamara, D.S., Dessus, P.: ReaderBench – automated evaluation of collaboration based on cohesion and dialogism. Int. J. Comput. Supported Collaborative Learn. **10**(4), 395–423 (2015)
5. Teasley, S.D.: Talking about reasoning: how important is the peer in peer collaboration? In: Resnick, L.B., Säljö, R., Pontecorvo, C., Burge, B. (eds.) Discourse, Tools and Reasoning. NATO ASI Series (Series F: Computer and Systems Sciences), vol. 160, pp. 361–384. Springer, Heidelberg (1997). https://doi.org/10.1007/978-3-662-03362-3_16
6. Dillenbourg, P.: What do you mean by collaborative learning? In: Dillenbourg, P. (ed.) Collaborative Learning: Cognitive and Computational Approaches, pp. 1–19. Elsevier, Oxford (1999)
7. Eckmann, J.-P., Kamphorst, O., Ruelle, D.: Recurrence plots of dynamical systems. EPL (Europhys. Lett.) **4**(9), 973 (1987)
8. Marwan, N., Kurths, J.: Nonlinear analysis of bivariate data with cross recurrence plots. Phys. Lett. A **302**(5–6), 299–307 (2002)
9. Rabinovich, M., Abarbanel, H.: The role of chaos in neural systems. Neuroscience **87**(1), 5–14 (1998)
10. de Wolf, N.J.G.: Visualising cross-recurrence patterns in natural language interaction, vol. Bsc. thesis. University of Amsterdam, Amsterdam, Netherlands (2014)

11. Galley, M., McKeown, K.: Improving word sense disambiguation in lexical chaining. In: 18th International Joint Conference on Artificial Intelligence (IJCAI 2003), pp. 1486–1488. Morgan Kaufmann Publishers, Inc., Acapulco (2003)
12. Landauer, T.K., Foltz, P.W., Laham, D.: An introduction to latent semantic analysis. Discourse Processes **25**(2/3), 259–284 (1998)
13. Mikolov, T., Chen, K., Corrado, G., Dean, J.: Efficient estimation of word representation in vector space. In: Workshop at ICLR, Scottsdale (2013)
14. Blei, D.M., Ng, A.Y., Jordan, M.I.: Latent Dirichlet allocation. J. Mach. Learn. Res. **3**(4–5), 993–1022 (2003)
15. Han, J., Pei, J., Kamber, M.: Data Mining: Concepts and Techniques. Elsevier, Amsterdam (2011)
16. Jermann, P., Nüssli, M.-A.: Effects of sharing text selections on gaze cross-recurrence and interaction quality in a pair programming task. In: Conference on Computer Supported Cooperative Work, pp. 1125–1134. ACM, Seattle (2012)
17. Jermann, P., Mullins, D., Nüssli, M.-A., Dillenbourg, P.: Collaborative gaze footprints: correlates of interaction quality. In: International Conference on Computer-Supported Collaborative Learning (CSCL 2011), pp. 184–191. International Society of the Learning Sciences, Honk Kong (2011)

Improving Short Answer Grading Using Transformer-Based Pre-training

Chul Sung[1]([✉]), Tejas Indulal Dhamecha[2], and Nirmal Mukhi[1]

[1] IBM Watson Education, Yorktown Heights, NY 10598, USA
{sungc,nmukhi}@us.ibm.com
[2] IBM Research, Bangalore, India
tidhamecha@in.ibm.com

Abstract. Dialogue-based tutoring platforms have shown great promise in helping individual students improve mastery. Short answer grading is a crucial component of such platforms. However, generative short answer grading using the same platform for diverse disciplines and titles is a crucial challenge due to data distribution variations across domains and a frequent occurrence of non-sentential answers. Recent NLP research has introduced novel deep learning architectures such as the Transformer, which merely uses self-attention mechanisms. Pre-trained models based on the Transformer architecture have been used to produce impressive results across a range of NLP tasks. In this work, we experiment with fine-tuning a pre-trained self-attention language model, namely Bidirectional Encoder Representations from Transformers (BERT) applying it to short answer grading, and show that it produces superior results across multiple domains. On the benchmarking dataset of SemEval-2013, we report up to 10% absolute improvement in macro-average-F1 over state-of-the-art results. On our two psychology domain datasets, the fine-tuned model yields classification almost up to the human-agreement levels. Moreover, we study the effectiveness of fine-tuning as a function of the size of the task-specific labeled data, the number of training epochs, and its generalizability to cross-domain and join-domain scenarios.

Keywords: Self-attention · Transfer learning · Student answer scoring

1 Introduction

Dialogue-based tutoring (DBT) platforms such as AutoTutor [6], Rimac [1], DeepTutor [24] and the Watson Tutor [28] have shown great promise in meeting individual student's needs. In such systems, the tutoring platform interacts with the student by asking questions and provides individual feedback based on all student answers. To provide appropriate feedback and rectify student mistakes, accurately understanding student answers is crucial. However, devising a generic short answer grading system that performs well across different questions and

© Springer Nature Switzerland AG 2019
S. Isotani et al. (Eds.): AIED 2019, LNAI 11625, pp. 469–481, 2019.
https://doi.org/10.1007/978-3-030-23204-7_39

domains of study is a challenge due to data distribution variations (differences in used language, length and depth of answers, use of non-sentential answers, among other issues).

Various Deep Learning (DL) based techniques have been explored for short answer grading [2,11,12,17,25]. However, availability of limited labeled data (reference and student answer pairs) often prohibits meaningful training; furthermore, due to domain discrepancy between the public corpora and short answer grading corpus, the utilization of the former by augmentation is not efficient. Lately, transfer learning has largely supplanted the use of the older DL techniques, and have had a substantial impact on the state of Natural Language Processing (NLP) [16]. The main concept within transfer learning is to apply the knowledge from one or more source tasks to a target task [18]. Broadly, a target task can use the knowledge of labeled data from other tasks or from unlabeled data called *self-taught learning* [21]. In NLP, word embedding is one of the most influential transfer models due to its capability of capturing semantic context of a word by producing vector representations of words from large unlabeled corpora such as Wikipedia and news articles [13].

As a transition of a robust transfer learning model, Peters *et al.* introduced contextualized word representations (called Embeddings from Language Models or ELMo) [19]. ELMo captured contextual information from word representations by combining the hidden states of multiple bidirectional LSTMs and initial embeddings. In 2018, diverse novel fine-tuning language models such as Universal Language Model Fine-tuning (ULMFiT) [9] and OpenAI's Generative Pre Training (GPT)[1] [20] were proposed followed by a robust transfer language model called Bidirectional Encoder Representations from Transformers, or BERT [5]. OpenAI's GPT and BERT adapted the Transformer architecture to learn the text representations, a novel and efficient language model architecture based on a self-attention mechanism [27]. However, while OpenAI's GPT used an unidirectional attention approach (the decoder in Transformer), BERT used a bidirectional one (the encoder in Transformer) to better understand the text context. BERT can be trained in two phases. In the *pre-training phase*, deep bidirectional representations inherited by the nature of the Transformer Encoder can use unlabeled huge corpora. In the *fine-tuning phase*, task-specific labeled data and parameter tuning is performed to optimize results for a specific problem, such as question answering or short answer grading.

In this work, we experiment with fine-tuning a pre-trained BERT language model and explore the following questions:

- How well do Transformer-based DL approaches (we use BERT as it is the latest iteration of such models) apply to short answer grading?
- How much does fine-tuning, involving the collection of domain-specific labeled answers, impact the results obtained?
- What is the amount of training (number of epochs) needed in order to produce an optimized model using this approach?
- How well does the same fine-tuned Transformer-based model work across different domains of study for the short answer scoring task?

---

[1] https://blog.openai.com/language-unsupervised/.

We begin with an overview of recent approaches in short answer grading, and an overview of BERT and the BERT model architecture, before presenting details on our experiments designed to answer these questions.

2 Related Work

Broadly speaking the literature pertaining to the problem of short answer grading can be categorized into two: (1) earlier approaches that relied heavily on hand-crafted features, and (2) recent deep learning approaches that require minimum, if not none at all, feature engineering.

2.1 Hand-Crafted Features

Mohler and Mihalcea [15] and Mohler *et al.* [14] are among the earliest research works towards automatic short answer grading. These approaches relied on various word similarity measures, corpus-based measures, and alignment of parses of reference and student answers. A benchmark in the field was established with the Student Response Analysis Challenge as part of SemEval-2013 [7]. Participating approaches relied on a range of hand-crafted features including corpus-based word similarities, WordNet based word similarities, part-of-speech tags, sentence parsing, and n-grams; one of the participants also explored domain adaptation. Broadly, the problem of Student Response Analysis is modeled as a special case of Textual Entailment or Semantic Textual Similarity. Ramachandran *et al.* [22] proposed to extract phrase patterns from reference answers to form basis of scoring approach. The approach improves over earlier approaches in that it explicitly extracts semantic information at sentence as opposed to earlier word similarity metrics. Ramachandran and Foltz [23] proposed a short answer grading based on text summarization.

2.2 Deep Learning Approaches

With the advances in deep learning approaches, various works leveraged these approaches. Sultan *et al.* [26] represented a sentence as sum of word embeddings [13] of its tokens in conjunction with other features. The approach uses word embeddings obtained by deep learning on large corpus; however, obtaining feature representations of a sentence as sum of word embeddings ignores the structural information. Thus, as a logical extension subsequent works have explored more sophisticated ways to obtain feature representations of answer sentences. Mueller and Thyagarajan [17] proposed a Long Short-Term Memory (LSTM) based Siamese network to compare student answer against reference answer. They observe that one of the major limitations in training LSTM networks is the lack of large amount of training data. They generate additional pairs of answers by replacing words in the original dataset. The extended dataset is used for training LSTM networks for short answer grading. The data intensive nature of deep learning approaches has emerged as an interesting issue for research, particularly in data-starved problems such as short answer grading.

Transfer Learning has evolved into a promising research direction to address this. It claims that a generic learning of natural language can be obtained from a data-rich generic task, which can be then *transferred* to downstream tasks which may have limited data. Research efforts to learn universal sentence embeddings for task-specific transfer have yielded impressive improvements on various benchmarks. Notable works include InferSent [4], ELMo [19], ULMFiT [9], GPT [20], and BERT [5]. Saha *et al.* [25] explored sentence embedding features from InferSent in conjunction with traditional token features. In another recent work, Marvaniya *et al.* [12] showed that short answer grading based on sentence embedding features can be further improved by leveraging their proposed scoring rubric approach. The current state of the short answer grading research has shown that transfer of sentence embeddings is useful, yet non-contextual approaches encounter their limitations at downstream tasks. In this study, we aim to demonstrate the ability and various characteristics of BERT (a latest and robust transfer language model) for short answer grading with limited domain-specific training data.

3   BERT for Short Answer Grading

The broad premise of BERT [5] is that there is a high-level language model that needs to be encoded into the network irrespective of the downstream task. The high-level language model is learned based on two semi-supervised objectives of (1) Masked Language Model (MLM) for a deep bidirectional representation and (2) Next Sentence Prediction (NSP) for understanding relationship between sentences; this training leverages multiple corpora. The resultant model, often called the pre-trained BERT model, forms the basis for downstream target tasks. For the task of short answer grading, we perform fine-tuning in the form of *Sentence Pair Classification*. This model allows to classify a pair of reference and student answers into desired categories of correct, incorrect, contradiction, and so on.

3.1   BERT Model Architecture

As described in Devlin *et al.* [5], BERT takes a single token sequence from a single text sentence for the MLM objective or from a pair of text sentences (adding [SEP] token between them as a separator) for the NSP objective. The special classification embedding [CLS] is added in front of each sequence and it is used as input to the classification-task layer. As shown in Fig. 1, the input representations are obtained by combining the token, segment, and *learned* position embeddings. The segment embeddings identify which sentence tokens are from and the position embeddings relative positioning of tokens. This is the input to the first Transformer Encoder layer and the output of this layer is fed into the next Transformer layer. BERT may have a stack of multiple Transformer layers. Each Transformer Encoder is composed of two major parts: a self-attention layer with multiple attention heads, followed by token-wise feed-forward layers.

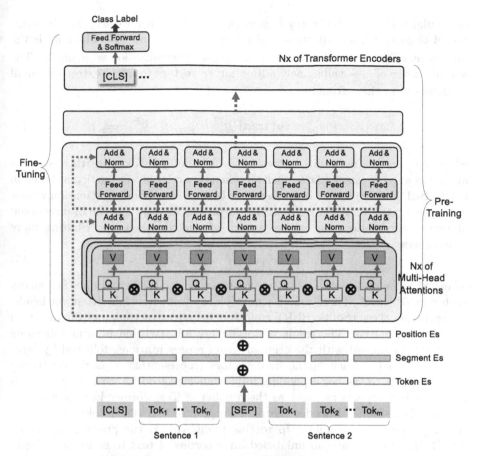

Fig. 1. BERT model architecture for short answer grading. We employed the *Sentence Pair Classification* task specific model using BERT. To describe the details of the model we used the same colors for the same representations as in [5,27].

Each attention head acts akin to a convolution in a convolutional neural network (ConvNet), except for a weighted average. As part of self-attention mechanism, BERT computes three vectors from each token (called query, key, and value) by multiplying three trainable weight matrices (W^Q, W^K, W^V respectively). The weight matrices emphasize different location values of the input as the role of kernels in ConvNet and they are adjusted for every head.

$$q_j^i = x_j W_i^Q \qquad k_j^i = x_j W_i^K \qquad v_j^i = x_j W_i^V \tag{1}$$

where, q_j^i, k_j^i, and v_j^i are the query, key, and value vectors (projections) respectively for jth token x_j in ith head. Then, with the query and key vectors BERT calculates attention weights by: (1) the dot product of the query vector of a particular token and all the key vectors ($\boldsymbol{k}_1^i...\boldsymbol{k}_n^i$ in ith head where n is the number of tokens), (2) an adjustment of the dot products by $\frac{1}{\sqrt{d_k}}$ where d_k is the dimension of the key vectors, and (3) a softmax normalization sequentially. The scaling factor of $\frac{1}{\sqrt{d_k}}$ helps finely adjust larger vectors to avoid extremely small gradients from the softmax.

$$aw_{j1}^i, ..., aw_{jn}^i = \mathtt{softmax}((\boldsymbol{q}_j^i \cdot \boldsymbol{k}_1^i, ..., \boldsymbol{q}_j^i \cdot \boldsymbol{k}_n^i)\frac{1}{\sqrt{d_k}})) \qquad (2)$$

where aw_{jk}^i is kth normalized attention weight for jth token in ith head. The attention weights capture how much all tokens are related to a particular token in head$_i$. BERT multiplies each value vector by the corresponding attention weight and sums up the weighted results. The output vector contains the bi-directional attention information, the value vectors of related tokens contributing more than others.

$$z_j^i = aw_{j1}^i \boldsymbol{v}_1^i + ... + aw_{jn}^i \boldsymbol{v}_n^i \qquad (3)$$

where z_j^i is the output of a self-attention layer for jth token and $\boldsymbol{v}_k^i$ is kth value vector in ith head. There may be multiple z_j from multiple attention heads. To aggregate these results, BERT concatenates all z_j vectors, multiplying them by a weight matrix. The result vector having all attention information along all heads is summed with the original token representations, followed by layer normalization [3]. Each of the final vectors (representing a particular token) discretely goes to the corresponding fully connected feed-forward network. This full procedure repeats as many as the number of Transformer Encoders and at the last Transformer Encoder the final output for the [CLS] token is used as the sequence representation. Up to this point, this is the pre-training model and BERT can leverage an unlabeled huge corpus of text to construct a high-level language model. Then, BERT adapts the labeled data for short answer grading not only for fine tuning the pre-training model but also constructing a classification model through the feed-forward classification layer on the pre-training model.

4   Experiments

We evaluated our proposed approach on two datasets:

1. **SCIENTSBANK-3way dataset of SemEval-2013** [7]: We used SCIENTS-BANK dataset for the 3-way task in SemEval 2013 challenge. The data consists of questions, reference answers, student answers, and three-way labels (CORRECT, INCORRECT, and CONTRADICTORY or in short CO, IC, and CD respectively) in the science domain. The SemEval 2013 challenge involves three classification subtasks on three given test sets: unseen answers (UA), unseen questions (UQ), and unseen domains (UD).

2. **Two psychology domain datasets:** The datasets contain a collection of questions, reference answers, student answers, and three-way labels (CORRECT, PARTIALLY-CORRECT, and INCORRECT or in short CO, PC, and IC respectively). These are based on student answers from two psychology-related textbooks (one is from behavioral physiology and has a lot of technical language and the other is from developmental psychology with mostly non-technical material). Each student response is manually annotated by three experts. Groundtruth is obtained as majority voting of the three annotations.

As shown in the Table 1, the class distribution of both datasets is highly skewed. Due to the class imbalance we select a macro-average-F1 method to observe how our proposed approach preforms overall across the latest other approaches. The macro-average-F1 computes the F1 score independently for each class and then takes the average of all F1 scores. Moreover, we report results in terms of accuracy and weighted-average-F1, but due to the class-imbalance in the datasets, these two metrics may provide biased evidences.

Table 1. Details of class distribution and train-test split protocols for SCIENTSBANK 3-way dataset of SemEval 2013 challenge and our psychology domain 1 and 2 datasets. The test set of SCIENTSBANK is divided into three different test sets for the three subtasks: unseen answers (UA), unseen questions (UQ), and unseen domains (UD).

Dataset	Class distribution			Train-test split	
				Training	Test
SemEval-2013 [7]	4,459 (CO)	5,307 (IC)	1,038 (CD)	4,969	540 (UA)
					733 (UQ)
					4,562 (UD)
Psychology domain 1	14,460 (CO)	3,845 (PC)	1,790 (IC)	16,076	4,019
Psychology domain 2	12,295 (CO)	2,495 (PC)	1,090 (IC)	12,704	3,176

4.1 Pre-training Setup

We chose **BERT$_{BASE}$, Uncased**[2] pre-trained model, which used the concatenation of BooksCorpus (800M words) and English Wikipedia (2,500M words) for pre-training. **Uncased** means that the text has been converted to lower-case before tokenization, dropping any accent markers. BERT uses WordPiece embeddings [29] using a 30,000 token vocabulary and up to 512 tokens are supported for the input sequence. The details of the BERT$_{BASE}$ model can be found in [5].

4.2 Fine-Tuning Setup

For fine-tuning the pre-trained BERT$_{BASE}$ model and a classification layer, we generated the two datasets in tab-separated values (TSV) files. We changed the learning rate of Adam optimizer to 2e–5 for SemEval-2013 and 3e–5 for two

---

[2] https://github.com/google-research/bert.

psychology domain datasets with the same batch size 32. We have also gradually reduced the training size up to 20% of the entire set to observe how many labeled data are required for fine-tuning. We changed the number of epochs from 4 to 12 to observe how many epochs the BERT and classifier are required to complete fine-tuning. For the fine-tuning process, we used two NVIDIA Tesla P100 GPUs (Graphics Card RAM 16 GB) and 120-GB memory.

Table 2. Performance on SciEntsBank Dataset of SemEval-2013 [7]. All results of ‡ are as reported in [25]. MEAD [23], Graph [23] and Marvaniya *et al.* [12] reported results on unseen answer protocol only as their approaches are designed for this scenario. Accuracy (Acc), macro-average-F1 (M-F1), and weighted-average-F1 (W-F1) are reported in percentage.

	Unseen answer			Unseen question			Unseen domain		
	Acc	M-F1	W-F1	Acc	M-F1	W-F1	Acc	M-F1	W-F1
Baseline [7]	55.6	40.5	52.3	54.0	39.0	52.0	57.7	41.6	55.4
ETS [8]	72.0	64.7	70.8	58.3	39.3	53.7	54.3	33.3	46.1
SOFTCAR [10]	65.9	55.5	64.7	65.2	46.9	63.4	63.7	48.6	62.0
MEAD [23]	-	42.9	55.4	-					
Graph [23]	-	43.8	56.7	-					
Sultan *et al.* [26]‡	60.4	44.4	57.0	64.3	45.5	61.5	62.7	45.2	60.3
Saha *et al.* [25]	71.8	66.6	71.4	61.4	49.1	62.8	63.2	47.9	61.2
Marvaniya *et al.* [12]	-	63.6	71.9	-					
Proposed BERT$_{\text{BASE}}$	**75.9**	**72.0**	**75.8**	**65.3**	**57.5**	**64.8**	**63.8**	**57.9**	**63.4**

4.3   Results and Analysis

We performed a set of experiments to study various aspects of the proposed BERT$_{\text{BASE}}$ model for the problem of short answer grading, including (1) performance comparison with published literature and human agreements, (2) sufficiency of fine-tuning in terms of supervised data requirement and the number of training epochs, (3) applicability of fine-tuned model on different domain, and (4) ability to jointly fine-tune for multiple domains. Based on the various experiments and their results presented on benchmark SciEntsBank dataset and our two psychology domain datasets, we make following key observations:

Table 3. Performance comparison of human agreements and the proposed method on our two psychology (psych.) domain datasets. Accuracy (Acc), macro-average-F1 (M-F1), and weighted-average-F1 (W-F1) are reported in percentage.

	Psych. domain 1			Psych. domain 2		
	Acc	M-F1	W-F1	Acc	M-F1	W-F1
Majority-vote vs. Human1	86.0	77.4	86.5	91.2	81.8	91.0
Majority-vote vs. Human2	89.4	81.1	89.6	88.9	80.9	89.1
Majority-vote vs. Human3	85.7	78.0	86.0	87.6	79.8	88.4
Proposed BERT$_{\text{BASE}}$	**91.8**	**85.7**	**91.8**	**91.0**	**82.2**	**91.0**

Effectiveness of Transfer Learning: As shown in Tables 2 and 3, on all the datasets the fine-tuned model yields impressive results. On SciEntsBank dataset, we establish state-of-the-art results. Compared to state-of-the-art, Saha *et al.* [25], which includes sentence embeddings of InferSent [4] along with token features, we report improvements ranging from 6% up to 10% in macro-average-F1. Note that, unsupervised pre-training of BERT helps to leverage a huge amount of existing natural language material. This puts the approach at an advantage over techniques such as InferSent [4] that requires large supervised (and therefore expensive and limited) corpus for pre-training.

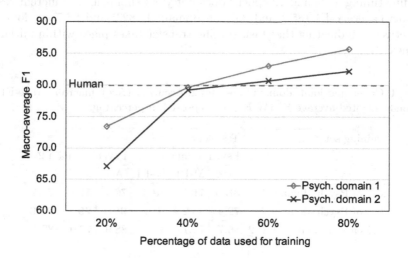

Fig. 2. Macro-average-F1 scores with different size of training sets of two domains, overlaid human performance. Evaluations are done on a held-out test set of 20%.

On our datasets, we obtain impressive macro-average-F1 of from 80% up to 85%, indicating the robustness of the model's transferability to the target task of short answer grading. On our datasets, we report human performance as a baseline against which the model can be compared. As outlined earlier, each student response is annotated by three experts. The variability in the annotation enables us to establish a human performance baseline. Table 3 lists each human annotation's comparison against the majority vote (MV) in terms of accuracy (Acc), macro-averaged-F1 (M-F1), and weighted-average-F1 (W-F1).

Effectiveness for Data-Starved Problems: Task-specific supervised fine-tuning is possible with small number of samples. On SciEntsBanks dataset, the training set includes ~5K samples; which yields results better than task-specific learning. To further study this property of the model, we design an experiment to train the model with small portions of training data. Figure 2 shows the performance in terms of macro-average-F1, when the training data is reduced from 80% of the whole set to mere 20%. Evaluation is done on a constant held-out test set consisting of 20% samples. Note the decrease in the slope as the training set expands, suggesting diminishing returns as training data is added. The increase in M-F1 is about 10% as the training set increases from 20% to 80%. For data-starved problems, a rather generous trade-off can be made to obtain a reasonably good performance with limited task-specific fine-tuning data. Interestingly, the M-F1 with 40% training data is in same range as human performance (shown in Table 3).

Effectiveness of Training Epoch on Fine-Tuning: We also performed experiments for fine-tuning BERT with varying number of epochs. We observed that fine-tuning for 4 and 12 epochs does not yield significantly different results on macro-average-F1 (85.7 and 85.4 on domain 1, 82.2 and 83.7 on domain 2 respectively), indicating that task-specific transfer takes place within initial few epochs only.

Table 4. Cross- and joint- domain fine-tuning. Accuracy (Acc), macro-average-F1 (M-F1), and weighted-average-F1 (W-F1) are reported in percentage.

Training set	Test set					
	Psych. domain 1			Psych. domain 2		
	Acc	M-F1	W-F1	Acc	M-F1	W-F1
40% of Psych. domain 1	**88.0**	**79.7**	**88.1**	76.4	51.1	75.5
40% of Psych. domain 2	72.4	48.2	70.0	**90.1**	**79.1**	**90.0**
40% each of domain 1 & 2	86.7	79.1	87.5	88.9	77.0	88.7

Effectiveness in Cross- and Joint- Domain Fine-Tuning: We further evaluated the fine-tuned model's ability to generalize to unseen domains. Table 4 reports the performance of fine-tuned models on both domains. It shows that the model fine-tuned using domain 1 yields very poor results on domain 2, and vice versa. This suggests that domain specific supervised data is indeed required for efficient fine-tuning. As a follow-up, we fine-tuned a model using a combined set of both domain data; which yields results relatively similar to domain specific tuning. It provides evidence that the model can be jointly fine-tuned for both models.

5   Conclusion

This paper conclusively demonstrates that Transformer-based pre-trained models push the state-of-the-art in short answer grading to a level that may be approaching the ceiling of what is possible. In comparison with human scorers, the model learns the "wisdom of the crowd", surpassing the performance of any individual human scorer on our datasets. The amount of fine-tuning needed is reasonable; even with just a few thousand labeled samples, we are able to get superior results. We also show that while applying a model fine-tuned on data associated with one domain cannot directly apply to grading other domains, it is possible to create a single model fine-tuned using data from multiple domains that works for each of them. Going forward, we expect to investigate whether adding an additional domain-specific text corpus to a pre-trained model improves the ability to process language for that domain. We will continue to experiment with ways to minimize the amount of fine-tuning (e.g., through characterization of what types of labeled samples yield the highest marginal improvement during fine-tuning, thus allowing for more efficient data collection for automated grading). Finally, work on model management, reuse of models, and devising efficient methods to add new labeled samples to existing fine-tuned methods will be of interest so that a model adapts over time.

Acknowledgements. We would like to thank Yoonsuck Choe (Texas A&M University) for helpful comments on an earlier version of this paper.

References

1. Albacete, P., Jordan, P., Katz, S.: Is a dialogue-based tutoring system that emulates helpful co-constructed relations during human tutoring effective? In: Conati, C., Heffernan, N., Mitrovic, A., Verdejo, M.F. (eds.) AIED 2015. LNCS (LNAI), vol. 9112, pp. 3–12. Springer, Cham (2015). https://doi.org/10.1007/978-3-319-19773-9_1

2. Alikaniotis, D., Yannakoudakis, H., Rei, M.: Automatic Text Scoring Using Neural Networks, June 2016. https://doi.org/10.18653/v1/p16-1068, https://arxiv.org/abs/1606.04289

3. Ba, J.L., Kiros, J.R., Hinton, G.E.: Layer Normalization, July 2016. http://arxiv.org/abs/1607.06450

4. Conneau, A., Kiela, D., Schwenk, H., Barrault, L., Bordes, A.: Supervised learning of universal sentence representations from natural language inference data. In: Proceedings of the 2017 Conference on Empirical Methods in Natural Language Processing, pp. 670–680 (2017)

5. Devlin, J., Chang, M.W., Lee, K., Toutanova, K.: BERT: Pre-training of Deep Bidirectional Transformers for Language Understanding, October 2018. http://arxiv.org/abs/1810.04805

6. D'mello, S., Graesser, A.: Autotutor and affective autotutor: learning by talking with cognitively and emotionally intelligent computers that talkback. ACM Trans. Interact. Intell. Syst. **2**(4), 23:1–23:39 (2013). https://doi.org/10.1145/2395123.2395128, http://doi.acm.org/10.1145/2395123.2395128

7. Dzikovska, M., et al.: Semeval-2013 task 7: the joint student response analysis and 8th recognizing textual entailment challenge. In: Second Joint Conference on Lexical and Computational Semantics (* SEM), Volume 2: Proceedings of the Seventh International Workshop on Semantic Evaluation (SemEval 2013), vol. 2, pp. 263–274 (2013)

8. Heilman, M., Madnani, N.: ETS: domain adaptation and stacking for short answer scoring. In: Second Joint Conference on Lexical and Computational Semantics (* SEM), Volume 2: Proceedings of the Seventh International Workshop on Semantic Evaluation (SemEval 2013), vol. 2, pp. 275–279 (2013)

9. Howard, J., Ruder, S.: Universal Language Model Fine-tuning for Text Classification (2018). http://arxiv.org/abs/1801.06146
10. Jimenez, S., Becerra, C., Gelbukh, A.: Softcardinality: hierarchical text overlap for student response analysis. In: Second Joint Conference on Lexical and Computational Semantics (* SEM), Volume 2: Proceedings of the Seventh International Workshop on Semantic Evaluation (SemEval 2013), vol. 2, pp. 280–284 (2013)
11. Kumar, S., Chakrabarti, S., Roy, S.: Earth mover's distance pooling over siamese lstms for automatic short answer grading. In: Proceedings of the Twenty-Sixth International Joint Conference on Artificial Intelligence, IJCAI-17, pp. 2046–2052 (2017). https://doi.org/10.24963/ijcai.2017/284
12. Marvaniya, S., Saha, S., Dhamecha, T.I., Foltz, P., Sindhgatta, R., Sengupta, B.: Creating scoring rubric from representative student answers for improved short answer grading. In: Proceedings of the 27th ACM International Conference on Information and Knowledge Management, pp. 993–1002. ACM (2018)
13. Mikolov, T., Sutskever, I., Chen, K., Corrado, G., Dean, J.: Distributed Representations of Words and Phrases and Their Compositionality, October 2013. http://arxiv.org/abs/1310.4546
14. Mohler, M., Bunescu, R., Mihalcea, R.: Learning to grade short answer questions using semantic similarity measures and dependency graph alignments. In: Proceedings of the 49th Annual Meeting of the Association for Computational Linguistics: Human Language Technologies-Volume 1, pp. 752–762. Association for Computational Linguistics (2011)
15. Mohler, M., Mihalcea, R.: Text-to-text semantic similarity for automatic short answer grading. In: Proceedings of the 12th Conference of the European Chapter of the Association for Computational Linguistics, pp. 567–575. Association for Computational Linguistics (2009)
16. Mou, L., et al.: How Transferable are Neural Networks in NLP Applications? March 2016. http://arxiv.org/abs/1603.06111
17. Mueller, J., Thyagarajan, A.: Siamese recurrent architectures for learning sentence similarity. In: AAAI, vol. 16, pp. 2786–2792 (2016)
18. Pan, S.J., Yang, Q.: A survey on transfer learning. IEEE Trans. Knowl. Data Eng. **22**(10), 1345–1359 (2010). https://doi.org/10.1109/TKDE.2009.191, http://dx.doi.org/10.1109/TKDE.2009.191
19. Peters, M.E., et al.: Deep contextualized word representations, February 2018. http://arxiv.org/abs/1802.05365
20. Radford, A., Narasimhan, K., Salimans, T., Sutskever, I.: Improving language understanding by generative pre-training (2018)
21. Raina, R., Battle, A., Lee, H., Packer, B., Ng, A.Y.: Self-taught learning: transfer learning from unlabeled data. In: Proceedings of the 24th International Conference on Machine Learning ICML 2007, pp. 759–766. ACM, New York (2007). https://doi.org/10.1145/1273496.1273592, http://doi.acm.org/10.1145/1273496.1273592
22. Ramachandran, L., Cheng, J., Foltz, P.: Identifying patterns for short answer scoring using graph-based lexico-semantic text matching. In: Proceedings of the Tenth Workshop on Innovative Use of NLP for Building Educational Applications, pp. 97–106 (2015)
23. Ramachandran, L., Foltz, P.: Generating reference texts for short answer scoring using graph-based summarization. In: Proceedings of the Tenth Workshop on Innovative Use of NLP for Building Educational Applications, pp. 207–212 (2015)

24. Rus, V., Stefanescu, D., Niraula, N., Graesser, A.C.: Deeptutor: towards macro-and micro-adaptive conversational intelligent tutoring at scale. In: Proceedings of the First ACM Conference on Learning @ Scale Conference L@S 2014, pp. 209–210. ACM, New York (2014). https://doi.org/10.1145/2556325.2567885, https://doi.acm.org/10.1145/2556325.2567885

25. Saha, S., Dhamecha, T.I., Marvaniya, S., Sindhgatta, R., Sengupta, B.: Sentence level or token level features for automatic short answer grading?: use both. In: Penstein Rosé, C., et al. (eds.) AIED 2018. LNCS (LNAI), vol. 10947, pp. 503–517. Springer, Cham (2018). https://doi.org/10.1007/978-3-319-93843-1_37

26. Sultan, M.A., Salazar, C., Sumner, T.: Fast and easy short answer grading with high accuracy. In: Proceedings of the 2016 Conference of the North American Chapter of the Association for Computational Linguistics: Human Language Technologies, pp. 1070–1075 (2016)

27. Vaswani, A., et al.: Attention is all you need. In: Guyon, I., et al. (eds.) Advances in Neural Information Processing Systems, vol. 30, pp. 5998–6008. Curran Associates, Inc. (2017). http://papers.nips.cc/paper/7181-attention-is-all-you-need.pdf

28. Ventura, M., et al.: Preliminary evaluations of a dialogue-based digital tutor. In: Penstein Rosé, C., et al. (eds.) AIED 2018. LNCS (LNAI), vol. 10948, pp. 480–483. Springer, Cham (2018). https://doi.org/10.1007/978-3-319-93846-2_90

29. Wu, Y., et al.: Google's neural machine translation system: Bridging the gap between human and machine translation. CoRR abs/1609.08144 (2016). http://arxiv.org/abs/1609.08144

Uniform Adaptive Testing Using Maximum Clique Algorithm

Maomi Ueno[1(✉)] and Yoshimitsu Miyazawa[2]

[1] The University of Electro-Communications, Tokyo, Japan
`ueno@ai.is.uec.ac.jp`
[2] The National Center for University Entrance Examinations, Tokyo, Japan

Abstract. Computerized adaptive testing (CAT) presents a tradeoff problem between increasing measurement accuracy and decreasing item exposure in an item pool. To address this difficulty, we propose a new CAT that partitions an item pool to numerous uniform item groups using a maximum clique algorithm and then selects the optimum item with the highest Fischer information from a uniform item group. Numerical experiments underscore the effectiveness of the proposed method.

Keywords: Computerized adaptive testing · e-testing ·
Item response theory · Maximum clique algorithm ·
Uniform test form assembly

1 Introduction

Computerized Adaptive Testing (CAT) selects and presents the optimal item that maximizes the test information (Fisher information measure) at the current estimated ability based on item response theory (IRT) from an item pool. After each response, the examinee's ability estimate is updated. Then the subsequent item is selected to have optimal properties at the new estimate. Adaptive item selection to each examinee can reduce the number of examined items so as not to decrease the test accuracy in comparison with the same fixed test. However, in conventional CATs, the same items tend to be presented to examinees who have similar ability. This property causes bias of the item exposure frequency in an item pool. Earlier studies [1] demonstrated that frequently exposed items deteriorate rapidly. To resolve this difficulty, Kingsbury and Zara (1989) proposed partitioning of an item pool into several groups of items and then selected the optimal item that maximizes Fisher information from the group minimizing item exposure [1]. Furthermore, van der Linden and et al. (1998,2004,2016) proposed a shadow-test approach that maximizes Fisher information under several constraints (e.g., test area and item exposure frequency) using integer programming [2–4]. Earlier methods mitigated the bias of item exposure frequency from an item pool. However, they encountered the difficulty that increases bias of measurement accuracy for examinees. In addition, this problem necessarily engenders a bias of examinees' required test lengths in CAT. Thus, a tradeoff exists between decreasing

© Springer Nature Switzerland AG 2019
S. Isotani et al. (Eds.): AIED 2019, LNAI 11625, pp. 482–493, 2019.
https://doi.org/10.1007/978-3-030-23204-7_40

item exposure and increasing measurement accuracy. Nevertheless, earlier methods do not address the tradeoff. To resolve that shortcoming, we propose a new framework that can control the balance between item exposure and measurement accuracy. More specifically, we use a state-of-the-art uniform test assembly technique to divide an item pool into several equivalent groups of items and thereby adjust the degree of item exposure. Regarding the uniform test forms, each form consists of a different set of items, but the forms have equivalent measurement accuracy(i.e., equivalent test information based on item response theory). Recent studies explored several techniques using AI technologies to generate numerous uniform test forms from an item pool [5–9]. Especially, among all methods, uniform test assembly using the maximum clique algorithm is known to generate the greatest number of uniform test forms [6–9]. This method formalized the uniform test assembly with overlapping items conditions as a maximum clique problem (MCP), where overlapping items represent common items among multiple test forms. Here it is noteworthy that the determined number of overlapping items increases the item exposure frequency. Determination of the number of overlapping items for the MCP method can therefore control the degree of item exposure.

The MCP has never been utilized for CATs. Therefore, this study proposes a new CAT method which reduces the degree of item exposure using the MCP. The proposed method partitions an item pool into numerous uniform item groups using the MCP method. Then, from a uniform item group, we select the optimum item with the highest Fischer information, which reflects the measurement accuracy.

Salient benefits of using the method are the following.

1. The proposed method solves the tradeoff between item exposure and the measurement accuracy (test length).
2. The proposed method decreases the bias of measurement accuracy (test length) for examinees, that's a uniform adaptive testing.

Experiments were conducted to compare the performances of the proposed method with conventional methods. The results show that the proposed method dynamically improves the measurement accuracy (reduces test length) without largely increasing item exposure. A particularly surprising finding is that increasing the item size of the uniform item group does not necessarily improve the measurement accuracy. Rather, an optimum item size exists for accuracy. This is the main reason why the proposed partition improves the measurement accuracy (reduces test length) without greatly increasing item exposure.

2 Computerized Adaptive Testing Based on Item Response Theory

2.1 Item Response Theory

In CAT, an examinee's ability parameter is estimated based on Item Response Theory (IRT) [10] to select the optimum item with the highest information.

In the two-parameter logistic model (2PLM), the most popular IRT model, the probability of a correct answer to item i by examinee j with ability $\theta \in (-\infty, \infty)$ is assumed as

$$p(u_i = 1|\theta) = \frac{1}{1 + exp[-1.7a_i(\theta - b_i)]}. \tag{1}$$

Therein, u_i is 1 when an examinee answers item i correctly, and 0 otherwise. Furthermore, $a_i \in [0, \infty)$ and $b_i \in (\infty, \infty)$ respectively denote the discrimination parameter of item i and the difficulty parameter of item i.

2.2   Fisher Information

The asymptotic variance of estimated ability based on the item response theory is known to approach the inverse of Fisher information [10]. Accordingly, item response theory usually employs Fisher information as an index representing the accuracy. In 2PLM, the Fisher information is defined when item i provides an examinee's ability θ using the following equations.

$$I_i(\theta) = \frac{[p'(u_i = 1|\theta)]^2}{p(u_i = 1|\theta)[1 - p(u_i = 1|\theta)]} \tag{2}$$

where

$$p'(u_i = 1|\theta) = \frac{\partial}{\partial\theta}p(u_i = 1|\theta).$$

Results imply that the examinee's ability can be discriminated using an item with high Fisher information $I_i(\theta)$. Accordingly, that ability estimation can be expected to be implemented by selecting items with the highest amount of Fisher information given an examinee's ability estimate $\hat{\theta}$.

The test information function $I_T(\theta)$ of a test form T is defined as $I_T(\theta) = \sum_{i \in T} I_i(\theta)$. The asymptotic error of ability estimate $\hat{\theta}$: $SE(\hat{\theta})$ can be obtained as the inverse of square root of the test information function at a given ability estimate $\hat{\theta}$ as $SE_T(\theta) = \frac{1}{\sqrt{I_T(\theta)}}$.

2.3   Computerized Adaptive Testing

In conventional CAT, adaptive items are selected from an item pool using the following procedures.

1. An examinee's ability is initialized to $\hat{\theta} = 0$.
2. An item maximizing Fisher information for given ability is selected from the item pool. It is presented to the examinee.
3. The examinee's ability estimate is updated from the correct/wrong response data to the item.
4. Procedures 2 and 3 are repeated until the update difference of the examinee's ability estimate decreases to a constant value ϵ or less.

Consequently, CAT can reduce the number of items examined so that it does not reduce the test accuracy in comparison to that of the same fixed test.

2.4   Constrained CAT with Item Exposure Control

In CAT, it is highly likely that the same set of items will be presented to examinees exhibiting similar abilities. Therefore, conventional CAT cannot be used practically in situations where the same examinee can take a test multiple times. Furthermore, because the ability variable follows the standard normal distribution, items with higher information around $\theta = 0$ tend to be exposed frequently. Therefore, bias of item exposure frequency occurs in an item pool. An earlier report [1] described that frequently exposed items tend to deteriorate rapidly.

To resolve this difficulty, Kingsbury and Zara (1989) proposed the partitioning of an item pool into several groups of items and then selected the optimal item maximizing Fisher information from the group minimizing item exposure [1,2,4]. The item-pool partitioning procedure is the following.

1. An item pool is partitioned into several groups of items.
2. The estimated ability of an examinee is initialized to $\hat{\theta} = 0$.
3. The group minimizing the number of exposure items is selected from an item pool.
4. The item maximizing Fischer information is selected from the group and is presented to the examinee.
5. After each response, the examinee's ability estimate is updated.
6. Procedures 2, 3, and 4 are repeated until the update difference of the estimated ability decreases to ϵ or less. ϵ is set to 0.01, which is used conventionally for actual computerized adaptive testing.

The number of groups was ascertained by comparing the respective performances of several numbers of groups.

Actually, van der Linden and et al. (1998, 2004, 2016) proposed a shadow-test approach that maximizes Fisher information under several constraints (e.g., test area and item exposure frequency) using integer programming [2–4]. The procedure used for constrained computerized adaptive testing is the following.

1. The estimated ability of an examinee is initialized to $\hat{\theta} = 0$.
2. The item set (shadow test with I items) maximizing Fischer information is then assembled using the integer programming shown below.

$$\text{maximize} \sum_{i=1}^{I} I_i(\theta) x_i \tag{3}$$

subject to

$$\sum_{i=1}^{I} x_i = n; \text{(test length)},$$

$x_i = 1$ if item i is included in the shadow test, $x_i = 0$ otherwise.

If the exposure count of item i is greater than R, then $x_i = 0$,

where R text is the upper bound of the exposure count by the user.

3. The item maximizing Fischer information is selected from the shadow test and is presented to an examinee.
4. After each response, the examinee's ability estimate is updated.
5. Procedures 2, 3, and 4 are repeated until the update difference of the estimated ability decreases to $\epsilon = 0.01$ or less.

Earlier methods mitigated the bias of item exposure frequency in an item pool. However, they led to the important difficulty of increased bias of measurement accuracies (errors) for examinees. Furthermore, this difficulty necessarily engenders a bias of examinees' required test lengths in CAT. In fact, a tradeoff exists between minimizing item exposure and maximizing the measurement accuracy (test information). Nevertheless, earlier methods do not adjust the tradeoff. For that reason, we propose a new CAT framework that can control the tradeoff.

3 Uniform Adaptive Testing Using Maximum Clique Algorithm

The proposed method partitions an item pool to numerous equivalent groups of items to adjust the degree of item exposure using the MCP method. The method then selects the optimum item with the highest Fischer information from a uniform partition of the item pool.

3.1 Uniform Partitioning of the Item Pool

To maximize the number of uniform tests with an overlap condition, Ishii et al. proposed the maximum clique problem for uniform test assembly [7]. The clique problem is a combinational optimization problem in graph theory. We apply this method to uniform partitioning of the item pool as described below.

Letting V be a finite set of vertexes, and letting E be a set of edges, the graph is represented as a pair $G = \{V, E\}$. The maximum clique problem seeks the clique which has the maximum number of vertexes in the given graph. Letting $G = \{V, E\}$ be a finite graph, and letting $C \subseteq V$ be a clique, then the maximum clique problem is formally defined as shown below:

$$
\begin{aligned}
&\textbf{maximize } |C| \\
&\textbf{subject to} \\
&\qquad \forall v, \forall w \in C, \{v, w\} \in E \\
&\qquad \text{(clique constraint)}.
\end{aligned}
\tag{4}
$$

Here, uniform partitioning of item pool has the following specifications:

1. Any item group in the uniform partition satisfies all partition constraints.
2. Any two item groups in a uniform partition comprise a different set of items (i.e., any two groups have fewer overlapping items than the number allowed in the overlapping constraint).

Accordingly, uniform partitioning of the item pool can be described as the maximum clique extraction from a graph:

$$V = \left\{ \begin{array}{l} s : s \in S, \text{Feasible item-group } s \\ \quad \text{satisfies all constraints} \\ \quad \text{excepting the overlapping} \\ \quad \text{constraint from a given} \\ \quad \text{item pool} \end{array} \right\}$$

$$E = \left\{ \begin{array}{l} \{s', s''\} : \text{The pair of } s' \text{ and } s'' \\ \quad \text{satisfies the} \\ \quad \text{overlapping constraint} \end{array} \right\}.$$

The test constraints include a constraint for the number of items, and the test information of the item group. Letting L_{θ_k} be a lower bound, and letting U_{θ_k} be an upper bound for test information related to $I_T(\theta_k)$, then a constraint for test information function is written as the following equation.

$$L_{\theta_k} \leq I_T(\theta_k) \leq U_{\theta_k} \tag{5}$$

Letting OC be the allowed number in the overlapping constraint and letting both s and s' be item groups, then the overlapping constraint is defined as the following equation:

$$\forall s, \forall s' \in V, \tag{6}$$
$$|s \cap s'| \leq \text{OC} \tag{7}$$

This maximum clique problem seeks the maximum set of feasible item groups in which any two groups satisfy the overlapping constraint. Therefore, this optimization problem theoretically maximizes the number of equivalent item groups. We apply the approximated MCP algorithm [9], which is a state-of-the-art algorithm, to obtain numerous uniform item groups.

3.2 Adaptive Item Selection from an Item Group

Using the obtained item groups, the proposed method selects and presents the optimal items to an examinee as explained below.

1. An arbitrary uniform item group is selected from a set of unused groups.
2. The optimal item maximizing Fischer information is selected from the group and presented to an examinee in Procedure 1.
3. The examinee's ability estimate is updated from his/her response.
4. Procedures 2 and 3 are repeated until the update difference of the estimated ability of the examinee reaches a constant value of $\epsilon = 0.01$ or less.

If a set of unused groups is empty in Procedure 1, then the algorithm resets it as a universal set of uniform item groups.

4   Numerical Evaluation

This section presents a comparison of the performance of the proposed method (designated as Proposal) to those of other computerized adaptive testing methods (conventional adaptive testing in 2.3 (designated as CAT), Kingsbury and Zara (1989) CAT in 2.4 (designated as KZ), and van der Linden's IP based CAT in 2.4 (designated as IP). For the proposed method and KZ, we construct item groups with 50 items (We write Proposal(50) and KZ(50)) and item groups with 100 items (We write Proposal(100) and KZ(100)) to investigate the effects of item sizes for the measurement accuracy and item exposure. Furthermore, we use two numbers of overlapping items for the proposed methods OC = 0 and OC = 10 to investigate the effects of OC for the measurement accuracy and item exposure. Additionally, we conduct experiments with $R = 50, 80, 90, 100$, and 150 as the upper bound exposure counts of IP. Also, L_{θ_k} and U_{θ_k} of the proposed method are determined as described in an earlier report [7]. We conducted the following two experiments using simulation data and actual data.

4.1   Simulation Experiment

We conducted a simulation experiment as described hereinafter.

1. Item pools with 500 and 1000 items are generated. The true parameters of each item are generated from $a_i \sim U(0,1)$ and $b_i \sim N(0,1)$.
2. The true abilities of examinees are sampled from $\theta \sim N(0,1)$.
3. Each adaptive testing method is conducted using each item pool. The examinees' response data are generated from $p(u_i \mid \theta)$. Correct response data are generated if $p(u_i \mid \widehat{\theta}) > 0.5$. Incorrect response data are generated otherwise. The convergence (Stopping) criterion is $\epsilon = 0.01$, which is used conventionally for actual computerized adaptive testing [14].
4. Procedures 2–3 are repeated 1000 times.

Table 1 presents the results. In Table 1, "overlapping items" represents the number of overlaps, "No. Item groups" denotes the number of generated (uniform for the proposal) item groups, "Avg. test length" stands for the average test length which reflects the measurement accuracy of the test (the standard error of test lengths in parenthesis), and "S.D. estimates error" expresses the standard deviation of asymptotic errors of estimates $\widehat{\theta}$. When the value of "S.D. estimates error" approaches to zero, the tests presented to the different examinees by the CAT have the same estimation accuracy.

Also, "Max. No. exposure item" shows the maximum number of exposure items. "Avg. exposure item" expresses the average exposure count of an item (the standard error of numbers of exposure items in parenthesis). For IP results, Table 1 shows only those results of R with the best accuracy (the smallest value of "Avg. test length") because of limitations of space. Consequently, $R = 150$ for 500 item pool size and $R = 80$ for 1000 item pool were selected.

Results obtained for "No. item groups" demonstrate that the proposed method generated numerous uniform item groups. "Avg. test length", which

Table 1. Results using simulated data

Item pool size	Methods	Over lapping items	No. item-groups	Avg. test length	S.D. estimates error	Max.No. exposure item	Avg. exposure item
500	AT	-	-	65.1 (7.29)	0.022	1000	130.1 (223.5)
	IP	-	-	73.2 (16.07)	0.588	150	146.4 (19.4)
	KZ(50)	0	10	57.6 (9.40)	0.031	649	115.2 (182.5)
	KZ(100)	0	5	62.2 (8.51)	0.029	751	124.4 (206.8)
	Proposal(50)	0	5	33.9 (7.82)	0.03	200	67.9 (84.7)
		10	136	34.3 (8.28)	0.032	227	68.6 (40.1)
	Proposal(100)	0	3	39.5 (7.35)	0.047	334	79.1 (123.1)
		10	99983	40.2 (6.86)	0.017	326	80.3 (78.5)
1000	AT	-	-	70.4 (7.11)	0.022	1000	70.4 (158.1)
	IP	-	-	79.5 (26.93)	0.48	80	79.5 (6.2)
	KZ(50)	0	20	62.4 (9.94)	0.034	450	62.4 (112.5)
	KZ(100)	0	10	66.1 (9.28)	0.031	593	66.1 (133.4)
	Proposal(50)	0	9	31.9 (7.13)	0.038	112	31.9 (46.3)
		10	8758	35.1 (8.19)	0.025	154	35.1 (21.8)
	Proposal(100)	0	4	39.4 (9.44)	0.063	200	39.4 (70.5)
		10	100000	42.5 (8.05)	0.015	227	42.9 (41.9)

Table 2. Result using actual data

Item pool size	Methods	Over lapping items	No. item-groups	Avg. test length	S.D. estimates error	Max.No. exposure item	Avg. exposure item
978	AT	-	-	65.5 (10.72)	0.016	1000	67 (163.2)
	IP	-	-	87.9 (18.22)	0.323	90	89.9 (1.2)
	KZ(50)	0	20	63.5 (12.30)	0.042	382	45.1 (88.3)
	KZ(100)	0	10	61.7 (13.11)	0.044	556	44.66 (108.6)
	Proposal(50)	0	7	41.2 (4.51)	0.043	143	42.1 (60.9)
		10	8669	40.7 (4.70)	0.043	136	41.6 (19.3)
	Proposal(100)	0	2	54.8 (8.84)	0.033	500	56 (139.3)
		10	7088	54.3 (7.67)	0.031	179	55.5 (42.6)

reflects the measurement accuracy of the corresponding CAT, is one of the most important indexes in this study because reducing the number of examined items so as not to increase item exposure solves the tradeoff between item exposure and the measurement accuracy. The results of "Avg. test length," which reflects the measurement accuracy of CAT, surprisingly show that the proposed method provides the best performance among all the CATs, although the item alternatives were constrained using the uniform item groups. Presenting items with extremely high information at the early stage of CAT is known to adversely cause the local solution for ability estimation and to interrupt the convergence to the true estimate because the estimate at the early stage is often far from the true one [2]. The proposed method has a uniform distribution of item characteristics over the whole ability area. It constrains the number of items with the uniform conditions so as not to select items with extremely high information for a specific ability area. This property shortens the test length. It is noteworthy that it mitigates item exposure because the proposed method presents less non-informative items to examinees. Additionally, it is notable that it yields the smallest standard deviation of test length. Therefore, the proposed method decreases the bias of measurement accuracy for examinees. No significant differences were found in the values of "S.D. estimates error" among the methods because all methods employ the same convergence criterion $\epsilon = 0.01$. Although the proposed method did not provide the lowest values of "Max. No. exposure item," the best values of IP result from the constraint of the upper bound R. The proposed method shows the lowest values of "Avg. exposure items" among all methods. Comparing the performances of the uniform item group sizes 50 and 100 for the proposed method and KZ, the performances with item size 50 are better than those with item size 100. The reason is that even the uniform item group is affected by the local solution problems when the item size is large. This result emphasizes that partitioning an item pool is effective for CAT. This result also suggests that there might be an optimal size of the uniform item group. In addition, comparing the performances of overlapping item sizes OC $= 0$ and OC $= 10$ for the proposed method, the number of generated uniform item groups with OC $= 50$ is much larger than that with OC $= 0$. However, contrary to expectations, performances with OC $= 0$ outperform those with OC $= 10$ for all criteria except for "No. item groups." The uniform item groups generated with OC $= 0$ have higher quality for uniform measurement accuracy and item exposure than those with OC $= 10$, although CAT must repeatedly use the same uniform item groups with OC $= 0$. Therefore, a tradeoff must exist between the quality of uniform item groups and the number of generated uniform item groups. Moreover, as the item pool size increases, the performances of CATs do not necessarily increase. In fact, the proposed method increases the performances as the item pool size increases only when the item group size is 50 because the item-groups can gather high quality items when the item size is large. From these results, it is recommended to develop a large size item pool and then assemble uniform item-groups with an optimum item size.

Conventional CAT shows the lowest values of "Avg. test length" and "S.D. estimates error" among all CATs because it repeatedly selects and presents the same items to different examinees. In addition, the values of "Max. No exposure items" and "Max. No. exposure item" of Conventional CAT have the worst results among all methods. Particularly "Max. No exposure items" demonstrates that conventional CAT presented the same item to all examinees.

Although IP decreased the values of "Max. No. exposure item" because of the upper bound R, it did not provide good performance for other criteria. However, a naive method, KZ, provided better performance than IP did, except for "Max. No. exposure item." The reason is that partitioning the item pool to several groups is highly effective for the same reason as that for the proposed method, although it uses naive random sampling.

The results demonstrate that partitioning an item pool to several groups with an appropriate number of items is highly effective for CAT.

4.2 Experiment Conducted Using Actual Data

This section presents evaluation of the effectiveness of the proposed method using actual data. An experiment was conducted using the item pool of real data, with 978 items, and a test constraint used in the synthetic personality inventory (SPI) examination (actual CAT style), which is a popular aptitude test in Japan [15].

Table 2 presents the results. For IP results, Table 2 presents only the results of $R = 90$ with the best accuracy (the smallest value of "Avg. test length") because of space limitations. The table shows almost identical results to those of the simulation experiment. Namely, the proposed method provides the best performances among all the methods. The generated uniform item groups with $OC = 10$ are much more numerous than those with $OC = 0$. In this case, performances with $OC = 10$ slightly outperform those with $OC = 0$ for all criteria. As described in Sect. 4.1, a tradeoff must exist between the quality of uniform item groups and the number of generated uniform item groups. In this experiment, the proposed method provides the best performance for item size of 50 and $OC = 10$.

5 Conclusions

This paper has demonstrated that CAT has a tradeoff problem between increasing measurement accuracy and decreasing item exposure in an item pool. To address this difficulty, we proposed a new CAT that partitions an item pool to numerous uniform item groups using a uniform test assembly based on the maximum clique algorithm. Then we select the optimum item with the highest Fischer information from a uniform item group.

Experiments were conducted to compare the performance of the proposed method with those of conventional methods. Results show that the proposed method dynamically improves the measurement accuracy (reduces test length) without greatly increasing item exposure. Contrary to our expectations, the results did not show that the proposed method using numerous uniform item groups necessarily outperforms that using a few groups. Results suggest that a tradeoff for the proposed method must exist between the quality of uniform item groups and the number of generated uniform item groups. We expect to investigate the means of solving this tradeoff problem as a subject of future work.

For this study, we used Fischer information measure as an item selection criterion. Although the Fischer information measure becomes accurate for late stage of CAT because it is an asymptotic approximation, recent studies have proposed more accurate information measures and the item selection algorithms [16–18]. We expect to apply the proposed uniform partition of item pool technique to the information measure and its applications [19,20] in future studies.

References

1. Kingsbury, G.G., Zara, A.R.: Procedures for selecting items for computerized adaptive tests. Appl. Measur. Educ. **2**(4), 359–375 (1989)
2. van der Linden, W.J., Reese, L.: A model for optimal constrained adaptive testing. Appl. Psychol. Measur. **22**(3), 259–270 (1998)
3. van der Linden, W.J., Veldkamp, B.: Constraining item exposure in computerized adaptive testing with shadow tests. J. Educ. Behav. Stat. **29**(3), 273–291 (2004)
4. Choi, S.W., Moellering, K.T., Li, J., van der Linden, W.J.: Optimal reassembly of shadow tests in CAT. Appl. Psychol. Measur. **40**(7), 469–485 (2016)
5. Songmuang, P., Ueno, M.: Bees algorithm for construction of multiple test forms in e-testing. IEEE Trans. Learn. Technol. **4**(3), 209–221 (2011)
6. Ishii, T., Songmuang, P., Ueno, M.: Maximum clique algorithm for uniform test forms assembly. In: Lane, H.C., Yacef, K., Mostow, J., Pavlik, P. (eds.) AIED 2013. LNCS (LNAI), vol. 7926, pp. 451–462. Springer, Heidelberg (2013). https://doi.org/10.1007/978-3-642-39112-5_46
7. Ishii, T., Songmuang, P., Ueno, M.: Maximum clique algorithm and its approximation for uniform test form assembly. IEEE Trans. Learn. Technol. **7**(1), 83–95 (2014)
8. Ishii, T., Ueno, M.: Clique algorithm to minimize item exposure for uniform test forms assembly. In: Conati, C., Heffernan, N., Mitrovic, A., Verdejo, M.F. (eds.) AIED 2015. LNCS (LNAI), vol. 9112, pp. 638–641. Springer, Cham (2015). https://doi.org/10.1007/978-3-319-19773-9_80
9. Ishii, T., Ueno, M.: Algorithm for uniform test assembly using a maximum clique problem and integer programming. In: André, E., Baker, R., Hu, X., Rodrigo, M.M.T., du Boulay, B. (eds.) AIED 2017. LNCS (LNAI), vol. 10331, pp. 102–112. Springer, Cham (2017). https://doi.org/10.1007/978-3-319-61425-0_9
10. Lord, F., Novick, M.R.: Statistical Theories of Mental Test Scores. Addison-Wesley, M.R. (1968)
11. Sun, K.T., Chen, Y.J., Tsai, S.Y., Cheng, C.F.: Creating IRT-based parallel test forms using the genetic algorithm method. Appl. Measur. Educ. **21**(2), 141–161 (2008)
12. Belov, D.I., Armstrong, R.D.: A constraint programming approach to extract the maximum number of non-overlapping test forms. Comput. Optim. Appl. **33**(2), 319–332 (2006)
13. van der Linden, W.J.: Linear Models for Optimal Test Design. Springer, New York (2005). https://doi.org/10.1007/0-387-29054-0
14. van der Linden, W.J., Glas, C.A.W.: Elements of Adaptive Testing. Springer, New York (2010). https://doi.org/10.1007/978-0-387-85461-8
15. Recruit, Synthetic Personality Inventory (SPI) (2014). http://www.spi.recruit.co.jp/
16. Ueno, M.: An extension of the IRT to a network model. Behaviormetrika **29**(1), 59–79 (2002)
17. Ueno, M. and Songmuang, P.: Computerized adaptive testing based on decision tree. In: The Tenth IEEE International Conference on Advanced Learning Technologies (ICALT), pp. 191–193 (2010)
18. Ueno, M.: Adaptive testing based on bayesian decision theory. In: Lane, H.C., Yacef, K., Mostow, J., Pavlik, P. (eds.) AIED 2013. LNCS (LNAI), vol. 7926, pp. 712–716. Springer, Heidelberg (2013). https://doi.org/10.1007/978-3-642-39112-5_95

19. Ueno, M., Miyasawa, Y.: Probability based scaffolding system with fading. In: Conati, C., Heffernan, N., Mitrovic, A., Verdejo, M.F. (eds.) AIED 2015. LNCS (LNAI), vol. 9112, pp. 492–503. Springer, Cham (2015). https://doi.org/10.1007/978-3-319-19773-9_49

20. Ueno, M., Miyazawa, Y.: IRT-based adaptive hints to scaffold learning in programming. IEEE Trans. Learn. Technol. **11**(4), 415–428 (2018)

Rater-Effect IRT Model Integrating Supervised LDA for Accurate Measurement of Essay Writing Ability

Masaki Uto[✉]

University of Electro-Communications, Tokyo, Japan
uto@ai.lab.uec.ac.jp

Abstract. Essay-writing tests are widely used in various assessment contexts to measure higher-order abilities of learners. However, a persistent difficulty is that ability measurement accuracy strongly depends on rater characteristics. To resolve this problem, many item response theory (IRT) models have been proposed that can estimate learners' abilities in consideration of rater effects. One remaining difficulty, however, is that measurement accuracy is reduced when few raters are assigned to each essay, a common situation in practical testing contexts. To address this problem, we propose a new rater-effect IRT model integrating a supervised topic model that can estimate abilities from raters' scores and the textual content of written essays. By reflecting textual content features in IRT-based ability estimates, the model can improve ability measurement accuracy when there are few raters for each essay. Furthermore, learners' abilities can be estimated using essay textual content alone, without ratings, when model parameters are known. Finally, scores for unrated essays can be estimated from textual content, so the model can be used for automated essay scoring. We evaluate the effectiveness of the proposed model through experiments using actual data.

Keywords: Item response theory · Latent Dirichlet allocation · Supervised topic model · Essay writing · Automated essay scoring

1 Introduction

The need to measure practical and higher-order abilities such as logical thinking, critical reasoning, and creative-thinking skills has recently increased in various assessment contexts, and essay-writing tests have attracted much attention as a way to measure such abilities [1,7,26,36,39]. Essay-writing tests evaluate learners' abilities from ratings given by multiple raters to essays of learners for some essay tasks. However, one difficulty is that ability measurement accuracy strongly depends on rater and task characteristics such as rater severity and task difficulty [7,13,14,23,31,46]. To address this difficulty, many item response theory (IRT) models that incorporate rater and task characteristic parameters have been proposed (e.g., [14,47]). These models can estimate learner abilities while

© Springer Nature Switzerland AG 2019
S. Isotani et al. (Eds.): AIED 2019, LNAI 11625, pp. 494–506, 2019.
https://doi.org/10.1007/978-3-030-23204-7_41

considering the effects of those characteristics. They can thereby provide more accurate ability measurement than either mean or total scores can provide [46].

One remaining difficulty, however, is that ability measurement accuracy falls when few raters are assigned to each essay, reducing the rating data for each learner (e.g., [43,45]). This situation is commonly encountered in practical essay-writing test situations because of the need for lower rater burdens and scoring costs [14,15].

To address this difficulty, we propose a new IRT model that can estimate learners' abilities using both rating data and the textual content of written essays. The proposed model is an IRT model with rater and task parameters that integrates latent Dirichlet allocation (LDA) [9], a representative topic model. Specifically, the model represents the relation between the ability estimate in IRT and the topic distribution of each essay, which is estimated from LDA, as a normal regression model. This formulation is inspired by supervised topic modeling [8], a state-of-the-art method for predicting response variables related to each text from topic information. The proposed model provides the following benefits:

1. By reflecting textual content features in IRT-based ability estimates, the model can improve ability measurement accuracy when there are few raters for each essay.
2. Learners' abilities can be estimated using essay textual content alone, without ratings, when model parameters are known.
3. Scores for unrated essays can be estimated from textual content, so the model can be used for automated essay scoring (AES).

This study demonstrates the effectiveness of the proposed model through experiments using actual data.

AES methods using machine learning, such as topic models or deep neural networks, have recently attracted attention in AI fields as another approach to reduce the burden of essay grading [2,3,10,16,20,41]. The accuracy of such methods is limited, however. Therefore, grading by raters is still required in medium-stakes and high-stakes assessments. As we describe below, our proposed model reduces grading burden while maintaining the accuracy of ability measurements. The proposed model is also unique as an AES method in that it can estimate essay scores with explicit consideration of the characteristics of raters and tasks. Furthermore, the proposed approach is expected to be useful to develop adaptive intelligent learning environments based on IRT because such environments are often adversely affected by data sparsity.

2   Data

This study assumes that rating data U obtained from an essay-writing test consist of a score $k \in \mathcal{K} = \{1, \cdots, K\}$ assigned by rater $r \in \mathcal{R} = \{1, \cdots, R\}$ to an essay of learner $j \in \mathcal{J} = \{1, \cdots, J\}$ for essay task $i \in \mathcal{I} = \{1, \cdots, I\}$.

Consequently, letting e_{ij} be the essay of learner j for task i, and letting U_{ijr} be the score assigned by rater r to e_{ij}, data U are defined as

$$U = \{U_{ijr} \in \mathcal{K} \cup \{-1\} \mid i \in \mathcal{I}, j \in \mathcal{J}, r \in \mathcal{R}\}, \tag{1}$$

with $U_{ijr} = -1$ denoting missing data. Although essay grading is not necessarily categorical, we use categorical grading because it is generally used in medium-stakes and high-stakes assessments.

Moreover, letting $\mathcal{V} = \{1, \cdots, V\}$ be a vocabulary list for essay collection $E = \{e_{ij} \mid i \in \mathcal{I}, j \in \mathcal{J}\}$, each essay $e_{ij} \in E$ is definable as a list of vocabulary words as

$$W_{ij} = \{W_{ijn} \in \mathcal{V} \mid n = \{1, \cdots, N_{ij}\}\}, \tag{2}$$

where W_{ijn} is the n-th word in essay e_{ij}, and N_{ij} is the number of words in essay e_{ij}.

This study was designed to measure learner ability accurately from rating data U and essays E. We use IRT and a topic model for that purpose.

3   Item Response Theory

IRT [27] is a test theory based on mathematical models. IRT represents the probability of a learner response to a test item as a function of latent learner ability and item characteristics such as difficulty and discrimination. IRT is widely used for educational testing because it offers many benefits such as: (1) It can measure the abilities of learners responding to different test items on the same scale. (2) It can also estimate learners' abilities while minimizing the effects of heterogeneous or aberrant items with low estimation accuracy.

Many IRT models are applicable to ordered categorical data [4,29,30,38] such as the essay-rating data used in this study. Those IRT models are applicable to two-way (learner × test item) data, but they are not directly applicable to rating data, which are usually given as three-way (learner × rater × task) data, as defined in Sect. 2. IRT models that incorporate rater parameters have been proposed to resolve this difficulty [32,33,42,46,47]. Note that those models treat item parameters in traditional IRT models as task parameters.

For those models, this study uses a model proposed in Ref. [48], which is expected to provide the most robust ability measurement when a large variety of rater characteristics is assumed to exist. The model defines the probability that rater r assigns score k to essay e_{ij} as

$$P_{ijrk} = \frac{\exp \sum_{m=1}^{k} [\alpha_r \alpha_i (\theta_j - \beta_i - \beta_r - d_{rm})]}{\sum_{l=1}^{K} \exp \sum_{m=1}^{l} [\alpha_r \alpha_i (\theta_j - \beta_i - \beta_r - d_{rm})]}, \tag{3}$$

where θ_j is the latent ability of learner j, α_i is a discrimination parameter for essay task i, β_i is the difficulty of task i, α_r is the consistency of rater r, β_r is the severity of rater r, and d_{rk} is the severity of rater r within category k. For model identification, we assume $\sum_{i=1}^{I} \log \alpha_i = 0$, $\sum_{i=1}^{I} \beta_i = 0$, $d_{r1} = 0$, and $\sum_{k=2}^{K} d_{rk} = 0$. These parameters can be estimated from rating data U.

As described in the Introduction, these IRT models provide higher ability measurement accuracy than do mean or total scores, because they can estimate learners' ability while considering various rater effects [46–48]. However, ability measurement accuracy is reduced when few raters are assigned to each essay, reducing rating data for each learner. To address this problem, we propose a method for measuring learners' ability θ_j that uses both rating data and textual information from written essays. This method uses a topic model to process the essay text.

4   Topic Model

Topic models estimate latent topics in a document from word occurrence frequencies, based on the assumption that certain words will appear depending on potential topics in the text. Example topic models include latent semantic analysis (LSA) [11], probabilistic latent semantic indexing (PLSI) [21], and latent Dirichlet allocation (LDA) [9]. We chose LDA for this study for its higher accuracy in estimating topics.

LDA introduces a latent variable that represents topic allocation for each word in each document. Specifically, $Z_{ijn} \in \mathcal{T} = \{1, \cdots, T\}$ (where T is the number of topics) denotes the topic allocation for word W_{ijn} in essay e_{ij}. Letting $\psi_{ij} = \{\psi_{ij1}, \cdots, \psi_{ijT}\}$ be parameters of a multinomial distribution over the T topics for e_{ij} (where $\psi_{ijt} \in \psi_{ij}$ is the occurrence probability of the t-th topic in essay e_{ij}), and letting $\phi_t = \{\phi_{t1}, \cdots, \phi_{tV}\}$ be parameters of a multinomial distribution over V vocabulary words for each topic t (where $\phi_{tv} \in \phi_t$ represents the probability of the v-th vocabulary word in the t-th topic), then LDA models a generative process for each word W_{ijn} and corresponding topic Z_{ijn} as

$$Z_{ijn} \sim Multi(\psi_{ij}), \quad W_{ijn} \sim Multi(\phi_{z_{ijn}}), \tag{4}$$

where $Multi(\cdot)$ represents a multinomial distribution with the given parameters. The topic distribution $Multi(\psi_{ij})$ shows topics that tend to be generated in each essay, and word distribution $Multi(\phi_t)$ shows how vocabulary words are used within each topic.

LDA assumes a Dirichlet prior distribution for each ψ_{ij} and ϕ_t. Letting γ and η be parameters of the Dirichlet prior distribution for ψ_{ij} and ϕ_t, their generative processes in LDA are

$$\psi_{ij} \sim Dir(\gamma), \quad \phi_t \sim Dir(\eta), \tag{5}$$

where $Dir(\cdot)$ denotes the Dirichlet distribution with a given hyperparameter.

LDA parameters can be estimated using a collapsed Gibbs sampler [19] from a bag-of-words representation of the text data $\boldsymbol{E}$.

Topic distribution parameters ψ_{ij} are regarded as representing an underlying semantic theme in each essay e_{ij} using a T-dimensional vector [8, 22, 50]. Since the topic information provides useful statistics for text collection, topic models have become a standard tool in machine learning, with many applications that transcend their original purpose of modeling textual data [12, 22, 25, 28, 44].

Texts are frequently associated with other variables, such as labels, tags, or ratings. Supervised topic models have attracted attention as a method for predicting such extra information using the topic information for texts, and the model proposed in this study was inspired by the supervised topic modeling approach. The next section introduces supervised LDA (sLDA) [8], a representative supervised topic model.

5   Supervised Topic Model

Assume a given response variable $y_{ij} \in \mathbb{R}$ associated with each essay e_{ij}. In this case, sLDA models the relation between y_{ij} and the topic information corresponding with essay e_{ij} as a regression model. When a normal linear model is used as the regression model, sLDA defines the generative process of y_{ij} as

$$y_{ij} \sim N(\boldsymbol{\omega}^{\mathrm{T}} \bar{\boldsymbol{Z}}_{ij}, \sigma_0^2), \tag{6}$$

where $N(\mu, \sigma^2)$ represents a normal distribution with mean μ and standard deviation σ, $\boldsymbol{\omega} = \{\omega_1, \cdots, \omega_T\}$ denotes topic weighting parameters for the response variable, and σ_0^2 is a hyperparameter representing variance of the response variable. Letting $\bar{\boldsymbol{Z}}_{ij} = \{\bar{Z}_{ij1}, \cdots, \bar{Z}_{ijT}\}$, $\bar{Z}_{ijt} \in \bar{\boldsymbol{Z}}_{ij}$ is defined as

$$\bar{Z}_{ijt} = \frac{\sum_{n=1}^{N_{ij}} \delta(Z_{ijn}, t)}{N_{ij}}, \tag{7}$$

where $\delta(a, b)$ is a function returning 1 if $a = b$ and 0 otherwise.

Supervised topic models including sLDA allow prediction of a response variable based on semantic themes in each text. They have thus achieved higher prediction accuracy than simple regression models that make direct predictions from word-frequency data [8,22,24,50], and thus have been applied with high performance to such prediction tasks (e.g., [18,24,25,35,49]). This study employs this approach to reflect essay topic information on ability measurements by the IRT model.

6   Proposed Model

The proposed model reflects essay topic information on ability measurement θ_j in the IRT model defined as Eq. (3), giving the ability distribution as

$$\theta_j \sim N(\boldsymbol{\omega}^{\mathrm{T}} \bar{\boldsymbol{Z}}_j, \sigma_0^2), \tag{8}$$

where $\boldsymbol{\omega} = \{\omega_1, \cdots, \omega_T\}$ denotes topic weighting parameters for the ability estimates and $\bar{\boldsymbol{Z}}_j = \{\bar{Z}_{j1}, \cdots, \bar{Z}_{jT}\}$. Here, $\bar{Z}_{jt} \in \bar{\boldsymbol{Z}}_j$ is defined as

$$\bar{Z}_{jt} = \frac{\sum_{i=1}^{I} \sum_{n=1}^{N_{ij}} \delta(Z_{ijn}, t)}{\sum_{i=1}^{I} N_{ij}}. \tag{9}$$

We assume that each learner writes multiple essays for multiple tasks and that the response variable is one ability value for each learner. Therefore, unlike sLDA as presented above, the proposed model defines $\bar{Z}_{jt}$ by summing the topic information of multiple essays. We furthermore use $\sigma_0^2 = 1.0$ in Eq. (8), because IRT generally uses a standard normal distribution as the ability distribution.

The proposed model is expected to provide higher accuracy in ability measurements than is possible from traditional IRT models with rater parameters, because it can estimate ability using both rating data and the textual characteristics of essays. Another advantage of the proposed model is that it can estimate abilities without rating data if the word distribution and parameters for raters, tasks, and topic weights are known. Furthermore, given these ability estimates, the model can automatically score unrated essays. Subsection 6.2 describes the procedures for ability estimation without rating data and for automated scoring of unrated essays.

6.1   Parameter Estimation Using MCMC

Representative parameter estimation methods in IRT include the marginal maximum likelihood estimation method using the EM algorithm and maximum posteriori estimation using the Newton–Raphson method [6]. However, for complicated IRT models such as those used in this study, the expected a posteriori (EAP) method using the Markov-chain Monte Carlo (MCMC) algorithm generally provides more accurate parameter estimates [17,46]. In LDA, EAP methods using a variational Bayesian (VB) inference [9] or using an MCMC algorithm [19] have been used. MCMC generally provides more robust estimation and is easier to implement than the VB method, but it entails higher computational costs [5].

MCMC for IRT models is generally conducted as a Metropolis–Hastings-within-Gibbs sampling method algorithm [33,46], while LDA generally uses a collapsed Gibbs sampler [19]. A collapsed Gibbs sampler improves the efficiency of MCMC sampling by marginalizing out certain model variables, and it is applicable to the proposed model. We thus estimate parameters in the proposed model by a Metropolis–Hastings-within-collapsed-Gibbs sampler algorithm. The algorithm marginalizes topic distribution parameters $\psi = \{\psi_{ij} | i \in \mathcal{I}, j \in \mathcal{J}\}$ and word distribution parameters $\phi = \{\phi_t | t \in \mathcal{T}\}$. Then, the topic allocation $Z = \{Z_{ijn} | i \in \mathcal{I}, j \in \mathcal{J}, n \in \{1, \cdots, N_{ij}\}\}$, the IRT parameters $\xi = \{\alpha_i, \beta_i, \alpha_r, \beta_r, d, \theta\}$, and weighting parameter ω are sampled from each marginalized full conditional posterior distribution. Here, $\alpha_i = \{\alpha_i | i \in \mathcal{I}\}$, $\beta_i = \{\beta_i | i \in \mathcal{I}\}$, $\alpha_r = \{\alpha_r | r \in \mathcal{R}\}$, $\beta_r = \{\beta_r | r \in \mathcal{R}\}$, $d = \{d_{rk} | r \in \mathcal{R}, k \in \mathcal{K}\}$, and $\theta = \{\theta_j | j \in \mathcal{J}\}$. Metropolis–Hastings method is used to draw samples from these distributions because they cannot be analytically determined. Due to space limitations, the details of the algorithm have been omitted.

6.2   Ability Estimation from Text Data and Automated Essay Scoring

As described above, the proposed model can estimate ability without rating data if the word distribution and rater, task, and topic weighting parameters are known. Specifically, they can be estimated by running the MCMC algorithm described above after replacing the sampling distribution of Z_{ijn} with that derived given the word distribution and the rater, task, and topic weighting parameters. Note that values of the given parameters are not sampled in MCMC.

Furthermore, the model can score unrated essays. Specifically, given abilities estimated from text data using the above procedure and given previously estimated rater and task parameters, the expected score $\hat{U}_{ij}$ for essay e_{ij} is calculable as

$$\hat{U}_{ij} = \sum_{r=1}^{R} \frac{1}{R} \sum_{k=1}^{K} k \cdot P_{ijrk}. \tag{10}$$

It is noteworthy that this method estimates the scores considering the characteristics of raters and tasks, unlike traditional AES methods do.

7   Experiments Using Actual Data

We used actual data to evaluate the effectiveness of the proposed model. For the evaluations, we gathered actual data through the following procedures:

1. 34 university students were recruited as subjects.
2. Subjects were asked to complete four essay-writing tasks created by the National Assessment of Educational Progress (NAEP) [34, 37].
3. After the participants completed all tasks, 10 raters evaluated all the essays using a rubric with five rating categories that was created based on NAEP writing assessment criteria for grade 12 [37].

We used the collected rating data and essay texts in the following experiments.

7.1   Evaluation of Ability Measurement Accuracy

We evaluated ability measurement accuracy by the proposed model for a situation in which there are few raters for each essay. We conducted the following experiment with varying numbers of topics T in the range $[1, 15]$.

1. We estimated parameters in the proposed model by the MCMC algorithm using all data. When $T = 1$, we set $\omega_1 = 0$ and omitted the sampling procedure for ω_1, because the topic distribution is meaningless. Prior distributions and hyperparameters were chosen to be consistent with values reported in related studies [40, 44, 48] as $\log \alpha_i \sim N(0.1, 0.4)$, $\log \alpha_r \sim N(0.0, 0.5)$, $\beta_i, \beta_r, d_{rk}, \omega_t \sim N(0.0, 1.0)$ $\eta = 1/T$, and $\gamma = 1/VT$. For the vocabulary words, we used nouns, verbs, adverbs, adjectives, and conjunctions extracted from the essays after removing stop words. Stop words are those appearing in only one or two essays and those appearing in more than half of the essays.

2. After randomly assigning $n \in \{1, 2\}$ raters to each essay, we set the ratings for essays without raters to missing values.
3. From the data with the missing values, we re-estimated learners' abilities by the MCMC method described in Subsect. 6.2. Here, we used the word distribution and the rater, task, and topic weighting parameters obtained in Step 1.
4 We calculated the root mean square error (RMSE) between the ability estimates obtained in Step 3 and those obtained in Step 1.
5 After repeating Steps 2–4 ten times, we calculated the mean RMSE value.

Figure 1 shows the results. The horizontal axis shows the number of topics T and the vertical axis shows the mean RMSE. The *One Rater* and *Two Raters* lines show the results for each number of assigned raters. Note that the proposed model with $T = 1$ is equivalent to the IRT model defined in Eq. (3).

Comparing the proposed model with the conventional IRT model ($T = 1$), we can confirm that the proposed model dynamically reduces the RMSE because it can use the semantic characteristics of the essays to estimate learners' abilities. The proposed model monotonically reduces the RMSE until the number of topics reaches $T = 4$ and provides similar performance for $T > 4$. When $T \geq 4$, the proposed model with $n = 1$ achieves similar RMSE as that in the conventional IRT model with $n = 2$. This suggests that the use of textual information for ability measurement can improve accuracy to the same extent as adding an additional rater in the conventional IRT model.

The above experiment shows that the proposed model effectively improves ability measurement accuracy when only few raters are assigned to each essay.

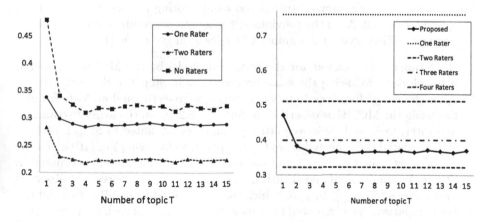

Fig. 1. Ability estimation errors.   **Fig. 2.** Score prediction errors.

7.2 Ability Measurement Accuracy Without Rating Data

We evaluated ability measurement accuracy by the proposed model for a situation in which estimations use text data but not rating data. We conducted the following experiment while varying the number of topics T in the range $[1, 15]$, as in the previous experiment.

1. We estimated parameters for the proposed model by the MCMC algorithm using all data, following the same procedure as in Step 1 in Subsect. 7.1.
2. After changing all ratings to missing values, we re-estimated all learners' abilities using the MCMC. This estimation was conducted following the procedures described in Subsect. 6.2, given the word distribution and rater, task, and topic weighting parameters estimated in Step 1.
3. We calculated RMSE between the ability estimates obtained in Step 1 and those obtained in Step 2.

The *No Raters* line in Fig. 1 shows the results, namely that although RMSE in the conventional IRT model ($T = 1$) is extremely large because it uses neither rating data nor textual information, the proposed model dynamically improves the RMSE. As in the previous experiment, the proposed model monotonically reduces RMSE until the number of topics reaches $T = 4$ and provides similar performance for $T > 4$. Furthermore, when $T \geq 4$, the proposed model with no raters achieves the lower RMSE than the conventional IRT model with $n = 1$ does. This suggests that by using only textual information, the proposed model can estimate learners' ability with the same or higher accuracy than does the conventional IRT model with one rater, demonstrating the effectiveness of the proposed model.

7.3   Accuracy of Automated Essay Scoring

We evaluated the accuracy of automated essay scoring for unrated data using the proposed model. As in the previous experiments, we conducted the following experiment while varying the number of topics T in the range $[1, 15]$.

1. We estimated parameters for the proposed model by the MCMC algorithm using all data, following the same procedure as in Step 1 in Subsect. 7.1.
2. After changing all ratings to missing values, we re-estimated all learner abilities using the MCMC as described in Subsect. 6.2, given the word distribution and rater, task, and topic weighting parameters estimated in Step 1.
3. We calculated the expected score $\hat{U}_{ij}$ for all essays following Eq. (10) given the abilities estimated in Step 2 and the rater and task parameters estimated in Step 1. We then calculated RMSE between the expected scores and observed mean scores $U_{ij} = \sum_r U_{ijr}/R$, which were computed using all rating data.
4. For comparison, we calculated the observed mean score of each essay using the ratings from $n \in \{1, \cdots, 5\}$ randomly selected raters, and computed RMSE between these scores and those computed using the full rating data. After repeating this procedure ten times, we calculated the mean RMSE value.

Figure 2 shows the results. The horizontal axis shows the number of topics T and the vertical axis shows the mean RMSE. The solid line (*Proposed*) indicates RMSEs between the scores predicted by the proposed model and the observed mean scores calculated using the full rating data. The other lines (n $Rater(s)$) show RMSEs between the observed mean scores as calculated using n the ratings of randomly selected raters and those calculated using the full rating data.

The results confirm a similar tendency with those of the previous experiments. Specifically, prediction error in the conventional IRT model ($T = 1$) is extremely large, while the proposed model dynamically improves RMSE, monotonically reducing it until the number of topics reaches $T = 4$ and providing similar performance for $T > 4$.

Comparing prediction error in the proposed model with that from the mean scores of n raters, the proposed model outperforms the mean score of 3 raters. This result suggests that the proposed model can appropriately score unrated essays.

8 Conclusion

We proposed a new IRT model that can use rater scores and essay texts to estimate learner abilities. The proposed model was formulated as an IRT model with rater and task parameters integrating supervised LDA. Through experiments using actual data, we demonstrated that the proposed model can provide higher accuracy of ability measurement than can the conventional IRT model when there are few raters for each essay because it can estimate abilities using both rating data and the semantic characteristics of the essays. We also showed that the proposed model can estimate learners' ability appropriately from essay texts without rating data, and that it can perform valid automated scoring for unrated essays.

Future studies must be conducted to analyze the topics estimated in actual data experiments and to examine methods to detect the optimal number of topics for the proposed model. Furthermore, we expect to evaluate the effectiveness of the proposed model using larger datasets and in comparison with the accuracy of automated essay scoring in earlier AES methods. We also expect to consider extensions of the proposed model using other topic models or deep neural network models in place of sLDA because improved representations of textual characteristics are anticipated to improve the performance of the proposed model.

Acknowledgment. This work was supported by JSPS KAKENHI Grant Numbers 17H04726 and 17K20024.

References

1. Abosalem, Y.: Beyond translation: adapting a performance-task-based assessment of critical thinking ability for use in Rwanda. Int. J. Second. Educ. **4**(1), 1–11 (2016)

2. Alikaniotis, D., Yannakoudakis, H., Rei, M.: Automatic text scoring using neural networks. In: Proceedings of the 54th Annual Meeting of the Association for Computational Linguistics, pp. 715–725. Association for Computational Linguistics (2016)
3. Amorim, E., Cançado, M., Veloso, A.: Automated essay scoring in the presence of biased ratings. In: Annual Conference of the North American Chapter of the Association for Computational Linguistics, pp. 229–237 (2018)
4. Andrich, D.: A rating formulation for ordered response categories. Psychometrika 43(4), 561–573 (1978)
5. Asuncion, A., Welling, M., Smyth, P., Teh, Y.W.: On smoothing and inference for topic models. In: Proceedings of International Conference on Uncertainty in Artificial Intelligence, pp. 27–34 (2009)
6. Baker, F., Kim, S.H.: Item Response Theory: Parameter Estimation Techniques. Statistics, textbooks and monographs. Marcel Dekker, New York (2004)
7. Bernardin, H.J., Thomason, S., Buckley, M.R., Kane, J.S.: Rater rating-level bias and accuracy in performance appraisals: the impact of rater personality, performance management competence, and rater accountability. Hum. Resour. Manag. 55(2), 321–340 (2016)
8. Blei, D.M., McAuliffe, J.D.: Supervised topic models. In: Proceedings of the 20th International Conference on Neural Information Processing Systems, pp. 121–128 (2007)
9. Blei, D., Ng, A., Jordan, M.: Latent Dirichlet allocation. J. Mach. Learn. Res. 3, 993–1022 (2003)
10. Dascalu, M., Westera, W., Ruseti, S., Trausan-Matu, S., Kurvers, H.: ReaderBench learns Dutch: building a comprehensive automated essay scoring system for Dutch language. In: André, E., Baker, R., Hu, X., Rodrigo, M.M.T., du Boulay, B. (eds.) AIED 2017. LNCS (LNAI), vol. 10331, pp. 52–63. Springer, Cham (2017). https://doi.org/10.1007/978-3-319-61425-0_5
11. Deerwester, S., Dumais, S.T., Furnas, G.W., Landauer, T.K., Harshman, R.: Indexing by latent semantic analysis. J. Am. Soc. Inf. Sci. 41(6), 391–407 (1990)
12. Duan, D., Li, Y., Li, R., Zhang, R., Wen, A.: Ranktopic: ranking based topic modeling. In: IEEE 12th International Conference on Data Mining, pp. 211–220 (2012)
13. Eckes, T.: Examining rater effects in TestDaF writing and speaking performance assessments: a many-Facet Rasch analysis. Lang. Assess. Q. 2(3), 197–221 (2005)
14. Eckes, T.: Introduction to Many-Facet Rasch Measurement: Analyzing and Evaluating Rater-Mediated Assessments. Peter Lang Pub. Inc., Frankfurt (2015)
15. Engelhard, G.: Constructing rater and task banks for performance assessments. J. Outcome Meas. 1(1), 19–33 (1997)
16. Farag, Y., Yannakoudakis, H., Briscoe, T.: Neural automated essay scoring and coherence modeling for adversarially crafted input. In: Proceedings of the 2018 Conference of the North American Chapter of the Association for Computational Linguistics: Human Language Technologies, pp. 263–271. Association for Computational Linguistics (2018)
17. Fox, J.P.: Bayesian Item Response Modeling: Theory and Applications. Statistics for Social and Behavioral Sciences. Springer, New York (2010). https://doi.org/10.1007/978-1-4419-0742-4
18. Gerrish, S.M., Blei, D.M.: Predicting legislative roll calls from text. In: Proceedings of International Conference on International Conference on Machine Learning, pp. 489–496 (2011)

19. Griffiths, T.L., Steyvers, M.: Finding scientific topics. Proc. Nat. Acad. Sci. **101**(Suppl. 1), 5228–5235 (2004)
20. Hastings, P., Hughes, S., Britt, M.A.: Active learning for improving machine learning of student explanatory essays. In: Penstein Rosé, C., et al. (eds.) AIED 2018. LNCS (LNAI), vol. 10947, pp. 140–153. Springer, Cham (2018). https://doi.org/10.1007/978-3-319-93843-1_11
21. Hofmann, T.: Probabilistic latent semantic indexing. In: Proceedings of Annual International ACM SIGIR Conference on Research and Development in Information Retrieval, pp. 50–57 (1999)
22. Jameel, S., Lam, W., Bing, L.: Supervised topic models with word order structure for document classification and retrieval learning. Inf. Retr. J. **18**(4), 283–330 (2015)
23. Kassim, N.L.A.: Judging behaviour and rater errors: an application of the many-Facet Rasch model. GEMA Online J. Lang. Stud. **11**(3), 179–197 (2011)
24. Li, F., Wang, S., Liu, S., Zhang, M.: SUIT: a supervised user-item based topic model for sentiment analysis. In: Proceedings of the Twenty-Eighth AAAI Conference on Artificial Intelligence, pp. 1636–1642 (2014)
25. Li, X., Ouyang, J., Zhou, X.: Supervised topic models for multi-label classification. Neurocomputing **149**, 811–819 (2015)
26. Liu, O.L., Frankel, L., Roohr, K.C.: Assessing critical thinking in higher education: Current state and directions for next-generation assessment. ETS Research Report Series 2014, 1, pp. 1–23 (2014)
27. Lord, F.: Applications of Item Response Theory to Practical Testing Problems. Erlbaum Associates, Hillsdale (1980)
28. Louvigné, S., Uto, M., Kato, Y., Ishii, T.: Social constructivist approach of motivation: social media messages recommendation system. Behaviormetrika **45**(1), 133–155 (2018)
29. Masters, G.: A Rasch model for partial credit scoring. Psychometrika **47**(2), 149–174 (1982)
30. Muraki, E.: A generalized partial credit model. In: van der Linden, W.J., Hambleton, R.K. (eds.) Handbook of Modern Item Response Theory, pp. 153–164. Springer, New york (1997). https://doi.org/10.1007/978-1-4757-2691-6_9
31. Myford, C.M., Wolfe, E.W.: Detecting and measuring rater effects using many-Facet Rasch measurement: Part I. J. Appl. Meas. **4**, 386–422 (2003)
32. Patz, R.J., Junker, B.W., Johnson, M.S., Mariano, L.T.: The hierarchical rater model for rated test items and its application to large-scale educational assessment data. J. Educ. Behav. Stat. **27**(4), 341–366 (1999)
33. Patz, R.J., Junker, B.: Applications and extensions of MCMC in IRT: multiple item types, missing data, and rated responses. J. Educ. Behav. Stat. **24**, 342–366 (1999)
34. Persky, H., Daane, M., Jin, Y.: The nation's report card: Writing 2002. Technical report, National Center for Education Statistics (2003)
35. Rodrigues, F., Ribeiro, B., Lourenço, M., Pereira, F.C.: Learning supervised topic models from crowds. In: Third AAAI Conference on Human Computation and Crowdsourcing, pp. 160–168 (2015)
36. Rosen, Y., Tager, M.: Making student thinking visible through a concept map in computer-based assessment of critical thinking. J. Educ. Comput. Res. **50**(2), 249–270 (2014)
37. Salahu-Din, D., Persky, H., Miller, J.: The nation's report card: Writing 2007. Technical report, National Center for Education Statistics (2008)

38. Samejima, F.: Estimation of latent ability using a response pattern of graded scores. Psychom. Monogr. **17**, 1–100 (1969)
39. Schendel, R., Tolmie, A.: Assessment techniques and students' higher-order thinking skills. Assess. Eval. High. Educ. **42**(5), 673–689 (2017)
40. Taddy, M.: On estimation and selection for topic models. In: Lawrence, N.D., Girolami, M.A. (eds.) Proceedings of International Conference on Artificial Intelligence and Statistics, vol. 22, pp. 1184–1193 (2012)
41. Taghipour, K., Ng, H.T.: A neural approach to automated essay scoring. In: Proceedings of the 2016 Conference on Empirical Methods in Natural Language Processing, pp. 1882–1891. Association for Computational Linguistics (2016)
42. Ueno, M., Okamoto, T.: Item response theory for peer assessment. In: Proceedings of IEEE International Conference on Advanced Learning Technologies, pp. 554–558 (2008)
43. Uto, M., Nguyen, T., Ueno, M.: Group optimization to maximize peer assessment accuracy using item response theory and integer programming. IEEE Trans. Learn. Technol. p. 1 (2019)
44. Uto, M., Louvigné, S., Kato, Y., Ishii, T., Miyazawa, Y.: Diverse reports recommendation system based on latent Dirichlet allocation. Behaviormetrika **44**(2), 425–444 (2017)
45. Uto, M., Thien, N.D., Ueno, M.: Group optimization to maximize peer assessment accuracy using item response theory. In: André, E., Baker, R., Hu, X., Rodrigo, M.M.T., du Boulay, B. (eds.) AIED 2017. LNCS (LNAI), vol. 10331, pp. 393–405. Springer, Cham (2017). https://doi.org/10.1007/978-3-319-61425-0_33
46. Uto, M., Ueno, M.: Item response theory for peer assessment. IEEE Trans. Learn.Technol. **9**(2), 157–170 (2016)
47. Uto, M., Ueno, M.: Empirical comparison of item response theory models with rater's parameters. Heliyon **4**(5), 1–32 (2018)
48. Uto, M., Ueno, M.: Item response theory without restriction of equal interval scale for rater's score. In: Penstein Rosé, C., Martínez-Maldonado, R., Hoppe, H.U., Luckin, R., Mavrikis, M., Porayska-Pomsta, K., McLaren, B., du Boulay, B. (eds.) AIED 2018. LNCS (LNAI), vol. 10948, pp. 363–368. Springer, Cham (2018). https://doi.org/10.1007/978-3-319-93846-2_68
49. Zheng, X., Yu, Y., Xing, E.P.: Linear time samplers for supervised topic models using compositional proposals. In: Proceedings of the 21st ACM SIGKDD International Conference on Knowledge Discovery and Data Mining, pp. 1523–1532 (2015)
50. Zhu, J., Ahmed, A., Xing, E.P.: MedLDA: maximum margin supervised topic models for regression and classification. In: Proceedings of the 26th International Conference on Machine Learning. pp. 1257–1264 (2009)

Collaboration Detection that Preserves Privacy of Students' Speech

Sree Aurovindh Viswanathan$^{(\boxtimes)}$ and Kurt VanLehn$^{(\boxtimes)}$

Arizona State University, Tempe, AZ 85281, USA
{sviswal0, kurt.vanlehn}@asu.edu

Abstract. Collaboration is a 21st Century skill as well as an effective method for learning, so detection of collaboration is important for both assessment and instruction. Speech-based collaboration detection can be quite accurate but collecting the speech of students in classrooms can raise privacy issues. An alternative is to send only whether or not the student is speaking. That is, the speech signal is processed at the microphone by a voice activity detector before being transmitted to the collaboration detector. Because the transmitted signal is binary (1 = speaking, 0 = silence), this method mitigates privacy issues. However, it may harm the accuracy of collaboration detection. To find out how much harm is done, this study compared the relative effectiveness of collaboration detectors based either on the binary signal or high-quality audio. Pairs of students were asked to work together on solving complex math problems. Three qualitative levels of interactivity was distinguished: Interaction, Cooperation and Other. Human coders used richer data (several audio and video streams) to choose the code for each episode. Machine learning was used to induce a detector to assign a code for every episode based on the features. The binary-based collaboration detectors delivered only slightly less accuracy than collaboration detectors based on the high quality audio signal.

Keywords: Collaborative learning · Machine learning · Learning analytics

1 Introduction

Collaboration is a 21st century skill as well as an effective method for learning [1]. However, learning to collaborate is not straightforward. Students may require feedback to develop collaboration skills [2]. For scaling up feedback and assessment of collaboration, automated methods for collaboration detection are required. Fortunately, when students interact via speech, collaboration can be differentiated from non-collaboration using current technology, as reviewed below.

However, monitoring collaboration in spoken conversations between students raises concerns about privacy. Although teachers are always entitled to hear the speech of their students, giving third parties access to student conversations may raise significant privacy concerns. An alternate approach would be to process the audio signal at the microphone by using a voice activity detector (also called speech activity detector), which converts the raw audio signal into a binary signal (1 = Speech, 0 = Silence). Once the audio signal is converted to a binary signal, it can be transmitted or stored for

© Springer Nature Switzerland AG 2019
S. Isotani et al. (Eds.): AIED 2019, LNAI 11625, pp. 507–517, 2019.
https://doi.org/10.1007/978-3-030-23204-7_42

collaboration analysis. Because the binary signal is incomprehensible, privacy is preserved. However, the loss of information may prevent effective classification of collaboration.

This paper compares the relative effectiveness of measuring collaboration based on either a high quality audio signal or its binary version. Since the total amount of information transmitted by high quality signal is many orders of magnitude greater than the binary version, we expected a large difference in classifier performance. The high quality audio signal was collected by headset microphones connected to tablets being used by small groups of students who were solving problems in a laboratory setting. Only low level analysis was performed on both signals. Spoken words were not used as part of the analysis. For analysis of high quality speech data, low level features such as pitch, shimmer and linear spectral features were used. For the binary signals, time series features such as absolute energy, approximate entropy and symmetry were used. All features were extracted by algorithms and no human coders were involved.

In order to create and evaluate collaboration detectors, the judgments of human coders were used as the 'gold standard' classification of the group's interactions. The coders had both high quality audio and several videos to aid their judgment. Collaboration detectors were then machine-learned from the human judgments. Their accuracies were measured using 10 fold cross validation.

2   Prior Work on Speech-Based Collaboration Detection

Many systems have explored automated analyses of interaction among group members [3]. Instead of speech, most such systems input typed text from students collaborating via forums, chat or email. Of the projects that used speech-based collaboration detectors [4–11], only 3 measured the accuracy of the classification. These 3 projects are the most similar to our project, so they will be reviewed here.

Just as we did for our high quality speech classifier, Gweon et al. [6, 13] used machine-learned classifiers based on low-level speech features. Although the amount of speech and silence were included as features, the temporal pattern of speech and silence were not considered in their analysis. Secondly, whereas collaboration was the focal code, Gweon et al. two projects chose different non-collaboration codes. This choice may impact accuracy, so we measured the accuracy of classifiers trained with different combinations of non-collaboration codes.

Just as we did for our binary-signal classifier, Martinez Maldonado et al. [9] used a voice activity detector to convert speech into binary. However, they used counting and proportions in their classifiers; they did not consider temporal patterns. Later, this group did consider temporal patterns [4, 8]. They used differential sequence mining to find temporal patterns of speech and silence that would reliably split groups into high and low collaborators. However they did not convert their findings into a collaboration detector and measure its accuracy.

Bassiou et al. [11] conducted studies that are quite similar to ours. They found that collaboration detectors induced from a binary signal were more accurate than collaboration detectors induced from low-level acoustic features. They also found that accuracy could be further improved by combining the two types of features. Their studies differed from ours in several ways. Most importantly, they used a coarser coding scheme for collaboration. The scheme merely indicated how many students in the group were participating actively in problem solving.

3 Data Collection

This section describes the context in which the data were gathered.

3.1 Task: Collaborative Writing

The subjects collaboratively solved a problem ("Boomerangs") that was developed by the Mathematics Assessment Project and appears on their site [14]. Like prior work on spoken collaboration detection, it is a math problem. However, unlike the prior work, students are required to write paragraph-long explanations (See Fig. 1). They were given solutions to an optimization problem done by 4 hypothetical students and were asked 3 questions about each solution. Thus, this task is actually a collaborative writing task. It clearly has a different cadence than most mathematical problem solving. In particular, there can be significant periods of time when one student is writing out an explanation developed by one or both students. Because collaborative writing is required in many tasks from outside mathematics, it is important to investigate the accuracy of speech-based collaboration detection while student are thus engaged.

3.2 Technology, Participants and Duration

Students worked together in pairs. Each student had their own tablet, a Samsung Galaxy Note 10.1. This tablet had active digitizer technology which allowed students to write easily and legibly on the tablet screen using a stylus. These tablets are connected to the Server via a Wireless network. The software used by participants is called FACT [22]. The FACT user interface mimicked a large poster on which students can write and draw content. Anything written by one group member was immediately visible to the other. Both members of the group could scroll and zoom independently of each other thus allowing them to focus on different parts of the poster.

The study was conducted in a laboratory setting. The participants were 38 graduate and undergraduate students from our university who were paid for their time. Prior to doing the collaborative writing task, students solved the optimization problem themselves in order to become familiar with it. All students were able to solve it easily. The overall task, including both problem solving and collaborative writing, took around 45 to 55 min to complete.

3.3 Raw Data Collection

The recording setup generated input streams from two unidirectional headset microphones, one omnidirectional microphone, two tablet screen and two Web cameras. A desktop screen recorder was used to combine these input signals for easier annotation by the human coders.

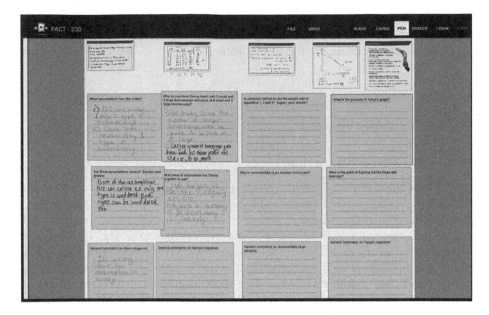

Fig. 1. Snapshot of students working on shared workspace

3.4 Coding Categories

This section describes the collaboration codes assigned by human coders. Because the coders could see all the input streams, whereas the collaboration detectors could only hear the audio from the two headset mics, the human coders' judgements were used as the "gold standard" against which the collaboration detectors were judged.

Several coding schemes for human judgement of collaboration have been devised [1, 11, 15–17]. The simplest merely count the number of members of the group that are actively participating. Thus, if both members of our groups were speaking or writing, then the group would be coded as collaborating.

However, even if all members of a group are participating, they may be interacting to varying degrees. On one end of this scale, they have split up the task and each member is working individually on a different subtask; this end of the scale is often called *cooperation*. On the other extreme, the members are essentially pursuing one line of reasoning or argumentation with all members contributing to it. That is, each member's utterance or action refers to and builds upon the prior contributions of other members. This end of the scale is often called *co-construction, transactivity* or just *collaboration*. The degree of interaction among group members manifests itself in different ways, so most coding schemes are multi-dimensional, where each dimension alerts coders to one way in which the cooperation/collaboration distinction can be observed.

When a collaboration detector is used in a classroom, it should help the teacher make a binary decision - whether to visit a group or not. Similarly, when it is used in a tutoring system, it should help the system make a binary decision – to intervene or not.

Different teachers and systems might have different concerns, but our sense of the literature is that most use cases can be covered if the collaboration detector outputs these three classifications:

1. *Interactive:* Students are working on the same part of the poster, and they contributions build upon each other. This is often called co-constructive or transactive behavior.
2. *Cooperation:* Both students are working, but they are working separately and independently, usually on different parts of the poster.
3. *Other:* At least one student is not working or contributes minimally to the task.

When the collaboration detector is used by a teacher who only cares that all students are participating, then the first two categories can be lumped together as "good" and the teacher only receives alerts about groups classified into the third category. On the other hand, if the teacher wants student to interact more intensely, then only the first category can be treated as "good" and the teacher receives alerts about groups classified into the second and third category.

This design matches the Chi's ICAP framework [1], which uses "Interactive" for category 1 (the I of ICAP). Category 2 means both students are "Constructive" (the C of ICAP). Category 3 means at least one student is Active, Passive or disengaged. The framework predicts that $I > C > A > P$ for learning gains, which implies $1 > 2 > 3$ for our categories. We used Chi's term "interactive" for the first category above.

Thus, we asked human judges to make the 3-way distinction above. However, during our analysis of accuracy, we considered the 3-way classification accuracy as well as several binary classifications obtained by lumping together 2 of the 3 categories.

Because our study was conducted in a camera-infused laboratory, we saw very little disengagement. Thus, the typical behavior of groups classified as "Other" was for one student to be doing all the work, while the other student watched and occasionally uttered brief agreements. In earlier work, we termed these categories "collaboration", "cooperation" and "asymmetric contribution" [21].

4 Analysis Methods

4.1 Audio Processing and Extraction of Audio Logs

The inputs to the collaboration detectors came only from the headset microphones; the other audio and video streams were seen only by the human coders. The first step in processing the headset mic audio was removal of background noise. Signal processing was carried out as proposed by Rafi et al. [18] along with few modifications. FFT windows was reduced to 0.25 from 0.5 and a soft mask by Wiener filtering method instead of a hard filter.

The binary version of the audio signal was obtained from the cleaned-up mic audio using a standard voice activity detector (WebRTC). The non-speech segments in the high quality audio signal was removed based on output from the voice activity detector.

4.2 Segmentation

As students answered the 12 questions, we noticed that they sometimes shifted their interaction patterns when transitioning from one question to the next. In order to avoid mixing up two different patterns of interactions, we placed segment boundaries between subtasks. More specifically, we used these criteria for placing a segment boundary: First, when there was a switch from one subtask to the other by any participant, a segment boundary was placed Second, if the particular activity took more than 4 min, then a segment boundary was placed immediately after the writing activity stopped for a brief time (like 5 s). This prevented overly long segments. Third, if a student went back to a different card and started writing, a segment boundary was placed. The average length of the segment was 110 s with standard deviation of 83 s.

4.3 Human Coding

Once the segmentation was performed, human annotators classified each segment as either interactive (I), cooperative (C) or other (O). The annotators used the audio-video stream obtained by the screen recorder and also used log data to understand various write events. If characteristics of multiple codes are found in the same segment, then the category with the greatest amount of time was assigned to the segment. Two human annotators tagged a sample of 80% of the overall segments. Inter rater agreement was considered acceptable with Cohen's kappa K = 0.76. For consistency across the whole dataset, the classifications of one annotator (the first author) were used in subsequent analysis.

4.4 Feature Extraction

In order to use standard machine learning algorithms to induce collaboration detectors for both the audio signal and the binary signal, the signals were represented as features.

The audio signal's features were from the OpenSMILE [20] audio feature set, which represents the state of the art in affect and paralinguistic recognition. Features were generated by a toolkit from the speech signal of each subject. The two individual subject feature vectors were then concatenated into one single feature vector for every segment.

The binary speech signal was characterized as a time series signal. This is obtained one per person. At a segment level, tsfresh [19] computes 794 time series features based on variations of both these signals with respect to time. The entire set of features obtained from tsfresh can be found in [19] and for OpenSMILE can be found in [20].

Group level features such as the duration of time when students spoke with each other (speech time per segment) and the duration of time when they did not speak (silence time) with each other were also extracted from the signal.

4.5 Feature Selection

Feature selection was performed because the number of features was greater than the number of observations. Pairwise correlations were performed on features likely to be redundant. Sets of highly correlated features (coefficient > 0.9) were reduced to a single feature chosen arbitrarily from the set.

For the high quality audio signal, recursive feature elimination is used to eliminate features that have low discriminative power across different classes.

For the binary signals, both the students' time series characterization along with its the collaboration class (I, C or O) was fed into tsFresh [19]. For each feature, it used Chi-square and other statistical tests to determine whether the features' value was reliability associated with the collaboration class. Only features whose p-value exceeded 0.05 were kept.

5 Results

As mentioned earlier, we developed collaboration detectors for several use cases. One pair of detectors distinguished all three categories (Interaction, Cooperation and Other). In addition, we generated binary classifiers focused on every single category.

5.1 Binary Classifier Focused on Cooperation

This section reports the accuracy of the binary classifier that was trained to discriminate Cooperation (C) from Non Cooperation (NC). We built classifiers using both high quality audio signal and the binary signal. Random Forests yielded the best results when compared to other algorithms such as logistic regression, bagging and boosting. The models were validated using tenfold cross validation. As Table 1 shows, the accuracy of the binary signal classifier (K = 0.53) was similar to the accuracy of the high quality audio classifier (K = 0.66).

Table 1. Confusion matrices for binary classifier focused on cooperation

High Quality Audio (F_1= 0.83, K= 0.66)

		Predictions	
		NC	C
True	NC	297	16
Class	C	15	38

Binary Logs (F_1= 0.76, K= 0.53)

		Predictions	
		NC	C
True	NC	294	19
Class	C	22	31

5.2 Binary Classifier Focused on Interaction

This section reports the accuracy of the binary classifier that trained to discriminate Interaction (I) from Non Interaction (NI). We built classifiers using both the high quality audio signal and the binary signal. Random Forests yielded the best results when compared to other algorithms. These models were validated using tenfold cross validation. As Table 2 shows, the accuracy of the binary signal classifier (K = 0.54) was similar to the accuracy of the high quality audio classifier (K = 0.62).

Table 2. Confusion Matrices of Binary Classifiers focused on Interaction

High Quality Audio (F_1= 0.80, K= 0.62)

		Predictions	
		I	NI
True	I	136	35
Class	NI	36	159

Binary Logs (F_1= 0.77, K= 0.54)

		Predictions	
		I	NI
True	I	129	42
Class	NI	40	155

5.3   Binary Classifier Focused on Other Category

This section reports the accuracy of the binary classifier that trained to discriminate Other (O) from Non Other (NO). We built classifiers using both high quality audio signal and the binary signal. Random Forests yielded the best results when compared to other algorithms. The models were validated using tenfold cross validation. As Table 3 shows, the accuracy of the binary signal classifier (K = 0.28) was similar to the accuracy of the high quality audio classifier (K = 0.36), but neither accuracy was high.

Table 3. Confusion Matrices for Binary Classifier Focused on Cooperation

High Quality Audio (F_1= 0.69, K= 0.36)

		Predictions	
		O	NO
True	O	82	60
Class	NO	50	174

Binary Logs (F_1= 0.64, K= 0.28)

		Predictions	
		O	NO
True	O	95	47
Class	NO	84	140

5.4   Ternary Classifier

This section reports the accuracy of the results of ternary classifier that is trained to discriminate three categories: Interaction (I), Cooperation (C) and Other (O). We built classifiers using both high quality audio signal and binary signal. Random Forests yielded the best results when compared to other types of algorithms. The models were validated using tenfold cross Validation. As Table 4 shows, the accuracy of the binary signal classifier (K = 0.44) was similar to the accuracy of the high quality audio signal classifier (K = 0.55).

Table 4. Confusion Matrices of Three way Classifiers

High Quality Audio (F_1 = 0.70, k=0.55)

		Predictions		
		I	O	C
True	I	94	31	17
Class	O	35	135	1
	C	17	0	36

Binary logs (F_1 = 0.63, k=0.44)

		Predictions		
		I	O	C
True	I	84	38	20
Class	O	46	123	2
	C	20	1	32

6 Discussion and Conclusion

When this project began, we did not think that detectors based on binary signals would perform well compared to collaboration detectors based on high quality audio signal. Against low expectations we got modest kappa scores - 0.53, 0.54, 0.44 except for the Other-focused binary classifier (0.28). To explain the Other classifier's inaccuracy, we can examine the results from the ternary classifier.

The contingency table of the ternary classifier (Table 4) shows that some of samples from the Interaction category are mistaken as Other category and vice versa. On the other hand, the Cooperation category was clearly separated from the Other category. This phenomenon is due to the fact that in some cases, when students worked alone with their partner watching, they continued to verbalize their writing and hence the machine learner assumed that they were Interacting with each other.

The binary classifier focused on Other category was minimally reliable since the classifier has to differentiate Other from Interaction and Cooperation combined together into one. The reliability was compromised since Interaction and Cooperation have entirely different audio characteristics and the Other category shares some character-istics of both. As a result, neither the binary nor full audio features could differentiate between them reliably. If the use case requires the Other category to be distinguished from non-Others, the ternary classifier should be used.

Although the collaboration detectors based on the binary signal performed rea-sonably well, there are a few caveats to consider. The first limitation is that synchro-nization of different data streams needs to be improved so that significant of manual labor can be avoided. Until this is achieved, our method cannot be used in classroom in real-time.

The second limitation is the usage of cards to automatically mark segment boundaries. This also allowed us to detect a change in subtask when subjects worked with each other. Although this helped in segmentation and improving reliability, we are not sure about the performance of the classifier when the boundaries are less salient.

Third, this study did not encounter off task behavior since it was a laboratory study recorded under a camera. Analysis of the semantic content of the speech may be necessary to detect off task behavior.

The fourth limitation is that the study was performed in artificial laboratory setting and it only involved two people at a time. This may have increased the accuracy of the full-audio classifier because there was no interference from other audio sources. This would also explain why Bassiou et al. [11] found that their collaboration detector based on a binary signal was more accurate than their collaboration detector based on a full audio signal: their speech was collected in a noisy classroom. Thus, their results combined with ours suggest a bright future for collaboration detection based on binary signals.

These finding suggest a clear direction for our future work, because they appear to solve several practical problems.

First, when a collaboration detection system needs high quality audio, the analog signal from the microphone must be sampled at a high bitrate. Transmitting these bitstreams wirelessly from 30 students can overload classroom radios, but using wires instead invites physical damage to the equipment when students or staff become entangled. In contrast, binary signals require less bandwidth, so 30 of them can probably be transmitted wirelessly reliably.

Finally, throat microphones can probably be used instead of headset microphones. These microphones detect noise only from the speaker and not ambient noise. Thus, the noise removal we performed prior to voice activity detection may not be necessary. Affordable throat microphones often produce somewhat distorted audio, so they would probably not work well for collaboration detection based on acoustic features.

Acknowledgements. This research was funded by the Diane and Gary Tooker chair for effective education in Science Technology Engineering and Math, by NSF grant IIS-1628782, and by the Bill and Melinda Gates Foundation under Grant OP1061281.

References

1. Chi, M.T.H., Wylie, R.: ICAP: a hypothesis of differentiated learning effectiveness for four modes of engagement activities. Educ. Psychol. **49**(4), 219–243 (2014)
2. Ladd, G.W., Kochenderfer-Ladd, B., Visconti, K.J., Ettekal, I., Sechler, C.M., Cortes, K.I.: Grade-school children's social collaborative skills: links with partner preference and achievement. Am. Educ. Res. J. **51**(1), 152–183 (2014)
3. Magnisalis, I., Demetriadis, S., Karakostas, A.: Adaptive and intelligent systems for collaborative learning support: a review of the field. IEEE Trans. Learn. Technol. **4**(1), 5–20 (2011)
4. Martinez-Maldonado, R., Kay, J., Yacef, K.: An automatic approach for mining patterns of collaboration around an interactive tabletop. In: Lane, H.C., Yacef, K., Mostow, J., Pavlik, P. (eds.) AIED 2013. LNCS (LNAI), vol. 7926, pp. 101–110. Springer, Heidelberg (2013). https://doi.org/10.1007/978-3-642-39112-5_11
5. Bachour, K., Kaplan, F., Dillenbourg, P.: An interactive table for supporting participation balance in face-to-face collaborative learning. IEEE Trans. Learn. Technol. **3**(3), 203–213 (2010)
6. Gweon, G., Jain, M., McDonough, J., Raj, B., Rosé, C.P.: Measuring prevalence of other-oriented transactive contributions using an automated measure of speech style accommodation. Int. J. Comput. Support. Collaborative Learn. **8**(2), 245 (2013)
7. Gweon, G.: Assessment and support of the idea co-construction process that influences collaboration (2012)
8. Martinez-Maldonado, R., Dimitriadis, Y., Martínez-Monés, A., Kay, J., Yacef, K.: Capturing and analyzing verbal and physical collaborative learning interactions at an enriched interactive tabletop. Int. J. Comput. Support. Collaborative Learn. **8**(4), 455 (2013)
9. Martinez-Maldonado, R., Yacef, K., Kay, J.: TSCL: a conceptual model to inform understanding of collaborative learning processes at interactive tabletops. Int. J. Hum. Comput. Stud. **83**, 62–82 (2015)
10. Roman, F., Mastrogiacomo, S., Mlotkowski, D., Kaplan, F., Dillenbourg, P.: Can a table regulate participation in top level managers' meetings? In: Proceedings of the 17th ACM International Conference on Supporting Group Work - GROUP 2012 (2012)
11. Bassiou, N., et al.: Privacy-preserving speech analytics for automatic assessment of student collaboration. In: Proceedings Interspeech, pp. 888–892 (2016)

12. Agrawal, P., Udani, M.: The automatic assessment of knowledge integration processes in project teams, Long papers, p. 462 (2011)
13. Gweon, G., Jain, M., McDonogh, J., Raj, B.: Predicting idea co-construction in speech data using insights from sociolinguistics. In: Proceedings of the Learning Sciences: The Future of Learning (2012)
14. http://map.mathshell.org/index.php
15. Meier, A., Spada, H., Rummel, N.: A rating scheme for assessing the quality of computer-supported collaboration processes. Comput.-Supported Collaborative Learn. **2**, 63–86 (2007)
16. Kahrimanis, G., Chounta, I.-A., Avouris, N.: Validating empirically a rating approach for quantifying the quality of collaboration. In: Daradoumis, T., Demetriadis, S., Xhafa, F. (eds.) Intelligent Adaptation and Personalization Techniques, pp. 295–310. Springer, Heidelberg (2012). https://doi.org/10.1007/978-3-642-28586-8_13
17. Gweon, G., Jain, M., McDonough, J., Raj, B., Rose, C.P.: Measuring prevalence of other-oriented transactive contributions using an automated measure of speech style accommodation. Int. J. Comput. Support. Collaborative Learn. **8**(2), 245–265 (2013)
18. Rafii, Z., Pardo, B.: Music/Voice separation using the similarity matrix, pp. 583–588 (2012)
19. Christ, M., Braun, N., Neuffer, J., Kempa-Liehr, A.W.: Time series FeatuRe extraction on basis of scalable hypothesis tests (tsfresh–A Python package). Neurocomputing (2018)
20. Eyben, F., Wöllmer, M., Schuller, B.: OpenSMILE: the Munich versatile and fast open-source audio feature extractor, pp. 1459–1462. ACM (2010)
21. Viswanathan, S.A., VanLehn, K.: Using the tablet gestures and speech of pairs of students to classify their collaboration. IEEE Trans. Learn. Technol. **11**, 230–242 (2018)
22. Cheema, S., VanLehn, K., Burkhardt, H., Pead, D., Schoenfeld, A.: Electronic posters to support formative assessment. In: Proceedings of the 2016 CHI Conference Extended Abstracts on Human Factors in Computing Systems, pp. 1159–1164 (2016)

How Does Order of Gameplay Impact Learning and Enjoyment in a Digital Learning Game?

Yeyu Wang[1]([✉]), Huy Nguyen[2], Erik Harpstead[2], John Stamper[2], and Bruce M. McLaren[2]

[1] University of Pennsylvania, Philadelphia, PA, USA
wangyeyu215@gmail.com
[2] Carnegie Mellon University, Pittsburgh, PA, USA

Abstract. When students are given agency in playing and learning from a digital learning game, how do their decisions about sequence of gameplay impact learning and enjoyment? We explored this question in the context of *Decimal Point*, a math learning game that teaches decimals to middle-school students. Our analysis is based on students in a high-agency condition, those who can choose the order of gameplay, as well as when to stop. By clustering student mini-game sequences by edit distance – the number of edit operations to turn one sequence into another – we found that, among students who stopped early, those who deviated more from a canonical game sequence reported higher enjoyment than those who did not. However, there were no differences in learning gains. Our results suggest that students who can self-regulate and exercise agency will enjoy the game, but the type and number of choices may also have an impact on enjoyment factors. At the same time, more investigation into the amount and means of delivering instruction to maximize learning efficiency within the game is necessary. We conclude by discussing digital learning game design lessons to create a game that more closely aligns with students' learning needs and affective states.

Keywords: Digital learning game · Decimal number · Edit distance · Clustering

1 Introduction

An important aspect of digital learning game design is deciding which gameplay elements the players (i.e., students) can control. In a typical game environment, players are offered a lot of agency - the capability to make their own decisions about how, what, and when they play. However, agency, which is often associated with engagement and enjoyment [41], may or may not be helpful to learning. Another nuance present in digital learning games is whether students should be given instructionally relevant choices, since young learners often have difficulty in making effective instructional decisions [33], in many cases resorting to unthoughtful choices [44].

One way to enhance students' experience and outcomes, while still giving them control over instructionally relevant aspects of gameplay, is to provide a recommendation

© Springer Nature Switzerland AG 2019
S. Isotani et al. (Eds.): AIED 2019, LNAI 11625, pp. 518–531, 2019.
https://doi.org/10.1007/978-3-030-23204-7_43

feature within the game that can suggest the (potentially) optimal next step, without reducing the students' sense of agency. To achieve this, an important step is examining the influences of different problem sequences and identifying those that are most beneficial in terms of learning, enjoyment, or ideally both. We examined this question in *Decimal Point*, a digital learning game composed of a variety of mini-games designed to help middle-school students learn decimals [30]. While the original version of the game features a canonical sequence of mini-games that aims at interleaving various problem types and visual themes, it is not designed to be optimal for both learning and enjoyment for all students. To build a recommender capability as outlined above, we would need to identify the features of a good sequence while, at the same time, noting that these features may vary based on individual students.

To tackle this issue, prior studies of *Decimal Point* have compared learning and enjoyment between a high- and low-agency condition [23, 34]. The high-agency group could play the mini-games in any order and also had the option to stop playing early or play extra games. In contrast, the low-agency group had to play all mini-games in a fixed order. Expanding on this work, we focused solely on the high-agency students and explored potential differences among them in our analysis. In other words, given that high-agency students can make their own choices about mini-game selection, how would different selection orders (i.e., *game sequences*) impact their experience? More specifically, we investigated the following research questions:

RQ1: *How do students' game sequences impact their self-reported enjoyment of the digital learning game?*

RQ2: *How do students' game sequences impact their learning outcomes from the digital learning game?*

2 Background

2.1 The *Decimal Point* Game

Decimal Point is a single-player game that helps middle-school students learn about decimal numbers and their operations (e.g., adding, ordering, comparing). The game is based on an amusement park metaphor (Fig. 1), where students travel to different areas of the park, each with a theme (e.g., Haunted House, Sports World), and play a variety of mini-games, each targeting a common decimal misconception [19, 25, 52].

In the original game [30], students were prompted to play the mini-games in a predefined, canonical sequence, according to the dashed line shown in Fig. 1A (starting from the upper left). This sequence was originally developed to maintain thematic cohesion and to interleave problem types, which has been shown to improve mathematics learning [37, 38]. However, it is unclear whether a different sequence could be more or less beneficial to students. A subsequent study by [34] further explored agency by comparing two versions of *Decimal Point*: high-agency and low-agency. In the high-agency condition, students could play the mini-games in any order, could stop halfway through (i.e., after 12 mini-games) or play extra rounds after finishing all 24 mini-games. In the low-agency condition, students played all mini-games in a fixed order, without the option to stop early or play more.

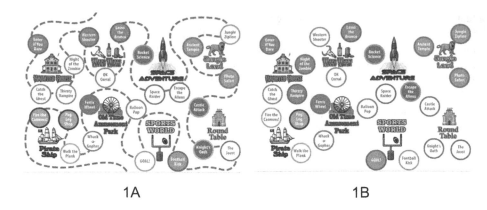

Fig. 1. The different game maps used in (A) low-agency and high-agency with line, and (B) high-agency without line. The filled circles denote completed mini-games.

The authors reported no differences in learning or enjoyment between the two conditions, and had two conjectures regarding the high-agency students. First, they may have been implicitly guided to follow the canonical sequence by the dashed line on the map (Fig. 1A), hence their experience was comparable to that of students in the low-agency condition. Second, high-agency students may not have felt that their specific mini-game choices were consequential, as they would either stop early or eventually end up having played all mini-games, same as the low-agency students. In other words, to these students, different game sequences may have seemed to result in the same outcome.

The first conjecture was confirmed by post-hoc analyses reported in [34] and in a follow-up study by [23]. [34] reported that 68% of high-agency students played only 24 mini-games, similar to those in the low-agency condition, in approximately the same order. The study in [23] introduced a new high-agency condition without the dashed line (Fig. 1B) and it was observed that students in this condition deviated from the canonical path significantly more than those in the original high-agency condition.

As the next step, in this paper, we investigate the second conjecture – whether different game sequences selected by students in the high-agency conditions (with and without the dashed line) can have an impact on learning and enjoyment.

2.2   Related Work

The high-agency version of *Decimal Point* has many characteristics of an exploratory learning environment (ELE) [1], where students are free to explore instructional materials rather than follow a predefined learning path. Other notable digital learning games of this type include *Physics Playground* [49], *iSTART-2* [50], *Quest Atlantis* [5] and *Crystal Island* [43]. Among these, *Crystal Island* has been the subject of an experimental manipulation similar to that of *Decimal Point*, with three agency conditions: (1) high-agency, where students could freely explore the game world and choose which activities to do and in what order, (2) low-agency, where students did the same activities but had to follow a fixed order, and (3) no-agency, where students only observed a video of an expert playing the game. Study results from [43] showed that low-agency students demonstrated the greatest learning gains but also exhibited undesirable behaviors such as a propensity for guessing, suggesting that some degree of agency may be beneficial, but too much is not.

An important task in ELEs is modeling students' learning to provide effective interventions based on fine-grained interactions with the learning environment [1]. A useful metric that can be derived from these sequential data is the *distance* - a measure of how similar two sequences are. Prior research has shown that in digital learning games, the distances from students' problem-solving sequences to an expert solution sequence are correlated with their learning gains [42] and test performance [20]. It is also possible to compute the distances among students' own sequences to cluster them. Analysis of the resulting clusters has been instrumental in several ELE assessment tasks: identifying player strategies in an algorithmic puzzle-based game [24], distinguishing between low and high achieving students in a problem-solving tabletop application [29], exploring the solution space in an open-ended physics game [22], and so on.

Another focus of the current work is student enjoyment and how it may be influenced by gameplay choices. In general, digital learning games are effective at promoting engagement and enjoyment by giving students control over the learning environment [45, 50, 51]. However, the effect of student choices is also subject to several nuances. First, it can vary based on individual students' self-regulation skills [32]. Second, students need to feel that their choices are meaningful and acquire a sense of agency (for a detailed discussion of agency within *Decimal Point*, refer to [23] and [34]). Third, the type and number of choices may affect their utility. In particular, choices that reflect personal interest will have the greatest effect, yet a large number of choices can become discouraging [36]. We will elaborate on these nuances in our later discussion.

3   Context

The work reported in this paper is a post-hoc analysis of data collected from two prior studies of *Decimal Point* [23, 34]. We briefly introduce the way these studies were conducted here before describing our analysis approach.

The two prior studies involved a total of 484 students. In this work, we focused on only 287 of those students in two conditions, high-agency with line (HAL) and high-agency without line (HANL), since these were the groups of students who could make their own mini-game selections, as opposed to those in the low-agency condition who could not make such choices. We further removed students who did not finish all of the pre- and posttest materials and evaluation surveys, which are used to measure learning and enjoyment outcomes, reducing the sample to 235 students (110 male, 125 female). The digital learning game and study materials included the following:

Pretest, Immediate Posttest and Delayed Posttest. The pretest, immediate posttest and delayed posttest (one week after the posttest), were administered online. The tests are isomorphic to one another (i.e., the same types of problems in the same order) and contain decimal items similar to those found in the game (e.g., ordering, comparing, and adding decimals). The tests were also counterbalanced across students (e.g., ABC, ACB, BAC, etc. for pre, post, and delayed). Learning gains from pretest to posttest and pretest to delayed posttest are used to measure learning outcome.

Intervention. Students playing the high-agency versions of the game were shown a game map depicted in Fig. 1A (for the HAL group) or B (for the HANL group), where they could make their mini-game selections. There is also a dashboard that provides information about the types of game activity, and shows current mini-game completion progress. After playing half of the mini-games, students would be notified that they could choose to stop playing at any time from this point. Once students finished all 24 mini-games, the map interface would be reset to allow each game to be played once more (with the same game mechanics but different question content). Hence, the number of mini-games played by each student ranges from 12 to 48.

Evaluation Questionnaire and Survey. After finishing the game, students were given an evaluation questionnaire and post-survey, which asked them to (1) rate their overall experiences using a 5-point Likert scale, with a variety of game enjoyment questions (e.g., "I liked doing this lesson"), (2) select their most favorite mini-game, and (3) reflect on their agency experience (e.g., "if you did this activity again, would you play fewer, the same, or more number of mini-games? Why?"). The scores from (1) are averaged to produce a measure of self-reported enjoyment.

4   Results

4.1   Game Sequence Clustering

Since there is no expert sequence in *Decimal Point* (as we previously mentioned, it is unclear if the canonical sequence is optimal), we did not measure deviation from expert solution like other studies [20, 42]. Instead, our goal was to look at trends in learning and enjoyment among students who played through the mini-games in a similar way. We took a clustering approach to create groups of students who played a similar sequence of mini-games and looked for differences between these groups. To be consistent with prior studies, and because it was shown to be useful for analyzing our type of sequential data [23, 34], we used the Damerau-Levenshtein edit distance [13] as a measure of similarity between sequences. This metric counts the minimal operations required to change one sequence to another using insertions, deletions, substitutions, and transpositions. The smaller the edit distance, the more similar two sequences are to one another. If the value is zero, the two sequences are identical; if the value is the sum of the two sequence lengths, they are completely distinct.

We then applied k-medoids clustering [6] with the pairwise edit distance matrix of all game sequences as input. In this way, students who played similar game sequences (i.e., have a smaller edit distance between one another) would be grouped within the same cluster. We experimented with different values of k (number of clusters) for k-medoids clustering. After searching from 2 to 20, we selected the optimal k value of 4, based on the best average Silhouette Coefficient [40]. The four cluster medoids are illustrated in Fig. 2. We named each cluster based on the key mnemonic features of its medoid. The first is Canonical Sequence (CS) with a medoid sequence identical to the canonical, following the dashed line in Fig. 1A. The second is Initial Exploration (IE) because students played a few mini-games out of order at the beginning of gameplay before returning to the canonical sequence. The third and fourth are Half on Top (HT) and Half on Left (HL) respectively because their medoids only span a portion of the game map (the top half and left half, respectively). Descriptive statistics for all clusters are included in Table 1.

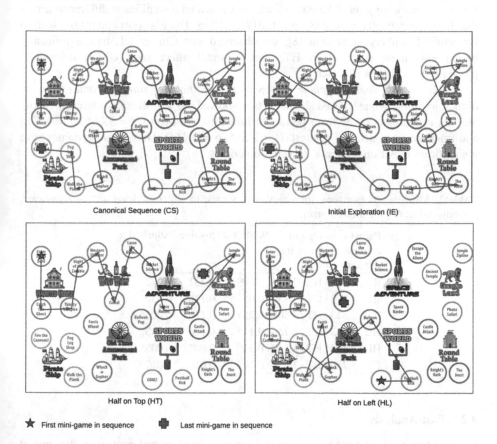

Fig. 2. Visualizations for the medoid game sequences in four clusters. Here the maps are shown without the line for clarity.

Table 1. Descriptive statistics for the four clusters.

Cluster	# of Students	# of Mini-games	Pretest	Immediate posttest	Delayed posttest	Enjoyment
CS	89	24.9 (4.6)	36.8 (13.1)	42.4 (10.3)	44.3 (10.1)	3.7 (0.8)
IE	14	26.9 (9.2)	37.3 (11.1)	42.4 (8.6)	42.9 (9.2)	3.8 (0.6)
HT	100	17.2 (6.3)	33.9 (12.9)	39.7 (11.7)	40.8 (12.3)	3.4 (1.0)
HL	32	19.6 (6.0)	40.0 (9.8)	43.3 (10.6)	44.3 (10.5)	3.8 (0.8)

To identify differences in learning and enjoyment across clusters, we conducted the Kruskal-Wallis test [14]. Kruskal-Wallis was chosen because our data did not satisfy the normality assumptions of an ANOVA. In the case of a significant difference, we used Dunn's post hoc [17] to perform pairwise comparisons between clusters. The effect size was also considered, following the thresholds of Cliff's Delta [39]. In this way, we examined our two research questions:

RQ1: *How do students' game sequences impact their self-reported enjoyment of the digital learning game?* Kruskal-Wallis test revealed a significant difference across the four clusters ($H = 10.248$, $p = 0.017$). Using Dunn's post hoc test with a Benjamini-Hochberg correction [8], we observed that Cluster HL had significantly higher enjoyment scores than HT, with a small effect size (Cliff's $d = 0.310$, $p = 0.007$), as shown in Table 2.

RQ2: *How do students' game sequences impact their learning outcomes from the digital learning game?* Kruskal-Wallis test showed no significant difference across clusters in gaining scores from pretest to immediate posttest ($H = 3.086$, $p = 0.378$) and from pretest to delayed posttest ($H = 2.585$, $p = 0.414$). Thus, clusters do not have a significant effect on students' learning outcomes.

Table 2. Multiple comparisons for enjoyment scores in different level (** - significant, $p <$ adjusted α; ^ - small effect size).

Pairwise median of enjoyment scores for different clusters				Cliff's d	p-value	Adjusted α
IE	3.813	HL	3.813	−0.096	0.731	0.042
IE	3.813	CS	3.750	0.055	0.748	0.050
IE	3.813	HT	3.500	0.257	0.130	0.025^
HL	3.813	CS	3.750	0.115	0.326	0.033
HL	3.813	HT	3.500	0.310	0.007	0.008**^
CS	3.750	HT	3.500	0.195	0.018	0.017^

4.2   Post Analysis

Prior studies have established agency as a sense of freedom and control by the student [54], and in the context of our digital learning game, the amount of deviation from the canonical path [23]. Given that there is a significant difference in enjoyment scores between two clusters, we further explored the relationship between game sequence, agency, and enjoyment through the following two metrics.

Theme Transition Frequency (TTF). We expected that students who exercised agency would look at the entire map and explore different theme areas, as opposed to selecting a mini-game nearest to their current location or staying within one theme. While students could stay within a theme that they liked, we believed they were unlikely to enjoy every theme; therefore, we still expected to see more exploration. To measure this behavior, we defined a new metric, called *theme transition frequency*, as the number of transitions between consecutive mini-games with different themes divided by the total number of transitions, for a given student. A value close to 1 means that the student tends to alternate between themes; a value close to 0 means that the student sticks to the same theme until all mini-games in that theme are completed. Next, we conducted Kruskal-Wallis test and found a significant difference in TTF across the four clusters (H = 52.421, p < 0.0005). To compare the TTF between pairs of clusters, we applied Dunn's post hoc test with Benjamini-Hochberg correction [8]. Cluster IE had significantly higher TTF than cluster CS, with a large effect size (Cliff's d = 0.527, p = 0.004). Cluster HL had significantly higher TTF than cluster CS with a large effect size (Cliff's d = 0.749, p < 0.0005) and higher than cluster HT with a small effect size (Cliff's d = 0.244, p < 0.022). Cluster HT had significantly higher TTF than cluster CS, with a medium effect size (Cliff's d = −0.454, p = 0.017) (Table 3).

Table 3. Multiple comparisons for TTF in different level (** - significant, p < adjusted α; ^ - small effect size, ^^ - medium effect size, ^^^ - large effect size)

Pairwise Median of TTF for Different clusters					Cliff's d	p-value	Adjusted α
IE	0.506	HL	0.675	−0.304	0.149	0.042
IE	0.506	CS	0.304	0.527	0.004	0.025**^^^
IE	0.506	HT	0.545	−0.041	0.996	0.050
HL	0.675	CS	0.304	0.749	< 0.0005	0.008**^^^
HL	0.675	HT	0.545	0.244	0.022	0.033**^
CS	0.304	HT	0.545	−0.454	< 0.0005	0.017**^^

Mini-game Preference. As the only difference in enjoyment we identified was among those who stopped early, in the HL and HT clusters, we conjectured that students may have had a stronger sense of enjoyment earlier in gameplay than towards the end. However, we did not have a mechanism to detect affective states over time. Therefore, as a proxy in examining this behavior, for each student, we looked at her self-reported favorite mini-game on the post-survey and where it occurred in her game sequence. More specifically, each student was labeled as one of three categories: (1) prefer one of the first three mini-games played, (2) prefer one of the last three mini-games played, and (3) prefer one between the first and last three mini-games. We then tested if the favorite mini-game is equally likely to appear in every part of the sequence. Since there are 24 mini-games in total, the null hypothesis is that the distribution of the three groups is 12.5%, 12.5% and 75% of the number of students respectively. We conducted a Chi-Square goodness of fit test [15] and found that this hypothesized distribution differs significantly from the empirical distribution of 30.2%, 59.6%, and 10.2% respectively (χ^2 = 67.42, df = 3, p < 0.0005). In particular, the first category, despite covering only the first three mini-games, accounted for almost one-third of the most favorite mini-game responses, much higher than its expected portion of 12.5%. This result implies that students tended to prefer their initial gameplay experience.

5   Discussion

In this work we explored the question of whether different game sequences lead to different learning and/or enjoyment outcomes for students in the high-agency condition who could decide on their mini-game selections. Across the four identified clusters of game sequences – CS, IE, HT, HL – we found no differences in learning, but Cluster HL had significantly higher enjoyment scores than Cluster HT. We discuss this key result, as well as our other results, in the following paragraphs.

With respect to learning, we saw that the varied numbers of mini-games played by students across the clusters did not result in learning differences. This outcome is consistent with [23], and the authors' proposed explanations are also applicable in our case. Students who stopped early may have been able to self-regulate their learning and learned as much as those who played all mini-games, resulting in more efficient learning [31]. Alternatively, it is possible that there is more instructional content than required for mastery in the game, so students who played all of the mini-games essentially over-practiced rather than being less efficient. There have been debates about the varying effects of over-practice; some researchers claim that it leads to decreased learning efficiency [10, 12], while others suggest it yields higher levels of fluency [27] and better long-term outcomes [16]. In our case, it appears that over-practice, if present, had a neutral effect, since students who potentially over-practiced achieved the same learning gains as those who did not. A step toward better understanding this would be to construct a knowledge component (KC) model of students' in-game learning [21] so that learning efficiency and over-practice can be validated through Bayesian knowledge tracing [12] and learning curve analysis [18, 28, 53]. Such a KC model could also be displayed to students to facilitate awareness of progress and self-regulation, in the form of an open learner model [9].

With respect to enjoyment, while students in HL and HT both played approximately half of the mini-games, the former played the most mini-games out of order, while the latter tended to follow the canonical sequence. This distinction, demonstrated by our analysis of theme transition frequency, suggests that the HL group exercised more agency and enjoyed the game more than HT. On the other hand, we expected that students in CS and IE would have more enjoyment than those in HL and HT, because the former group also had the option to stop early yet chose to continue playing. However, we did not observe this difference. One possible explanation is that students in CS and IE did not stop early because they were not good at self-regulating their learning, rather than because they were enjoying the game more. This idea is supported by the observation from Fig. 2 that the game sequences in CS and IE are close to the canonical sequence, suggesting that students did not exercise agency in mini-game selection. A second explanation is that the novelty of the game environment may wear off towards the end, i.e., students may have experienced a "burnout effect" with diminished feeling of progression [11], which influenced their rating of overall enjoyment. Survey responses of mini-game preference did in fact show that students tended to favor the initial mini-games. A potential reason for this phenomenon is the nature of choices in *Decimal Point*. According to [7], engaging in choices or self-control is effortful and draws on limited resources. Therefore, a large number of choices can become overwhelming [26, 46], and making several independent choices in a limited time may result in fatigue or ego-depletion [36]. In the high-agency condition, students first have to select one of the 24 mini-games, then one of the 23 remaining mini-games, and so on. Those who played all mini-games had to make 24 such selections within the timeframe of the study, so they may have experienced ego-depletion, which resulted in reduced enjoyment. Also, towards the end of gameplay, students do not have as many options to pick from because the completed mini-games are blocked from re-selection; however, this lack of choice may instead lead to decreased sense of agency. [36] suggested that there is an optimum number of choices that balances between the cognitive load from too many choices and the lack of agency from too few. Identifying this number for *Decimal Point* is left for future work.

In summary, we derived the following game design lessons from our analyses. First, one should aim for just the right amount of instructional content so that students can master the materials yet not incur the potential negative effects of over-practice. It can be difficult to initially estimate how much content is sufficient, but educational data mining techniques (e.g., learning curve analysis [21, 28]) can help revise and improve the materials in subsequent iterations. In addition, like *Decimal Point*, a game could allow students to control how much practice they are given, with proper scaffolding to assist them in self-regulating (e.g., an open learner model [9]). Second, when providing students with instructionally relevant choices, one should take into account factors such as agency, burnout and ego-depletion in designing the type and number of choices [11, 36]. Third, when collecting data from survey questions, one should note that students tend to report on their most recent experience, near the end of gameplay, rather than the overall experience.

Finally, we should point out that in this work, posttest scores and survey responses were used to measure the impacts of game sequence clusters. While these metrics are consistent with our prior studies [23, 34], it is possible that more fine-grained measures, for example those taken after each in-game action or mini-game played, would provide a better understanding of the influences of game sequences. In particular, we can use moment-by-moment learning models [3] to understand whether immediate or delayed learning takes place, and learning curve analysis [12] to track students' performance over time. For enjoyment, we will analyze learner affect by integrating automated affect detectors [2, 4, 35] in our data collection and analysis procedures, which can yield more reliable results than survey responses alone. This direction is consistent with the view of digital learning game researchers that students' learning and enjoyment should be assessed by in-game data rather than external measures [47, 48].

6   Conclusion

Our work investigated the effects of game sequences in *Decimal Point*. There were no differences in learning across sequence clusters, However, among students who chose to stop playing early, at around half of the mini-games, those who deviated more from the canonical order and switched between theme areas reported higher enjoyment scores. These results lead to important questions about the amount of instructional content, the nature of choices, and the interplay of various engagement factors in the context of digital learning games. We intend to investigate these questions in future work to better understand the dynamics of students' game experience. This, in turn, will help us develop better AI techniques to personalize the game for increased enjoyment and learning.

Acknowledgements. This work was supported by NSF Award #DRL-1238619. The opinions expressed are those of the authors and do not represent the views of NSF. Special thanks to J. Elizabeth Richey for offering valuable feedback. Thanks to Scott Herbst, Craig Ganoe, Darlan Santana, Farias, Rick Henkel, Patrick B. McLaren, Grace Kihumba, Kim Lister, Kevin Dhou, John Choi, and Jimit Bhalani, all of whom made important contributions to the development of the Decimal Point game.

References

1. Amershi, S., Conati, C.: Combining unsupervised and supervised classification to build user models for exploratory. JEDM J. Educ. Data Min. **1**, 18–71 (2009)
2. Baker, R.S.J.d., et al.: Towards sensor-free affect detection in cognitive tutor algebra. In: Proceedings of the 5th International Conference on Educational Data Mining, pp. 126–133 (2012)
3. Baker, R.S., Goldstein, A.B., Heffernan, N.T.: Detecting learning moment-by-moment. Int. J. Artif. Intell. Educ. **21**, 5–25 (2011)
4. Baker, R.S., Inventado, P.S.: Educational data mining and learning analytics. In: Larusson, J. A., White, B. (eds.) Learning Analytics, pp. 61–75. Springer, New York (2014). https://doi. org/10 1007/978-1-4614-3305-7_4
5. Barab, S., Pettyjohn, P., Gresalfi, M., Volk, C., Solomou, M.: Game-based curriculum and transformational play: designing to meaningfully positioning person, content, and context. Comput. Educ. **58**, 518–533 (2012)
6. Bauckhage, C.: Numpy/scipy recipes for data science: k-medoids clustering. Researchgate. Net, February 2015
7. Baumeister, R.F., Bratslavsky, E., Muraven, M.: Ego depletion: is the active self a limited resource? In: Self-Regulation and Self-Control, pp. 24–52. Routledge, New York (2018)
8. Benjamini, Y., Hochberg, Y.: Controlling the false discovery rate: a practical and powerful approach to multiple testing. J. R. Stat. Soc. Ser. B Methodol. **57**, 289–300 (1995)
9. Bull, S., Nghiem, T.: Helping learners to understand themselves with a learner model open to students, peers and instructors. In: Proceedings of Workshop on Individual and Group Modelling Methods that Help Learners Understand Themselves, International Conference on Intelligent Tutoring Systems, pp. 5–13. Citeseer (2002)
10. Cen, H., Koedinger, K.R., Junker, B.: Is over practice necessary?-Improving learning efficiency with the cognitive tutor through educational data mining. Front. Artif. Intell. Appl. **158**, 511 (2007)
11. Cook, D.: What are game mechanics. Lost Gard (2006)
12. Corbett, A.T., Anderson, J.R.: Knowledge tracing: modeling the acquisition of procedural knowledge. User Model. User-Adapt. Interact. **4**, 253–278 (1994)
13. Damerau, F.J.: A technique for computer detection and correction of spelling errors. Commun. ACM **7**, 171–176 (1964)
14. Daniel, W.W.: Kruskal–Wallis one-way analysis of variance by ranks. Appl. Nonparametric Stat. 226–234 (1990)

15. Dodge, Y.: The Concise Encyclopedia of Statistics. Springer, New York (2008). https://doi.org/10.1007/978-0-387-32833-1
16. Doroudi, S., Holstein, K., Aleven, V., Brunskill, E.: Towards understanding how to leverage sense-making, induction and refinement, and fluency to improve robust learning. In: Proceedings of the 8th International Conference on Educational Data Mining, pp. 376–379 (2015)
17. Dunn, O.J.: Multiple comparisons using rank sums. Technometrics 6, 241–252 (1964)
18. Eagle, M., et al.: Estimating individual differences for student modeling in intelligent tutors from reading and pretest data. In: Micarelli, A., Stamper, J., Panourgia, K. (eds.) ITS 2016. LNCS, vol. 9684, pp. 133–143. Springer, Cham (2016). https://doi.org/10.1007/978-3-319-39583-8_13
19. Glasgow, R., Ragan, G., Fields, W.M., Reys, R., Wasman, D.: The decimal dilemma. Teach. Child. Math. 7, 89 (2000)
20. Hao, J., Shu, Z., von Davier, A.: Analyzing process data from game/scenario-based tasks: an edit distance approach. JEDM-J. Educ. Data Min. 7, 33–50 (2015)
21. Harpstead, E., Aleven, V.: Using empirical learning curve analysis to inform design in an educational game. In: Proceedings of the 2015 Annual Symposium on Computer-Human Interaction in Play, pp. 197–207. ACM (2015)
22. Harpstead, E., MacLellan, C.J., Koedinger, K.R., Aleven, V., Dow, S.P., Myers, B.: Investigating the solution space of an open-ended educational game using conceptual feature extraction (2013)
23. Harpstead, E., Richey, J.E., Nguyen, H., McLaren, B.M.: Exploring the subtleties of agency and indirect control in digital learning games. In: Proceedings of the 9th International Conference on Learning Analytics & Knowledge, pp. 121–129. ACM (2019)
24. Horn, B., Hoover, A.K., Barnes, J., Folajimi, Y., Smith, G., Harteveld, C.: Opening the black box of play: Strategy analysis of an educational game. In: Proceedings of the 2016 Annual Symposium on Computer-Human Interaction in Play, pp. 142–153. ACM (2016)
25. Isotani, S., McLaren, B.M., Altman, M.: Towards intelligent tutoring with erroneous examples: a taxonomy of decimal misconceptions. In: Aleven, V., Kay, J., Mostow, J. (eds.) ITS 2010. LNCS, vol. 6095, pp. 346–348. Springer, Heidelberg (2010). https://doi.org/10.1007/978-3-642-13437-1_66
26. Iyengar, S.S., Lepper, M.R.: Rethinking the value of choice: a cultural perspective on intrinsic motivation. J. Pers. Soc. Psychol. 76, 349 (1999)
27. Koedinger, K.R., Corbett, A.T., Perfetti, C.: The knowledge-learning-instruction framework: bridging the science-practice chasm to enhance robust student learning. Cogn. Sci. 36, 757–798 (2012)
28. Koedinger, K.R., Stamper, J.C., McLaughlin, E.A., Nixon, T.: Using data-driven discovery of better student models to improve student learning. In: Lane, H.C., Yacef, K., Mostow, J., Pavlik, P. (eds.) AIED 2013. LNCS (LNAI), vol. 7926, pp. 421–430. Springer, Heidelberg (2013). https://doi.org/10.1007/978-3-642-39112-5_43
29. Maldonado, R.M., Yacef, K., Kay, J., Kharrufa, A., Al-Qaraghuli, A.: Analysing frequent sequential patterns of collaborative learning activity around an interactive tabletop. In: Educational Data Mining 2011 (2010)
30. McLaren, B.M., Adams, D.M., Mayer, R.E., Forlizzi, J.: A computer-based game that promotes mathematics learning more than a conventional approach. Int. J. Game-Based Learn. IJGBL 7, 36–56 (2017)
31. McLaren, B.M., Lim, S.-J., Koedinger, K.R.: When and how often should worked examples be given to students? New results and a summary of the current state of research. In: Proceedings of the 30th Annual Conference of the Cognitive Science Society, pp. 2176–2181 (2008)

32. McNamara, D.S., Shapiro, A.M.: Multimedia and hypermedia solutions for promoting metacognitive engagement, coherence, and learning. J. Educ. Comput. Res. **33**, 1–29 (2005)
33. Metcalfe, J., Kornell, N.: The dynamics of learning and allocation of study time to a region of proximal learning. J. Exp. Psychol. Gen. **132**, 530 (2003)
34. Nguyen, H., Harpstead, E., Wang, Y., McLaren, B.M.: Student agency and game-based learning: a study comparing low and high agency. In: Penstein Rosé, C., et al. (eds.) AIED 2018. LNCS (LNAI), vol. 10947, pp. 338–351. Springer, Cham (2018). https://doi.org/10.1007/978-3-319-93843-1_25
35. Ocumpaugh, J.: Baker Rodrigo Ocumpaugh monitoring protocol (BROMP) 2.0 technical and training manual. New York, NY Manila Philippines. Teachers College, Columbia University and Ateneo Laboratory for the Learning Sciences, vol. 60 (2015)
36. Patall, E.A., Cooper, H., Robinson, J.C.: The effects of choice on intrinsic motivation and related outcomes: a meta-analysis of research findings. Psychol. Bull. **134**, 270 (2008)
37. Patel, R., Liu, R., Koedinger, K.R.: When to block versus interleave practice? Evidence against teaching fraction addition before fraction multiplication. In: CogSci (2016)
38. Rohrer, D., Dedrick, R.F., Burgess, K.: The benefit of interleaved mathematics practice is not limited to superficially similar kinds of problems. Psychon. Bull. Rev. **21**, 1323–1330 (2014)
39. Romano, J., Kromrey, J.D., Coraggio, J., Skowronek, J.: Appropriate statistics for ordinal level data: should we really be using t-test and Cohen's d for evaluating group differences on the NSSE and other surveys. In: Annual Meeting of the Florida Association of Institutional Research, pp. 1–33 (2006)
40. Rousseeuw, P.J.: Silhouettes: a graphical aid to the interpretation and validation of cluster analysis. J. Comput. Appl. Math. **20**, 53–65 (1987)
41. Ryan, R.M., Rigby, C.S., Przybylski, A.: The motivational pull of video games: a self-determination theory approach. Motiv. Emot. **30**, 344–360 (2006)
42. Sawyer, R., Rowe, J., Azevedo, R., Lester, J.: Filtered time series analyses of student problem-solving behaviors in game-based learning. In: Proceedings of the 11th International Conference on Educational Data Mining, pp. 229–238 (2015)
43. Sawyer, R., Smith, A., Rowe, J., Azevedo, R., Lester, J.: Is more agency better? The impact of student agency on game-based learning. In: André, E., Baker, R., Hu, X., Rodrigo, Ma.M. T., du Boulay, B. (eds.) AIED 2017. LNCS (LNAI), vol. 10331, pp. 335–346. Springer, Cham (2017). https://doi.org/10.1007/978-3-319-61425-0_28
44. Schneider, W.: The development of metacognitive knowledge in children and adolescents: major trends and implications for education. Mind Brain Educ. **2**, 114–121 (2008)
45. Schønau-Fog, H., Bjørner, T.: "Sure, I Would Like to Continue" a method for mapping the experience of engagement in video games. Bull. Sci. Technol. Soc. **32**, 405–412 (2012)
46. Schwartz, B.: Self-determination: the tyranny of freedom. Am. Psychol. **55**, 79 (2000)
47. Shaffer, D.W., Gee, J.P.: The right kind of GATE: computer games and the future of assessment. In: Mayrath, M.C., Clarke-Midura, J., Robinson, D.H., Schraw, G. (eds.) Technology-Based Assessments for 21st Century Skills: Theoretical and Practical Implications From Modern Research. Information Age Publications, Charlotte, NC, 211–228 (2012)
48. Shute, V.J., Ventura, M.: Stealth Assessment: Measuring and Supporting Learning in Video Games. MIT Press, Boca Raton (2013)
49. Shute, V.J., Ventura, M., Kim, Y.J.: Assessment and learning of qualitative physics in newton's playground. J. Educ. Res. **106**, 423–430 (2013)
50. Snow, E., Jacovina, M., Varner, L., Dai, J., McNamara, D.: Entropy: a stealth measure of agency in learning environments. In: Educational Data Mining 2014 (2014)

51. Spires, H.A., Rowe, J.P., Mott, B.W., Lester, J.C.: Problem solving and game-based learning: effects of middle grade students' hypothesis testing strategies on learning outcomes. J. Educ. Comput. Res. **44**, 453–472 (2011)
52. Stacey, K., Helme, S., Steinle, V.: Confusions between decimals, fractions and negative numbers: a consequence of the mirror as a conceptual metaphor in three different ways. In: PME Conference, pp. 4–217 (2001)
53. Stamper, J.C., Koedinger, K.R.: Human-machine student model discovery and improvement using dataShop. In: Biswas, G., Bull, S., Kay, J., Mitrovic, A. (eds.) AIED 2011. LNCS (LNAI), vol. 6738, pp. 353–360. Springer, Heidelberg (2011). https://doi.org/10.1007/978-3-642-21869-9_46
54. Wardrip-Fruin, N., Mateas, M., Dow, S., Sali, S.: Agency reconsidered. In: DiGRA Conference. Citeseer (2009)

Analyzing Students' Design Solutions in an NGSS-Aligned Earth Sciences Curriculum

Ningyu Zhang[1]([✉]), Gautam Biswas[1], Jennifer L. Chiu[2], and Kevin W. McElhaney[3]

[1] Department of EECS, Institute for Software Integrated Systems, Vanderbilt University, 1025 16th Avenue South, Nashville, TN 37212, USA
{ningyu.zhang,gautam.biswas}@vanderbilt.edu
[2] Department of Curriculum, Instruction and Special Education, University of Virginia, PO Box 400273, Charlottesville, VA 22904, USA
jlc4dz@virginia.edu
[3] Education Division, SRI International, 333 Ravenswood Avenue, Menlo Park, CA 94025, USA
kevin.mcelhaney@sri.com

Abstract. This paper analyzes students' design solutions for an NGSS-aligned earth sciences curriculum, the Playground Design Challenge (PDC), for upper-elementary school (grade 5 and 6) students. We present the underlying computational model and the user interface for generating design solutions for a school playground that has to meet cost, water runoff, and accessibility constraints. We use data from the pretest and posttest assessments and activity logs collected from a pilot study run in an elementary school to evaluate the effectiveness of the curriculum and investigate the relations between students' behaviors and their learning performances. The results show that (1) the students' scores significantly increased from pretest to posttest on engineering design assessments, and (2) students' solution-generation and testing behaviors were indicative of the quality of their design solutions as well as their pre-post learning gains. In the future, tracking such behaviors online will allow us to provide adaptive scaffolds that help students improve on their engineering design solutions.

Keywords: Technology-enhanced learning · NGSS · Engineering design · Learning analytics

1 Introduction

Design activities provide learners a supportive, authentic, and effective context to experiment with and develop an understanding of real-world scenarios using models of scientific processes [12,15]. Design-based learning activities, especially complex design problems, have shown great potential and promise in benefiting K-12 students' learning [5,6,15]. The Next Generation Science Standards

© Springer Nature Switzerland AG 2019
S. Isotani et al. (Eds.): AIED 2019, LNAI 11625, pp. 532–543, 2019.
https://doi.org/10.1007/978-3-030-23204-7_44

(NGSS) of the United States include engineering disciplinary core ideas and practices within the three-dimensional performance expectations (PEs) of disciplinary core ideas, science and engineering practices, and crosscutting concepts as early as kindergarten [7]. There are also efforts from the engineering and science education community to promote engineering design activities in elementary classrooms to strengthen science education [15]. However, previous studies have also highlighted the challenges and barriers to integrating engineering concepts and practices into elementary school curricular settings [5,9].

This paper reports on the Playground Design Challenge (PDC), an NGSS-aligned curricular unit for upper elementary school students (age 11 to 12). PDC integrates the Earth science and engineering domains through (1) scientific investigations that involve physical experiments on the absorption of different surface materials, (2) building conceptual models to understand the concepts of water absorption and runoff after a rainfall; (3) an engineering design challenge, where students can design playground models that meet specified constraints, and evaluate the construction cost and total water runoff of a designed playground [3]; and (4) use of computational models implemented in NetsBlox [1] that enable students to test different design solutions.

In the rest of this paper, we present the learning environment and the underlying computational model used in the PDC. We describe the data collected from a pilot study to evaluate the effectiveness of the curriculum and investigate the relations between students' behaviors and their pre-post learning gains. More specifically, we investigated the following research questions:

RQ 1: How effective was the intervention in improving students engineering design proficiency?

RQ 2: How well did the students' engineering design solutions correlate with their pre- to posttest learning gains?

RQ 3: How did students' behaviors of exploring the *problem space* and generating their engineering designs align with the performances on the NGSS PEs?

2   Engineering Design and the K-12 Curricula

Engineering design involves complex cognitive processes such as (1) understanding the problem, (2) generating ideas, (3) learning new concepts necessary for solving problems, (4) developing and testing models, and (5) analyzing and revising solutions [9]. Design-based instruction is more accessible to elementary school students as younger learners tend to have "less apprehension toward design challenges" compared to elder learners [15, p. 515]. In addition, design activities have great potential and promise to benefit science learning because scientific scenarios can be contextualized into compelling design problems [5,6,15].

Whereas science learning through problem-solving has received a lot of attention in secondary school curricula *e.g.,* [6] there is much less focus on *science-through-design* learning for younger pupils [15] (exceptions are Penner *et al.* [12] and Wendell *et al.*'s [15] work). Penner *et al.*'s study with third-grade students involved designing models of the human elbow. Students engaged in a series of design-related activities such as building, testing, and evaluating models. The students then used the elbow models to explore the biomechanics of the human body [12]. Wendell *et al.* implemented a LEGO^{TM} design challenge for elementary-grade students that created a synergy between science learning knowledge and engineering design [15]. Both studies demonstrated how students' problem-solving processes in a *problem space* [10] could provide engaging teaching and learning strategies.

3   Methods

3.1   The Engineering Design Environment

We created a computational model to simulate the effect of rainfall on water runoff from a playground to surrounding areas. The total runoff rate is calculated by the Rational Equation, a widely-applied method in hydraulic engineering to estimate the peak discharge of a small watershed [14]. The equation for peak discharge volume is $Q = c \times i \times A$, where c is a unit-less runoff ratio, i is the rain intensity, and A is the drainage area. To make the equation more understandable to elementary school students, we simplified the equation by assuming that the playground had a unit area, thereby eliminating the area variable from the model. As a result, the runoff coefficient is interpreted as the amount of discharge per unit of rainfall intensity (measured in inches) on the playground.

In this pilot study, students were provided access to a pre-built, interactive computational model implemented in NetsBlox [1], which they used to construct and test playground designs[1]. Students could combine seven surface materials for constructing their playground: (1) concrete, (2) natural grass, (3) artificial turf, (4) engineered wood chips, (5) sand, (6) rubber tiles, and (7) poured rubber. As part of their design task, students chose materials that were appropriate for the different parts of the playground (this was specified as requirements for specific play areas, e.g., soccer field, basketball court, swing sets, etc.) and met the runoff and cost constraints; in addition, they had to make sure that the field was wheelchair-accessible. The cost, runoff ratio, and accessibility of each material were provided to the students to help them design the playground.

Students constructed their playground by clicking on the squares and selecting from the seven available materials. When students chose a material, the square's look reflected the choice of material, and its cost was added to the total playground cost. The total runoff from the playground for a specified amount of rainfall was also updated using the Rational Equation.

The left part of Fig. 1 shows a playground design built with materials such as natural grass, artificial turf, sand, concrete, and rubber tiles to allow for a soccer field (four squares), a basketball court (two squares), and a play area with swings and other equipment (two squares). The three icons on the top of the UI on Fig. 1 represent the control buttons for the simulation. Students could click on

---

[1] Students did not program the computational model in this pilot study, however, we have added programming activities in NetsBlox for future studies.

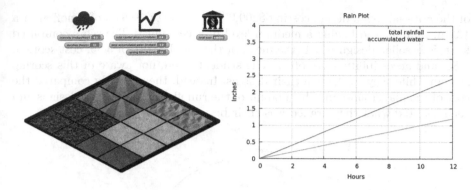

Fig. 1. The user interface of the playground design and a rain plot.

the dark cloud icon to open a dialog box to select the intensity and duration of the rain. After a simulation run, students could view a plot of the results (rainfall amount and runoff by the hour), and check the cost of the current playground under the bank icon.

The system logged five types of actions as students interacted with the computational model: (1) adding/removing the surface material for a square; (2) resetting all squares on the playground to the initial empty state, and (3–5) clicking on the 3 control buttons. The values of the model variables (*e.g.*, the choice of the surface materials, the total cost, and runoff rate) were also logged for *post hoc* analyses.

3.2   Playground Design Criteria and Scoring

The students were informed that a satisfying playground design must meet three criteria: (1) runoff ≤ 0.5 in. after 1.2 in. of total rainfall in 4 h, (2) cost $\leq \$200,000$ for the playground, and (3) having sufficient accessibility for students in wheelchairs. The accessibility criterion was not quantified in the design specifications, but for our *post hoc* analysis of the students' designs, we assigned scores of 1.0, 2.0, and 3.0 to low, medium, and high accessibility materials. Because the values associated with the three design criteria had widely different scales, we applied a simple transformation to each criterion to reduce them to a value between 1.0 and 5.0 to ensure that each criterion was given equal weight in our assigned evaluation score for a student's design. Specifically, because the playground cost varied between \$40,000 and \$600,000, we applied an inverse linear scaling to convert the actual playground cost into a score in the range [1.0, 5.0], where 1.0 represented \$600,000 and 5.0 represented \$40,000. Similarly, the runoff values were scaled to the same range with 1.0 representing a 0.96-in. runoff design (the maximum possible) and 5.0 representing a 0.24-in. runoff (the minimum possible) after the 1.2-in. rainfall.

We then used the mean of the 3 sub-scores as the score of a playground design. Figure 2 presents a visualization of a baseline design that just meets all of the criteria, *i.e.*, a design costing $200,000, resulting in 0.5 in. of runoff after a 1.2-in. rainfall, and having a medium level of accessibility. The score computed for this baseline design is 3.4 (the mean of the runoff score of 3.35, cost score of 3.85, and accessibility score of 3). The students were not aware of this scoring system while they worked on their designs. Instead, they directly compared the cost of the playground and the amount of the runoff for a number of designs and then selected what they argued was their best solution.

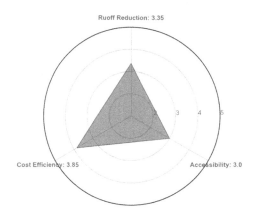

Fig. 2. The scores of the baseline design calculated by post hoc analysis

3.3   Assessment of Integrated Science and Engineering Proficiency

We designed a summative assessment that included 3 tasks measuring students' proficiency with NGSS upper elementary engineering design Performance Expectations [7]. One task assessed students' ability to define a design problem (3-5-ETS1-1). The second and third tasks assessed students' ability to generate and compare multiple possible solutions (3-5-ETS1-2) [8]. The assessment modality included multiple choice and constructed response questions. All three tasks were designed around the scientific concepts of water runoff. Task rubrics rewarded the extent to which students could make valid engineering decisions and whether these decisions were informed by the underlying scientific concept of water absorption and runoff. Sample questions and more detailed discussions of the development of the 3-dimensional assessment, its alignment to the NGSS, and the description of the grading rubrics have been presented in [3,8].

A total of 397 students (123 fifth-graders and 274 sixth-graders) from an upper-elementary school in the United States participated in the 4-week pilot study (about 1 h per school day). The study was led by science teachers with researchers playing the role of observers. The school district's STEM coordinator and one participating teacher were closely involved in the development and implementation of the curriculum. However, the teachers had not taught the PDC curriculum prior to this implementation. The summative assessment was administered as a pre- and posttest at the beginning and the end of the study.

4   Results and Discussions

4.1   Learning Gains from the Curriculum Unit

The pre- and posttest scores of a subset of 107 students were graded at the time of this analysis. We did not include students who missed a pretest or a posttest, leaving us with 88 students. We confirmed with a Kolmogorov-Smirnov test [4] that the pre-post scores were not normally distributed, and then used the non-parametric Wilcoxon rank-sum test [4] to examine if the differences in pre- to posttest scores were significant. Table 1 reports the test statistics of the overall scores and their breakdown.

Table 1. Learning gains (N = 88).

Test	Points	Pre score (std)	Post score (std)	p-value	z-score	Effect size
Total	18	4.72 (3.52)	6.50 (3.61)	<0.001	4.68	0.35
Def. problem	6	1.44 (1.11)	2.03 (1.33)	<0.001	3.77	0.28
Gen. solution	4	1.59 (1.45)	2.02 (1.52)	0.012	2.50	0.19
Comp. solutions	8	1.68 (1.78)	2.44 (1.76)	0.008	2.67	0.20

The students' learning performance showed statistically significant improvements in all aspects. This helps answer RQ 1, *i.e.*, that there appeared to be a positive association between the curriculum and the students' improved proficiency in engineering design tasks. However, the effect sizes were small, and there was a considerable gap between the posttest scores students attained the maximum possible score. Therefore, there is room for students to improve their engineering design abilities, which we hope to achieve by refining the current curriculum. This result also matches the literature that engineering design is challenging for elementary school students [15].

4.2   Playground Design Behaviors and Design Scores

As discussed in Sect. 3.1, the system logs five types of actions as students experiment and design their playgrounds. During the study, it recorded a total of 79,003 actions from 357 students. In this paper, we focus on two types of measures of the log data that relate to evaluating design solutions: (1) the number of tests conducted by a student; and (2) the scores assigned to a student-generated playground design. Both measures indicate how the students searched the solution space [10] and how well the generated solutions met the design criteria.

Table 2 presents the definition and descriptive statistics related to the number of tests the students conducted and the solution they chose. The relatively large variance in the number of test actions can be explained by classroom observations that some students worked in pairs to generate these solutions. This was expected because students were encouraged to work with each other and discuss their solutions with others in the classroom.

Table 2. Descriptive statics of test data.

Variable name	Description	Mean (std)	Range
Num test	Number of tested designs	7.64 (7.19)	1–36
Satisfy designs	Number of tested designs satisfying all criteria	3.28 (4.53)	0–27
Best score	Highest score of all tested designs	3.86 (0.15)	2.85–4.03
Last score	Score of the last (temporally) tested design	3.76 (0.17)	2.85–4.03
Submitted score	Score of the design submitted to WISE	3.77 (0.14)	3.42–3.97
Score diff	Difference between submitted and best scores	0.08 (0.18)	−0.30–0.47

Students submitted their final playground design to the Web-based Inquiry Science Environment (WISE) [13], where they also participated in a number of instructional activities. The submitted designs were scored by the method discussed in Sect. 3.2. A negative value for the submitted score difference (the last row of Table 2) implies that some students submitted a design that was better than what they tested during their computational modeling experiments. This discrepancy can be partly explained by the fact that students collaborated for some of the time and the solution reported may have resulted not from individual work but the collaboration, which produced better solutions than the individual efforts. On the other hand, classroom observations and interviews also indicated that some students arbitrarily reported designs that they thought looked good, although they did not actually test these solutions. The second situation echoed reports in the literature that students' focus during design activities may be diverted by personal aesthetics [5].

We compared the scores of the students' *submitted* designs to the highest scores of *tested* designs and found that less than 10% of the students reported their best design on WISE. A Mann-Whitney U-test [4] showed a significant difference in the scores. The average submitted score was 3.77 (stdev = 0.14) and the average best solution score generated in the NetsBlox environment was 3.86 (stdev = 0.15). This difference was significant (p-value < 0.001) with a large effect size of 3.48. This result indicates that although the students were able to generate satisfying design solutions, they did not make much of an attempt to compare the different solutions they had generated. More importantly, the difference in the reported solution and the best solution provides insight into students' understanding and learning to generate optimal playground design solutions. We discuss the implications in Sect. 4.3.

4.3   Correlation Analyses

Table 3 reports the correlation coefficients (Spearman's ρ) of the performance and behavioral measures from the 88 students whose pretest and posttest scores were available. Statistically significant correlations are marked with *s. We present a few observations from the correlation analysis and discuss how they can help answer research questions 2 and 3.

Table 3. Correlation coefficients of measures (*: p-value < 0.05, **: p-value < 0.01).

	Pre score	Post score	Learning gain	Num test	Satisfy designs	Best score	Last score	Sbmtd. score
Post score	0.64**							
Learning gain	−0.25*	0.53**						
Num test	0.12	0.13	0.05					
Satisfy designs	0.12	0.00	−0.13	0.75**				
Best score	0.17	0.06	−0.11	0.80**	0.81**			
Last score	0.06	0.04	−0.10	0.66**	0.65**	0.53**		
Submitted score	−0.09	−0.09	0.49**	0.31*	0.17	−0.04	0.23*	
Score diff	0.11	0.04	−0.41**	0.23*	0.06	0.06	0.05	0.18

First, the students' pre- and posttest scores are highly correlated ($\rho = 0.64$). This is an expected result—studies [2, 11] suggest that a learner's prior knowledge in a domain facilitates further learning in the domain. Surprisingly, the weak negative correlation between pretest score and learning gains ($\rho = -0.25$) implies those who had high prior knowledge did not learn as much from the intervention. The small negative correlation between learning gains and the playground design scores ($\rho = -0.11$) implies that the students learned about design criteria, but may have not applied them in an effective way to generate their design solutions. However, the correlation is not significant, implying there may be no true effect between the two variables.

Second, the learning gain is correlated with the submitted design scores ($\rho = 0.49$). We expected such a correlation because we believed that the students' performance in the engineering design activities should contribute to their improvement of the engineering proficiency (as evaluated by the pre-post assessments). This observation provides insights into RQ 2 that the students' engineering design solutions are indicative of their learning gains.

Third, the number of tests conducted by the students is correlated with (1) the number of satisfying designs ($\rho = 0.75$), (2) the best design scores ($\rho = 0.80$), and (3) the submitted design scores ($\rho = 0.31$). Additionally, the number of satisfying designs also correlated with the highest design scores ($\rho = 0.81$). It suggests that the students who committed more effort on systematically creating and testing design solutions were more likely to find better playground designs.

Fourth, the *difference* between the submitted design scores and the best design scores is moderately and negatively correlated with the learning gain ($\rho = -0.41$). In addition, when we analyzed the correlation of this variable and the learning gains of each individual assessment items (*i.e.*, defining problem, generating solutions, and comparing solutions – results not reported in Table 3 due to the space limitations), we found that the learning gains for the *comparing solutions* sub-task is strongly and negatively correlated with the score difference ($\rho = -0.72$). This evidence suggests that students' ability to discern better design solutions in the learning environment is strongly indicative of the NGSS PE of comparing solutions (3-5-ETS1-2).

Fifth, the submitted scores had a larger correlation to the score of the *last* tested design ($\rho = 0.23$) than the *best* design ($\rho = -0.04$). Because the submitted scores seem to be independent of the best design scores, it is reasonable to believe that a large number of students simply reported the results of their last generated design solution rather than the best design solutions. Nonetheless, these observations provide some evidence to answer RQ 3 that students' learning behaviors and performances directly link to the NGSS engineering performance.

4.4   A Case Description

In this subsection, we present a case study using the log data from one student to illustrate his/her playground design processes. The student was among the most successful students in the pilot study based on their learning improvement and design performances. The student tested the designs 29 times (at the 98th percentile, abbreviated as % later), and 13 tested designs satisfied all design criteria (95%). The student had the highest design score of 3.92 (67%), which is also the design submitted to WISE (85%). The student's overall pretest score, posttest score, and standardized learning gain were 5 points, 8 points, and 0.42 (59%, 99%, and 90%, respectively). Figure 3 provides a visualization of the students' playground design projected onto a 3-dimension space. The three axes of the figure correspond to the runoff, cost, and accessibility aspects of the design criteria. Each dot on the 3-D plot marks a tested design. The shaded region stands for a satisfying solution space, *i.e.*, all dots contained in the solution space mark a satisfying design.

The student's initial design used poured rubber (the most expensive material) on 4 squares and had a total cost of $255,000, failing to satisfy the cost criterion. On the second try, the student replaced poured rubber on 2 squares with less expensive materials and made a satisfying design. Despite succeeding early on, the student continued to explore additional solutions in an apparent effort to further improve the solution. The student tried other designs using more concrete, a less absorbent material that caused more runoff, resulting in a few designs that again failed the runoff criteria (designs 3–7). After addressing the runoff problem, the student made the best design at the 10th attempt and kept experimenting. Later, the student replaced half of the concrete squares with natural grass, which in turn caused the playground not being accessible anymore (designs 14–16). Then the student tried a new design with artificial

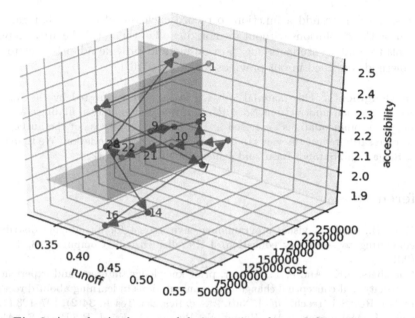

Fig. 3. A student's playground design projected on a 3-dimension space.

turf and concrete, raising the total cost over the limit again (No. 17). Finally, after exploring a few other inexpensive designs consisting most of natural grass and concrete, the student created a satisfying final solution. By this time, the student had experimented with all 7 surface materials.

This case study presents the trajectory of a successful designer who also achieved high learning gains. More importantly, it shows how we can derive features such as (1) the transitions between non-satisfying designs and satisfying designs and (2) the changes between the designs (visualized as the arrows in Fig. 3). These features will provide a great opportunity to use data-driven methods to characterize students' learning behaviors and provide feedback to help them improve their design proficiency over time.

5 Conclusions

In this paper, we introduced the computational model and the learning activities that the students engaged in the Playground Design Challenge. We presented results from the data collected from a pilot study and discussed how the students' behaviors in the design activity could influence the performance of the design, which in turn linked to their learning performance as evaluated by NGSS-aligned pre-post assessments. For future work, (1) we have integrated the computational modeling activities in the latest version of the curriculum and planned to investigate the synergistic effect between the scientific modeling activities and engineering design activities. (2) To assist the engineering design process, we plan to add a function to record each tested design that can ease comparing these solutions without memorizing all of them. (3) We are also building tools to analyze students' log data online and provide adaptive scaffolding with methods outlined in our previous work, *e.g.*, [16,17].

Acknowledgments. This material is based upon work supported by the National Science Foundation under Grant No. DRL-1742195. Any opinions, findings, and conclusions or recommendations expressed in this material are those of the authors and do not necessarily reflect the views of the National Science Foundation. We thank Ron Fried, Reina Fujii, Satabdi Basu, and James Hong for their work.

References

1. Broll, B., et al.: A visual programming environment for introducing distributed computing to secondary education. J. Parallel Distrib. Comput. **118**, 189–200 (2018)
2. Chambers, S.K., Andre, T.: Gender, prior knowledge, interest, and experience in electricity and conceptual change text manipulations in learning about direct current. J. Res. Sci. Teach. Off. J. Natl. Assoc. Res. Sci. Teach. **34**(2), 107–123 (1997)
3. Chiu, J., McElhaney, K.W., Zhang, N., Biswas, G., Fried, R., Basu, S., Alozie, N.: A principled approach to NGSS-aligned curriculum development integrating science, engineering, and computation: a pilot study. Paper presented at the 2019 NARST Annual International Conference (2019)
4. Conover, W.J., Conover, W.J.: Practical Nonparametric Statistics. Wiley, New York (1980)
5. Kafai, Y.B., Ching, C.C.: Affordances of collaborative software design planning for elementary students' science talk. J. Learn. Sci. **10**(3), 323–363 (2001)
6. Kolodner, J., Crismond, D., Fasse, B., Gray, J., Holbrook, J., Puntembakar, S.: Putting a student-centered learning by design curriculum into practice: lessons learned. J. Learn. Sci. **12**(4), 495–547 (2003)
7. NGSS Lead States: Next Generation Science Standards: For States, By States (2013)
8. McElhaney, K.W., Basu, S., Wetzel, T., Boyce, J.: Three-dimensional assessment of NGSS upper elementary engineering design performance expectations. Paper presented at the 2019 NARST Annual International Conference (2019)
9. Mehalik, M.M., Doppelt, Y., Schuun, C.D.: Middle-school science through design-based learning versus scripted inquiry: better overall science concept learning and equity gap reduction. J. Eng. Educ. **97**(1), 71–85 (2008)
10. Newell, A., Simon, H.A.: Human Problem Solving. Prentice-Hall Englewood Cliffs, Upper Saddle River (1972)
11. Osman, M.E., Hannafin, M.J.: Effects of advance questioning and prior knowledge on science learning. J. Educ. Res. **88**(1), 5–13 (1994)
12. Penner, D.E., Lehrer, R., Schauble, L.: From physical models to biomechanics: a design-based modeling approach. J. Learn. Sci. **7**(3–4), 429–449 (1998)
13. Slotta, J.D., Linn, M.C.: WISE Science: Web-Based Inquiry in the Classroom. Teachers College Press, New York (2009)
14. Thompson, D.B.: The rational method. Civil Engineering Department Texas Tech University, pp. 1–7 (2006)

15. Wendell, K.B., Rogers, C.: Engineering design-based science, science content performance, and science attitudes in elementary school. J. Eng. Educ. **102**(4), 513–540 (2013). https://doi.org/10.1002/jee.20026

16. Zhang, N., Biswas, G.: Understanding students' problem-solving strategies in a synergistic learning-by-modeling environment. In: Penstein Rosé, C., et al. (eds.) AIED 2018. LNCS (LNAI), vol. 10948, pp. 405–410. Springer, Cham (2018). https://doi.org/10.1007/978-3-319-93846-2_76

17. Zhang, N., Biswas, G., Dong, Y.: Characterizing students' learning behaviors using unsupervised learning methods. In: André, E., Baker, R., Hu, X., Rodrigo, M.M.T., du Boulay, B. (eds.) AIED 2017. LNCS (LNAI), vol. 10331, pp. 430–441. Springer, Cham (2017). https://doi.org/10.1007/978-3-319-61425-0_36

Hierarchical Reinforcement Learning for Pedagogical Policy Induction

Guojing Zhou, Hamoon Azizsoltani, Markel Sanz Ausin, Tiffany Barnes, and Min Chi[(✉)]

Department of Computer Science, North Carolina State University,
Raleigh, NC 27695, USA
{gzhou3,hazizso,msanzau,tmbarnes,mchi}@ncsu.edu

Abstract. In interactive e-learning environments such as Intelligent Tutoring Systems, there are pedagogical decisions to make at two main levels of granularity: whole problems and single steps. Recent years have seen growing interest in data-driven techniques for such pedagogical decision making, which can dynamically tailor students' learning experiences. Most existing data-driven approaches, however, treat these pedagogical decisions *equally, or independently*, disregarding the long-term impact that tutor decisions may have across these two levels of granularity. In this paper, we propose and apply an offline, off-policy Gaussian Processes based Hierarchical Reinforcement Learning (HRL) framework to induce a hierarchical pedagogical policy that makes decisions at both problem and step levels. In an empirical classroom study with 180 students, our results show that the HRL policy is significantly more effective than a Deep Q-Network (DQN) induced policy and a random *yet reasonable* baseline policy.

Keywords: Hierarchical Reinforcement Learning · Pedagogical policies

1 Introduction

Interactive e-Learning Environments such as Intelligent Tutoring Systems (ITSs) and educational games have become increasingly prevalent in educational settings. In domains like math and science, solving a problem often requires producing one or multiple steps, each of which is the result of applying a domain principle or rule. For example, $2x + 5 = 9$ can be solved for x in two steps: (1) subtract the same term 5 from both sides of the equation; and (2) divide both sides by 2. Tutoring in such domains is thus often structured as a two-loop procedure [35]: the outer loop makes *problem* level decisions, such as problem selection; while the inner loop controls *step* level decisions, such as whether or not to give hints or give a feedback. As a result, there are decisions to make and opportunities to give at *different levels of granularity*, such as hints, worked examples, immediate feedback, or suggested subgoals, and some are more important or impactful than others. Human decision-makers treat these distinct levels of granularity differently and are capable of selecting between them [7,12].

© Springer Nature Switzerland AG 2019
S. Isotani et al. (Eds.): AIED 2019, LNAI 11625, pp. 544–556, 2019.
https://doi.org/10.1007/978-3-030-23204-7_45

Data-driven approaches, and especially reinforcement learning (RL), have been shown to improve the effectiveness of ITSs [4,5,9,10,19,28,29,39]. However, most prior applications of RL for pedagogical policy induction treat all system decisions *equally or independently* and do not account for the long-term impact of higher-level actions or the interaction of decisions made at different levels. In this paper, we propose and apply an offline, off-policy Gaussian Processes-based (GP-based) Hierarchical Reinforcement Learning (HRL) framework to induce a hierarchical pedagogical policy at two levels of granularity: *problem* and *step*. More specifically, our HRL policy will first make a problem-level decision and then make step-level decisions based on the problem-level decision. In this study, for example, our HRL policy first decides whether the next *problem* should be a worked example (WE), problem solving (PS), or a faded worked example (FWE). In WEs, students observe how the tutor solves a problem; in PSs students solve the problem themselves; in FWEs, the students and the tutor *co-construct* the solution. Based on the problem-level decision, the HRL policy then makes step-level decisions on whether to elicit the next solution step from the student, or to show it to the student directly. We refer to such decisions as *elicit/tell* decisions. If WE is selected, an all-tell step policy will be carried out; if PS is selected, an all-elicit policy will be executed; finally if FWE is selected, the tutor will decide whether to elicit or to tell a step based on the corresponding induced step-level policy. Both WE and PS can be seen as two extreme ends of FWEs. Therefore, one non-hierarchical way to make decisions would be to focus on step-level decisions alone.

In a classroom study, we compared the HRL induced hierarchical policy (HRL) with two step-level policies: a Deep Q-Network induced policy (DQN) and a *random yet reasonable* (Random) policy because both *elicit* and *tell* are always considered to be *reasonable* educational interventions in our learning context. 180 students were randomly assigned to three conditions and our results showed that the HRL policy was significantly more effective than the DQN and Random policies, and no significant difference was found between the two latter policies. For time on task, no significant difference was found between the HRL condition and Random but the former (HRL) spent significantly more time than DQN. Finally, the induced HRL policy is more likely to select PS and FWE than WE, which confirmed our hypothesis that HRL would provide the right balance to pedagogical decision making, targeting WEs and tells to just those problems and steps that need them.

2   Background and Related Work

2.1   Previous Research on Applying RL to ITSs

Generally speaking, RL approaches can be classified as online, where the agent learns a policy in real time by interacting with the environment, or offline, where the agent learns from pre-collected training data. RL approaches can also be divided into on-policy vs. off-policy, based on the relationship between their behavior and estimation policies [32]. In on-policy RL, the behavior policy used to control how the agent explores the environment (online), or collects training data (offline), is the same as the estimation policy being learned. In off-policy methods, these two policies may be unrelated. Both online and offline RL approaches have been used for pedagogical policy induction in recent years; among them, prior research mainly took an off-policy RL approach [3,4,9,10,13,19,28,36,39]. Next, we will describe prior RL work from the online vs. offline perspective.

Online RL research to induce pedagogical policies has often relied on simulations or simulated students. As a consequence, the success of these approaches is heavily dependent on the accuracy of the simulations. Beck et al. [3] applied temporal difference, with off-policy ϵ-greedy exploration, to induce pedagogical policies that would minimize the students' time on task. Iglesias et al. applied another common online, off-policy approach named Q-learning to induce policies for efficient learning [9,10]. More recently, Rafferty et al. applied POMDP with off-policy tree search to induce policies for faster learning [19]. Wang et al. applied an online, off-policy Deep-RL approach to induce a policy for adaptive narrative generation in educational game [36]. All of the models described above were evaluated via simulations or classroom studies, yielding improved student learning and/or behaviors as compared to baseline policies.

Offline RL approaches, on the other hand, "take advantage of previous collected samples, and generally provide robust convergence guarantees" [25]. The success of offline RL is thus often heavily dependent on the quality of the training data. One common convention is to collect an exploratory corpus by training students on an ITS that makes random yet *reasonable* decisions and then apply RL to induce pedagogical policies from that corpus. Shen et al. applied value iteration and least square policy iteration on a pre-collected training corpus to induce pedagogical policies aimed at improving students' learning performance [27,28]. Chi et al. applied policy iteration to induce a pedagogical policy aimed at improving students' learning gains [4]. Mandel et al. [13] applied an offline POMDP approach to induce a policy which aims to improve student performance in an educational game. In classroom studies, most models above were found to yield certain improved student learning relative to a baseline policy.

Despite these successes, the necessity for accurate simulations (online) or large training corpora (offline) has limited the wide use of RL for policy induction. Additionally, prior research on both online RL and offline RL has not taken the granularity of decisions into account when applying RL techniques for the induction of pedagogical policies. In the remainder of the paper, we will refer to these approaches as flat RL to differentiate them from our new HRL approach.

It has been widely shown that HRL can be more effective and data-efficient than flat RL approaches [6,11,18,22,37]. HRL generally breaks down a large decision-making problem into a hierarchy of small sub-problems and induces a policy for each of them. Since the sub-problems are small, they usually require less data to find the optimal policies. For example, Cuayhuitl et al. induced navigation policies [6] at 3 levels: buildings, floors, and corridors, showing that

HRL converged to an optimal policy in much fewer iterations. Peng et al. showed success using temporal HRL to induce locomotion control policies for path following and soccer dribbling while flat policies could not complete these tasks [18]. Although promising, the use of hierarchy requires additional information, such as the transitions and rewards at different levels of granularity, to induce a policy, and this may be hard to get from pre-collected data. Therefore, most existing HRL applications have been online, but here, we propose and apply an offline, off-policy HRL approach. To the best of our knowledge, this is the first attempt to apply HRL to induce pedagogical policies.

2.2   WE, PS and FWE

Prior research has investigated the effectiveness of WE, PS, FWE, and their various combinations [14–17,21,23,26,31,33]. When focusing on PS and WE, Mclaren et al. found no significant difference in learning performance between studying WE-PS pairs and doing PS-only, but the former spent significantly less time than the PS-only [16]. In a subsequent study, Mclaren et al. compared three conditions: WE-only, PS-only and WE-PS pairs [15]. Similarly, no significant differences were found among them in terms of learning gains, but the WE condition spent significantly less time than the other two; and no significant time on task difference was found between PS-only and WE-PS pairs.

Several studies were conducted comparing different combinations of WE, PS, and FWE. Renkl et al. compared WE-FWE-PS with WE-PS pairs, and the former significantly outperformed the latter on student learning performance while no significant difference was found between them on time on task [21]. Similarly, Najar et al. compared adaptive WE/FWE/PS with WE-PS pairs [17]. They found that the former significantly outperformed WE-PS pairs in terms of learning outcomes and the former also spent significantly less time on task than the latter. For adaptive WE/FWE/PS, they used expert rules to make decisions based on student learning states. Finally, Salden et al. compared three conditions: WE-FWE-PS, FWE, and PS-only [23]. Their results showed that FWE outperformed WE-FWE-PS, which in turn outperformed PS-only, and no significant time on task difference was found among the three conditions. Note that in their study, the order of WE, FWE, and PS were fixed in WE-FWE-PS; while in FWE, the tutor used an adaptive pedagogical policy, expert rules combined with data-driven student models. In short, previous studies have shown that alternating among WE, PS, and FWEs can be more effective than only alternating between WE and PS; however, it is not clear whether the former can be more effective than only using FWEs. On the other hand, prior research either used a fixed policy (WE-FWE-PS) or hand-coded expert rules combined with data-driven student models to make decisions. In this work, we applied an offline, off-policy HRL framework to derive a hierarchical pedagogical policy directly from empirical data. Its effectiveness is directly compared against another data-driven FWE policy induced by applying one of the state-of-the-art flat RL methods: Deep Q-Network.

3   Policy Induction

In this work, both our proposed HRL framework and DQN are offline, off-policy in that they induce policies from a historical dataset $\mathcal{D}$ collected by training students on the ITS that makes random yet reasonable decisions. RL focuses on inducing effective decision making policies for an agent with the goal of maximizing the agent's cumulative rewards. In many domains, RL is applied with immediate rewards. In an automatic call center system, for example, the agent can receive an immediate reward for every question it asks because the impact of each question can be assessed instantaneously [38]. Immediate rewards are generally more effective than delayed rewards for RL-based policy induction. This is because it is easier to assign appropriate credit or blame when the feedback is tied to a single decision. The more we delay the rewards or punishments, the harder it becomes to assign credit or blame properly. The availability of immediate rewards is especially important for HRL approaches. On the other hand, the most appropriate reward to use in ITSs is student learning gains, which are typically unavailable until the entire training process is complete. This is due to the complex nature of the learning process which makes it difficult to assess students' learning moment by moment and more importantly, many instructional interventions that boost short-term performance may not be effective over the long-term. Therefore, we first proposed and applied a Gaussian Processes based (GP-based) approach to infer "immediate rewards" from the delayed rewards and then applied HRL and DQN to induce the corresponding hierarchical or step-level policies based on the inferred immediate rewards. In the following, we will briefly describe: (1) our proposed GP-based approach to infer immediate rewards, (2) our offline, off-policy GP-based HRL framework, and (3) DQN. We now present a few critical details of the process, but many have been omitted to save space.

3.1   GP-Based Approach for Immediate Reward Inference

Our historical dataset $\mathcal{D}$ consists of student-ITS interaction trajectories with different lengths. Each trajectory d can be viewed as: $s_1 \xrightarrow{a_1, r_1} s_2 \xrightarrow{a_2, r_2} \cdots s_n \xrightarrow{a_n, r_n}$. Here $s_i \xrightarrow{a_i, r_i} s_{i+1}$ indicated that at the i_{th} turn in d, the learning environment was in state s_i, agent executed action a_i and received reward r_i, and then the learning environment transferred into state s_{i+1}. Because our primary interest is to improve students' final learning, we used Normalized Learning Gain (NLG) as the reward because it measures students' gain *irrespective of their incoming competence*. $NLG = \frac{posttest - pretest}{\sqrt{1 - pretest}}$ where *pretest* and *posttest* refer to the students' test scores before and after the ITS training respectively and 1 is the maximum score. Given that a student's NLG will not be available until the entire training is completed, only terminal states have non-zero rewards. Thus for a trajectory d, $r_1 \cdots, r_{n-1}$ are all equal to 0, and only the final reward r_n is equal to the student's $NLG \times 100$, which is in the range of $(-\infty, 100]$.

To infer the immediate rewards from the final delayed reward for each trajectory, we applied Gaussian Processes (GP) to learn a distribution function f for the expected values and the standard deviations of all of the immediate rewards. More specifically, a prior probability is given to each possible function before observation. Then, higher probabilities are given to the functions where the sum of the generated immediate rewards is close to the observed delayed reward. In other words, the immediate rewards inside each trajectory were inferred by minimizing the mean square error (MMSE) of additive Gaussian distributions [8]. The immediate rewards were distributed inside each trajectory by assuming that they follow Gaussian distributions and that these rewards add up to the delayed reward for each trajectory. Following the Gaussian Process Regression [1,20] and the shared mutual information existed in the feature representation, we can thus infer the immediate rewards from delayed rewards.

3.2 An Offline, Off-policy GP-Based HRL for Policy Induction

Most HRL research is based upon an extension of Markov Decision Processes (MDPs) called Discrete Semi-Markov decision processes (SMDPs) and the central idea behind the HRL approach is to transform the problem of inducing effective pedagogical policies into one of computing an optimal policy for choosing actions in SMDPs. An MDP describes a stochastic control process and formally corresponds to a 4-tuple: <S,A,T,R>. When inducing pedagogical policies, the states S are vector representations composed of relevant learning environment features such as the difficulty level of a problem, percentage of the correct entries a student entered so far and so. In this study, we have a total of 142 state features to describe the learning environment; the actions A are selected from $\{WE, FWE, PS\}$ for problem-level decisions and from {elicit, tell} for steps; the reward function R is calculated from the system's success measures: students NLG. Once the $\{S, A, R\}$ has been defined, the transition probabilities T are estimated from the training corpus, $\mathcal{D}$. Once a complete MDP is constructed, calculation of an optimal policy via policy iteration is straightforward.

SMDPs extend the existing MDP framework with the addition of a set of complex activities [2] or options [30], each of which can invoke other activities recursively, thus allowing for hierarchical policy functions. The *complex* activities are distinct from the primitive actions in that a complex activity may contain multiple *primitive* actions. In our applications, WE, PS and FWE are complex activities while elicit and tell are primitive actions. A complex activity consists of three elements: a policy π that maps states to each available option, a termination condition, and an initiation set. A solution to the SMDP mentioned above is an optimal policy (π^*), a mapping from state to complex activities or primitive actions, that maximizes the expected discounted cumulative reward for each state.

The complex activities in SMDPs can take a variable number of low-level activity (or actions) to execute across multiple time steps. This makes it necessary to extend the state-transition function to take into account the activity length. If an activity a takes t' time steps to be executed in state s, then the state transition probability function given s and a is defined by the joint distribution of the result state s' and the number of time steps t' when action a is performed in the state s: $P(s', t'|s, a)$. The expected reward function is also extended to accumulate over the waiting time in s given action a. More specifically, the Q-value function $Q(s, a)$ represents the expected discounted reward the agent will gain if it takes an action a in a state s and follows the policy to the end and for SMDP, the Bellman equation can be re-written as:

$$Q(s, a) = R(s, a) + \sum_{s', t'} \gamma^{t'} P(s', t'|s, a) \max_{a' \in A} Q(s', a') \tag{1}$$

$0 \leq \gamma \leq 1$ is a discount factor. If γ is less than 1, then it will discount rewards obtained later. For HRL, learning occurs at multiple levels. The global learning generates a policy for the top level decision and local learning generates a policy for each complex activity. This process retains the fundamental assumption of RL: that goals are defined by their association with reward, and thus that the objective is to discover actions that maximize the long-term cumulative reward. Local learning focuses not on learning the best policy for the overall task but the best policy for the corresponding complex activity.

In our offline off-policy HRL framework, both problem- and step-level policies were learned by recursively using the Gaussian Processes to estimate the Q-value function in Eq. 1. Using an actor-critic policy iteration framework, we iteratively update the policy. This process continues until the Q-value function and the induced policy converged. We assume that the Q-value function follows a prior distribution and by combining the prior of Q-value function and the inferred immediate rewards, the Gaussian Process Regression can provide the posterior distribution of the Q-value function approximation in a tractable way. In this work, our training corpus contains a total number of 1118 students' interaction logs collected from a series of seven prior studies which followed the identical procedure and learning materials as the students in this study described below. To induce the hierarchical policy, we defined a problem-level semi-MDP for determining whether the next problem should be WE, PS or FWE and for each of the training problems, we defined a step-level semi-MDP for inducing a step-level policy to determine elicit vs. tell if a complex activity FWE is selected for that training problem.

3.3   DQN for Policy Induction

A Double DQN approach [34] with the prioritized experience replay technique [24] was applied to induce the DQN step-level policy. A multi-layer perceptron neural network was used to approximate the Q-function. The inputs to the neural network were the last 3 step observations of a student and the outputs were the Q values for each possible step level action (in our case, elicit and tell). The network consists of two 64-unit layers with the rectified linear unit (ReLU) activation function (except that the output layer has no activation function). As a convention for this algorithm, an experience replay buffer and a target network were used to stabilize the training. The data and immediate rewards used for DQN policy induction were identical to those used for HRL.

4   Empirical Experiment

Participants. This study was conducted in the undergraduate Discrete Mathematics course at the Department of Computer Science at North Carolina State University in the Fall of 2018. The study was given as one of the regular homework assignments; students had one week to complete it and were graded based upon their demonstrated effort rather than performance. Students ($N = 180$) were randomly assigned into three conditions (60 in each of HRL, DQN, and Random). Due to preparations for exams and the length of the experiment, 140 students completed the study. 3 students who scored perfectly in the pre-test were excluded from our subsequent analysis. In addition, 9 students who completed the study in groups were excluded. The remaining 128 students were distributed as follows: $N = 44$ for HRL, $N = 45$ for DQN, and $N = 39$ for Random. A χ^2 test shows that the participants' completion rate did not differ by condition: $\chi^2(2) = 1.03, p = 0.598$.

Pyrenees is a web-based ITS that teaches students a general problem solving strategy and 10 major principles of probability, such as the Complement Theorem and Bayes' Rule. It provides students with step-by-step instruction, immediate feedback, and on-demand help. Specifically, the help is provided via a sequence of increasingly specific hints. The last hint in the sequence, i.e., the bottom-out hint, tells student exactly what to do. Except for the decision granularity, the remaining components of the tutor, including the GUI interface, the training problems, and the tutorial support were identical for all students.

Procedure. All three conditions went through the same four phases: (1) textbook, (2) pre-test, (3) training on the ITS, and (4) post-test. The only difference among them was the policy employed by the ITS. During **textbook**, all students read a general description of each principle, reviewed some examples, and solved some training problems. The students then took a **pre-test** which contained a total of 14 single- and multiple-principle problems. Students were not given feedback on their answers, nor were they allowed to go back to earlier questions (this was also true for the post-test). During **training on the ITS**, all three conditions received the same 12 problems in the same order. Each domain principle was applied at least twice. Finally, all students took the 20-problem **post-test**; 14 of the problems were isomorphic to the pre-test. The remainder were non-isomorphic multiple-principle problems.

Grading Criteria. The pre- and post-test problems required students to derive an answer by writing and solving one or more equations. We used three scoring rubrics: binary, partial credit, and one-point-per-principle. Under the binary rubric, a solution was worth 1 point if it was completely correct or 0 if not. Under the partial credit rubric, each problem score was defined by the proportion of correct principle applications evident in the solution. A student who correctly applied 4 of 5 possible principles would get a score of 0.8. The One-point-per-principle rubric in turn gave a point for each correct principle application. All of the tests were graded in a double-blind manner by a single experienced grader.

Table 1. Learning performance and time on task

Condition	Pre	Iso post	Full post	Adj post	NLG	Time (hours)
HRL(44)	66.4(18.8)	85.8(14.6)	75.3(16.9)	77.7(10.3)	14.3(19.2)	2.19(.64)
DQN(45)	73.9(13.6)	85.2(13.1)	74.2(14.6)	71.2(12.0)	−2.2(29.4)	1.81(.58)
Random(39)	66.3(18.9)	80.5(19.5)	69.0(19.6)	71.4(13.8)	−0.1(35.0)	1.97(.52)

The results presented below were based upon the partial-credit rubric but the same results hold for the other two. For comparison purposes, all test scores were normalized to the range of $[0, 100]$.

5   Results

Despite of random assignment, a one-way ANOVA analysis on the pre-test score showed a marginally significant difference among the three conditions: $F(2, 125) = 2.805$, $p = 0.064$, $\eta = 0.043$. Subsequent contrast analysis showed that DQN scored significantly higher than HRL: $t(125) = 2.06$, $p = 0.042$, $d = 0.46$ and Random: $t(125) = 2.01$, $p = 0.046$, $d = 0.46$; but there is no significant difference between HRL and Random: $t(125) = 0.02$, $p = 0.986$, $d = 0.00$. The results suggest that while our random assignment indeed balanced the HRL and Random conditions' incoming competence, it did not do so for the DQN condition. Therefore, we mainly focus on comparing learning performances that consider the pre-test differences, that is, adjusted post-test and NLG especially the latter because it is the reward we used for policy induction.

Table 1 shows the mean and standard deviation (SD) of students' learning performance and total training time results across three conditions. From left to right, it shows the condition with the number of students in parentheses, pre-test (Pre), isomorphic post-test (Iso Post), full post-test (Full Post), adjusted post-test (Adj Post), Normalized Learning Gain (NLG), and the total training time on the ITS in hours (Time).

Isomorphic Post-test. To measure students' learning improvement, we compared their isomorphic post-test scores with their pre-test scores. A repeated measures analysis using test type (pre-test vs. isomorphic post-test) as a factor and test score as the dependent measure showed a main effect for test type: $F(1, 127) = 158.63$, $p < 0.0001$, $\eta = 0.555$ in that students scored significantly higher in the isomorphic post-test than in the pre-test. More specifically, all three conditions scored significantly higher in the isomorphic post-test than in the pre-test: $F(1, 43) = 110.74$, $p < 0.0001$, $\eta = 0.720$ for HRL, $F(1, 44) = 34.73$, $p < 0.0001$, $\eta = 0.441$ for DQN, and $F(1, 38) = 38.47$, $p < 0.0001$, $\eta = 0.503$ for Random. This showed that the basic practice and problems, domain exposure, and interactivity of our ITS effectively help students acquire knowledge, even when the decisions are made randomly yet reasonably.

Table 2. Step level tutor decisions

Condition	Elicit	Tell	Pct Tell
HRL	309.0(60.4)	88.7(66.1)	22.025(15.870)
DQN	205.8(51.6)	188.9(53.0)	47.794(12.974)
Random	200.5(15.9)	203.5(17.4)	50.354(2.482)

Adjusted Post-test. To comprehensively evaluate students' final performance, we performed analysis on the full post-test score which has an additional six multiple-principle problems. An ANCOVA analysis on the post-test using the pre-test score as a covariate showed a significant difference among the three conditions: $F(2, 124) = 3.86$, $p = 0.024$, $\eta = 0.030$. Subsequent contrast analysis on the adjusted post-test score showed that the HRL condition scored significantly higher than the DQN condition: $t(125) = 2.53$, $p = 0.013$, $d = 0.57$ and the Random condition: $t(125) = 2.36$, $p = 0.020$, $d = 0.52$. No significant difference was found between DQN and Random. The results suggest that the HRL policy is significantly more effective than the DQN policy and the Random policy.

NLG. Similarly, a one-way ANOVA analysis on the NLG showed that there is a significant difference among the three conditions: $F(2, 125) = 4.39$, $p = 0.014$, $\eta = 0.066$. Subsequent contrast analysis showed that the HRL condition scored significantly higher than the DQN condition: $t(125) = 2.75$, $p = 0.007$, $d = 0.66$ and the Random condition: $t(125) = 2.30$, $p = 0.023$, $d = 0.52$. Again, no significant difference was found between DQN and Random. The results suggest again that the HRL policy significantly outperformed the DQN policy and the Random policy.

Time on Task. A one-way ANOVA analysis on time on task showed a significant difference among the three conditions: $F(2, 125) = 4.74$, $p = 0.010$, $\eta = 0.071$. More specifically, the HRL condition spent significantly more time than the DQN condition: $t(125) = 3.07$, $p = 0.003$, $d = 0.62$ and marginally significantly more time than the Random condition: $t(125) = -1.75$, $p = 0.082$, $d = 0.39$.

Tutor Decisions. Our preliminary log analysis revealed that for the HRL condition, the average number of problem-level decisions students received are: .95(1.16) for WE, 5.07(2.58) for PS and 3.98(2.49) for FWE. Thus the HRL policy was more likely to choose PS and FWE than WE. Table 2 shows the number of step-level decisions students received across the three conditions. The first column shows the condition followed by the number of elicit and tell and finally the percentage of tell. Our preliminary step-level log analysis results showed that the HRL condition received more elicit than tell; while the other two conditions received a relatively balanced amount. A one-way ANOVA analysis on the percentage of tell revealed a significant difference among the three conditions: $F(2, 125) = 71.47$, $p < 0.0001$, $\eta = 0.533$. Subsequent contrast analysis showed that the HRL condition received significantly less tell than the DQN condition: $t(125) = -10.00$, $p < 0.0001$, $d = 1.78$ and the Random condition: $t(125) = -10.60$, $p < 0.0001$, $d = 2.42$. In addition, the HRL and the DQN condition had a much higher SD on tell percentage. This suggests that the HRL policy and the DQN policy made more personalized decisions than the Random policy.

6   Conclusion and Discussion

In this study, we proposed and applied an offline, off-policy GP-based HRL framework to induce a hierarchical pedagogical policy. The policy makes decisions first at the problem level and then the step level. At the problem level, it decides whether the next problem should be WE, PS or FWE. If FWE is selected, a corresponding step-level policy will be activated to decide whether the next step should be elicit or tell. In an empirical classroom study, we compared the HRL policy with a DQN induced step-level policy and a Random step-level policy. Our results showed that the HRL policy was significantly more effective than the DQN policy and the Random policy and no significant difference was found between the latter two policies. For time on task, there was no significant difference between the HRL condition and the Random condition, but the former spent significant more time than the DQN condition. Finally, the HRL policy was more likely to choose PS and FWE than WE.

The results suggest that HRL can be more effective than flat RL in pedagogical policy induction. One possible explanation is that HRL has an explicit problem-level vision. At the problem level, HRL views a problem as an atomic action, and this abstraction has two potential advantages: (1) it aggregates the effects of all steps in a problem and (2) it converts a long step-level sequence into a short problem-level sequence. The aggregation of steps across a problem may provide HRL with a better estimation of the effect of taking a series of steps; while the problem sequence may give HRL a better view of the long-term effects of each problem. Theoretically, flat RL could learn the impact of a problem by aggregating step-level information, but there is no guarantee that it would. Our results confirm the intuition that HRL should outperform flat RL on pedagogical policy induction because it can simultaneously learn at two levels of granularity - the problem level outer loop and the step level inner loop.

References

1. Azizsoltani, H., Sadeghi, E.: Adaptive sequential strategy for risk estimation of engineering systems using gaussian process regression active learning. Eng. Appl. Artif. Intell. **74**(July), 146–165 (2018)
2. Barto, A.G., Mahadevan, S.: Recent advances in hierarchical reinforcement learning. Discrete Event Dyn. Syst. **13**(1–2), 41–77 (2003)
3. Beck, J., Woolf, B.P., Beal, C.R.: ADVISOR: a machine learning architecture for intelligent tutor construction. AAAI/IAAI **2000**(552–557), 1–2 (2000)
4. Chi, M., VanLehn, K., Litman, D., Jordan, P.: Empirically evaluating the application of reinforcement learning to the induction of effective and adaptive pedagogical strategies. User Model. User Adap. Inter. **21**(1–2), 137–180 (2011)
5. Clement, B., Oudeyer, P.Y., Lopes, M.: A comparison of automatic teaching strategies for heterogeneous student populations. In: EDM 2016–9th International Conference on Educational Data Mining (2016)

6. Cuayáhuitl, H., Dethlefs, N., Frommberger, L., Richter, K.-F., Bateman, J.: Generating adaptive route instructions using hierarchical reinforcement learning. In: Hölscher, C., Shipley, T.F., Olivetti Belardinelli, M., Bateman, J.A., Newcombe, N.S. (eds.) Spatial Cognition 2010. LNCS (LNAI), vol. 6222, pp. 319–334. Springer, Heidelberg (2010). https://doi.org/10.1007/978-3-642-14749-4_27
7. Evens, M., Michael, J.: One-on-One Tutoring by Humans and Computers. Psychology Press (2006)
8. Guo, D., Shamai, S., Verdú, S.: Mutual information and minimum mean-square error in Gaussian channels. IEEE Trans. Inf. Theor. **51**(4), 1261–1282 (2005)
9. Iglesias, A., Martínez, P., Aler, R., Fernández, F.: Learning teaching strategies in an adaptive and intelligent educational system through reinforcement learning. Appl. Intell. **31**(1), 89–106 (2009)
10. Iglesias, A., Martínez, P., Aler, R., Fernández, F.: Reinforcement learning of pedagogical policies in adaptive and intelligent educational systems. Knowl. Psychol. Press-Based Syst. **22**(4), 266–270 (2009)
11. Kulkarni, T.D., Narasimhan, K., Saeedi, A., Tenenbaum, J.: Hierarchical deep reinforcement learning: integrating temporal abstraction and intrinsic motivation. In: Advances in Neural Information Processing Systems, pp. 3675–3683 (2016)
12. Lajoie, S.P., Derry, S.J.: Motivational techniques of expert human tutors: lessons for the design of computer-based tutors. In: Computers as Cognitive Tools, pp. 83–114. Routledge (2013)
13. Mandel, T., Liu, Y.E., Levine, S., Brunskill, E., Popovic, Z.: Offline policy evaluation across representations with applications to educational games. In: Proceedings of the 2014 International Conference on Autonomous Agents and Multi-agent Systems, pp. 1077–1084. International Foundation for Autonomous Agents and Multiagent Systems (2014)
14. McLaren, B.M., van Gog, T., Ganoe, C., Yaron, D., Karabinos, M.: Exploring the assistance dilemma: comparing instructional support in examples and problems. In: Trausan-Matu, S., Boyer, K.E., Crosby, M., Panourgia, K. (eds.) ITS 2014. LNCS, vol. 8474, pp. 354–361. Springer, Cham (2014). https://doi.org/10.1007/978-3-319-07221-0_44
15. McLaren, B.M., Isotani, S.: When is it best to learn with all worked examples? In: Biswas, G., Bull, S., Kay, J., Mitrovic, A. (eds.) AIED 2011. LNCS (LNAI), vol. 6738, pp. 222–229. Springer, Heidelberg (2011). https://doi.org/10.1007/978-3-642-21869-9_30
16. McLaren, B.M., Lim, S.J., Koedinger, K.R.: When and how often should worked examples be given to students? New results and a summary of the current state of research. In: CogSci, pp. 2176–2181 (2008)
17. Najar, A.S., Mitrovic, A., McLaren, B.M.: Adaptive support versus alternating worked examples and tutored problems: which leads to better learning? In: Dimitrova, V., Kuflik, T., Chin, D., Ricci, F., Dolog, P., Houben, G.-J. (eds.) UMAP 2014. LNCS, vol. 8538, pp. 171–182. Springer, Cham (2014). https://doi.org/10.1007/978-3-319-08786-3_15
18. Peng, X.B., Berseth, G., Yin, K., Van De Panne, M.: DeepLoco: dynamic locomotion skills using hierarchical deep reinforcement learning. ACM Trans. Graph. (TOG) **36**(4), 41 (2017)
19. Rafferty, A.N., Brunskill, E., Griffiths, T.L., Shafto, P.: Faster teaching via POMDP planning. Cogn. Sci. **40**(6), 1290–1332 (2016)
20. Rasmussen, C.E.: Gaussian processes in machine learning. In: Bousquet, O., von Luxburg, U., Rätsch, G. (eds.) ML -2003. LNCS (LNAI), vol. 3176, pp. 63–71. Springer, Heidelberg (2004). https://doi.org/10.1007/978-3-540-28650-9_4

21. Renkl, A., Atkinson, R.K., Maier, U.H., Staley, R.: From example study to problem solving: smooth transitions help learning. J. Exp. Educ. **70**(4), 293–315 (2002)
22. Ryan, M., Reid, M.: Learning to fly: an application of hierarchical reinforcement learning. In: Proceedings of the 17th International Conference on Machine Learning. Citeseer (2000)
23. Salden, R.J., Aleven, V., Schwonke, R., Renkl, A.: The expertise reversal effect and worked examples in tutored problem solving. Instr. Sci. **38**(3), 289–307 (2010)
24. Schaul, T., Quan, J., Antonoglou, I., Silver, D.: Prioritized experience replay. arXiv preprint arXiv:1511.05952 (2015)
25. Schwab, D., Ray, S.: Offline reinforcement learning with task hierarchies. Mach. Learn. **106**(9–10), 1569–1598 (2017)
26. Schwonke, R., Renkl, A., Krieg, C., Wittwer, J., Aleven, V., Salden, R.: The worked-example effect: not an artefact of lousy control conditions. Comput. Hum. Behav. **25**(2), 258–266 (2009)
27. Shen, S., Ausin, M.S., Mostafavi, B., Chi, M.: Improving learning & reducing time: a constrained action-based reinforcement learning approach. In: Proceedings of the 26th Conference on User Modeling, Adaptation and Personalization, pp. 43–51. ACM (2018)
28. Shen, S., Chi, M.: Reinforcement learning: the sooner the better, or the later the better? In: Proceedings of the 2016 Conference on User Modeling Adaptation and Personalization, pp. 37–44. ACM (2016)
29. Stamper, J.C., Eagle, M., Barnes, T., Croy, M.: Experimental evaluation of automatic hint generation for a logic tutor. In: Biswas, G., Bull, S., Kay, J., Mitrovic, A. (eds.) AIED 2011. LNCS (LNAI), vol. 6738, pp. 345–352. Springer, Heidelberg (2011). https://doi.org/10.1007/978-3-642-21869-9_45
30. Sutton, R.S., Precup, D., Singh, S.: Between MDPs and semi-MDPs: a framework for temporal abstraction in reinforcement learning. Artif. Intell. **112**(1–2), 181–211 (1999)
31. Sweller, J., Cooper, G.A.: The use of worked examples as a substitute for problem solving in learning algebra. Cogn. Instr. **2**(1), 59–89 (1985)
32. Thomas, P., Brunskill, E.: Data-efficient off-policy policy evaluation for reinforcement learning. In: International Conference on Machine Learning, pp. 2139–2148 (2016)
33. Van Gog, T., Kester, L., Paas, F.: Effects of worked examples, example-problem, and problem-example pairs on novices learning. Contemp. Educ. Psychol. **36**(3), 212–218 (2011)
34. Van Hasselt, H., Guez, A., Silver, D.: Deep reinforcement learning with double Q-learning. In: AAAI, vol. 2, p. 5. Phoenix, Nairobi (2016)
35. Vanlehn, K.: The behavior of tutoring systems. IJAIED **16**(3), 227–265 (2006)
36. Wang, P., Rowe, J., Min, W., Mott, B., Lester, J.: Interactive narrative personalization with deep reinforcement learning. In: Proceedings of the Twenty-Sixth International Joint Conference on Artificial Intelligence (2017)
37. Wang, X., Chen, W., Wu, J., Wang, Y.F., Yang Wang, W.: Video captioning via hierarchical reinforcement learning. In: Proceedings of the IEEE Conference on Computer Vision and Pattern Recognition, pp. 4213–4222 (2018)
38. Williams, J.D.: The best of both worlds: unifying conventional dialog systems and POMDPs. In: Interspeech, pp. 1173–1176 (2008)
39. Zhou, G., Wang, J., Lynch, C., Chi, M.: Towards closing the loop: bridging machine-induced pedagogical policies to learning theories. In: EDM (2017)

Author Index

Printed in the United States
By Bookmasters